Precision Metal Additive Manufacturing

Precision Metal Additive Manufacturing

Edited by
Richard Leach and Simone Carmignato

CRC Press
Taylor & Francis Group
Boca Raton London New York

CRC Press is an imprint of the
Taylor & Francis Group, an **informa** business

First edition published 2021
by CRC Press
6000 Broken Sound Parkway NW, Suite 300, Boca Raton, FL 33487-2742

and by CRC Press
2 Park Square, Milton Park, Abingdon, Oxon, OX14 4RN

Library of Congress Cataloging-in-Publication Data

Names: Leach, R. K., editor. | Carmignato, Simone, editor.
Title: Precision additive metal manufacturing / edited by Richard Leach and Simone Carmignato.
Description: First edition. | Boca Raton, FL : CRC Press, 2020. | Includes bibliographical references and index.
Identifiers: LCCN 2020013553 (print) | LCCN 2020013554 (ebook) | ISBN 9781138347717 (hardback) | ISBN 9780429436543 (ebook) | ISBN 9780429791284 (adobe pdf) | ISBN 9780429791260 (mobi) | ISBN 9780429791277 (epub)
Subjects: LCSH: Metal-work. | Additive manufacturing.
Classification: LCC TS205 .P68 2020 (print) | LCC TS205 (ebook) | DDC 671--dc23
LC record available at https://lccn.loc.gov/2020013553
LC ebook record available at https://lccn.loc.gov/2020013554

ISBN: 978-1-138-34771-7 (hbk)
ISBN: 978-0-429-43654-3 (ebk)

Typeset in Palatino
by Deanta Global Publishing Services, Chennai, India

Contents

Acknowledgements

The editors would like to thank the EU H2020 MSCA project PAM², 'Precision Additive Metal Manufacturing' (Grant Agreement #721383), for providing funding for four of the contributing authors (Markus Baier, Amal Charles, Rajit Ranjan and Amrozia Shaheen), for facilitating joint research on precision AM between many of the contributing authors of this book and for enabling significant progress beyond the state of the art in precision AM.

Editors

Richard Leach is Professor in Metrology at the University of Nottingham and heads up the Manufacturing Metrology Team. Prior to this position, he was at the National Physical Laboratory from 1990 to 2014. His primary love is instrument building; from concept to final installation, and his current interests are the dimensional measurement of precision and additive manufactured structures. His research themes include the measurement of surface topography, development of methods for measuring 3D structures, development of methods for controlling large surfaces to high resolution in industrial applications and traceability of X-ray computed tomography. He is a leader of several professional societies and a visiting professor at Loughborough University and the Harbin Institute of Technology.

Simone Carmignato is Professor of Manufacturing Technologies at the University of Padua. His main research activities are in the areas of precision manufacturing, dimensional metrology and industrial computed tomography. He is the author of books and hundreds of scientific papers, and is an active member of leading technical and scientific societies. He has been chairman, organiser and keynote speaker for several international conferences, and has received national and international awards, including the 'Taylor Medal' from CIRP, the International Academy for Production Engineering.

Contributors

Can Ayas is assistant professor at the Delft University of Technology in the Precision and Microsystems Department. He has had research positions in University of Groningen, Eindhoven University of Technology and the University of Cambridge. His research interests are concentrated on the development of computational tools in the field of mechanics of materials and structural design, such as AM processes, topology optimisation for AM, plasticity at small scales, hydrogen embrittlement and design of lattice materials.

Markus Baier is a researcher at the University of Padua and has been a Marie Curie research fellow within the European project 'Precision Additive Metal Manufacturing'. His research activities are mainly focused on the accuracy enhancement of metal AM, applying advanced measurement techniques and X-ray computed tomography.

Martin Baumers is assistant professor of Additive Manufacturing Management at the Centre for Additive Manufacturing at the University of Nottingham. He focuses on the optimisation of workflows for efficient manufacturing execution in AM, developing solutions to capacity utilisation problems, the estimation of build time and process energy consumption and cost, as well as computational shape complexity measurement.

Francois Blateyron is vice president of Research and Metrology at Digital Surf, a company that develops metrological software specialising in surface texture and microscopy image analysis. He has been involved in the development of metrological algorithms for almost thirty years, and in standardisation at ISO/TC213 for twenty years. He collaborates with many research laboratories around the world involved in dimensional and surface metrology, and has contributed to several journals and conferences.

David Butler is a reader in Manufacturing Metrology at the University of Strathclyde. Prior to joining Strathclyde in 2017, he spent the previous eighteen years in Singapore as an associate professor at the Nanyang Technological University and as founding director of the Advanced Remanufacturing and Technology Centre – a 250-strong translational research centre. His research interests are mainly industry driven and include the development of traditional and non-conventional abrasive removal processes for aerospace and semiconductor applications. From 2017 to 2019, he also held a joint appointment as the Science Area Leader for Additive Manufacturing at the National Physical Laboratory.

Amal Charles is a researcher at the Institute for Automation and Applied Informatics, Karlsruhe Institute of Technology. His research focuses on improving the precision of metal AM processes, specifically though the optimisation of laser based powder bed fusion processes using statistical and computational methods. His specific interest is in improving the as-built quality of overhanging surfaces through localised parameter optimisation in down-facing areas.

Bianca Maria Colosimo is professor in the Department of Mechanical Engineering at Politecnico di Milano, where she is co-leading the AddMe lab, a laboratory equipped with all the relevant technologies for metal AM. Her main research interest is in the area of industrial data modelling, monitoring and control, with special focus on (big) data (for example, functional data, 3D point clouds, surfaces, signals, images and video-images) for advanced manufacturing applications, including AM. She is editor-in-chief of the *Journal of Quality Technology* and a senior member of several societies in quality and manufacturing.

Ben Dutton is a scientist with 17 years' experience in non-destructive evaluation (NDE). As a technical specialist at the Manufacturing Technology Centre (MTC), he is leading projects involving quality for AM. He supports the MTC's Net Shape and Additive Manufacturing group on several projects which require in-process and post-build NDE. He is actively leading the development of several international NDE ISO/ASTM standards for AM, and he is part of the ASTM AM Centre of Excellence team at the MTC.

Ahmed Elkaseer is a senior research associate at Karlsruhe Institute of Technology. He has more than 15 years' experience in experimental, modelling, simulation and optimisation-based studies of advanced manufacturing processes, with a recent emphasis on additive manufacturing and Industry 4.0 applications. Dr Elkaseer has several publications in advanced manufacturing technologies and serves as editorial board member for a number of journals.

David Gilbert is a senior research engineer within the Non-Conventional Machining Team at MTC, leading the Automated Post-Processing Team. Prior to working at the MTC, he undertook research in pulsed-laser micro-machining applications for the aerospace industry with Rolls-Royce and the University of Nottingham. His interests lie in robotics and automation, specifically in applications where flexible/non-contact machining process are required. The research conducted within his team revolves around process optimisation for a wide range of surface finishing techniques, and automation of those processes, including automated surface texture measurement and analysis for closed-loop process feedback.

Marco Grasso is assistant professor in the Department of Mechanical Engineering of Politecnico di Milano. The general framework of his research consists of statistical process monitoring of manufacturing processes and production systems via signal and video-image data analysis and modelling. His most recent research is carried out in the framework of metal AM, with a focus on in-situ sensing and monitoring of laser and electron beam powder bed fusion processes.

Jesper H. Hattel is professor in modelling of manufacturing processes at the Department of Mechanical Engineering, DTU. His research interests are modelling of processes such as casting, joining, composites manufacturing and AM. This involves the use of computational methods within the disciplines of heat transfer, fluid dynamics, solid mechanics and materials science. Applications range from microelectronics to automotive industry to wind turbines.

Matthijs Langelaar is associate professor at Delft University of Technology in the Structural Optimization and Mechanics Group, which is part of the Mechanical Engineering Faculty. His research focus is on topology optimisation, ranging from large deformation active mechanisms to multi-physics design problems. In recent years, he has been active in the development of new manufacturing constraints for topology optimisation, in particular related to AM.

Sankhya Mohanty currently holds a researcher position in numerical modelling of metal AM processes at the Department of Mechanical Engineering, DTU. His research interests are multi-physics modelling of manufacturing processes, through-process modelling of material behaviour and statistical and numerical methods for optimisation.

Shawn Moylan is a mechanical engineer and project leader at the National Institute of Standards and Technology. His research career has focused on precision measurements of manufacturing equipment, ranging from micromachining to multi-axis coordinated motion, and, more recently, to AM machines. He is an internationally recognised leader in the advanced manufacturing standards community, especially ASTM Committee F42 on Additive Manufacturing Technology, ISO/TC261 Additive Manufacturing and ASME B46 Surface Texture. He is also on the advisory board for the Additive Manufacturing Standardization Collaborative, a joint effort among ANSI and America Makes.

Lewis Newton is a research fellow at the University of Nottingham within the Manufacturing Metrology Team. His research was initially in the development of methods for measuring and characterising the surface texture of as built and finished AM surfaces: investigating good practice for surface texture measurement instrumentation and feature-based characterisation pipelines of features specific to the AM process. Currently, his research is in the development of photogrammetric systems for the measurement of aesthetic surfaces and further investigations into feature-based characterisation pipelines for novel applications.

Rajit Ranjan is a researcher at Delft University of Technology in the Structural Optimization and Mechanics Group, which is part of the Mechanical Engineering Faculty. His research focuses on numerical modelling of heat transfer phenomena during the powder bed fusion process and inclusion of thermal models within the topology optimisation scheme. He has worked on several industrial projects with Hitachi, GE and Boeing, where improvements in manufacturing technologies (conventional and additive) were key aspects.

Elia Sbettega is researcher at the University of Padua. His main research interests are in the area of precision engineering, with focus on coordinate metrology and X-ray computed tomography.

Steffen G. Scholz is the head of the research team at Karlsruhe Institute of Technology. He is also the principal investigator in the Helmholtz funded long-term programs 'Digital System Integration' and 'Printed Materials and Systems'. He has more than 15 years' research and development experience in the field of system integration and automation, flexible scalable production, polymer micro- and nano-replication, process optimisation and control, with a special focus on 3D printing and Industry 4.0 concepts and applications. He is a chair of different international conferences within the scope of manufacturing technologies and sustainability and a visiting professor at the Vellore Institute of Technology.

Nicola Senin is associate professor of Manufacturing Technologies and Systems at the Department of Engineering, University of Perugia and an associate researcher at the University of Nottingham. His primary research interests are in industrial metrology (surface and dimensional) and in the area of in-process sensing, monitoring and control for advanced manufacturing processes. He is part of the editorial board of *Surface Topography:*

Metrology and Properties and member of international scientific and technical societies in manufacturing and precision engineering.

Amrozia Shaheen is currently a Marie Skłodowska-Curie research fellow in the Manufacturing Metrology Team, University of Nottingham. Her research is about optical form measurement of complex rough surfaces to improve the precision of AM parts, including the development of multi-view fringe projection systems.

Bethan Smith is a technology manager leading the Non-Conventional Machining Team at MTC. Prior to leading the team, she was involved in directing and growing the Post-Processing for Additive Manufacturing research portfolio at MTC. Her interests lie in the development and application of new and novel techniques to overcome post-processing challenges within additive parts, in particular for features with no direct line-of-sight. More recently, she has been exploring new approaches in post-processing for AM, such as the use of multiple complementary processes for controlling the surface texture and the use of laser technology for creating functional surfaces for improved performance.

Fred van Keulen is the Antoni van Leeuwenhoek chair at the Technical University of Delft. From 2005 to 2010 he was the first chairman of the new Department PME at TU Delft and was scientific director of the Delft Centre for Mechatronics and Microsystems. His research activities focus on mechanics and optimisation with typical applications such as micro/nano devices and the associated high-performance tools and instruments. In his research, manufacturing aspects play a key role.

Wilson Vesga is a materials and structural integrity researcher. He has worked on varied projects gaining extensive experience, both practical and analytical, in materials and mechanical characterisation, NDE techniques and structural integrity. In 2016, he joined MTC in the Metrology and NDT Group. He is actively involved in R&D projects to assess quality and integrity of parts, both post-build and in-process for AM, welding, net shape and composites, applying NDE technologies. He is also working and collaborating to develop standards for AM (ISO/ASTM JG 59 & JG 60) and composites (ASTM WK69139) for application in the aerospace and automotive sectors.

Peter Woolliams is a senior research scientist at the National Physical Laboratory. He has spent twenty years working on developing novel measurement systems in a range of areas, including colorimetry, fibre optics, biophotonics and materials science as well as providing measurement support to companies. He is currently working on developing measurement assurance for the AM industry and is a member of the ISO and ASTM Additive Manufacturing Standards committees, ensuring measurement good practice is included.

Filippo Zanini is assistant professor at the University of Padua, with teaching assignments in Additive Manufacturing and Computer-Integrated Manufacturing. His research experience is in the areas of precision manufacturing and coordinate metrology, including traceability of X-ray computed tomography for advanced manufacturing applications.

1

Introduction to Precision Metal Additive Manufacturing

Martin Baumers, Simone Carmignato and Richard Leach

CONTENTS

1.1 Introduction to Additive Manufacturing

The advent of additive manufacturing (AM), also referred to as '3D printing' (which, however, is a less appropriate term, mostly used in non-technical contexts – see Section 1.2.1), has received considerable attention from technology observers and manufacturing professionals over recent decades. A number of interesting possibilities are associated with the technology: it is seen as an opportunity to realise new designs and design techniques, switch to manufacturing approaches that are more responsive to demand, customise products' cost-efficiently and generally 'digitise' manufacturing (Berman 2012, D'Aveni 2015, Manyika et al. 2013, Segars 2018, Baumers and Holweg 2019). Despite the claims of novelty, AM technology is in reality based on implementations that reach back to the 1980s – the first commercial AM system was brought to the market in 1986 by 3D Systems (see Leach et al. 2019 for a short but relevant history). Since then, AM has entered a long – and still ongoing – process of evolving into a credible and accepted general-purpose manufacturing approach. However, it is also worth considering that humans have been using cutting and forming technologies for thousands of years, so AM is still relatively new.

One motivation for writing this book is to reinforce that, at present, the AM journey is far from complete. The industrial community, and also academic engineering research, is currently investing significant effort and resource into turning AM from the 'next big thing' into a reality that, above all, delivers manufacturing value for its users and ultimately for the consumer. The premise of this book is that the ability to deliver metallic components in a precise and reliable way is central to this significant industrial project.

As defined more specifically in the following parts of this introduction, AM can be thought of as a process that builds objects from three-dimensional (3D) digital models, usually by adding material in a layer-by-layer way. This 'additive' method of processing material sets AM apart from conventional machining approaches, such as casting, forging, moulding or machining processes. While often perceived as a single type of technology, AM is in fact an umbrella term covering a group of quite disparate material deposition technologies (see Section 1.2.2), each with its own characteristics, advantages and drawbacks. Further complicating matters, the range of available build materials is broad, including metals, polymers, ceramics and biomaterials, and has expanded significantly over recent years.

A very brief summary of the history of AM is highly instructive as a backdrop for this book as a treatment of the current technological state of the art, especially with its focus on precision manufacturing applications. Back in the 1990s, AM technologies, known then under the label 'rapid prototyping' (see Section 1.2.1), were developed as a means to automate the manual fabrication of various types of prototypes, for instance in the automotive industry (see, for example, Jacobs 1992). Matching this clear industrial requirement, the early generations of the technology were considered suitable only for the production of prototypes and design studies. Since then, AM, and the ambition of its proponents, has significantly expanded: at the time of writing of this book (early 2020) the precision, repeatability and material range has increased to the point that some AM processes are considered viable as industrial-production technologies. The Taniguchi plot shown in Figure 1.1 illustrates how precision manufacturing capability has evolved over time and tentatively locates current AM and micro-scale AM (including processes such as two-photon

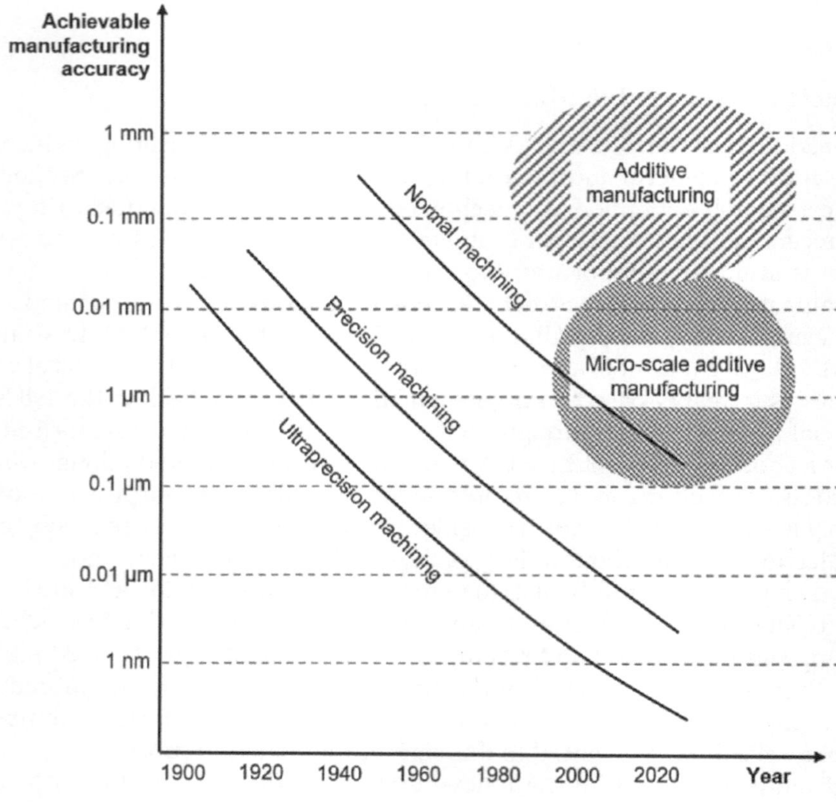

FIGURE 1.1
Taniguchi plot indicating the development of precision manufacturing capability and AM as a function of time. (Adapted from Taniguchi 1974.)

polymerisation) within this framework. As can be seen, AM processes still require considerable improvements in manufacturing accuracy to be on par with conventional machining technology.

Successes in a spectrum of niche applications demonstrate the progress AM has made so far. Most notably, in the hearing-aid and dental-replacement industry, AM has dethroned other manufacturing processes to become the new de facto standard way of making such products. Similar dethroning is happening in the world of prosthetics. These successes have encouraged some observers to claims that AM will displace conventional tooled manufacturing processes in the near future en masse: 'within the next five years we will have fully-automated, high-speed, large-quantity additive manufacturing systems that are economical even for standardized parts' (D'Aveni 2015). AM technology is thus being perceived as 'ready to emerge from its niche status and become a viable alternative to conventional manufacturing processes in an increasing number of applications' (Cohen et al. 2014).

It is important to note, however, that only a few examples of the successful commercial adoption of AM in the manufacture of standard parts have been reported to date. Perhaps the most frequently cited example of such an application is the fuel injector nozzle adopted in the most recent generation of CFM International's Leading Edge Aviation Propulsion (LEAP) turbofan aeroengines (see also Chapter 2, Figure 2.1). Allowing the consolidation of a significant number of metal parts into a highly complex component manufactured via laser powder bed fusion, the decisive advantage offered by AM in this application is that it allows a redesign of the component's complex internal structure, leading to a slower build-up of fuel residue, in turn leading to a longer service life (Shields and Carmel 2013).

As illustrated by this example, applications of AM usually draw on the technology's specific advantages that stem from its tool-less operating principles. These advantages include the elimination of the need to invest in costly tooling and the ability to introduce geometric changes to individual units at little or no additional process cost, as well as a general capability of realising highly complex product geometries.

One question that is frequently overlooked, however, is whether the characteristics of AM technology make it a suitable substitute for traditional tool-based manufacturing processes in high-volume manufacturing settings. To be successful in such contexts, AM must not only be able to achieve the required product geometries in principle, and be competitive on a unit cost basis, but also deliver product and process characteristics that are compatible with other supporting processes found in such settings. This includes design, product engineering and material supply activities occurring before the manufacturing stage and, perhaps most crucially, post-processing and product-qualification processes following the AM operation. This remains an important question that has not received sufficient attention within the engineering community, with most existing work conceptually outlining AM's capability of challenging traditional manufacturing supply chains (see, for example, Cotteleer and Joyce 2014, D'Aveni 2015, De Jong and De Bruijn 2013, Laplume et al. 2016, Tuck et al. 2008, Weller et al. 2015).

Returning to the question of the 'next big thing', the considerations outlined in this introduction will be important in determining how 'big' exactly AM can become in the future. Having steadily grown into an industry worth in excess of $9 billion annual sales revenue over the last three decades (Wohlers 2019), some technology observers predict accelerated growth in the years ahead as AM is adopted into more and more applications. As forecast by the United Kingdom's AM National Strategy (Additive Manufacturing UK 2017), the global AM industry is likely to reach an industry size of between $22 billion and in excess of $70 billion in the year 2025. The large discrepancy between the lower and upper estimates originate from the lack of clarity as to how compatible AM is with established underpinning and supporting processes in current manufacturing practice. Such issues are also at the heart of the prospects of AM in precision manufacturing applications. For this reason, this book provides a much-needed and timely contribution to help establish a clearer understanding of AM.

1.2 Basic Definitions

The basic general AM terms used in this book are defined in the following. Each definition includes an indication of the reference providing the relevant terminology and definition. Further definitions, pertaining to specific domains, are given in the main chapters of this book (for example, definitions of geometrical parameters and metrology terms are given in Chapter 10).

1.2.1 General Terms

- *Additive manufacturing (AM)* – process of joining materials to make parts from 3D model data, usually layer upon layer, as opposed to subtractive manufacturing and formative manufacturing methodologies (ISO/ASTM FDIS 52900 2019).
- *AM system* – machine and auxiliary equipment used for AM (ISO/ASTM FDIS 52900 2019).
- *AM machine* – section of the AM system including hardware, machine-control software, required setup software and peripheral accessories necessary to complete a build cycle for producing parts (ISO/ASTM FDIS 52900 2019).
- *Rapid prototyping (RP)* – application of AM intended for reducing the time needed for producing prototypes. Historically, RP was the first commercially significant application for AM and has, therefore, been commonly used as a general term for this type of technology. (ISO/ASTM FDIS 52900 2019).
- *Rapid tooling (RT)* – application of AM intended for the production of tools or tooling components with reduced lead times as compared to conventional tooling manufacturing (ISO/ASTM FDIS 52900 2019).
- *3D printing* – this term is often used in a non-technical context synonymously with AM; however, *3D printing* is a more specific term, referring to the fabrication of objects through the deposition of a material using a print head, nozzle or other printer technology (ISO/ASTM FDIS 52900 2019).
- *STL* – file format for model data describing the surface geometry of an object as a tessellation of triangles used to communicate 3D geometries to machines in order to build physical parts. The STL file format was originally developed as part of the CAD package for the early STereoLithography Apparatus, thus referring to that process. It is sometimes also described as 'Standard Triangulation Language' or 'Standard Tessellation Language', though it has never been recognised as an official standard by any standards developing organisation (ISO/ASTM FDIS 52900 2019).
- *AM file format (AMF)* – file format for communicating AM model data including a description of the 3D surface geometry with native support for colour, materials, lattices, textures, constellations and metadata. Similar to STL, the surface geometry is represented by a triangular mesh, but in AMF the triangles can also

be curved. AMF can also specify the material and colour of each volume and the colour of each triangle in the mesh (ISO/ASTM FDIS 52900 2019). The standard specification of AMF is given in ISO/ASTM 52915 (2016).

- *Feedstock* – Bulk raw material supplied to the AM building process. For AM building processes, the bulk raw material is typically supplied in various forms, such as liquid, powder, suspensions, filaments and sheets (ISO/ASTM FDIS 52900 2019).

1.2.2 Process Categories

- *Binder jetting (BJT)* – AM process in which a liquid bonding agent is selectively deposited to join powder materials (ISO/ASTM FDIS 52900 2019).
- *Directed energy deposition (DED)* – AM process in which focused thermal energy is used to fuse materials by melting as they are being deposited. 'Focused thermal energy' means that an energy source (for example, laser, electron beam or plasma arc) is focused to melt the materials being deposited (ISO/ASTM FDIS 52900 2019).
- *Material extrusion (MEX)* – AM process in which material is selectively dispensed through a nozzle or orifice (ISO/ASTM FDIS 52900 2019).
- *Material jetting (MJT)* – AM process in which droplets of feedstock material are selectively deposited. Example feedstock materials for material jetting include photopolymer resin and wax (ISO/ASTM FDIS 52900 2019).
- *Powder bed fusion (PBF)* – AM process in which thermal energy selectively fuses regions of a powder bed (ISO/ASTM FDIS 52900 2019).
- *Sheet lamination (SHL)* – AM process in which sheets of material are bonded to form a part (ISO/ASTM FDIS 52900 2019).
- *Vat photopolymerisation (VPP)* – AM process in which liquid photopolymer in a vat is selectively cured by light-activated polymerisation (ISO/ASTM FDIS 52900 2019).

1.2.3 Other Terms

- *Slicing process* – pre-manufacturing stage of an AM process, involving slicing the facet (volume) model into several successive layers and recording the information contained within each layer (ISO 17296-4 2014).
- *Build chamber* – enclosed location within the AM system where the parts are fabricated (ISO/ASTM FDIS 52900 2019).
- *Build platform* – base which provides a surface upon which the building of the parts is started and supported throughout the build process (ISO/ASTM FDIS 52900 2019).
- *Curing* – change of the physical properties of a material by means of a chemical reaction. An example of an important curing operation in AM is the changing of a polymer resin from liquid to solid by light-activated cross-linking of molecule chains (ISO/ASTM FDIS 52900 2019).
- *Hatch distance* – separation between two consecutive tracks along the scan path, measured by a distance from the centre of one track to the centre of the next one (Masood 2014).
- *Staircase effect* – discretisation phenomenon due to the layer-wise nature of AM processes, causing edges of layers to be visible on the surfaces, especially on slanted or curved surfaces (Gibson et al. 2015).

- *Support* – structure separate from the part geometry that is created to provide a base and anchor for the part during the building process. Supports are typically removed from the part prior to use. For certain processes such as MEX and MJT, the support material can be different from the part material and deposited from a separate nozzle or print head. For certain processes, such as metal PBF, auxiliary supports can be added to serve as an additional heat sink for the part during the building process (ISO/ASTM FDIS 52900 2019).
- *Lattice structure* – 3D geometric arrangement composed of connective links between vertices (points) creating a functional structure (ISO/ASTM FDIS 52900 2019).
- *Hybrid manufacturing* – Manufacturing methods combining advantages of AM (for example, reduced material wastage and geometrical freedom) and advantages of subtractive manufacturing (for example, dimensional accuracy and surface finish) (Srivastava et al. 2019).
- *Post-processing* – process step, or a series of process steps, taken after the completion of an AM build cycle in order to achieve the desired properties in the final product (ISO/ASTM FDIS 52900 2019).

1.3 Towards Precision Additive Manufacturing

Precision engineering has been, and continues to be, one of the disciplines needed to enable future technological progress, especially in the area of manufacturing, and there is no reason this will not be the case for AM. Demands for increasing precision have also evolved in almost all other industrial sectors, ranging from planes, trains and automobiles to printers (printing electronics and optical surfaces), scientific and analytical instruments (microscopes and telescopes to particle accelerators), medical and surgical tools, and traditional and renewable power generation. At the root of all of these technologies are increasingly advanced machines and controls, and, in manufacturing, advanced products. This ubiquitous need for increased precision has resulted in the growth of an international community, with precision engineering societies in Asia, Europe and America gathering to exchange information and ideas focused on conceptual and technological solutions for the creation of increasingly precise processes. The European Society for Precision Engineering and the American Society for Precision Engineering have joined forces to run an annual conference on the subject of 'Advancing Precision in Additive Manufacturing', where key players in the precision community meet those from the AM community and exchange ideas and recent progress. An outcome of these societal activities has been a growing awareness in the AM community of the underlying tools and approaches for the design of mechanical processes to optimise and evaluate instrument and machine performance. These principles and design concepts, applied in the AM community, form the basis of this book.

From its history, it might be supposed that a simple, clear definition of precision engineering would be apparent (see Winchester 2018 for an in-depth historical account of precision engineering). However, unlike many fields of study for which there is a targeted and, therefore, limited scope, precision engineering spans many scientific and engineering disciplines, and an enquirer is likely to get different answers from practitioners focusing on different aspects in this field, such as machine designers, instrumentation developers or those involved with interpretation of measurement data. One possible definition is to determine the ratio of the value of accuracy, repeatability, resolution, uncertainty or some other term that quantifies error with the range of an instrument or machine. It has been stated that engineering starts to become precision engineering when these ratios approach 1 part in 10^6 (Leach and Smith 2018). Therefore, a diamond-turning machine with a range of 100 mm in its axes and a repeatability of 100 nm would be considered a precision machine. The average metal laser powder bed fusion AM machine, on the other hand, may have a cube volume of 100 mm, but the repeatability of the machine is perhaps 100 µm, i.e. a ratio of 1 part in 10^3. By the above definition, this is not considered precision engineering. Precision engineering has also been promoted as the pursuit of determinism in manufacturing processes (Bryan 1993). In this view, the lack of precision that often comes from a lack of repeatability is considered a lack of attention to causal effects within the process. Precision tolerances have been achieved with metal AM in some examples (discussed throughout this book), but this has been less by following the principles of precision engineering and more to do with 'throwing it hard enough until it sticks'. Considerable effort is expended to optimise the design and process, and then a vast amount of post-process inspection is carried out and multiple finishing operations are applied. A better way to approach this would be to apply precision principles at all stages of the manufacturing: design, the AM machine construction and control, the various measurement processes, the finishing operations, functional testing and every other stage. This is the focus of this book: how can the principles of precision engineering be applied to improve the determinism of the AM manufacturing of a product.

Precision engineering might also be viewed as an endeavour to identify and quantify the causes leading to non-repeatability and utilise the principles of precision engineering to reduce, and preferably eliminate, the effect. For example, with machine tools, common sources of non-repeatability include thermal effects, loose joints, vibration, limited measurement and actuator resolution, contamination and friction. Again, approaches for addressing these limitations to precision in the AM of a product are the focus of this book.

The foundational concepts in precision engineering are covered in detail elsewhere (see, for example, *Basics of Precision Engineering* [2018], which can be seen as a companion text to this book, and whose Chapter 1 is particularly suggested to the reader interested in the fundamentals). *Precision Metal Additive Manufacturing* takes each step in the current AM process chain and investigates where the precision aspects can be enhanced. There are many existing comprehensive books on the AM process, so this book tries not to repeat the basic material if possible and simply supplies appropriate references. Each chapter presents the state of the art in its subject area, how precision can be used and enhanced and what are the remaining research challenges to address. Chapter 2 looks at the AM design process, specifically at topology optimisation and how precision principles can be applied to enhance the design and, ultimately, the product. Specifically, critical overhang elimination, overheating prevention and distortion reduction are considered. Chapter 3 discusses the various AM processes and particularly provides insight into the development of precision AM processes. The setting of priorities, significant process parameters and performance indicators for the AM process, as well as reasons for lack of process precision, are presented. This is followed by a discussion on data-driven process development. Chapter 4 presents the various techniques used to model the metal AM processes. The concepts of length and time scales of different physical phenomena are discussed, followed by potential applications of the different models towards reducing defects and increasing the precision of AM parts. The chapter presents the concepts of uncertainty in models and the need for proper verification and calibration procedures – key features of any precision engineering design. Chapter 5 presents the background and basic principles of secondary finishing of AM components, as well as the range of technical solutions available. Chapter 6 gives an overview of the status with specification standards for AM. Cost implications are always at the heart of a design process and form the basis of Chapter 7. The specification of a robust cost model for precision AM, extending existing approaches through the inclusion of a sub-model reflecting intensive inspection activities, is presented for the first time. Machine performance evaluation is vital for objectively demonstrating that a machine can meet specifications, standards and quality management systems and for allowing users to make informed decisions on how to best use their resources; this is the focus of Chapter 8. Although AM offers relatively new ways to produce parts, the methods used to evaluate the performance of the machines apply expertise developed in precision engineering, especially in machine tool metrology. Chapter 9 presents the various non-destructive testing methods that are being adapted or developed to investigate the integrity of AM parts, and it includes a detailed summary of the various defects and manufacturing anomalies. Chapters 10, 11 and 12 present off-line metrology principles and techniques: coordinate metrology, surface metrology and X-ray computed tomography, respectively. The benefit of the design freedom of AM can be outweighed by the complexity in measuring and verifying the resulting parts. Chapter 13 describes on-machine sensing, monitoring and control methodologies that can be used in metal AM, and particularly in powder bed fusion processes. Special attention is devoted to intelligent data-driven methodologies for on-machine measurement, monitoring and control methods.

References

Additive Manufacturing UK. 2017. *National Strategy 2018 – 25 Leading Additive Manufacturing in the UK*. Coventry, UK: UK Additive Manufacturing Steering Group.

Baumers, M., and Holweg, M. 2019. On the economics of additive manufacturing: Experimental findings. *Journal of Operations Management* 65:794–809.

Berman, B. 2012. 3-D printing: The new industrial revolution. *Business Horizons* 55:155–162.

Bryan, J. B. 1993. The deterministic approach in metrology and manufacturing. *ASME 1993 International Forum on Dimensional Tolerancing and Metrology*, June, Michigan, USA.

Cohen, D., Sargeant, M., and Somers, K. 2014. 3-D printing takes shape. *McKinsey Quarterly* 1:1–6.

Cotteleer, M., and Joyce, J. 2014. 3D opportunity for production: Additive manufacturing makes its (business) case. *Deloitte Review* 15:146–161.

D'Aveni, R. 2015. The 3-D printing revolution. *Harvard Business Review* 93:40–48.

De Jong, J. P., and De Bruijn, E. 2013. Innovation lessons from 3-D printing. *MIT Sloan Management Review* 54:43.

Gibson, I., Rosen, D. W., and Stucker, B. 2015. *Additive Manufacturing Technologies*. Berlin, Germany: Springer.

ISO 17296 part 4. 2014 Additive manufacturing – General principles – Part 4: Overview of data processing. Geneva, Switzerland: International Organization of Standardization.

ISO/ASTM FDIS 52900. 2019. Additive manufacturing – General principles – Part 1: Vocabulary and fundamental concepts. Geneva, Switzerland: International Organization of Standardization.

ISO/ASTM 52915. 2016. Specification for additive manufacturing file format (AMF) Version 1.2. Geneva, Switzerland: International Organization of Standardization.

Jacobs, P. F. 1992. *Rapid Prototyping & Manufacturing: Fundamentals of Stereolithography*. Dearborn, MI: Society of Manufacturing Engineers.

Laplume, A. O., Petersen, B., and Pearce, J. M. 2016. Global value chains from a 3D printing perspective. *Journal of International Business Studies* 47:595–609.

Leach, R. K., and Smith, S. T. 2018. *Basics of Precision Engineering*. Boca Raton, FL: CRC Press.

Leach, R. K., Bourell, D., Carmignato, S., Donmez, A., Senin, N., and Dewulf, W. 2019. Geometrical metrology for metal additive manufacturing. *CIRP Annals* 68:677–700.

Manyika, J., Chui, M., Bughin, J., Dobbs, R., Bisson, P., and Marrs, A. 2013. *Disruptive Technologies: Advances That Will Transform Life, Business, and the Global Economy* (Vol. 180). San Francisco, CA: McKinsey Global Institute.

Masood, S. 2014. Advances in additive manufacturing and tooling. In: Hashmi, S. (ed.), *Comprehensive Materials Processing*. Amsterdam, the Netherlands: Elsevier ScienceDirect.

Segars, A. H. 2018. Seven technologies remaking the world. *MIT Sloan Management Review*.

Shields, M., and Carmel, J. 2013. Turbofan engine technology upgrades: How should suppliers react? Avascent report. http://www.avascent. com/2013/09/turbofan-engine-technology-upgrades-suppliers-react/(accessed 2 October 2019).

Srivastava, M., Rathee, S., Maheshwari, S., and Kundra, T. K. 2019. *Additive Manufacturing – Fundamentals and Advancements*. Boca Raton, FL: CRC Press.

Taniguchi, N. 1974. On the basic concept of nanotechnology. *Proceedings of the International Conference on Production Engineering*, 2:18–23, Tokyo, Japan.

Tuck, C. J., Hague, R. J., Ruffo, M., Ransley, M., and Adams, P. 2008. Rapid manufacturing facilitated customization. *International Journal of Computer Integrated Manufacturing* 21:245–258.

Weller, C., Kleer, R., and Piller, F. T. 2015. Economic implications of 3D printing: Market structure models in light of additive manufacturing revisited. *The International Journal of Production Economics* 164:43–56.

Winchester, S. 2018. *The Perfectionists: How Precision Engineers Created the Modern World*. New York, NY: Harper Collins USA.

Wohlers, T. 2019. *Wohlers Report*. Fort Collins, CO: Wohlers Associates Inc.

2

Topology Optimisation Techniques

Rajit Ranjan, Can Ayas, Matthijs Langelaar and Fred van Keulen

CONTENTS

2.1 Introduction

In recent years, additive manufacturing (AM) has evolved from a prototyping technique into an established industrial manufacturing process. One of the primary reasons for this rapid rise in popularity is the increased design freedom that AM offers. Due to the inherent layer-by-layer material addition approach in AM, geometric complexity is not associated with extra cost (see Chapter 7 for an analysis of the cost implications of AM). This is in contrast to conventional manufacturing methods, such as milling, drilling and casting, where the manufacturing cost is directly correlated with geometric complexity. The increased shape freedom associated with AM thus allows manufacturing of highly complex and efficient engineering structures, for example, suitable for high-technology precision applications. However, the large degree of design freedom, particularly in the context of multi-physics fields, demands for advanced computational design tools.

The re-design of a General Electric (GE) fuel nozzle, shown in Figure 2.1, is a representative case where AM has enabled an increase in the overall product efficiency, along with significant cost savings (Grunewald 2016). However, the process of (re-)designing for AM

FIGURE 2.1
The re-design of the GE fuel nozzle printed using the direct metal laser sintering process. The nozzle provides functional integration of twenty separate parts while providing 25% reduction in mass. These improvements translate into a saving of $3 million per aircraft, per year. (From GE Additive 2018. With permission.)

is complex and typically involves multi-disciplinary computational and experimental trials, which obviously leads to increased overall costs. Therefore, the concept of 'design for additive manufacturing' (DFAM) emerged and refers to the large set of design principles for developing optimised designs that exploit the capabilities of AM.

The challenges associated with DFAM can be broadly attributed to two factors. The first DFAM challenge arises from the fact that, typically, AM parts are designed to operate in a complex multi-physics environment where precise functionality is desired. For example, in the case of the GE fuel nozzle, optimal mixing of fuel and air is of paramount importance for the efficient operation of the gas turbine, but determining the optimal nozzle design to achieve this is non-trivial. Another industrial example of an AM re-design is the ASML conditioning ring shown in Figure 2.2 (Loncke 2014). Here, an improvement in thermal control was achieved by re-designing the cooling channels that were manufactured by AM. The AM shape freedom enables complex channel geometries, but this also increases the demands on the design process. Both of these are examples where geometrically more complex designs outperform their relatively simpler predecessors. However, the complications associated with the design process also increase significantly. Typically, in such cases, multi-physics and multi-scale simulations are required to evaluate the functional performance of the designs. This is far from trivial, making the design process complex and computationally involved. Furthermore, the dimensional tolerance requirements for such precision applications are typically tight.

The second challenge associated with DFAM is related to the AM process itself. Laser powder bed fusion (L-PBF) processes are typically used for fabricating complex, high-technology metal AM parts. However, it is well known that L-PBF processes are not free

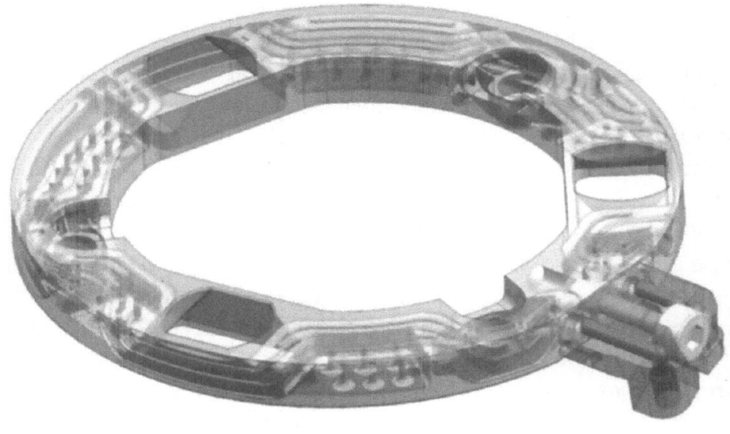

FIGURE 2.2
Re-design of the ASML conditioning ring. AM facilitates design of more complex integrated cooling channels that improve the thermal performance by a factor of six. (From Loncke 2014. With permission.)

from limitations and, if overlooked, these limitations can cause various types of defects in the final parts (see Chapter 13 for a description of typical defects in L-PBF). Defects lead to increased costs, as additional post-processing steps are required to address them. Moreover, in certain cases, defects lead to complete build failure (Kastsian and Reznik 2017) or a poor-quality part. This issue is even more relevant for precision parts. Therefore, due account for the process limitations should be made during the design stage, especially for precision parts. The above summarised challenges call for powerful computational design techniques. In particular, topology optimisation (TO) is highly beneficial for AM products that rely on a delicate interaction of multiple fields. Before introducing TO, AM-related limitations which are relevant for precision AM parts are discussed briefly.

The most well-known example of one such limitation is the building of overhanging features. Downfacing surfaces should be avoided if the overhang angle θ measured between the surface and base-plate is smaller than a critical value, as shown in Figure 2.3 (Mertens et al. 2014a). This is a typical example of a geometrical restriction. There are other relatively more complex issues, such as local overheating or distortion during the fabrication process, which also compromise the quality of the part. Inclusion of such complex criteria during the design stage is non-trivial as it requires a thorough knowledge of L-PBF process physics.

L-PBF is essentially a multi-scale process as it typically involves high-energy laser beams with radii of the order of 50 μm (micro-scale) for fabricating parts that have dimensions up to 30 cm (macro-scale). Also, it is a multi-physics process in which heat transfer, fluid dynamics, solid mechanics and phase transformations play critical roles (Schoinochoritis et al. 2017). In order to address the intricate physical interactions during manufacturing, as well as the potentially multi-domain aspects of the products during service, AM designers

FIGURE 2.3
The overhang angle θ as measured between part surface and base-plate should remain above a critical value θ^{cr}.

need a radical shift from traditional heuristic design techniques. In fact, exploiting the full design freedom offered by AM for a part relying on the interactions of the multiple physical domains for its operations is too complex for 'manual' design. Computational design tools can greatly help in this context as they can assist designers in developing complex new designs that meet certain functional criteria, while also considering the limitations of the AM process.

TO is a computational design method which, in the context of AM, can address functionality and manufacturability constraints in a mathematically rigorous way (Bendsøe and Sigmund 2003). A unique feature of TO is its capability to provide detailed material layouts, which can be manufactured exclusively with AM. Thus, it is *the* design tool to exploit the freedom offered by AM. A brief description of TO is given in Section 2.2. Since TO leads to typically complex designs, the limitations imposed by AM can be violated. This would potentially lead to manual design modifications and thereby deterioration of the performance. Therefore, TO methods that address limitations associated with precision AM are discussed in Section 2.3. These methods lead to computer-generated designs which not only satisfy AM requirements, but also outperform manual designs and are highly competitive. Lastly, challenges and an outlook are presented in Section 2.4.

2.2 Topology Optimisation

TO is a computational design method which aims to answer a fundamental engineering question: where to place material inside a design domain to get the best performance? The seminal paper on TO (Bendsøe 1989) demonstrated the method by solving mechanical design problems, for example, finding the optimal material distribution having maximum stiffness for a predefined load. Over time, TO has found other application fields, for example, fluid dynamics, heat transfer, optics, electromagnetics and their combinations. Nevertheless, the basic idea of TO remains the same: casting a given design problem into a material distribution problem that can be solved using gradient-based mathematical optimisation algorithms (Sigmund and Maute 2013).

In the past thirty years, the TO method evolved in multiple directions. These developments can be classified into two main groups. The first group is based on discretising the domain into a set of finite elements and then defining the resulting material distribution by allocating material or voids to each of these elements (see Figure 2.4a). The second group treats the problem as a classic shape optimisation and describes the material distribution with iterative boundary movements to evolve the part geometry (see Figure 2.4b). An additional step for hole nucleation may be included to easily increase the complexity of the layout. Prominent examples are the TO schemes involving a level-set-based description of the geometry. The reader is referred to the review of Dijk et al. (2013) for a detailed discussion on this method.

In both groups of TO methods, there are many variants which have been studied, see (Deaton and Grandhi 2014) for an overview. However, the most prominent TO method in both the commercial software domain and in academic research is the density-based method, which is part of the material distribution group. Consequently, in the remainder of this chapter, the focus will be on this method and, in particular, how these popular TO techniques can account for AM limitations that are relevant for producing precision parts.

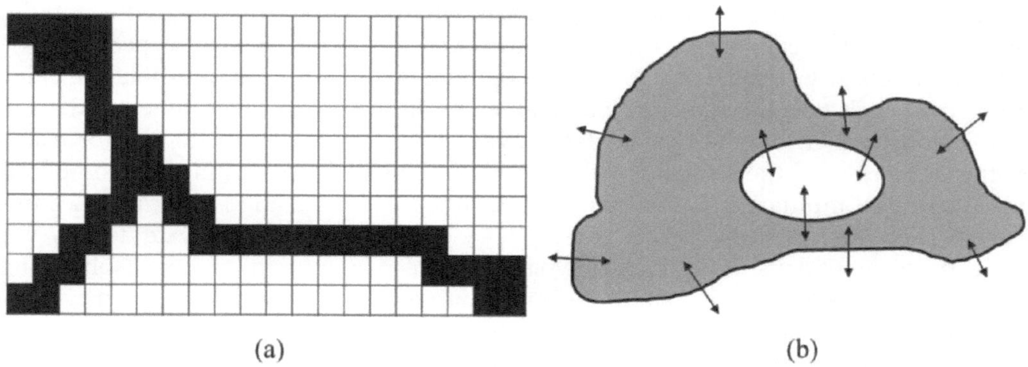

FIGURE 2.4
(a) Optimal material distribution is found by dividing the design domain into a set of finite elements followed by iterative allocation of material/void (black/white) to each element. (b) Iterative boundary movement and hole nucleation for finding optimal shapes.

2.2.1 Density-Based TO Method

The density-based TO formulation is introduced here using an example design problem of stiffness maximisation (compliance minimisation). The seminal paper on TO (Bendsøe 1989) focused on this problem, and since then, stiffness maximisation has been commonly used for TO. Density-based TO involves finding the stiffest material distribution inside a design domain for a given set of loads and boundary conditions, while the part volume is restricted. Figure 2.5a shows an example of a two-dimensional (2D) design domain with loading and kinematic boundary conditions, i.e. a point load acts at the lower right vertex of the design domain while the left edge remains fixed.

The process starts with discretising the design domain into a set of finite elements. In Figure 2.5b, 4-node bi-linear square elements are used as they lead to a structured mesh for which the formulation is relatively simple. Alternatively, an unstructured mesh can be used for spatial discretisation. Each element is then assigned a design variable ρ which

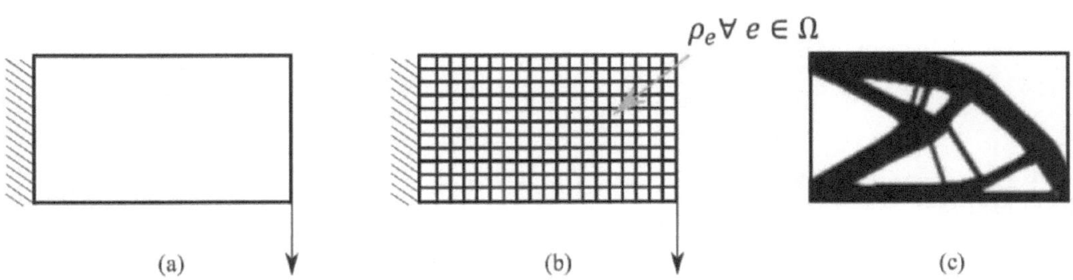

FIGURE 2.5
A 2D example for TO of a cantilever beam. (a) Design domain Ω with kinematic loading conditions, i.e. point load at lower right vertex, while left edge remains fixed. (b) Discretised domain (schematic, actual mesh is finer) with each finite element assigned a design variable ρ_e (subscript 'e' denotes the element number). (c) Optimal material distribution for this problem found using the density-based TO formulation.

represents the density of the element normalised with the physical density of the material. A density value thus can be either zero or one, representing void or solid respectively.

The aim of density-based TO is to find the optimal layout with densities that are either zero or one, such that the structure has maximum stiffness. However, this is a discrete optimisation problem where design derivatives of the response functions – for example, the derivative of the stiffness with respect to density field – cannot be computed. Consequently, computationally efficient gradient-based optimisation algorithms cannot be used. Therefore, the formulation is relaxed by allowing for intermediate density values, i.e. the element densities can take any value between zero (void) and one (solid). This transforms the problem into a continuous optimisation problem. The optimal material distribution is then determined in an iterative manner, where element densities are updated in each iteration, making use of the corresponding design derivatives such that the overall stiffness of the structure increases. Once the stiffness of the structure converges to a stable maximum value, the TO terminates.

Although allowing for intermediate densities enables the use of gradient-based optimisation, a crisp so-called black and white design, i.e. a design that exclusively contains void and solid areas with clear boundaries, is still preferred, as intermediate densities may be difficult to realise physically. Although, in recent years, researchers have started capitalising on intermediate densities by using lattice structure–based designs which become manufacturable due to AM (Amir and Mass 2018), this may not be suitable for all applications. Hence, the above formulation is combined with a scheme which puts a penalty on the intermediate densities and drives the final design towards a solid/void (one/zero or black/white) configuration. The most commonly used approach to achieve a crisp black/white design is referred to as solid isotropic material with penalisation (SIMP) (Bendsøe and Sigmund 1999). Typically, the SIMP approach in combination with finite element analysis (FEA) is used for numerical computation of the optimisation objective and constraints. In the following section, the mathematical formulation for TO is presented.

2.2.1.1 Problem Formulation

The discretised version of the design domain Ω is shown in Figure 2.5(b), where each finite element (FE) is characterised by a design variable, i.e. ρ_e for the eth element. The SIMP-based TO problem, where the aim is to minimise compliance c, can be written as

$$\min : c(\rho) = \mathbf{F}^{\mathrm{T}}\mathbf{U} = \mathbf{U}^{\mathrm{T}}\mathbf{K}\mathbf{U} = \sum_{e=1}^{N} (\rho_e)^p \, \mathbf{u}_e{}^T \mathbf{k}_e \mathbf{u}_e, \tag{2.1}$$

which is subject to

$$\mathbf{K}\mathbf{U} = \mathbf{F}, \tag{2.2}$$

$$g(\rho) = \frac{V(\rho)}{V_0} - f \leq 0, \tag{2.3}$$

$$0 < \rho_{\min} \leq \rho_e \leq 1, \ \forall \ e \in \Omega, \tag{2.4}$$

where $\boldsymbol{\rho}$ is the vector of all the design variables ρ_e, $\mathbf{U}$ and $\mathbf{F}$ are the nodal degrees of freedom and load vectors respectively, $\mathbf{K}$ is the global stiffness matrix, $\mathbf{u}_e$ and $\mathbf{k}_e$ are the element displacement vector and stiffness matrix respectively, N is the total number of elements, ρ_{min} is the minimum physical density (non-zero to avoid singularity of $\mathbf{K}$), V and V_0 are the volume of solid region and design domain respectively and f is the prescribed maximum volume fraction of material allowed in the design domain. The equilibrium equation for the discretised FE problem is given by Equation (2.2). Finally, the SIMP penalisation is imposed as a power law on the density values given by the penalty factor $p > 1$, as shown in Equation (2.1). The optimised cantilever design found using the presented formulation is shown in Figure 2.5(c).

2.2.1.2 Sensitivity Analysis

The next step is the calculation of sensitivities, i.e. the design derivatives of objective c and constraint $g(\boldsymbol{\rho})$ with respect to design variable ρ_e. The design derivative of Equation (2.3) can be easily found as it depends directly on design variable ρ_e, but for Equation (2.1), the derivatives of the nodal degrees of freedom with respect to the design variables are required when direct differentiation is applied. Typically in TO, there are a large number of design variables and a comparatively small number of response functions (i.e. objectives and constraints); therefore, it is logical to avoid explicit computation of the derivatives of the nodal degrees of freedom for computational tractability. For this purpose, the adjoint method is used, where an augmented compliance response is defined by adding the governing equation (Equation 2.2) in zero form, multiplied with a Lagrange multiplier, thus

$$c = \mathbf{F}^{\mathrm{T}}\mathbf{U} + \boldsymbol{\lambda}^{\mathrm{T}}\left(\mathbf{F} - \mathbf{K}\mathbf{U}\right). \tag{2.5}$$

Next, Equation (2.5) is differentiated with respect to the design variable ρ_e and the terms containing the derivative with respect to displacements are collected, i.e

$$\frac{\partial c}{\partial \rho_e} = \left(\mathbf{F}^{\mathrm{T}} - \boldsymbol{\lambda}^{\mathrm{T}}\mathbf{K}\right)\frac{\partial \mathbf{U}}{\partial \rho_e} - \boldsymbol{\lambda}^{\mathrm{T}}\frac{\partial \mathbf{K}}{\partial \rho_e}\mathbf{U}. \tag{2.6}$$

Equation (2.6) can be simplified to

$$\frac{\partial c}{\partial \rho_e} = -\boldsymbol{\lambda}^{\mathrm{T}}\frac{\partial \mathbf{K}}{\partial \rho_e}\mathbf{U}, \tag{2.7}$$

such that λ satisfies the adjoint equation

$$\mathbf{F}^{\mathrm{T}} - \boldsymbol{\lambda}^{\mathrm{T}}\boldsymbol{K} = 0. \tag{2.8}$$

It should be noted that Equation (2.8) resembles the form of equilibrium Equation (2.2). Therefore, for the choice $\boldsymbol{\lambda} = \mathbf{U}$, the displacement derivatives vanish from Equation (2.6). Consequently, the sensitivity of the objective with respect to the design variable can be efficiently evaluated using

$$\frac{\partial c}{\partial \rho_e} = -\mathbf{U}^{\mathrm{T}}\frac{\partial \mathbf{K}}{\partial \rho_e}\mathbf{U}. \tag{2.9}$$

It is worth mentioning that sensitivity computation is relatively simple for compliance minimisation problems as the adjoint variable turns out to be equal to the displacement. Consequently, the compliance minimisation problem is also referred to as *self-adjoint*. However, this is not the case in general. Typically, an additional linear adjoint equation has to be solved to compute the design derivative for each response. Also, note that design independent load is assumed here; therefore, the derivative of **F** with respect to the design variable ρ_e vanishes, which might not be the case for other more complex problems.

2.2.1.3 Filtering Techniques

If the presented TO method is used in its current form, it is likely to encounter numerical difficulties, such as mesh-dependency and artefacts known as *checkerboard patterns* (Bendsøe and Sigmund 2003). For instance, the TO result of the cantilever beam problem shown in Figure 2.6(a) exhibits classic checkerboard patterns. This is a numerical artefact as the patterns achieve artificially high but unrealistic stiffness due to the inability of the employed low-order FEs to represent the true structural response of alternating solid void elements, reminiscent of a checkerboard. In order to avoid these undesired oscillatory density patterns, spatial filtering techniques are used.

The density filtering technique suggested by Bruns and Tortorelli (2001) is one of the most commonly used approaches. It essentially applies a low-pass filter to an input density field, producing a filtered field where density oscillations below a certain length scale have been removed. While this eliminates the checkerboard artefacts from the solution space, it also serves to impose a minimum feature size on the design and thereby allows for mesh-independent solutions. The so-called density filter determines filtered element densities as

$$\tilde{\rho}_e = \frac{\sum_{j \in N_e} W_{e,j} \rho_e}{\sum_{j \in N_e} W_{e,j}} \tag{2.10}$$

where N_e is the neighbourhood of an element e located at position $\mathbf{r}_e$ and $W_{e,j}$ defines the weight of a control variable at spatial position $\mathbf{r}_j$, using, for example, a linearly decaying distance function: $W_{e,j} = \max\left(0, R - \left\|\mathbf{r}_j - \mathbf{r}_e\right\|\right)$, with filter radius R. The neighbourhood is defined as

$$N_e = \left\{ j \ : \text{dist}\left(e, j\right) \leq R \right\} \tag{2.11}$$

FIGURE 2.6
(a) Checkerboard patterns obtained in the TO result of a cantilever problem. (b) Optimal design found using density filtering technique with filter radius R set as 2 elements wide.

where the operator dist(e,j) is the centre-to-centre distance between elements e and j. This essentially means that the filtered element density $\tilde{\rho}_e$ for element e is defined as a weighted mean of densities of neighbouring elements located within a distance R from the centre of element e. It is important to note that, with the introduction of filters, the original densities ρ_e are now referred to as the *design variables*, while the filtered densities $\tilde{\rho}_e$ represent the actual physical densities on which the performance is determined. The optimisation problem thus becomes to find a set of design variables $\boldsymbol{\rho}$ for which compliance $c(\tilde{\boldsymbol{\rho}})$ is minimised. The sensitivity of compliance with respect to physical densities $\tilde{\rho}_e$ is still given by Equation (2.9), while the consistent sensitivity with respect to design variable ρ_e is obtained by the chain rule

$$\frac{\partial c}{\partial \boldsymbol{\rho}} = \frac{\partial c}{\partial \tilde{\boldsymbol{\rho}}} \frac{\partial \tilde{\boldsymbol{\rho}}}{\partial \boldsymbol{\rho}}. \tag{2.12}$$

Figure 2.6(b) presents an optimal solution of the same cantilever problem, solved using the concept of the density filter, given by Equations (2.10)–(2.12). Filter radius R is set as two elements wide, and the resulting design is free from checkerboard patterns.

Many other filtering techniques have been proposed to control certain aspects of TO solutions, including manufacturing constraints. For further details, the reader is referred elsewhere (Sigmund 2007, Vatanabe et al. 2016).

2.2.1.4 Solution Approaches

A given TO problem can be solved using gradient-based optimisation techniques. Commonly used techniques are the optimality criteria (OC) method, sequential linear programming (SLP) methods and the method of moving asymptotes (MMA). Among these, MMA and its variant, globally convergent MMA, are the most popular (Svanberg 1987).

2.2.1.5 Application Domains

In this chapter, a structural problem as an example to explain the TO method was used because it is most common and relatively simple. For further understanding, readers can obtain Matlab code for this problem from Andreassen et al. (2011), and a 3D Matlab implementation can be obtained from Liu and Tovar (2014). However, the TO method can be used for any optimisation problem governed by differential equations. For this reason, TO has been used in design problems comprising thermal, fluid flow, optics and/or acoustics-based partial differential equations. The goal remains to find the optimal material distribution for a well-defined objective and a set of constraints.

2.3 Topology Optimisation for Precision Metal AM

The symbiotic relationship between TO and AM is well recognised (Gibson et al. 2014). Designs produced using TO are typically geometrically complex, and AM acts as the enabling technology for realising such designs. On the other hand, TO is an ideal design tool for AM as it allows exploitation of the shape freedom offered by AM. The first study recognising the potential of combining TO and AM was published in 2011 (Brackett et al. 2011).

Since then, numerous TO for AM approaches have been presented, and readers are referred to Liu et al. (2018) for a comprehensive overview. In this chapter, the discussion is restricted to methods which aim to enhance the dimensional precision of AM parts. In this context, three AM-associated manufacturing aspects that are detrimental for the dimensional precision of the parts are incorporated into TO methods: critical overhanging features, local overheating and thermal distortion. Illustrative examples are given followed by industrial case studies to showcase the benefits of the methods.

2.3.1 TO Methods for Avoiding Overhangs in Precision AM Parts

As discussed earlier, L-PBF is the most commonly used AM technique for manufacturing precision metal parts. In L-PBF, metal powder is selectively melted in a layer-by-layer manner using a laser or an electron beam. In this process, fabrication of downfacing (overhanging) surfaces requires the overhanging feature to be supported by the underlying loose powder. Insufficient stiffness and poor conductivity of the powder can lead to dimensional inaccuracies in the final part. Design guidelines have been formulated that recommend use of a critical overhanging angle θ^{cr} which depends on the process parameters, machine specifications and part material. This implies that the angle measured between the part surface and the base-plate, as shown in Figure 2.3, should be greater than θ^{cr}. A number of experimental studies have found the critical overhang angle typically ranges from 40° to 50° (Cloots et al. 2017, Wang et al. 2013). When overhangs with shallower angles appear in the design, the surface quality of the downfacing surfaces becomes unacceptable. Moreover, for designs with complex geometrical features (which are common in the precision industry), this design guideline can be too restrictive. Hence, sacrificial supports are printed beneath downfacing surfaces. These supports are later removed; thus, they are waste material. Support removal also adds to post-processing time (Strano et al. 2013). Moreover, interfaces where the part and the support meet remain rough and require further finishing in order to meet the desired geometric tolerances (Ranjan et al. 2017a).

There are a number of prevalent techniques in industry for reducing the number of supports, such as the use of scaffold and cellular support structures (Vanek et al. 2014). Moreover, finding a build orientation that requires the least number of support is also suggested (Zhang et al. 2017). However, all of these considerations are preventive measures that can be applied after the design is finalised. Inclusion of these constraints within TO offers the possibility of addressing the support reduction directly within the optimisation framework. In this context, a number of studies are presented in the literature (Gaynor and Guest 2016, Van de Ven et al. 2018b), and here, the idea of the AM filter presented by Langelaar (2017) is discussed. Among others, this is one of the simplest and easy-to-implement approaches which addresses the overhang issue and delivers designs free from acute overhangs. The approach has been already adopted by industrial TO software Simulia's Tosca, and others.

2.3.1.1 Two-Dimensional Overhang Control

The basic idea behind AM filter is shown in Figure 2.7 in a 2D setting for clarity. The grid represents a selection of the FEs of a design domain similar to the one shown in Figure 2.5 (b), and $\rho_{\text{blueprint}}$ represents the 'blueprint' density of element 4 before the AM filter is applied. $\rho_{\text{blueprint}}$ basically refers to $\tilde{\rho}_e$ found after density filtering in Equation (2.10). Here, element 4 of layer i is printable if it is properly supported by the elements lying directly beneath it, i.e. elements 1, 2 and 3 in layer $(i-1)$. This support neighbourhood S for element

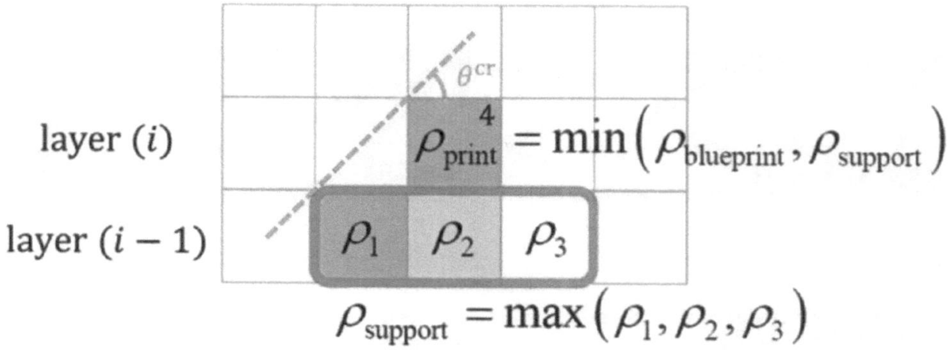

FIGURE 2.7
The concept of the AM filter in 2D. A supporting region S, highlighted by the blue boundary, is defined underneath Element 4. The printed density ρ_{print} is calculated for each element using min. and max. operations. The critical angle θ^{cr} can be controlled by changing the aspect ratio of the elements. (From Langelaar 2017. With permission.)

4 is highlighted with the blue box in Figure 2.7. Simply put, the idea behind the AM filter is that the density of element 4 should not exceed the maximum density in its support neighbourhood S, and this condition is imposed as a (simplified) AM process. Note that if square elements are used for implementing the AM filter (as shown in Figure 2.7), then features less than $\theta^{cr} = 45°$ are prohibited. The method can be easily extended to address other values of θ^{cr} by modifying the aspect ratios of elements (see Figure 2.7).

If this rule is imposed for all the elements, in a sequential layer-by-layer fashion reminiscent to the actual AM process, the resulting design will be fully printable. Max and Min operators are used for imposing this rule mathematically, i.e.

$$\rho_{support} = \max(\rho_1, \rho_2, \rho_3), \tag{2.13}$$

and

$$\rho_{print} = \min(\rho_{blueprint}, \rho_{support}), \tag{2.14}$$

where ρ_{print} represents the density of element 4 after the AM filter is applied. However, Max and Min operators are non-differentiable and, as discussed in Section 2.2.1, impede the sensitivity computation, which is a critical step for TO. Therefore, these operators are replaced by their continuous, smooth approximations, enabling consistent sensitivity analysis. The details of the sensitivity calculation are not presented here for brevity, and interested readers are referred to Langelaar (2017). The AM filter is applied during each TO iteration, and ρ_{print} is calculated for each element in the design domain. ρ_{print} represents the final printable density field, and the design performance (for example, compliance) is evaluated on this field.

An illustrative result is presented in Figure 2.8 where optimised designs for the cantilever problem introduced in Figure 2.5(a) are shown. The designs obtained with standard TO and TO with the AM filter are shown in Figure 2.8(a) and 2.8(b), respectively. The build direction coincides with the y-axis, which is also indicated by the blue base-plate in Figure 2.8(b). The optimised design with the AM filter is fully printable for a critical overhang angle of $\theta^{cr} = 45°$, unlike the standard design, which has long overhanging features even with $\theta = 0$. However, the compliance of the design optimised using an AM filter is 6%

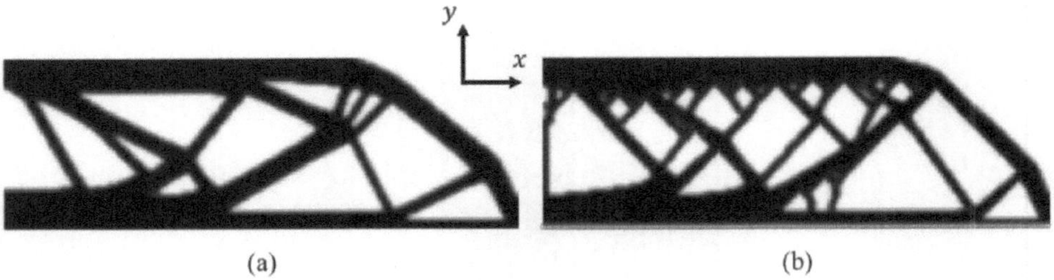

FIGURE 2.8
Results for cantilever case discretised using 180 and 60 elements in x and y direction, respectively. (a) Standard TO, $c = c_{ref}$. (b) TO with AM filter, $c/c_{ref} = 106\%$. (From Langelaar 2017. With permission.)

higher (less stiff) than that of the standard design, as some fraction of the available material is invested in making the top overhangs printable.

2.3.1.2 3D Overhang Control

The AM filter can be easily extended to 3D with a support neighbourhood that includes five elements instead of three, as illustrated in Figure 2.9 (Langelaar 2016). A 3D cantilever problem is schematically illustrated in Figure 2.10(a). In Figure 2.10(b), the standard TO result is given. Optimal designs for three different AM build orientations are presented in Figure 2.10(c) to 2.10(e). It is clear that the AM filter avoids all the not-permitted overhangs in the resulting designs. For example, small connections are introduced in Figure 2.10(d) that support the part from the base-plate, thus avoiding overhanging features. These connections by and large do not contribute to the stiffness of the part but are added by the AM filter just to meet the overhang constraint. Therefore, this design has the highest compliance (lowest stiffness) as compared to all other designs.

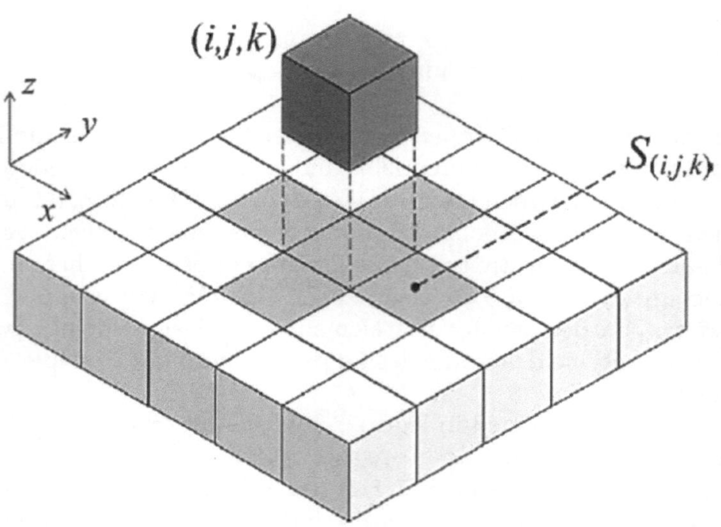

FIGURE 2.9
Concept of AM filter extended in 3D. The support neighbourhood $S_{(i,j,k)}$ here includes five elements beneath the element characterised by position (i,j,k). (From Langelaar 2016. With permission.)

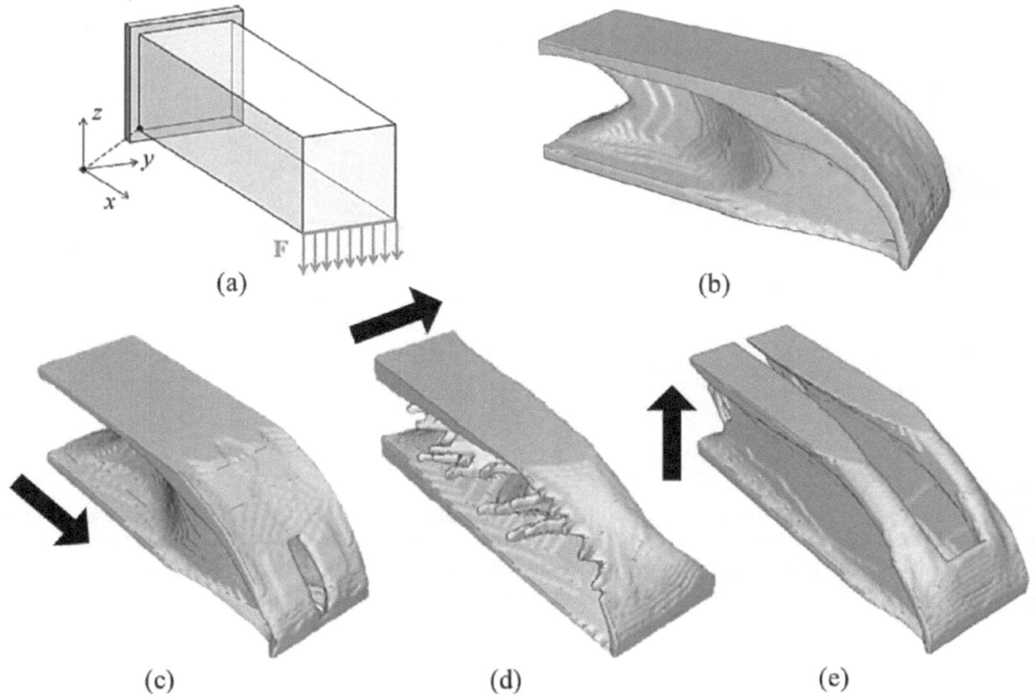

FIGURE 2.10
(a) Cantilever beam problem definition discretised with 150, 50 and 50 elements in x, y and z, respectively (b) Solution found using standard TO, $c = c_{ref}$ (c) Solution with AM filter with build orientation in $+x$ direction, c/c_{ref} = 101% (d) Solution with AM filter with build orientation in $+y$ direction, c/c_{ref} = 107% (e) Solution with AM filter with build orientation in $+z$ direction, c/c_{ref} = 99%. (From Langelaar 2016. With permission.)

2.3.1.3 Support Inclusion

The AM filter described above can result in completely self-supporting designs. However, complete prohibition of supports is rather restrictive and can have an adverse effect on the stiffness performance. Therefore, a relatively more holistic approach is presented in Langelaar (2019), where user-defined costs are associated with printing supports, and a cost for machining effort required for support removal is also considered. Using these costs, optimisation evaluates the trade-off in reducing part performance against support addition. Furthermore, for a part that undergoes finishing operations (for example, drilling of high-precision holes after printing), constraints are included to guarantee the required degree of stiffness for precision drilling/machining.

An industrial case study of optimising an air-brake component is presented. The problem definition is shown in Figure 2.11(a) where the green region represents the design domain, the blue region represents clamping points when the component is in service and F_m and F_s are service and machining loads, respectively.

The result of TO is shown in Figure 2.11(b), and it is interesting to observe the evolution of 'tree-like' supports where a single strut starts from the base-plate and splits into different directions to support different areas of the component. These supports offer material savings compared to conventional pillar supports. Note that there is a tendency to minimise the support/part interface since there is a cost associated with support removal in

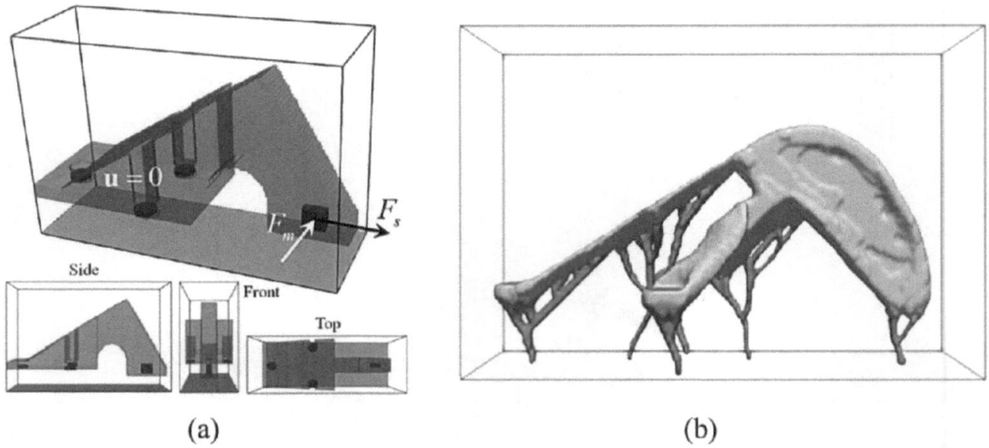

FIGURE 2.11
(a) The design domain is represented in green. The design is clamped at the blue regions. Service and machining loads are indicated by F_s and F_m, respectively, applied on the red region. (b) Optimal design found using the novel TO where component (grey) and supports (green) are presented. The shown design is for machining tolerance of u = 100 μm, i.e. a deformation of 100 μm is allowed for the machining load. (From Langelaar 2019. With permission.)

the objective. More complex issues – for example, supports for maximum heat evacuation – can also be achieved using TO.

Lastly, in general, the build orientation for printing of an AM part is decided once a design has been finalised. For example, for the design shown in Figure 2.8(a), the preferred orientation would be either the $+x$ or $-x$ direction, as less supports will be needed. However, build orientation is fixed before the optimisation starts in the approaches presented in this section. Therefore, another extension of the method is presented where combined optimisation of the part, support and build orientation is performed. The details of this implementation are not included here, and interested readers are referred to Langelaar (2018).

2.3.2 TO Methods for Preventing Overheating in Precision AM Parts

Local overheating during the L-PBF process is a major issue compromising precision in the final product. The unmolten powder, which has significantly lower conductivity compared to the bulk material, does not allow for proper heat evacuation for a newly deposited layer (Mertens et al. 2014b). This leads to an increase in the size of the melt pool, which could result in surface defects such as dross and balling (Sames et al. 2016). A vacuum seal manufactured by KMWE is shown in Figure 2.12, where dross formation is highlighted (Gramsma and Laheij 2016). Dross formation leads to high levels of surface texture in the built part, and significant post-processing effort is needed to meet the dimensional tolerances.

Typically, the issue of overheating is associated with overhangs, and support structures are thus used for conducting away the excess heat. However, the geometric overhang control methods, as described in Section 2.3.1, do not necessarily guarantee overheating prevention. For example, a specimen manufactured by Adam and Zimmer (2014) demonstrates that various features within a part with identical overhang angle can have different overheating behaviours. Moreover, for high-quality AM parts, precise control of material microstructure is desired, and local overheating has an adverse effect on microstructural

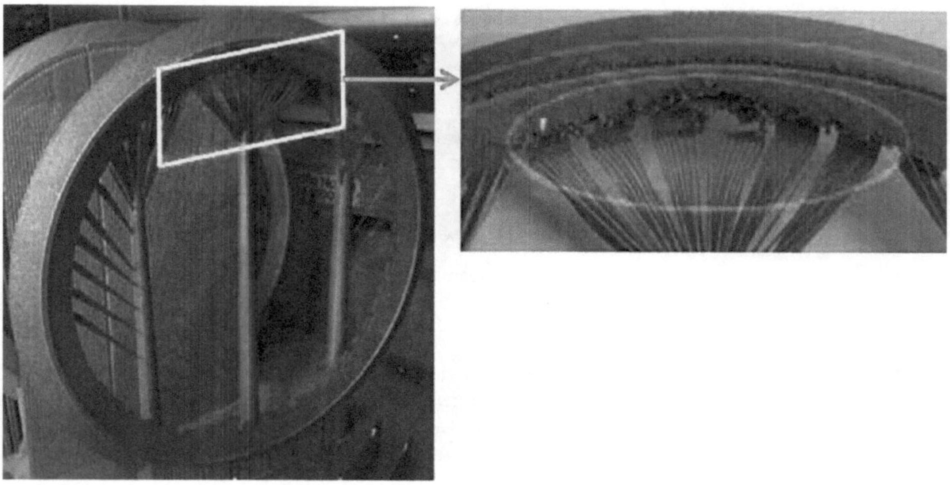

FIGURE 2.12
A vacuum seal manufactured by KMWE using L-PBF. The final printed part has heavy dross defects in the overhanging region. Thin supports were used to conduct away the excess heat, but they are not effective enough. (From Gramsma and Laheij 2016. With permission.)

control. Another problem that arises due to localised overheating is more pronounced deformations which compromise precision. In the study presented by Kastsian and Reznik (2017), deformations in the build direction cause collision with the re-coater blade leading to complete build failure. Consequently, addressing the issue of overheating during an AM process is important when designing precision components.

A novel TO method to prevent AM-associated overheating is presented by Ranjan et al. (2017b), which constrains AM related overheating within the density-based TO framework. The rationale behind the approach is that local topology determines whether a given region is prone to overheating or not. Therefore, a local conductivity test is performed. As a first step, the geometry is decomposed into a number of overlapping slabs. Here, the term *slab* refers to a set of subsequent AM layers. Each slab is subjected to thermal boundary conditions similar to that of a layer in a typical AM process, and the temperature field is obtained by performing a steady-state thermal FEA. A steady-state analysis is preferred as it is found to be capable of identifying 'hotspots' while providing significant computational advantages, making it possible to integrate it with TO. The final temperature field is obtained for the entire geometry by selecting a maximum value for each FEA node and regions with relatively high temperature values are identified as hotspots. Figure 2.13 presents a schema of the process in which a wedge-shaped geometry with a hole is divided into overlapping slabs. The assembled temperature field shows that the region just above the circular hole particularly tends to accumulate heat and, thus, is identified as a hotspot.

The hotspot detection scheme is integrated with a density-based TO process through a hotspot constraint. In each TO design iteration, the hotspot detection algorithm is applied on the entire design domain and temperature information is obtained. An aggregation scheme is then used to specify a maximum allowable hotspot temperature T^{cr} as a constraint. The critical overhang angle θ^{cr} for the applicable AM process is used to calculate the corresponding T^{cr} using the hotspot detection method described above. The MMA (Svanberg 1987) is used for optimisation, and the required sensitivity information is computed by the adjoint method.

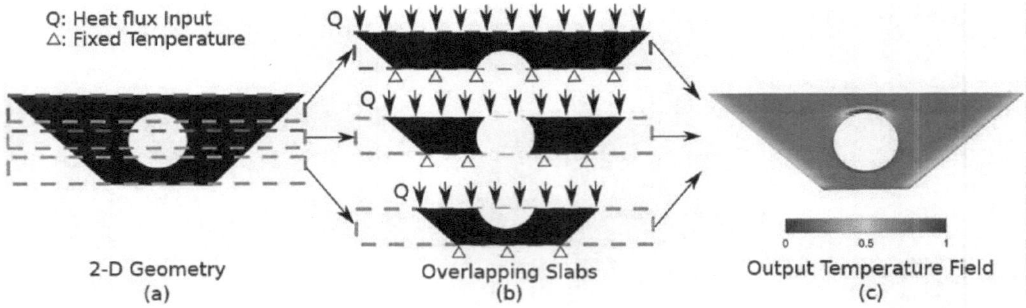

FIGURE 2.13
Hot spot detection method: (a) a wedge-shaped geometry decomposed into a set of overlapping slabs (b) individual slabs with applied boundary conditions (c) temperature field obtained by assembling temperature information from all the slabs and then picking the maximum value at each location. Temperatures are normalised using maximum value as normalisation constant. (From Ranjan et al. 2017. With permission.)

Representative TO results are shown in Figure 2.14, where the method is applied on the aforementioned cantilever loading case and designs for both standard TO and hotspot-based TO are depicted in Figure 2.14(a) and 2.14(b) respectively. The temperature fields obtained after applying the hotspot detection method on the final designs are also superimposed. For comparison, temperature fields for both designs are normalised to a common scale, using the maximum temperature in Figure 2.14(a) for normalisation. It can be seen that hotspot temperatures significantly reduce in the design obtained using the hotspot-based TO method. This is primarily achieved by avoiding the long overhangs present in the standard TO case which, if fabricated, would cause significant dross, leading to rough surfaces and loss of precision. A critical overhang angle of $\theta^{cr} = 40°$ is used for the presented result, i.e. $T^{cr} = f(\theta^{cr})$. A 4% decrease in stiffness is found. A green-coloured strip in Figure 2.14 indicates the substrate plate that dictates the build orientation for which optimisation was performed.

An application of this hotspot-based TO approach in the context of the precision industry is presented by Sinico et al. (2019). TO was performed on a metallic mould insert, which is to be produced by L-PBF of maraging 300 steel material. The design of the mould insert

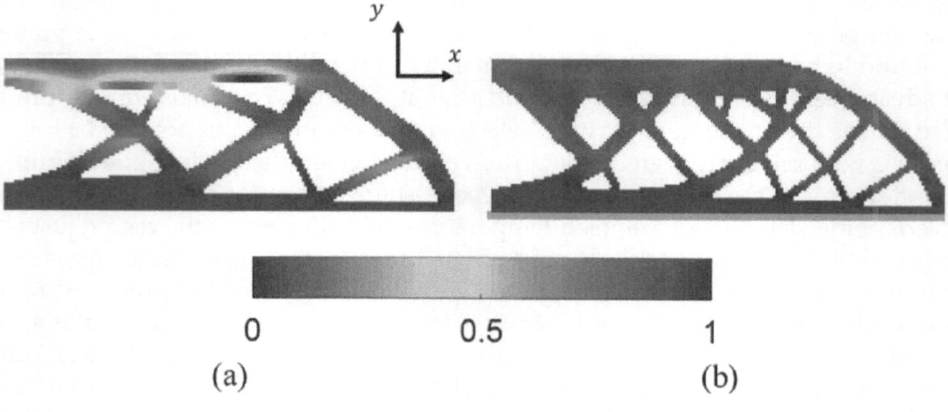

FIGURE 2.14
Results for cantilever case discretised using 180 and 60 elements in x and y direction, respectively. (a) Standard TO, $c = c_{ref}$ (b) Hotspot based TO, $c/c_{ref} = 104\%$. Temperature fields are normalised to a common scale for comparison.

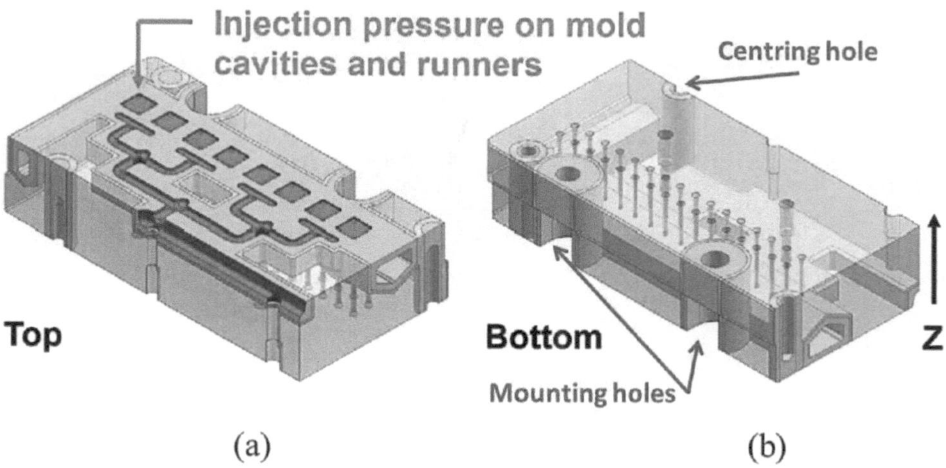

FIGURE 2.15
Design space for TO of an injection mould. Yellow features indicate regions that should remain intact during the TO process, i.e. so-called non-design space. Pressure due to plastic injection is applied on the runner cavities. The surfaces of mounting and centring holes are considered fixed. (From Sinico et al. 2019. With permission.)

had already been partially optimised for AM, with a simple conformal cooling channel running beneath the mould cavities. The minimisation of mass for the purpose of reducing L-PBF production time and material use, however, had not yet been considered at the design stage. TO was, therefore, selected as an ideal approach to perform this last step before the fabrication of the insert.

Using the double symmetry of the original mould design, only one quarter of the mould is optimised, and the isometric views of the top and bottom of the quarter mould are shown in Figure 2.15(a) and 2.15(b), respectively. Here, the yellow regions represent the 'keep-in' features, i.e. this region has been excluded from the TO design space and referred to as non-design domain. The design space, represented in pink, is the domain where the TO could optimise the material layout for the given set of loads, boundary conditions and constraints, with the objective of maximising the stiffness of the mould.

The loading condition mimicked the injection pressure load on the runners and mould cavities using the maximum possible pressure multiplied by a safety factor of 1.5. The water pressure in the channel was considered negligible. Mounting holes and centring holes (highlighted in Figure 2.15b) were set as non-moving surfaces. The mechanical properties of maraging 300 steel were used. The black arrow in Figure 2.15(b) represents the build orientation.

The optimised designs using standard and hotspot TO are shown in Figure 2.16(a) and 2.16(b), respectively. The presented results are for half of the mould. The mass has been reduced to 50% of the initial design, while the production time is reduced by 43% for the design shown in Figure 2.16(a). Lastly, the hotspot temperatures were found to be 30% lower in the design found using hotspot TO compared to that of standard TO.

This case study demonstrates that adopting enhanced TO methods at an early design stage can assist in conceptualising efficient, low-mass, easy-to-manufacture designs for precision applications. This can directly lead to savings in cost and time, and can even make the difference between a manufacturable, economically feasible solution or an infeasible one.

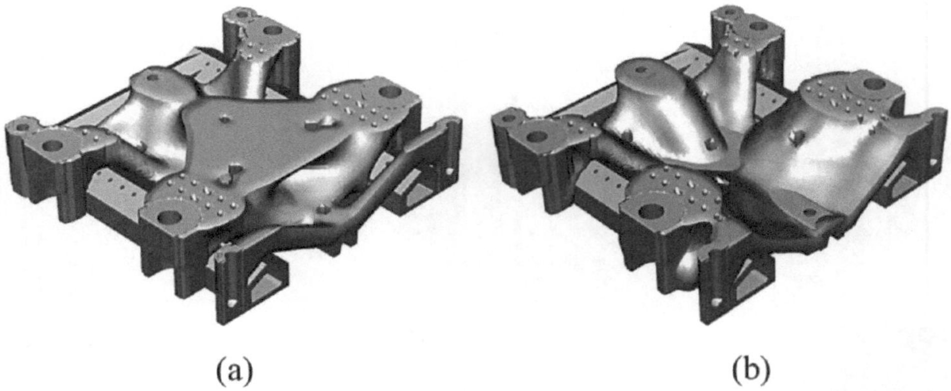

FIGURE 2.16
(a) Design obtained using standard TO (b) Design obtained using hotspot TO. (From Sinico et al. 2019. With permission.)

2.3.3 Towards TO Methods for Avoiding Distortion in Precision AM Parts

Every point in an AM part undergoes cyclic thermal loading during the L-PBF process, where temperatures rise and fall multiple times (Yang et al. 2018). The first temperature peak occurs when powder is melted using the laser or electron beam. Then, the temperature drops while the beam is scanning other distant zones within the same layer. However, when the laser beam scans a region in close proximity, another peak occurs. Finally, when the entire layer has been scanned, there is a dwell period in which the substrate plate moves downward, and the re-coater spreads a new layer of powder. Temperatures fall significantly during this period. The next peak in temperature is experienced when material point just above, in the freshly laid layer, goes through the same cycle. These heating/cooling cycles cause thermal expansion and contraction, eventually leading to thermal distortion during the process (Simson et al. 2017). Also, the part is clamped on the substrate plate which induces residual stresses and, when unclamped, leads to significant distortions in the final part. Figure 2.17 depicts printed beams where cracks are visible due to stress

FIGURE 2.17
The build-up of residual stresses in the part during the AM process results in part distortion and, in extreme cases, part failure. (From O'Neal 2018. With permission.)

build-up. Residual stress is detrimental to the final part quality, particularly in the case of precision AM components.

A thermo-mechanical L-PBF process model can be used to predict final stresses and distortions for a given design. However, part-scale coupled thermo-mechanical models are computationally expensive (see Chapter 3). For example, Denlinger et al. (2014) reported a simulation time of forty hours when simulating a large AM part (3810 mm × 457 mm × 25 mm). In the context of TO, it becomes even more challenging as design evaluation, process simulation and the associated sensitivity computations have to be conducted for every design iteration, and typically repeated hundreds of times. Therefore, a priority for current research is to first develop suitable approximate L-PBF models that can adequately capture process-induced distortions in practical timeframes.

In the context of predicting final distortions in AM parts, the inherent strain method is a promising candidate. Originally developed for predicting residual stresses in welding, the inherent strain method provides an option for fast prediction of final part distortion without simulating the thermal evolution during L-PBF (Keller and Ploshikhin 2014). The inherent strain method requires a calibration step where residual distortions are first quantified through experiments or high-fidelity models. This needs to be carried out only once, and then, when the inherent strain is known for a particular material and an associated set of machine parameters, it can be used for mechanical simulations of other parts. Experimental validation suggests that the inherent strain method accurately predicts distortion, and hence, it has been used in a number of research studies (Liang et al. 2018, Setien et al. 2018). Nevertheless, the computational cost of the inherent strain method is still too high to loop it iteratively within an optimisation framework within a practical timeframe. Note that there are recent studies which attempt to integrate the full inherent strain method with TO (Cheng et al. 2019), in spite of the substantial computational cost.

An approach recently presented by Munro et al. (2019) establishes the idea of decoupling layers during the linear elastic simulations of a layer-by-layer L-PBF model based on the inherent strain method. This allows parallelisation of analysis steps, which provides a major reduction in the computational time needed for the analysis. Figure 2.18(a) and

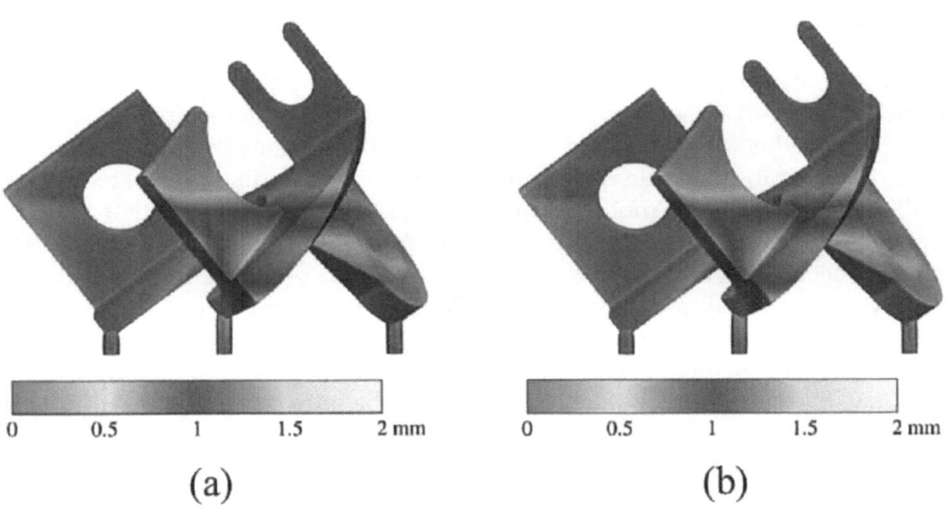

FIGURE 2.18
Deformation fields found for L-PBF using inherent strain method (a) sequential nonlinear simulation (b) stepwise parallel simulation. (From Munro et al. 2019. With permission.)

2.18(b) show deformation fields obtained by sequential non-linear and step-wise linearised parallel simulation, respectively, for an industrial test part. It is reported that a reduction of approximately 86% has been achieved in computational time, while 95% of the deformations agree within a margin of 10 μm (with largest distortions approaching 2 mm). More importantly, this example demonstrates that the linearised inherent strain method can accurately predict deformation for a real size part within a timeframe of a few minutes. This enables future integration of the aforementioned modelling approach with TO, to allow for distortion minimisation by re-design of the part, the support structures or both.

2.4 Challenges and Outlook

The field of DFAM is continuously evolving along with the AM technology itself. The idea of dedicated TO formulations for AM is relatively new but developing at a rapid pace. In the context of addressing AM-related issues within TO, geometric design guidelines have been the main focus of research, for example, overhang criteria, cavity avoidance (Liu et al. 2015) and accessibility for ease of support removal (Van de Ven et al. 2018a). Among these examples, overhang criteria have been investigated most intensively, and most commercial TO software packages now offer this as an option (for example, Tosca, Siemens NX). However, in the context of precision components, it is becoming more evident with growing research that geometry-based guidelines are not sufficient to avoid defects in the final part. Moreover, in an optimisation setting, geometry-based approaches can also become overly restrictive (Ranjan et al. 2017, Wildman and Gaynor 2017). Therefore, to more directly and accurately evaluate the limits of the process, a physics-based approach that mimics the L-PBF is needed.

In theory, a highly detailed process model, which captures the complex physical interactions during the L-PBF process, can be coupled with TO. However, the computational burden associated with such a model makes this an unrealistic approach for practical and industrial applications. Due to this limitation, researchers have started to explore the idea of 'simplified' part-scale AM models, which introduce approximations in order to reduce the computational burden. In general, the usability of a simplified model depends on the aim of the analysis, as simplifying assumptions may miss some aspects of the process while capturing others. For example, a number of studies consider only conduction as the mode of heat transfer while neglecting convection and radiation (Badrossamay and Childs 2007, Paul et al. 2014). Another common simplification is to simulate the deposition of an entire layer or number of layers in one step (Afazov et al. 2017, Peng et al. 2016). The thermal and mechanical AM process models presented in Section 2.3.2 and Section 2.3.3, respectively, are also examples of such simplified schemes. An open research question is how to determine the most appropriate simplifications for a given purpose. Another challenge is the effective inclusion of simplified models within a TO framework, where possibly different levels of accuracy can be considered to balance computational effort during the optimisation process.

Finally, it is important to note that the computational complexity increases even more in the case of designing precision components. Typically, such components have tight tolerances along with features with intricate functionalities which again involve complex multi-physics simulations, for example, the industrial cases discussed in Section 2.1. Therefore, in such cases, both the functionality and manufacturability of the component add to the computational expenses, making the DFAM process even more challenging. Nevertheless, the current state of TO and AM research is promising and developing at a rapid pace, and the need for DFAM solutions is clearly recognised.

References

Adam, G. A. O., and Zimmer, D. 2014. Design for additive manufacturing-element transitions and aggregated structures. *CIRP-J. Manuf. Sci. Tec.*, 7(1), 20–28.

Afazov, S., Denmark, W. A. D., Lazaro Toralles, B., Holloway, A., and Yaghi, A. 2017. Distortion prediction and compensation in selective laser melting. *Addit. Manuf.*, 17, 15–22.

Amir, O., and Mass, Y. 2018. Topology optimization for staged construction. *Struct. Multidiscip. Optim.*, 57(4), 1679–1694.

Andreassen, E., Clausen, A., Schevenels, M., Lazarov, B. S., and Sigmund, O. 2011. Efficient topology optimization in MATLAB using 88 lines of code. *Struct. Multidiscip. Optim.*, 43(1), 1–16.

Badrossamay, M., and Childs, T. H. C. 2007. Further studies in selective laser melting of stainless and tool steel powders. *Int. J. Mach. Tool. Manu.*, 47(5), 779–784.

Bendsøe, M. P. 1989. Optimal shape design as a material distribution problem. *Struct. Multidiscip. Optim*, 1(4), 193–202.

Bendsøe, M. P., and Sigmund, O. 1999. Material interpolation schemes in topology optimization. *Arch. Appl. Mech.*, 69(9–10), 635–654.

Bendsøe, M. P., and Sigmund, O. 2003. *Topology Optimization: Theory, Methods and Applications*. Berlin Heidelberg: Springer-Verlag.

Brackett, D., Ashcroft, I., and Hague, R. 2011. Topology optimization for additive manufacturing. *Proc. 22nd Int. SFF Conf.* Austin, TX, 348–362.

Bruns, T. E., and Tortorelli, D. A. 2001. Topology optimization of non-linear elastic structures and compliant mechanisms. *Comput. Method Appl. M.*, 190(26–27), 3443–3459.

Cheng, L., Liang, X., Bai, J., Chen, Q., Lemon, J., and To, A. 2019. On utilizing topology optimization to design support structure to prevent residual stress induced build failure in laser powder bed metal additive manufacturing. *Addit. Manuf.*, 27, 290–304.

Cloots, M., Zumofen, L., Spierings, A. B., Kirchheim, A., and Wegener, K. 2017. Approaches to minimize overhang angles of SLM parts. *Rapid Prototyp. J.*, 23(2), 362–369.

Deaton, J. D., and Grandhi, R. V. 2014. A survey of structural and multidisciplinary continuum topology optimization: Post 2000. *Struct. Multidiscip. Optim.*, 49(1), 1–38.

Denlinger, E. R., Irwin, J., and Michaleris, P. 2014. Thermomechanical modeling of additive manufacturing large parts. *J. Manuf. Sci. E.*, 136(6), 1–8.

Dijk, N. P. Van, Maute, K., Langelaar, M., and Keulen, F. van. 2013. Level-set methods for structural topology optimization: A review. *Struct. Multidiscip. Optim.*, 48(3),437–472.

Gaynor, A. T., and Guest, J. K. 2016. Topology optimization considering overhang constraints: Eliminating sacrificial support material in additive manufacturing through design. *Struct. Multidiscip. Optim.*, 54(5),1157–1172.

GE Additive. 2018. New manufacturing milestone: 30,000 additive fuel nozzles. Available at: https://www.ge.com/additive/stories/new-manufacturing-milestone-30000-additive-fuel-nozzles.

Gibson, I., Rosen, D., and Stucker, B. 2014. *Additive Manufacturing Technologies: 3D Printing, Rapid Prototyping, and Direct Digital Manufacturing*. New York: Springer-Verlag.

Gramsma, A., and Laheij, D. 2016. Exploring the 3D printing of metal vacuum seals. *Mikroniek*, Issue 2, 32–35.

Grunewald, S. J. 2016, GE Additive. 2018. New manufacturing milestone: 30,000 additive fuel nozzles. Available at: https://www.ge.com/additive/stories/new-manufacturing-milestone-30000-additive-fuel-nozzles

Kastsian, D., and Reznik, D. 2017. Reduction of local overheating in selective laser melting. *Proc. SIM-AM*, Munich, Germany.

Keller, N., and Ploshikhin, V. 2014. New method for fast predictions of residual stress and distortions of AM parts. *Proc. 25th Int. SFF Conf*, Austin, TX, 1229–1237.

Langelaar, M. 2016. Topology optimization of 3D self-supporting structures for additive manufacturing. *Addit. Manuf.*, 12, 60–70.

Langelaar, M. 2017. An additive manufacturing filter for topology optimization of print-ready designs. *Struct. Multidiscip. Optim.*, 55(3), 871–883.

Langelaar, M. 2018. Combined optimization of part topology, support structure layout and build orientation for additive manufacturing. *Struct. Multidiscip. Optim.*, 57(5), 1985–2004.

Langelaar, M. 2019. Integrated component-support topology optimization for additive manufacturing with post-machining. *Rapid Prototyp. J.*, 25(2), 255–265.

Liang, X., Cheng, L., Chen, Q., Yang, Q., and To, A. C. 2018. A modified method for estimating inherent strains from detailed process simulation for fast residual distortion prediction of single-walled structures fabricated by directed energy deposition. *Addit. Manuf.*, 23, 471–486.

Liu, J., Gaynor, A. T., Chen, S., Kang, Z., Suresh, K., Takezawa, A., Li, L., Kato, J., Tang, J., Wang, C. C. L., Cheng, L., Liang, X., and To, A. C. 2018. Current and future trends in topology optimization for additive manufacturing. *Struct. Multidiscip. Optim.*, 57(6), 2457–2483.

Liu, K., and Tovar, A. 2014. An efficient 3D topology optimization code written in Matlab. *Struct. Multidiscip. Optim.*, 50(6), 1175–1196.

Liu, S., Li, Q., Chen, W., Tong, L., and Cheng, G. 2015. An identification method for enclosed voids restriction in manufacturability design for additive manufacturing structures. *Front. Mech. Eng.*, 10(2), 126–137.

Loncke, D. 2014. Following Moore's Law. *Mikroniek*, Issue 6, 17–22.

Mertens, R., Clijsters, S., Kempen, K., and Kruth, J.-P. 2014a. Optimization of scan strategies in selective laser melting of aluminum parts with downfacing areas. *J. Manuf. Sci. E.*, 136(6), 61012–61017.

Mertens, R., Clijsters, S., Kempen, K., and Kruth, J.-P. 2014b. Optimization of scan strategies in selective laser melting of aluminum parts with downfacing areas. *J. Manuf. Sci. E.*, 136(6), 061012.

Munro, D., Ayas, C., Langelaar, M., and van Keulen, F. 2019. On process-step parallel computability and linear superposition of mechanical responses in additive manufacturing process simulation. *Addit. Manuf.*, 28, 738–749.

O'Neal, B. 2016. Swanson School of Engineering & Aerotech partner to refine metal additive manufacturing with fast computational modeling. *3DPrint*. Available at: https://3dprint.com/146259/swanson-aerotech-metal-am/ (accessed December 15, 2019).

Paul, R., Anand, S., and Gerner, F. 2014. Effect of thermal deformation on part errors in metal powder based additive manufacturing processes. *J. Manuf. Sci. E.*, 136(3), 031009.

Peng, H., Khouzani, M. G., Gong, S., Attardo, R., Ostiguy, P., Gatrell, B. A., Budzinski, J., Tomonto, C., Neidig, J., Shankar, M. R., Billo, R., Go, D. B., and Hoelzle, D. 2016. Part-scale model for fast prediction of thermal distortion in DMLS additive manufacturing; Part 2: A quasi-static thermomechanical model. *Proc.27th Int. SFF Conf.*, Austin, TX, 382–397.

Ranjan, R., Samant, R., and Anand, S. 2017a. Integration of design for manufacturing methods with topology optimization in additive manufacturing. *J. Manuf. Sci. E.*, 139(6), 1–14.

Ranjan, R., Yang, Y., Ayas, C., Langelaar, M. and Van Keulen, F. 2017b. Controlling local overheating in topology optimization for additive manufacturing. *Proc. Special Interest Group Meeting on Dimensional Accuracy and Surface Finish in Additive Manufacturing, euspen*, 17–19, Leuven, Belgium.

Sames, W. J., List, F. A., Pannala, S., Dehoff, R. R., and Babu, S. S. 2016. The metallurgy and processing science of metal additive manufacturing. *Int. Mater. Rev.*, 61(5), 315–360.

Schoinochoritis, B., Chantzis, D., and Salonitis, K. 2017. Simulation of metallic powder bed additive manufacturing processes with the finite element method: A critical review. *Proc. Inst. Mech. Eng. B*, 231(1), 96–117.

Setien, I., Chiumenti, M., van der Veen, S., San Sebastian, M., Garciandía, F., and Echeverría, A. 2018. Empirical methodology to determine inherent strains in additive manufacturing. *Comput. Math. Appl.*, 78(7), 2282–2295.

Sigmund, O. 2007. Morphology-based black and white filters for topology optimization. *Struct. Multidiscip. Optim.*, 33(4–5), 401–424.

Sigmund, O., and Maute, K. 2013. Topology optimization approaches: A comparative review. *Struct. Multidiscip. Optim.*, 48(6), 1031–1055.

Simson, T., Emmel, A., Dwars, A., and Böhm, J. 2017. Residual stress measurements on AISI 316L samples manufactured by selective laser melting. *Addit. Manuf.*, 17, 183–189.

Sinico, M., Ranjan, R., Moshiri, M., Ayas, C., Langelaar, M., Witvrouw, A., and Van Keulen, F. 2019. A mold insert case study on topology optimized design for additive manufacturing. *30th Int. SFF Conf.*, Austin, TX.

Strano, G., Hao, L., Everson, R. M., and Evans, K. E. 2013. A new approach to the design and optimisation of support structures in additive manufacturing. *Int. J. Adv. Manuf. Tech.*, 66(9–12), 1247–1254.

Svanberg, K. 1987. The method of moving asymptotes – a new method for structural optimization. *Int. J. Numer. Meth. Eng.*, 24(2), 359–373.

Van de Ven, E., Langelaar, M., Ayas, C., Maas, R., and Van Keulen, F. 2018a. Topology optimization with overhang filter considering accessibility of supports. *10th ESMC Conf.*, Bologna, Italy.

Van de Ven, E., Maas, R., Ayas, C., Langelaar, M., and Van Keulen, F. 2018b. Continuous front propagation-based overhang control for topology optimization with additive manufacturing. *Struct. Multidiscip. Optim.*, 57(5), 2075–2091.

Vanek, J., Galicia, J. A. G., and Benes, B. 2014. Clever support: Efficient support structure generation for digital fabrication. *Comput. Graph. Forum*, 33(5), 117–125.

Vatanabe, S. L., Lippi, T. N., Lima, C. R. d., Paulino, G. H., and Silva, E. C. N. 2016. Topology optimization with manufacturing constraints: A unified projection-based approach. *Adv. Eng. Softw.*, 100, 97–112.

Wang, D., Yang, Y., Yi, Z., and Su, X. 2013. Research on the fabricating quality optimization of the overhanging surface in SLM process. *Int. J. Adv. Manuf. Tech.*, 65(9–12), 1471–1484.

Wildman, R. A., and Gaynor, A. T. 2017. *Topology Optimization for Reducing Additive Manufacturing Processing Distortions*. Aberdeen, MD: Weapons and Materials Research Directorate.

Yang, Y., Knol, M. F., Van Keulen, F., and Ayas, C. 2018. A semi-analytical thermal modelling approach for selective laser melting. *Addit. Manuf.*, 21, 284–297.

Zhang, Y., Bernard, A., Harik, R., and Karunakaran, K. P. 2017. Build orientation optimization for multi-part production in additive manufacturing. *J. Intell. Manuf.*, 28(6), 1393–1407.

3

Development of Precision Additive Manufacturing Processes

Ahmed Elkaseer, Amal Charles and Steffen G. Scholz

CONTENTS

3.1 Introduction

This chapter discusses the development of precision in metal additive manufacturing (AM), and mainly focuses on the laser powder bed fusion (L-PBF) method, since it is the dominant technique for fabricating three-dimensional (3D) metal AM parts. The precision of AM parts is an issue that many technology developers face, especially when metallic components are manufactured (DebRoy et al. 2018, Leach and Smith 2018). Lack of process precision can be due to various causes, such as the dispersion of material properties, the variability and interdependent effects of process parameters, machine and equipment accuracies, lack of in-process control systems and the thermal nature of the process (Liu and Shin 2019, Ríos et al. 2018). Manufacturing industries producing components and equipment for aerospace, marine and automotive applications have realised that some geometrical deficiencies and high manufacturing costs could be overcome if large/customised components with intricate 3D features are manufactured by AM (Cabanettes et al. 2018, Liu et al. 2019). Given the size and long processing times of AM components, precision is critical, for example, to the transport sector, where personal safety depends on the reliability of manufactured structural components, or in the medical industry, for the production of patient specific implants, prosthetics, etc. With metal AM parts, small geometrical defects are amplified over time as layers are deposited one upon another and controlled deposition allows this issue to be overcome (Michopoulos et al. 2018, Thompson et al. 2016).

Although AM techniques have recently seen increasing adoption by various industrial sectors, the precision of metal manufactured parts remains the main barrier to the full potential of AM processes, gaining in increased market acceptance and penetration (Witvrouw et al. 2018). In this context, this chapter pieces together the roadmap to further develop precision in metal AM processes. Following the introduction, Section 3.2 describes the state of the art of precision AM processes and gives insights into their development. This is followed by a discussion of the significant process parameters and AM performance indicators in Sections 3.4 and 3.5, respectively. Data-driven process improvement methods are discussed in Section 3.6, including different design of experiments (DoEs), modelling of process performance (quantifying input/output process relationships) and process optimisation. The development of precision AM processes in the domain of Industry 4.0 is discussed in Section 3.7, encompassing real-time monitoring of AM processes and artificial intelligence (AI)–powered decision-making systems for digital quality control. Future perspectives on precision metal AM processes are given in Section 3.8. Finally, Section 3.9 provides this chapter's conclusions.

3.2 State of the Art and Insight into Precision Process Development

In addition to its minimal material waste, L-PBF (see Figure 3.1) offers relatively high accuracy and geometrical flexibility, making this technology widely adoptable by various industrial sectors (Zavala-Arredondo et al. 2019). The process of fabricating near-net-shape geometries using AM techniques, followed by post-processing such as heat treatment and machining (see Chapter 5), yields more efficient production results than traditional processes with specific materials and in specific markets. The main challenge with AM

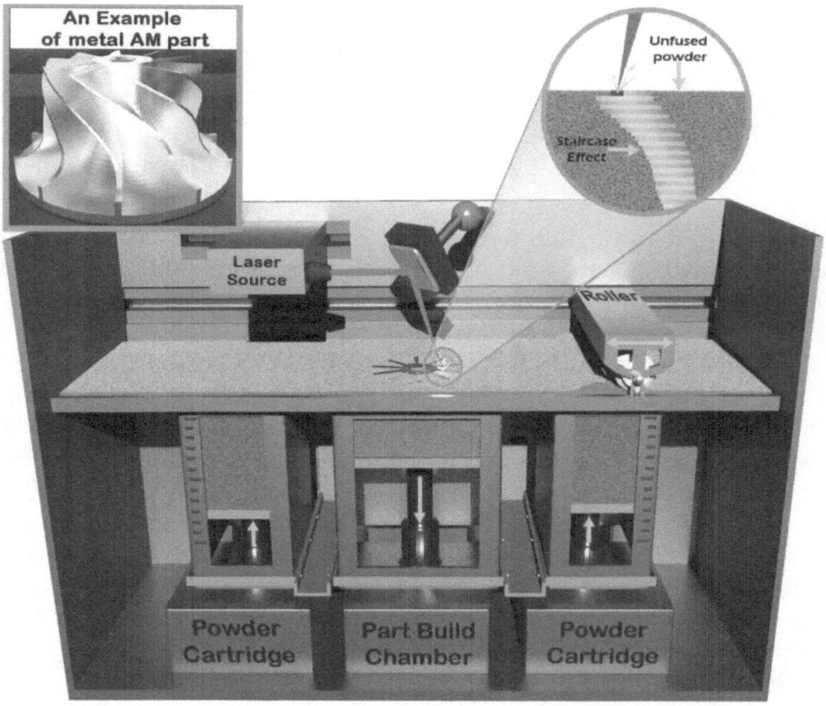

FIGURE 3.1
Illustration of the L-PBF process.

technologies is to ensure defect-free, structurally sound AM parts through high-precision processes.

L-PBF uses a laser beam to selectively melt metallic powder particles in a powder bed. The process steps can be explained as seen in Figure 3.1 and include the following:

1. The first step for any AM process begins with the creation of the model of the part to be printed using computer-aided design (CAD) software (see Chapter 2).
2. The CAD data of the required part is then translated into the STL format, the most commonly used file format for AM.
3. The STL file is then transferred to AM pre-processing software where the various process conditions and parameters, such as laser power, scan speed and layer thickness, are set.
4. The part is digitally sliced and the file transferred to the L-PBF system ready to start printing.
5. During the active printing process, the desired part is built by selectively melting successive layers of powder on top of a build platform. The laser beam irradiates the powder, and once sufficient power is applied, the powder melts and forms a molten pool.
6. This molten pool forms solid material in the shape of the product upon cooling.
7. Once the printing of each layer is finished, the build platform is lowered by the previously defined layer thickness.

8. A new powder layer is applied and levelled by the coaters onto the powder bed.
9. The laser then begins to scan and melt the next layer of the part.
10. This process continues successively until the desired part has been printed.
11. Following the completion of the print, the build platform is removed from the build chamber for further heat treatment, part/material removal and other finishing processes.
12. Excess powder remaining within the build chamber is removed using a vacuum cleaner. The powder is then sieved and can be reused a number of times for printing.

Compared to conventional manufacturing processes, AM processes provide a number of advantages, such as lower lead times, near-net-shape production, design flexibility and high material efficiency rates. An important driver for the aerospace industry is the buy-to-fly ratio, which is the ratio of the mass of the total material required to produce a part and the mass of the final component itself. In the aerospace industry, it is not rare to see buy-to-fly ratios of twenty to thirty due to the high requirement for mass optimisation of parts. Due to the ability to only utilise the necessary amount of material, AM has shown its capacity for producing components with buy-to-fly ratios almost approaching unity, representing large savings in terms of cost, energy and materials (Achillas et al. 2015, Conner et al. 2014, Mellor et al. 2014).

Improving the dimensional accuracy (which is one of the main issues regarding the quality of as-built AM components) will help develop the precision of metal AM processes (Thompson et al. 2016). There has been considerable research to establish repeatable and predictable AM processes; some examples are given below. A mathematical model to predict the final geometrical quality of as-built parts was proposed (Dantan et al. 2017). This model was used to simulate the processes and thus to optimise the design of the CAD file for higher precision AM parts in terms of the resulting surface texture and dimensional deviations. The accuracy of the defect characterisation method, manufacturing signature and form defects were found to be significant when attempting to control the geometrical variation and to improve the quality of AM parts. The effects of process parameters and material properties on the mechanical performance, part porosity and density of 304L stainless steel parts were experimentally examined (West et al. 2017). The results revealed that the natural frequency of the printed part is highly correlated to the yield and ultimate strength and, to a lesser extent, to the part density. A mathematical modelling approach to mitigate and compensate distortion of the built parts induced by material shrinkage and residual stresses in the powder bed fusion (PBF) process was proposed (Afazov et al. 2017a). The developed model utilised areal optical measurement data for correcting the design geometry and was experimentally validated by manufacturing a turbine blade and an impeller of Inconel 718. The results demonstrated the feasibility of the distortion compensation methodology to develop a precision AM process and to minimise distortion of the as-built part (Afazov et al. 2017b).

3.3 Setting Priorities

Similar to other technologies that are currently driving Industry 4.0, AM requires the 'zoom out, zoom in' approach to optimisation as proposed by John Hagel (Davis and Schwab 2018).

This dual approach can be useful when discussing AM process optimisation. Firstly, the optimisation of the AM technology itself, which deals with the overall improvement of all facets of AM processes, such as the optimisation of process parameters, improvement of energy efficiency and sustainability development for improved material efficiency. Secondly, the optimisation of the complete system to which the AM processes belongs, investigating the impact of AM technologies on established global process chains, the potential for AM to disrupt conventional manufacturing systems, as well as the potential for AM to create a wide variety of new and innovative products and services. When it comes to setting priorities, it is, therefore, important to visualise both the process level and the system-wide circumstances. In particular, while optimising AM processes, it is important to investigate the effect of this optimisation on the entire AM system. In addition, system-wide performance requirements for AM systems, such as lead time or takt time, can in turn help drive technology and process development. A holistic approach is, therefore, necessary for developing both the technology as well as the system proactively. In the following sections, technological aspects of AM processes as well as system-wide aspects will be introduced and briefly discussed.

3.4 Significant Process Parameters

A large number of process parameters can control the L-PBF process. Figure 3.2 shows an Ishikawa diagram for L-PBF processes development. In order to develop precision processes, it is critical to understand the effect of each process parameter. However, various research groups have concluded that L-PBF process parameters exhibit a high degree of interdependent and interacting effects on the quality of as-built parts (Charles et al. 2019c, Khorasani et al. 2019, Shipley et al. 2018). These effects are, therefore, the primary reasons for the difficulty in achieving precision in powder-bed AM processes.

The most significant L-PBF process parameters can be classified into four broad groups, as follows (Aboulkhair et al. 2014):

1. Laser-related parameters.
2. Scan-related parameters.
3. Powder-related parameters.
4. Build chamber–related parameters.

These parameters are explained in more details in Section 3.4.1–3.4.4, and Figure 3.3 lists the governing parameters within these four categories.

Figure 3.4 shows a visual representation of the various process parameters in L-PBF processes.

3.4.1 Laser-Related Process Parameters

Properties of the laser beam are defined as follows:

- *Laser power* P – the optical power output of the laser beam. P has units of watts and is the nominal power from a continuous wave laser or the average power of a pulsed or modulated laser.

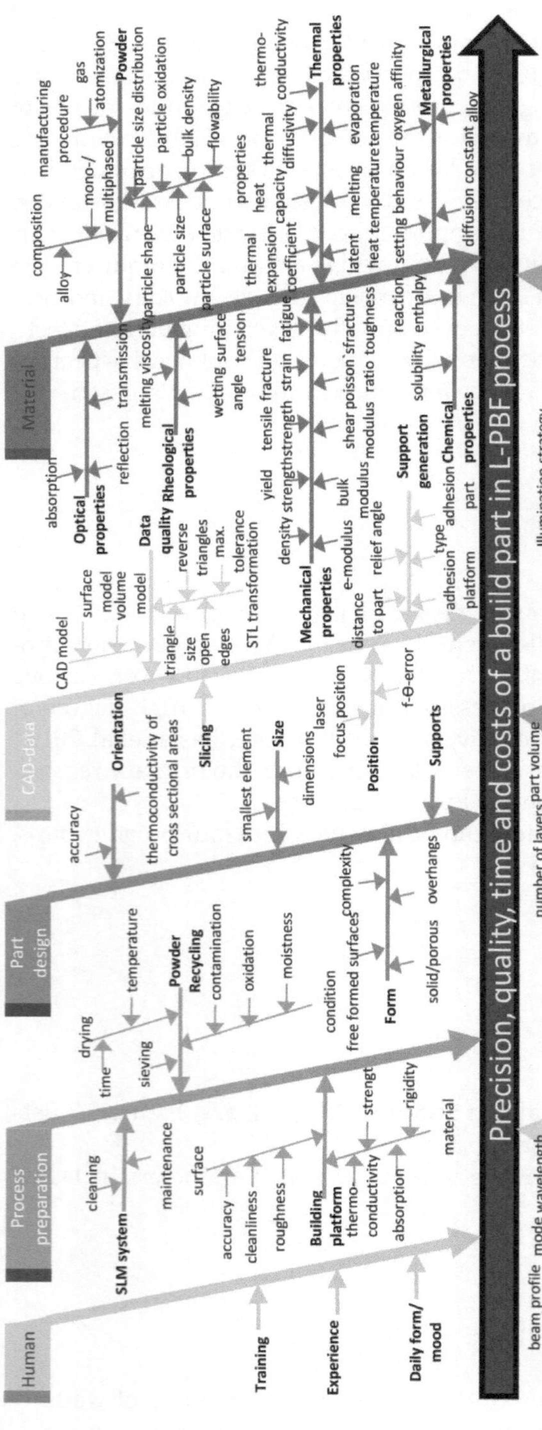

FIGURE 3.2
Ishikawa diagram for L-PBF process development. (Modified from Rehme, 2010.)

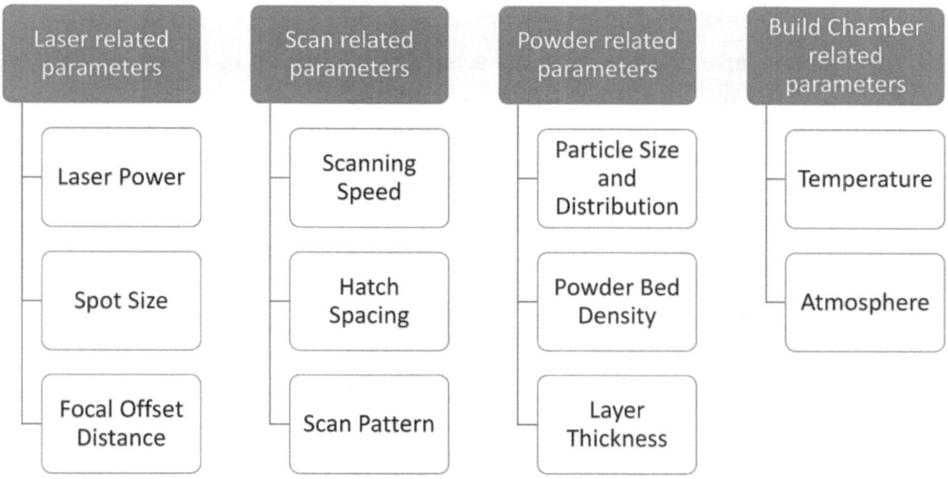

FIGURE 3.3
The most significant parameters in the L-PBF technique.

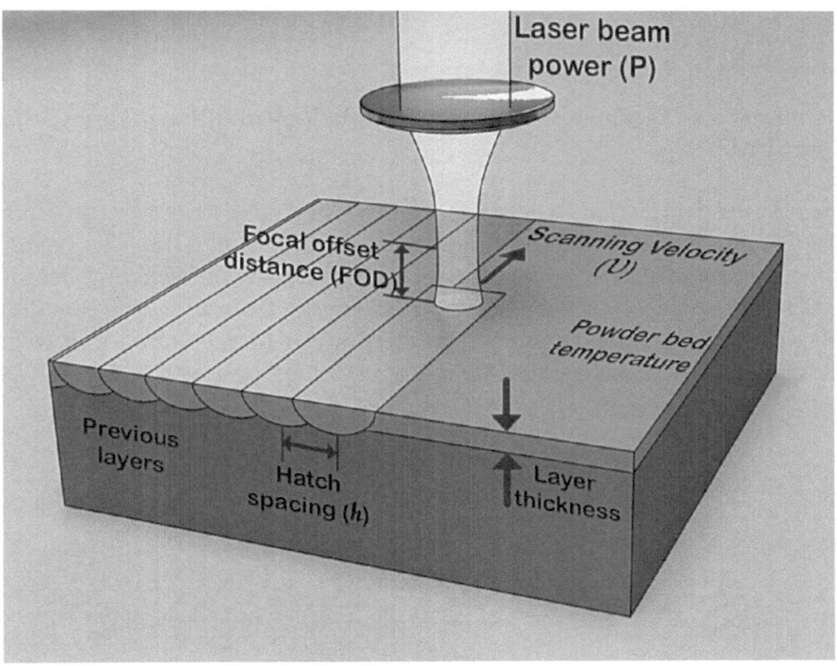

FIGURE 3.4
Process parameters in L-PBF.

- *Spot size d* – the smallest diameter of the laser beam at its focal plane when focussed by a lens.
- *Focal offset distance* – the distance between the focal plan of the laser and the printing surface.

3.4.2 Scan-Related Process Parameters

Scan-related process parameters control the scanning properties of the laser beam on the powder bed. The most significant of them are as follows:

- *Scanning speed v* – the translational speed of the laser beam while scanning the metal powder. The scanning speed and the laser power are often used together as combined process parameters (see Section 3.4.5) to control the L-PBF process.
- *Hatch spacing h* – the distance between the centres of two successive scan tracks; see Figure 3.4.
- *Scan (track) pattern* – the motion followed by the laser beam on the surface of the powder bed. Different machine tool manufacturers and pre-processing software allow the use of various scan track patterns. Figure 3.5 depicts three of the most commonly used scanning pattern examples: hexagonal, rectangular and striped. For the hexagonal scan pattern (Figure 3.5a), the laser scans the outline of a hexagon and then fills the inside while following the shape of the hexagon. In the rectangular cell scanning strategy (Figure 3.5b), the laser first scans the outline of a rectangular cell and fills each cell with alternating scan directions (90° rotation). The last image, Figure 3.5c, shows the stripes scan pattern where the laser scans in one direction and fills adjacent sections consecutively.

3.4.3 Powder-Related Process Parameters

Powder-related process parameters affect the morphology and the quality of the powder. They are described below.

- *Particle size and distribution* – the standard powder size distribution, as used in the L-PBF process, is between 15 μm and 45 μm (Sinico et al. 2018). The particle size and the size distribution significantly affect the packing density. Smaller particles provide larger surface area for absorption of energy from the laser; however, they can also cause larger gap formation in the powder bed, creating porosity in the final part. Smaller particles also increase agglomeration and reduce flowability (Sun et al. 2017).

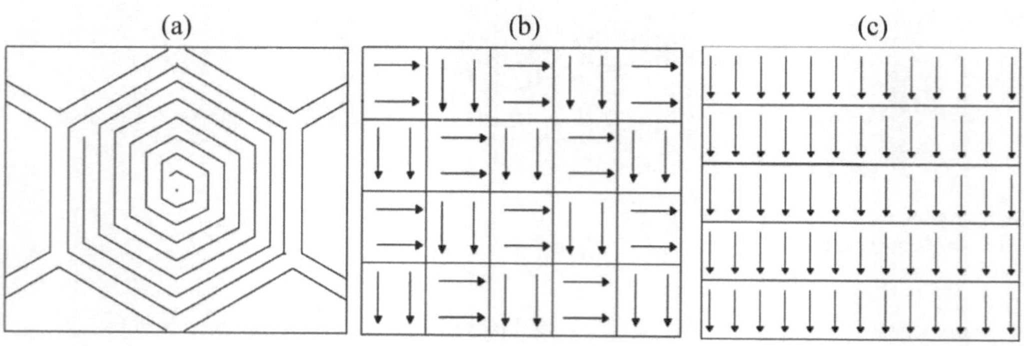

FIGURE 3.5
Different scanning strategies in L-PBF: (a) the hexagonal cell scanning strategy, (b) the rectangular cell scanning strategy and (c) the strips scanning strategy.

- *Powder bed density* – high packing densities are generally preferred as they lower internal stresses and porosity and improve the surface texture. Spherical particle morphology greatly improves the flowability of powder, thereby improving the packing density (Sun et al. 2017). The powder bed packing density also affects the thermal conductivity of parts, as a higher density increases the number of contact points between powder particles and thereby improves the thermal conductivity of the powder bed.
- *Layer thickness t* – the distance between each successive layer in the PBF process; see Figure 3.4. In practice, a 30 µm layer thickness is commonly used for fine prints with higher strength, while a 60 µm–90 µm layer thickness is used for faster but coarser prints (Sufiiarov et al. 2017).

3.4.4 Build Chamber-Related Parameters

- *Temperature* – the temperature inside the build chamber is an important process parameter and depends to a large extent on the material used for printing. Temperature can directly affect the surface integrity of the printed parts as well as their mechanical properties, such as tensile strength (Khorasani et al. 2019). The build chamber temperature needs to be controlled to ensure that it is in an optimal range for the material to be printed whilst also minimising thermal gradients. Lowering the temperature gradient between the build chamber and the printed material decreases the risk of internal stress formation within the part, which is a potential cause for printing defects, ultimately resulting in build failures.
- *Atmosphere* – due to the high temperatures inside the build chamber of L-PBF machines, the presence of oxygen inside the chamber can cause oxidation of the used material. Certain metals used in L-PBF are especially vulnerable to oxidation processes, therefore requiring a matching selection of the desired inert atmosphere inside the chamber. Furthermore, reduction of oxygen inside the build chamber minimises the risk of porosity within parts, thereby improving the mechanical properties of the part. Commonly used gases inside build chambers include argon and helium (Schmidt et al. 2017).

3.4.5 Combined Processing Parameters

A number of combined process parameters are used by researchers trying to understand the effect of various energy densities on the quality of components. Some of these parameters are presented in the following, as compiled by (Sun et al. 2017):

Volumetric energy density E_V [J mm^{-3}]:

$$E_V = \frac{P}{vht} \tag{3.1}$$

Linear input energy density E_I [J mm^{-3}]:

$$E_I = \frac{4P}{\pi v d^2}. \tag{3.2}$$

Surface energy density E_S [J mm^{-2}]:

$$E_S = \frac{P}{vd}. \tag{3.3}$$

Linear energy density E_L [J mm^{-1}]:

$$E_L = \frac{P}{v}. \tag{3.4}$$

3.5 Additive Manufacturing Performance Indicators

Researchers focussing on optimisation of metal AM processes have considered various performance indicators as optimisation criteria (Allaire et al. 2017, Qin et al. 2017, Shi et al. 2017, Shipley et al. 2018, Smith et al. 2016). The main performance indicators for a part produced by AM are listed in the following:

3.5.1 Mechanical Properties

Commonly used methods for testing mechanical properties of components include testing the tensile strength and hardness, which can also be utilised for classifying the performance of AM components. Fortunately, a number of standards already exist that describe the testing methodology and guidelines for testing AM components in particular. These standards are described in detail in Chapters 6 and 9.

3.5.2 Dimensional Accuracy

In the case of any manufacturing or machining process, the dimensional accuracy of the final part is of utmost importance and is one of the most critical defining characteristics in judging the stability and quality of the process. Dimensional accuracy is defined as the closeness of the dimensions of a produced part with the dimensions of an ideal part (represented, for example, by the CAD design of the part).

Dimensional accuracy in L-PBF processes is affected primarily by the high-energy densities imparted onto the powder and the high operating temperatures involved in the process, which cause large thermal gradients in the solid material just below the laser spot. The high temperatures in the upper layers cause them to expand, while the much colder layers beneath them restrict this expansion. This thermal gradient, therefore, induces compressive stresses on the upper layers. When these upper layers cool, the compressive stress is converted to residual tensile stresses, which are the major reason for warping and cracking of printed metal parts in L-PBF processes (Kempen et al. 2011). This mechanism of residual stress formation is depicted in Figure 3.6.

Various research groups have, therefore, focussed on different methods for reducing residual stresses in order to improve dimensional accuracy (Kruth et al. 2012, Mohanty and Hattel 2016, Robinson et al. 2018). In addition, several other process-dependent factors contributing to dimensional deviation and accuracy exist, such as part orientation and location, support structures and post-processing steps.

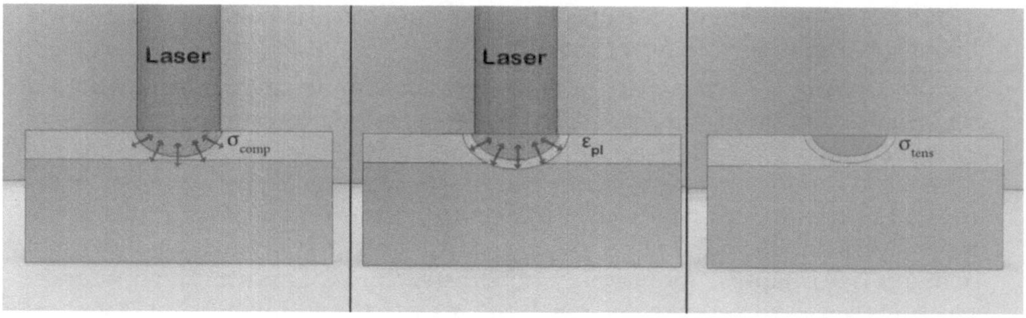

FIGURE 3.6
Mechanism of formation of internal stresses in the L-PBF processes.

- *Part orientation and location* – since the L-PBF process is a layer-by-layer fabrication process, the orientation of the printed part will greatly affect the final dimensional accuracy of the part (Mukherjee 2019). Parts with holes or cylindrical features are printed with their axis parallel to the build direction, as shown in Figure 3.7, to avoid the staircase effect and to ensure maximum accuracy. In metal PBF processes, another major factor in determining dimensional accuracy is dross formation phenomenon, which occurs in down-facing surfaces where the laser causes the melt-pool to form directly on top of loose powder (Han et al. 2018). This phenomenon results in the formation of an overheated zone and, therefore, melting of more material than required, which is visible as dross. Therefore, parts with

FIGURE 3.7
Example of the effect of part orientation on the quality of the final part.

cylindrical or overhanging features need to be oriented such as to maximise the dimensional accuracy (Zhang et al. 2017).

- *Support structures* – when part orientation alone cannot guarantee minimum dimensional deviation, support structures are required to print parts that have excessive down-facing surfaces (Cheng et al. 2019). Apart from providing the mechanical support that prevents the bending or warping of parts, support structures also play an important role in transferring the excess heat generated at the upper layers into the build plate, thereby contributing to the reduction of thermal gradients (see Chapter 2 for more on support structures). The inclusion of support structures causes drawbacks, such as the increase of the overall build time, as support structures necessitate the inclusion of a separate and additional process step for support structure removal. Some research also indicates that the act of support removal actually contributes to further dimensional inaccuracies and surface defects on the printed part in the form of burrs, which then necessitate a further finishing process (Triantaphyllou et al. 2015).
- *Post-processing* – because of the high thermal gradients present during the printing process, as discussed above, internal residual stresses are present in the final printed parts (Kruth et al. 2012), causing major deformations, warping and sometimes catastrophic failure, as shown in Figure 3.8. After printing the parts, and prior to the removal from the build platform, it is, therefore, common practice to perform a heat-treatment step to reduce the presence of these residual stresses. Depending on the material, the parameters for this stress relief differ. The simulation and optimisation of these heat-treatment parameters are major focuses of current research in the field of metal AM (De Baere et al. 2018).

3.5.3 Surface Texture

For the manufacturing industry, surface texture is an important performance indicator and is discussed in detail in Chapter 11.

FIGURE 3.8
Failure due to internal residual stresses during the AM process.

3.5.4 Part Density

The density of parts produced by L-PBF is largely affected by various defects present within the volume of the part, such as cracks, discontinuities, foreign materials and porosity, that in turn affect the mechanical properties and dimensional accuracy of the components (Shipley et al. 2018). Many researchers, therefore, focus on achieving 'fully dense' parts. A fully dense part is defined as having a density of over 99% (Eylon and Froes 1990), although there is no widely accepted definition of the term 'fully dense'. Various investigations have been conducted or are still ongoing regarding the minimum requirements for part density of components produced by L-PBF processes, as well as establishing process windows for maximising the density of printed parts (Cunningham et al. 2017, Han et al. 2017, Kasperovich et al. 2016).

3.5.5 Total Build Time

The total build time is an important performance indicator. Metal AM processes are considered to be expensive and time consuming, which hinders their mainstream application in industrial settings. Since build time directly affects the lead time for delivery of parts, this can cause disruption of existing process chains and supply chains, as well as logistical plans (see Chapter 7). Numerous researchers have started to focus on build-time optimisation, such as through a meta-heuristic optimisation technique that demonstrated the benefits of build-time minimisation using a case study (Hallmann et al. 2019). Accurate models for build-time prediction for extrusion-based AM by taking into account the acceleration and deceleration of the machine head have been developed (Komineas et al. 2018). However, build-time optimisation is an area of AM research that has developed over the last few years, and further research is required in the field of metal AM.

3.5.6 Energy Consumption

In recent years, many companies have adopted strategies towards improving sustainable practices and reducing the carbon footprint on the environment and atmosphere. The role of sustainability in AM and the reasons why manufacturers adopt AM technologies are discussed elsewhere (Niaki et al. 2019). AM already contributes to sustainability in one way as it promotes material efficiency, which has already attracted some companies to adopt AM (Sauerwein et al. 2019). Environmental assessment studies of AM technologies in the automotive industry have been conducted using a lifecycle assessment for L-PBF processes (Böckin and Tillman 2019). Their conclusions indicate that L-PBF processes can improve lifecycle environmental performance by mass optimisation of parts. However, further investigations are needed to understand the total costs of ownership as well as lifecycle assessments, energy consumption and recycling (Charles et al. 2019a, Fauth et al. 2019). These research topics could further justify and promote investment in AM technologies by industrial companies.

3.5.7 System-Wide Performance Indicators

Manufacturing industries employ key performance indicators (KPIs) to judge their operational efficiency and outputs. These KPIs can cover a large range of topics, from improvements in efficiency to improvements in customer experience, and are, therefore, used in many production engineering research areas (Barnes-Schuster et al. 2006, Hällgren et al. 2016). Some of the KPIs used by manufacturing industry are as follows:

1. *On time delivery* – the fraction of time that a completed product is delivered to a customer on time, as per the schedule that was promised to the customer.
2. *Cycle time* – the time required for the manufacturing of a product starting from initiation of a production order until the finished product. Process parameters such as build time directly affect the cycle time of a product.
3. *Changeover time* – the time necessary for a machine tool or a production line to change from producing one product to another. In AM processes, changeover times reflect the time that has elapsed between the end of one build and the initiation of another.
4. *Yield* – the fraction of products that are manufactured correctly the first time without the need for scrap or rework.
5. *Throughput* – the measure of the number of products being produced by a machine, cell, line, etc. over a specified period.
6. *Overall equipment efficiency (OEE)* – the overall efficiency and effectiveness of a machine tool, production line or cell, given by

$$\text{OEE} = \text{Availability} \times \text{Performance} \times \text{Quality}. \tag{3.5}$$

7. *Manufacturing cost per unit excluding materials* – the cost of all controllable manufacturing costs that go into the production of a given product.

The above KPIs are just a few of the many currently in use. In most industrial settings, the KPIs are subject to continuous optimisation procedures. During process optimisation and technological development of AM processes, it is important to consider the impact of KPIs on the overall AM value chain; thus, using these metrics to judge process improvements could be highly beneficial. These KPIs have been used in many cases for judging other conventional machining and manufacturing processes (Abu Qudeiri et al. 2018, Elkaseer et al. 2018a, Elkaseer et al. 2014, Fauth et al. 2019, Gyulai et al. 2018, Lingitz et al. 2018, Salehi et al. 2016, Sivakumar and Manivel 2020), but their use for AM is still limited and they need to be further included in AM optimisation and improvement studies. The inclusion of system-wide KPIs is especially relevant as AM is starting to leave laboratories and R&D facilities and moving towards implementation by manufacturers within their existing process chains. When mainstream industrialisation is the goal, system-wide KPIs for AM will play an important role, along with process-based performance indicators such as dimensional accuracy and surface quality, in order to judge and consider AM as an asset to manufacturing process chains.

3.6 Data-Driven Process Improvement

Manufacturing companies are aware of the importance of data collection and data analytics to develop their process knowledge and to remain competitive. Nevertheless, data analytics has recently been gaining more attention, since it plays a significant role in streamlining manufacturing operations to be faster and more efficient, especially in the domain of smart manufacturing. In particular, the benefits of data processing/analytics are not limited to visualising the status of the manufacturing processes but also empower artificial intelligence (AI) algorithms to produce smart solutions and concurrent corrective actions (see Section 3.7.2). Currently, manufacturing data is utilised to model AM processes and to facilitate the optimisation of the process parameters. Data can be gathered via sensors in real time (see Section 3.7.1 and Chapter 13) or from off-line experimental measurements (see Section 3.6.1, 3.6.2 and Chapters 10, 11 and 12).

A data-driven process improvement is structured based on five steps (collection of data, sharing, analysis, optimisation and feedback), as shown in Figure 3.9.

In a given AM process, data can be collected manually via an operator or automatically via smart sensors. AM data is highly heterogeneous as it is collected from different sources, such as sensors, cameras, databases and measurements of experimental tests. The format of the collected data has to be digitised and normalised to render it compatible with the modelling and optimisation algorithms. In addition, the digital format makes the data shareable between all smart assets in the manufacturing system (server machine, control units, monitoring tools, inspection operator and process documentation). Data collection and data sharing are followed by a further data analysis step to extract and visualise the mathematical model/objective function of the process. The data-analysis algorithms need to conduct an additional step for data cleaning/deleting in order to prevent data redundancy and data inconsistency (DeCastro-García et al. 2018, Rahm and Do 2000). Machine learning algorithms can be utilised to simplify the analysis via parallel processing of the data without necessitating a high level of computing power. Machine learning algorithms also reduce the bandwidth for sharing data and handle issues such as missing information and communication failures (Dean and

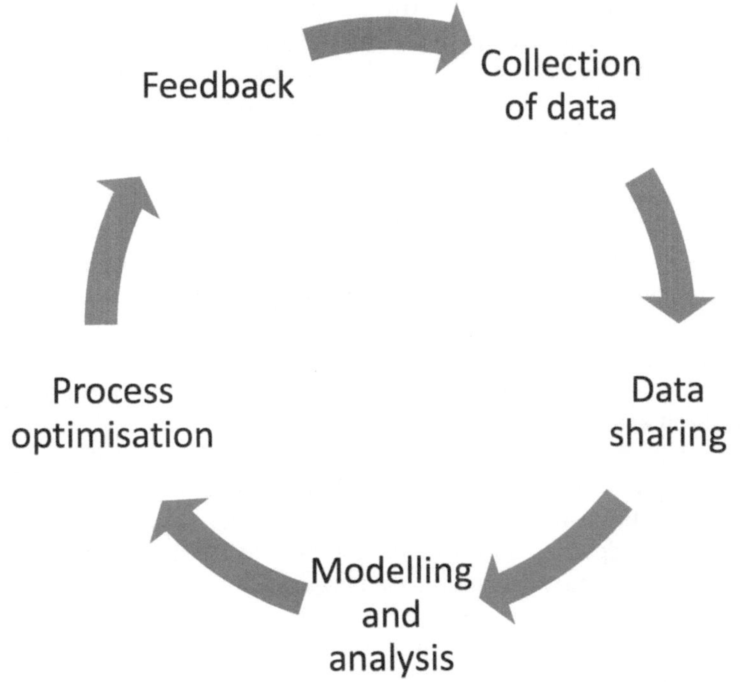

FIGURE 3.9
Data-driven process improvement steps. (Modified from Buer et al. 2018.)

Ghemawat 2004). The analysis step produces a deterministic mathematical function that models the relationship between all processed data. This model/mathematical function is used to predict/describe the behaviour of the AM process under various conditions (Banerjee et al. 2014, Mertens et al. 2014, Tapia et al. 2016). Following the mathematical modelling, the optimisation step identifies the optimal values of the process parameters for the best possible performance of the AM process. In particular, the optimisation step adjusts the parameters of the AM model in order to maximise the precision/quality of the product or minimise manufacturing errors/costs, depending on the objective function. Finally, the feedback step feeds the machine/operator the corrective actions/parameters in order to continuously improve the process.

3.6.1 Design of Experiments

Statistical design of experiment (DoE) techniques systematically quantify the correlations and effects of governing input parameters on output responses of a process. DoE is a systematic method for evaluating a set of experimental trials. Although DoE does not attempt to describe the underlying physical phenomenon of the process, it enables the user to identify the interrelationships between different process variables to determine their optimal values (Siebertz et al. 2010). In particular, DoE offers sufficient knowledge and information about the process with the least possible experimental effort (Montgomery 2019). For this purpose, the critical parameters of the process need to be identified and rigorously examined.

Every DoE approach generally encompasses factors, levels and output variables (process responses). Factors are separate input parameters that are altered on different levels (alternative values for each factor). Ideally, the factors and their interactions influence the output variables, whereby it is possible to draw conclusions from the experimental results.

A DoE can be set up in various ways, differing in structure, purpose and, therefore, complexity and elaboration. The most intuitive option would be to combine and test all factors at each level, leading to a full factorial design. Unfortunately, when investigating several factors at many different levels, this design expands exponentially, according to

$$n_r = n_l^{n_f} \tag{3.6}$$

where n_r is the number of runs in a full factorial design, n_f is the number of factors to be examined and n_l is the number of levels for each factor.

This high number of runs and the resulting high expenditure of time and resources is the reason why fractional factorial designs (FFDs) were developed. Fractional factorial DoEs enable detailed insight, while at the same time keeping the number of trials as low as possible (Montgomery 2019).

The FFD uses two principles: the sparsity principle and the hierarchical ordering principle. These two principles help to reduce the number of trials by examining only the important effects and interactions and by prioritising lower-order effects (Fang et al. 2018, 16f.).

Two types of FFD are described in this section. First, central composite design (CCD) is constructed of three building components (see Figure 3.10), namely eight corner (factorial) experiments, six symmetrically arrayed axial experiments and replicated centre experiments (at least three replicated experiments to quantify the robustness and validity of the CCD model); hence the approach is termed a *composite* design (Carlson 2001, Siebertz et al. 2010). To investigate problems with higher-order polynomials, the axial experiments can be expanded to consider five levels for each parameter.

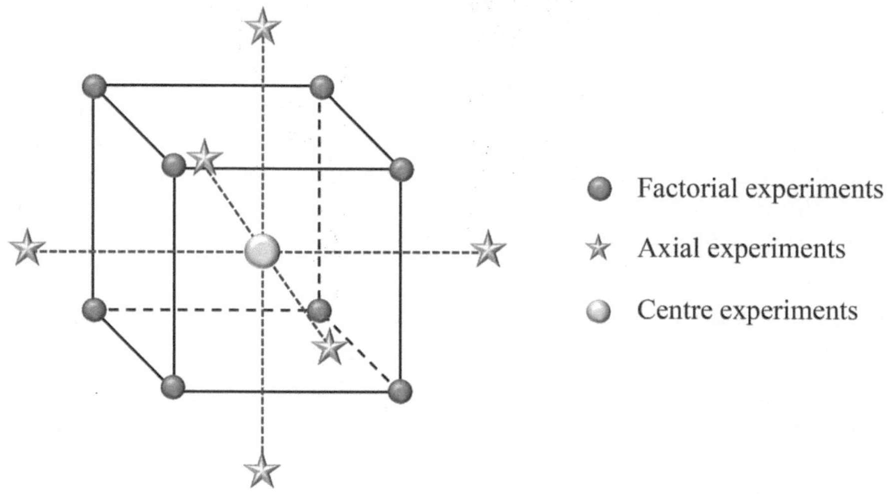

FIGURE 3.10
Structure of central composite design.

The second type of FFD is orthogonal arrays (OAs), also known as Taguchi designs, which are defined by a $n \times m$ matrix and noted by $L_n(s^m)$, where n is the number of runs, m is the number of factors and s is the number of levels (Montgomery 2019). Moreover, n can be understood as a fraction of the full factorial s^m. The speciality of OA is that each factor pair exists an equal number of times, creating a balanced repetition of levels. Taguchi also designed mixed-level OA, featuring factors with different numbers of levels (Mitra 2011). In OA designs, various sizes of test series are available, such as L25, L52 and L125, presenting OA designs entailing 25, 52 and 125 trials, respectively. A compromise must be determined between the degree of detail and the time/resources required/available for the experimental trials.

3.6.2 Modelling of Process Performance (Quantifying Input/Output Process Relationships)

Physical testing and experimentation costs raw materials, time and human resources. On the other hand, the development of reliable process models enables accurate prediction of the performance of the manufacturing process with the least experimentation effort and waste of resources (Abu Qudeiri et al. 2019, Elkaseer et al. 2016). In AM processes, the modelling of the manufacturing processes aims to consider the effect of fabrication process parameters to predict some features of the AM part, such as obtainable part shape and dimensions, surface texture and materials strength. However, a proposed AM modelling approach has to take into consideration the complex relationships between the process parameters, properties of raw materials, machine capability/features and process uncertainty as well as the impact on the overall process/value chain and the system-wide KPIs. In some cases, the modelling approach has to integrate advanced inspection solutions in order to study the effect of heat distribution on the topography of the manufactured product (Mies et al. 2016).

The accuracy of modelling software depends on the experimental design, the algorithm design, the comprehensive data collected during the AM process and other physical factors that can incite a physical response in the shape/structure of the fabricated product

(Jurrens et al. 2013). In order to develop AM process modelling, analytical and experimental-based approaches (statistical regression and computational machine learning, among others) are widely utilised. Although analytical approaches have seen increasing adoption in modelling different AM processes, the large body of reported attempts are not detailed enough to properly model the process due to its high complexity. For this reason, this section will focus on experimental-based modelling approaches only (see Chapter 4 for more on AM modelling). In particular, statistical regression and artificial neural networks for modelling of experimental data will be discussed.

3.6.2.1 Regression and Statistical Analysis

Statistical analysis is a methodology for systematically examining DoE results in order to categorise the most significant process parameters, to evaluate the effect of these parameters on the process responses and to identify the proper process parameters for optimal responses (Baturynska 2018, Grasso et al. 2018). Among statistical analysis techniques, regression is an effective methodology, utilising the least-squares method for determining the degree of correlation between process parameters and process responses, or the experimental results–based model. The proposed model can be used for process prediction and as an objective function in the optimisation of the process in a later step. Simple regression analysis is applied when a process response depends on a single process parameter, while multiple regression analysis is used in the case of dependence on multiple process parameters (Noryani et al. 2019). Regression analysis is developed based on a pre-defined relationship between the process parameters and process responses (Stulp and Sigaud 2015), such as linear, logarithmic and polynomial.

Statistical analysis has been used by several researchers to optimise process parameters, such as laser power, scan speed, powder thickness, hatch spacing and scan strategy for the L-PBF process. The Taguchi method has been previously used to optimise these process parameters and analyse their results using analyses of variance (ANOVA). These regression equations showed a linear relationship between the density of the AM fabricated part and laser power, scan speed, powder thickness and the scan strategy applied (Sun et al. 2013). In another study, a rapid method for parameter optimisation of the L-PBF process by using linear regression to predict geometric characteristics was conducted (Shi et al. 2017). Similarly, a statistical analysis–based experimental study to examine the effects of a range of process parameters and their interactions on the resultant surface texture of down-facing surfaces in L-PBF was conducted. The effects of the laser power, the scan speed and the scan spacing on the obtainable surface quality of down-facing surfaces were identified. It was concluded that the performance of the L-PBF process is highly dependent on the energy absorbed by the powder (Charles et al. 2019b).

3.6.2.2 Artificial Neural Network Modelling

An artificial neural network (ANN) is a computational algorithm used to mimic the behaviour of a biological neural network, and was developed to describe and solve complex problems (Abiodun et al. 2018). The single processing unit of the ANN is a neuron that is connected to other neurons by weights. Data processing is performed via these connections (weights and interconnected neurons), which also connect adjacent layers. Each layer is a group of neurons collectively functioning at the same processing step within a neural network. The ANN model consists of an input layer, hidden layers and an output layer. Figure 3.11 shows the schematic structure of an ANN where input data is fed to the input

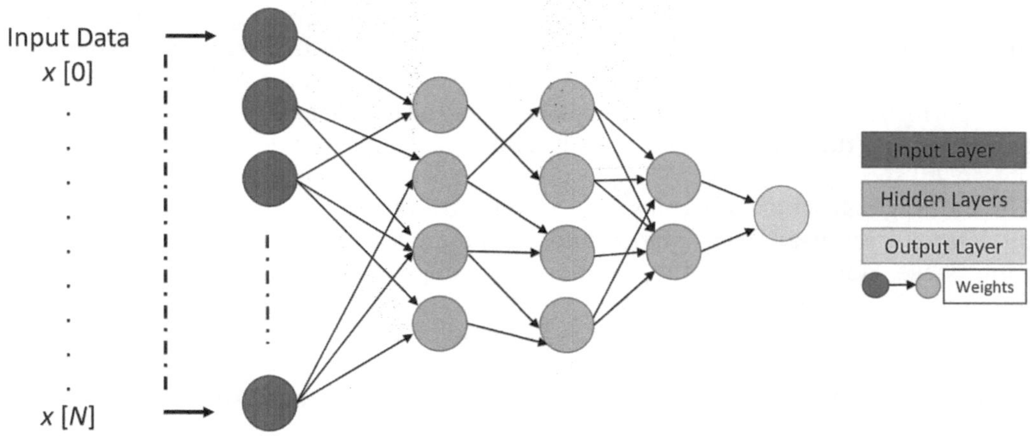

FIGURE 3.11
Structure of artificial neural network.

layer (first layer), and the number of the neurons equals the number of input parameters. The number and structure of the hidden layers are determined during the design phase of the ANN and are closely associated with the problem to be modelled (Mathieu et al. 2016). In each layer, the neurons process the input transferred from previous neurons via the associated weight to generate the output based on its mathematical formula, termed an *activation function* (LeCun et al. 2015). The output layer can consist of only one neuron or multiple neurons, depending on the process responses to be considered.

ANNs can utilise different learning methods, such as supervised, unsupervised, semi-supervised and reinforcement learning, depending on the problem to be modelled. The supervised learning approach is utilised for labelled datasets (collected data that is tagged with a certain label). In this case, the dataset is used to generate a model that processes the input features to predict or classify the output information. If the dataset contains unlabelled examples only, the unsupervised learning method is employed to create a model that processes the input features to solve a specific problem, such as clustering. In a semi-supervised learning technique, the dataset consists of both labelled and unlabelled examples (Shukla et al. 2018). The objective of semi-supervised learning is similar to that of supervised learning, but the addition of unlabelled data enhances the learning process (LeCun et al. 2015). In reinforcement learning, the algorithm perceives and interacts with the environment in order to execute actions based on its observed state. In particular, the algorithm performs different actions and receives corresponding rewards/penalties (Abbeel et al. 2007). Based on the received rewards/penalties, the algorithm then takes consequential action. After a number of iterations, the algorithm will learn the rules of the environment, which are also known as 'policies', and apply them for future computations accordingly. From this wide range of learning strategies, ANNs are able to solve numerous non-linear complex problems (Bengio et al. 2013).

As a first step in the development of an ANN model, the structure of the ANN must be designed by deciding the number of neurons and hidden layers. This is followed by the learning step, in which the optimal values for the learning parameters, such as weights, are identified. The learning step helps to minimise the difference between the actual data and the results predicted by the ANN model. In order for the ANN model to perform the learning process, the dataset must be split into three categories: (1) training dataset, (2) validation dataset and (3) testing dataset. The training dataset is used to build up the model and identify the learning parameters of the network, whereas the validation dataset is utilised to tune the learning parameters of the ANN in case of inadequate performance for the validation datasets. The testing dataset is employed post-validation to assess the efficiency of the ready-to-be-used algorithm with an unseen dataset (Nielsen 2015).

The training dataset usually comprises around 70% of the total dataset, with 15% of the total dataset being used for validation and the remaining 15% of the dataset for the final testing phase. However, this categorisation percentage can be adjusted according to the individual ANN design requirements.

The workflow of the training process is as follows. First, ANN training is started by a random initialisation of weights in the neural network. The training dataset is then fed into the network (Figure 3.12 [1–2]). Next, the ANN estimates the output (Figure 3.12 [3]) and calculates the error of the neural network by comparing the predicted output with the actual result (Figure 3.12 [4]). The calculated error is then propagated back to the network to update the weight of every neuron in the network (Figure 3.12 [5–6]). Steps 1–6 (see Figure 3.12) are continuously repeated until all training data has been fed into the ANN and a high estimated performance output has been achieved.

Having finished the training steps with the training dataset, the ANN will continue to process the validation dataset to enhance the model (steps 9–12, Figure 3.12). During the validation phase, overfitting problems can occur when an ANN model generates results that are not as good as those obtained during the training phase (see Figure 3.12 [8]). This could be due to overfitting of training datasets, including noise and other unwanted features. In order to prevent overfitting, it is recommended to increase the amount of training data. Dropout regularisation is also a potential solution, where some neurons could be

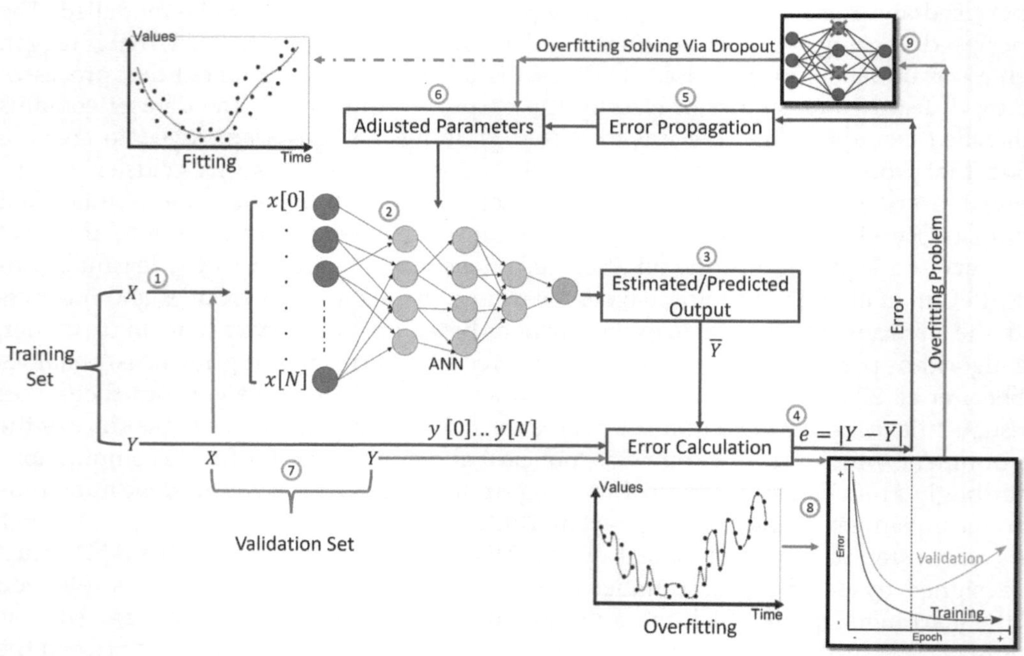

FIGURE 3.12
Training artificial neural network.

randomly dropped out during the training phase, thus enabling other neurons to learn better (LeCun et al. 2015). In addition, cross-validation is a powerful methodology for avoiding overfitting problems; it works by partitioning the training data into small groups in order to generate test partition from this data to tune the ANN model (Pereyra et al. 2017, Srivastava et al. 2014).

Recently, ANN modelling has been adopted for a number of different manufacturing solutions (Meireles et al. 2003), such as for control and robotics (Aylor et al. 1992, Fu et al. 2019, Lipson and Pollack 2000). In addition, ANN has been used for tool-wear prediction in machining processes (Charles et al. 2019b, D'Addona and Teti 2013, Elkaseer et al. 2019, Nassef et al. 2018). In the case of AM, an ANN modelling algorithm for fused deposition modelling (FDM) processes was developed. The ANN model predicts the surface texture of the AM product in order to improve the surface quality. It was found that the accuracy of the ANN model is better than other analytical methods (Vahabli and Rahmati 2016).

An ANN algorithm to optimise the counterbalance of thermal shrinkage and shape deformation for L-PBF fabricated parts was developed (Chowdhury and Anand 2016). An ANN framework to optimise the L-PBF process parameters for minimal fabrication time for the printed part was developed (Marrey et al. 2019). In another study, the combination between ANN and finite element simulation of AM parts was developed to enable the accurate simulation and optimisation of the part's structure, considering microstructure features (Koeppe et al. 2018).

3.6.3 Process Optimisation

Optimisation is a technique for identifying the optimal solutions among different possible alternatives for a certain problem, considering a set of constraints. Mathematical and computational methods are utilised to deal with different deterministic and stochastic problems. To some extent, a number of manufacturing problems are relatively straightforward and are considered single-objective problems that can be addressed using deterministic optimisation techniques. However, the trade-off between the quality, cost and productivity of the manufacturing process, in addition to the advent of AM, have introduced multi-objective problems that necessitate advanced optimisation techniques (Mukherjee and Ray 2006).

Examples of the well-known deterministic techniques are first-order and second-order derivative techniques. For instance, gradient descent is a first-order method, which utilises the gradient information of the problem distribution to find the optimum solution. However, one of the disadvantages of first-order methods is that their requirement for high accuracy necessitates small steps, thus requiring a large number of iterations and long calculation times. Accordingly, second-order optimisation methods can be used to optimise simple problems using a quadratic approximation method. Nevertheless, with the increase in the degree of complexity of the problem distribution, the computational time increases exponentially (Kochenderfer and Wheeler 2019).

For complex problems, stochastic optimisation techniques, such as simulated annealing and population algorithms, have been developed. In particular, the simulated annealing algorithm offers a methodology for finding the global optima of a problem that contains a large number of local optima. Simulated annealing is one of the stochastic optimisation methods inspired by metallurgy. It uses an initial randomisation strategy followed by repetitive adaptive direct searches per iteration to explore the entire problem distribution and determine the optimal point (Ghosh et al. 2019). It is worth stating that all of the aforementioned methods utilise individual single points moving across the entire distribution to reach the optimum point, which can be time-consuming. This has led to the emergence of population optimisation methods, such as genetic algorithms (GAs) and particle swarm algorithms (PSOs), inspired by biological science (Wilson and Mantooth 2013). In particular, GAs exploit the population of multiple individual points to scout out the entire distribution and thus to increase the chance of reaching the optimal point and to avoid fixing to local optimum points (Blum et al. 2012). This benefit has made GAs suitable for handling complex problems (McCall 2005).

The GA comprises five main steps: initial population, fitness function, selection, crossover and mutation, as shown in Figure 3.13. The GA process begins with the generation of a random set of individuals (Figure 3.13 [1]), known as the *population*. The fitness function (Figure 3.13 [2]) then determines the fitness score of each individual, where the individuals with the highest fitness level will be selected in the selection step (Figure 3.13 [3]) to be regenerated. In particular, among the selected individuals, the algorithm creates a crossover of the pairs of individuals with the highest score, resulting in an individual with an even higher fitness level (Figure 3.13 [4]). Following this, the mutation will be applied to the output by randomly flipping several binary values of the produced individual (Figure 3.13 [6]) to form a new generation (Figure 3.13 [7]), and the process is terminated if the problem is solved (see Figure 3.14). Otherwise, the algorithm will repeat the process starting from the selection step.

Similar to GA, PSO solves complex problems utilising a population of candidate solutions. However, it uses what is known as the *momentum*, which enables individuals to accelerate speed in a preferable direction to reach the optimal point.

Algorithms such as the GA or PSO have been utilised to find the optimal process parameters for the L-PBF process. In particular, an ANN model has been developed to correlate the effects of L-PBF process parameters (layer thickness, layer power, hatch spacing,

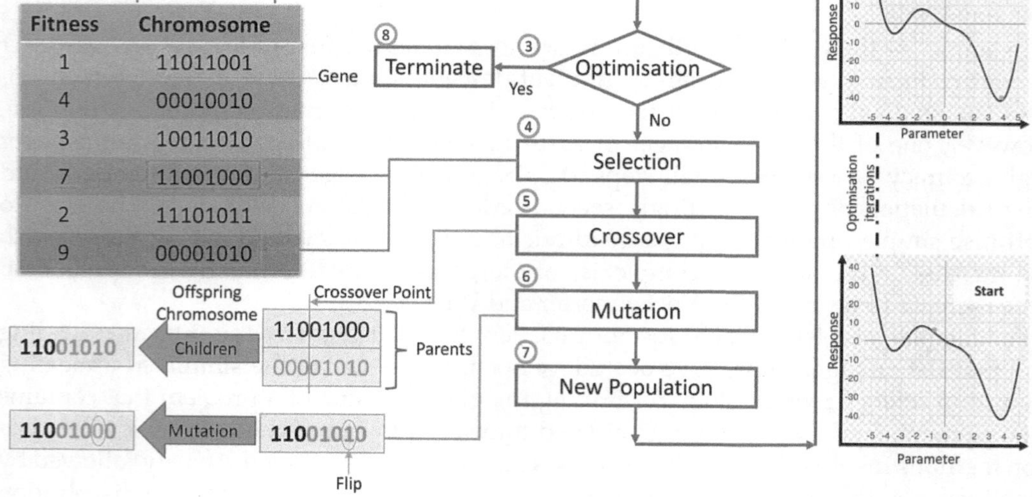

FIGURE 3.13
Main steps involved in optimisation using the genetic algorithm.

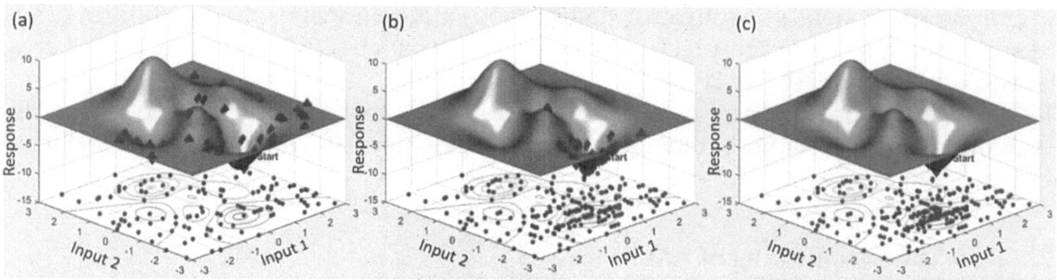

FIGURE 3.14
Optimisation process using genetic algorithm (a) initialisation of random population, (b) near optimal solutions and (c) optimal solution is identified.

scanning speed, interval time, surroundings temperature and scanning) and the process performance in terms of final part shrinkage, and fed into a GA to identify the optimal parameters for minimal shrinkage (Rong-Ji et al. 2008). In another study, a GA was used to optimise the hatch direction to improve the homogeneity of AM parts. Data of microstructure changes were collected via scanning electron microscopy (Ning et al. 2005). GA can also be utilised to solve multi-objective problems. In particular, a non-dominated sorting GA was utilised to optimise the strength and volumetric shrinkage in FDM processes (Gurrala and Regalla 2014). A GA was adopted to improve the reproducibility of the additive metal deposition (AMD) process to fabricate Ti6Al4V AM parts. In particular, the non-dominate sorting GA was applied to identify the optimal process parameters for minimal dimensional error of printed walls (Möller et al. 2016). The use of evolutionary optimisation algorithms in AM systems has been discussed in detail along with future prospects and trends (Leirmo and Martinsen 2019).

3.7 Precision Processes in the Domain of Industry 4.0

Conventional manufacturing businesses are in the throes of a recently emerging digital transformation (Schlaepfer and Koch 2015). The advent of the fourth industrial revolution, or Industry 4.0, has inspired digitalisation in manufacturing industries (Yeo et al. 2017), which facilitates smart real-time interactions between different manufacturing assets (Uhlemann et al. 2017). The application of Industry 4.0 concepts to manufacturing technologies requires real-time monitoring, industrial internet of things (IIoT) solutions, cloud computing, AI algorithms, process simulation techniques and digital quality control in order to realise cyber-physical systems (CPS) (Dilberoglu et al. 2017). The introduction of such smart CPSs into manufacturing technologies is predicted to pave the way for digital solutions to the transformation of existing manufacturing technologies (Schlaepfer and Koch 2015), where the cyber and physical worlds interact profitably. Although digital twins (DTs) and CPSs share the same essential concepts of a comprehensive and seamless linkage between the physical and digital worlds, DT is a subset of CPS. In particular, a DT is characterised by developing high-fidelity virtual models of the physical system to simulate their performance in the real world and provide feedback (Tao et al. 2019). A DT of AM has been recently developed to provide almost instantaneous acquisition of physical issues and precise prediction of manufacturing outcomes and to enhance the capabilities of the AM systems to produce high-precision products in a cost-effective manner (DebRoy et al. 2017, Mukherjee and DebRoy 2019).

Real-time monitoring of processes and AI-powered decision-making techniques for digital quality control are among the main pillars of Industry 4.0. This section will focus on these topics.

3.7.1 Real-Time Monitoring of AM Processes

In a CPS, the physical world has to be continuously and digitally defined by capturing the conditions of all manufacturing assets, such as actuators, materials, stock status in the warehouse, manufacturing environment, power consumption, fabrication process parameters, responses of the fabrication process and product quality attributes (see Chapter 13 for an in-depth treatment of in-process measurement). The capturing step can be performed using intelligent sensors, which can be invasive or non-invasive and interact with a smart system in order to visualise the physical world in a digital format (Elkaseer et al. 2018b, Salem et al. 2019); see Figure 3.15. The step from capturing to digitalisation is known as *data acquisition* and is considered as the first phase in the CPS workflow that facilitates real-time monitoring of the manufacturing operations (Lee et al. 2015). Note that in this chapter, the term *monitoring* applies to the whole manufacturing process, whereas in Chapter 13, it is mainly used in the context of on-machine or in-process measurement.

3.7.2 Artificial Intelligence and Decision-Making Systems for Digital Quality Control

Recently, AI and decision-making/decision support systems have started to play an important role in Industry 4.0 concepts because they enable the manufacturing processes to be continuously optimised in order to create zero-defect products and first-time-right processing. In addition to off-line analysis, AI enables analysis of real-time monitored data and accurate prediction of the part properties to meet customer/market demands. In addition, AI enables handling of big data to extract features and support decisions (Duan et al. 2019). Furthermore, AI algorithms can help conduct pre-process simulation (Mourtzis et al. 2019) of the manufacturing process in order to estimate the quality of the AM part and thus to optimise the fabrication parameters before real production. These capabilities help

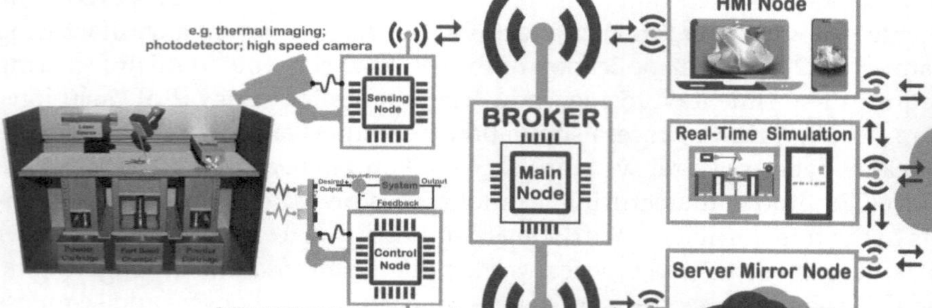

FIGURE 3.15
Real-time monitoring for digital quality control of L-PBF.

implement flexible and customisable manufacturing scenarios (Scholz et al. 2019) to fabricate new products with low cost and short lead times. In terms of AM technologies, an AI framework for a CPS of L-PBF was developed, which was used to identify and control the most significant parameters for the L-PBF process (Amini et al. 2019).

Another approach in digital quality control is the concept of machine learning (ML) or deep learning (DL) techniques. In these technologies, classifiers are used to recognise/classify the defects in the AM product. The detection/recognition methods can be performed using convolution neural networks (CNNs) or support vector machines (SVMs) (Deris et al. 2011, Karpathy et al. 2014, Pfrommer et al. 2018). However, DL has an advantage over ML due to its capability for defining/extracting the features autonomously without the need for feature engineering, which has enabled DL to effectively deal with the AM parts features and issues.

A max-pooling CNN was developed for detecting manufacturing defects in steel surfaces (Masci et al. 2013). The results showed that the performance of the proposed max-pooling CNN algorithm was better than the SVM algorithm. A vision system to recognise surface defects in fabricated metallic parts has been developed. This vision system was developed using a CNN (Tao et al. 2018). The computer vision algorithm inspects the surface to detect various defects, determine their locations and then classify them into groups via the CNN. For the L-PBF process, a DL algorithm has been developed to perform classification for melt-pool images with respect to the laser power applied during the process. The results showed that the part quality and manufacturing defects of the part are linearly proportional to the density of the AM part (Kwon et al. 2020).

An in-situ quality monitoring and control algorithm for AM has been developed using acoustic emission. The algorithm analyses the acoustic emission generated in the L-PBF process using a spectral CNN. This analysis helps to detect the optimal process parameters for the best possible performance of the process and to develop a high precision process (Shevchik et al. 2018). More examples of the use of AI in AM applications are given in Chapter 13.

3.8 Future Perspectives for Precision AM Processes

AM technologies are set to cause disruptions in the way that everyday products are made. The extent of the impact of these disruptions depends on the extent to which they are adopted by manufacturing companies. The three primary areas where research effort is needed in order to promote industrial uptake of AM – the costs, the comparative benefits of AM over subtractive machining processes and the rate at which these benefits occur – have been theoretically discussed (Tofail et al. 2018). Others have tried to develop frameworks for implementing AM within conventional manufacturing systems (Elkaseer et al. 2018c). Another study conducted fourteen case studies with AM companies and used the empirical results to develop a general four-component industrial AM system (Eyers and Potter 2017). Similarly, a structural model for implementing AM motivated by the lack of consideration of social-technical studies in this field has also been presented (Mellor et al. 2014). Currently, though, the major application of AM within companies is still as a tool for prototyping. However, the application of AM in the industrial environment and for actual part production is already ramping up, with many organisations considering adopting AM for low-volume production.

The Deloitte University press (Cotteleer and Joyce 2014) lists four paths through which companies can prepare themselves for implementing AM within their process chains.

Path I: This is a low-risk path for companies that are tentative about the advantages that AM can provide to their business and consists of using AM as a value addition tool to their process chains, such as for rapid prototyping, custom tooling or low-rate production. This has been the most commonly followed path within companies attempting to use AM over the last thirty years. Using this path, companies can improve specific areas of their business, such as the speed of the design process, thereby acquiring cost benefits. Companies do not seek radical alterations in either supply chains or products but may also explore AM technologies to improve value delivery for current products within existing supply chains.

Path II: Companies take advantage of scale economics offered by AM as a potential enabler of supply-chain transformation for the products they offer. This path focusses on using the increased responsiveness and flexibility that AM provides for supply chains, such as manufacturing close to point-of-use and reduction in inventory. This path allows companies, both large and small, to capitalise on the ability of AM to deliver products faster (shorter lead times), with reduced costs and with better quality than competitors.

Path III: This path completely relies on the capacity of AM to create innovations through improved part designs or functionalities for the products offered by companies. It includes implementing additional benefits for the customer's sake, such as reducing the number of parts within components or embedding different sensors within components that track the health of the products. Companies benefit from the 'advantages of scope' offered by AM technologies to achieve new levels of performance or innovation in their products. This path focusses on aspects such as customised production, functional printing and zero cost of increased complexity as provided by AM processes, and has a higher associated risk compared with the other paths, since it mostly depends on customer acceptance of any added benefits and any added costs that may arise due to this.

Path IV: Companies alter both supply chains and products in the pursuit of new business models. This is a high-risk methodology that will empower mass customisation and customer empowerment, and seeks to combine the tactics and benefits of both paths II and III, thereby creating new pathways for adding value into products while impairing the ability of competitors to compete.

Focussing on improving the precision of AM processes and thus creating a robust AM process will further facilitate the adoption of AM by companies. A robust process can be standardised and will have a highly repeatable performance. This is an important criterion for industrial companies, especially in the automotive sector, where production takes place in large numbers.

Current trends in AM are further assisted by developments in technologies such as photogrammetry, 3D scanning and functional printing. Increased integration of research efforts focussing on design for AM, process development, metrology and standardisation efforts have contributed to steady progress being made towards mainstream adoption of AM techniques. Developments in the field of multi-material printing show promise for creating flexible electronics and sensors (Emon et al. 2019). Also, '4D printing' technology has arrived, bringing expectations for application in smart devices, smart packaging, metamaterial and biomedical applications (Kuang et al. 2019).

3.9 Conclusions

This chapter has presented AM as a technology for the fourth industrial revolution, containing the simultaneous 'zoom out, zoom in' approach necessary for optimisation, and combining the optimisation of process and technological aspects of AM technology. A holistic development of process-wide aspects of industrial engineering performance indicators, such as yield, throughput and cycle time, has also been discussed. It is, therefore, important to investigate the impact of process parameters and variables on the overall process chain in order to meet industrial demands for quality, productivity and capability. This is important for the successful integration of AM technologies into conventional manufacturing processes.

This chapter also introduced both process-based and system-wide aspects of AM technologies followed by short summaries of common AM performance indicators. The later sections of this chapter introduced various data-driven process improvement approaches for process modelling and process optimisation. Finally, this chapter presented some insights into the use of other Industry 4.0 technologies such as AI and decision support systems for further optimising AM technologies. Building precision into AM processes will contribute to improving its predictability and robustness, which are key enablers and, therefore, necessary for successful industrial adoption.

It can be concluded that AM technologies are set to cause wide changes in the way that everyday objects are manufactured. The extent of the impact of these changes depends on the manner in which they are adopted by manufacturing companies. Currently, the major application of AM is as a tool for prototyping. Conventional manufacturing processes have the benefits of decades of research and development, and in order for AM to catch up in a relatively short period of time, it must leverage its improved technological capabilities. By incorporating Industry 4.0 concepts and applications, aspects of precision engineering can be fully developed in AM processes for widespread adoption by industry.

Acknowledgements

The authors would like to acknowledge the support provided by the Karlsruhe Nano Micro Facility (KNMF-LMP, http://www.knmf.kit.edu/), a Helmholtz research infrastructure at KIT, as well as the support of the Helmholtz Research Program STN (Science and Technology of Nanosystems) at KIT.

References

Abbeel, P., Coates, A., Quigley, M. and Ng, A. 2007. An application of reinforcement learning to aerobatic helicopter flight. In: Schölkopf, B., Platt, J. C. and Hoffman, T. (eds.), *Advances in Neural Information Processing Systems 19*. Cambridge, MA: MIT Press.

Abiodun, O. I., Jantan, A., Omolara, A. E., Dada, K. V., Mohamed, N. A. and Arshad, H. 2018. State-of-the-art in artificial neural network applications: A survey. *Heliyon* 4:e00938.

Aboulkhair, N. T., Everitt, N. M., Ashcroft, I. and Tuck, C. 2014. Reducing porosity in ALSi10mg parts processed by selective laser melting. *Addit. Manuf.* 1–4:77–86.

Abu Qudeiri, E. J., Saleh, A., Ziout, A., Mourad, I. A.-H., Abidi, H. M. and Elkaseer, A. 2019. Advanced electric discharge machining of stainless steels: Assessment of the state of the art, gaps and future prospect. *Materials* 12:907.

Abu Qudeiri, J. E., Mourad, A.-H. I., Ziout, A., Abidi, M. H. and Elkaseer, A. 2018. Electric discharge machining of titanium and its alloys: Review. *Int. J. Adv. Manuf. Tech.* 96:1319–1339.

Achillas, C., Aidonis, D., Iakovou, E., Thymianidis, M. and Tzetzis, D. 2015. A methodological framework for the inclusion of modern additive manufacturing into the production portfolio of a focused factory. *J. Manuf. Syst.* 37:328–339.

Afazov, S., Denmark, W. A. D., Lazaro Toralles, B., Holloway, A. and Yaghi, A. 2017a. Distortion prediction and compensation in selective laser melting. *Addit. Manuf.* 17:15–22.

Afazov, S., Okioga, A., Holloway, A., Denmark, W., Triantaphyllou, A., Smith, S.-A. and Bradley-Smith, L. 2017b. A methodology for precision additive manufacturing through compensation. *Precis. Eng.* 50:269–274.

Allaire, G., Dapogny, C., Estevez, R., Faure, A. and Michailidis, G. 2017. Structural optimization under overhang constraints imposed by additive manufacturing technologies. *J. Comput. Phys.* 351:295–328.

Amini, M., Chang, S. and Rao, P. 2019. A cybermanufacturing and AI framework for laser powder bed fusion (LPBF) additive manufacturing process. *Manuf. Lett.* 21:41–44.

Aylor, S., Rabelo, L. and Alptekin, S. 1992. Artificial neural networks for robotics coordinate transformation. *Comput. Ind. Eng.* 22:481–493.

Banerjee, S., Kulesha, G., Kester, M. and Mont, M. 2014. Emerging technologies in arthroplasty: Additive manufacturing. *J. Knee Surg.* 27:185–191.

Barnes-Schuster, D., Bassok, Y. and Anupindi, R. 2006. Optimizing delivery lead time/inventory placement in a two-stage production/distribution system. *Eur. J. Oper. Res.* 174:1664–1684.

Baturynska, I. 2018. Statistical analysis of dimensional accuracy in additive manufacturing considering STL model properties. *Int. J. Adv. Manuf. Technol.* 97:2835–2849.

Bengio, Y., Courville, A. and Vincent, P. 2013. Representation learning: A review and new perspectives. *ITPAM* 35:1798–1828.

Blum, C., Chiong, R., Clerc, M., De Jong, K., Weise, T., Neri, F. and Michalewicz, Z. 2012. Evolutionary optimization. In: Chiong, R., Weise, T. and Michalewicz, Z. (eds.), *Variants of Evolutionary Algorithms for Real-World Applications*. Springer-Verlag Berlin Heidelberg.

Böckin, D. and Tillman, A.-M. 2019. Environmental assessment of additive manufacturing in the automotive industry. *J. Cleaner Prod.* 226:977–987.

Buer, S.-V., Fragapane, G. I. and Strandhagen, J. O. 2018. The data-driven process improvement cycle: Using digitalization for continuous improvement. *Proc. IFAC*, Bergamo, Italy, 1035–1040.

Cabanettes, F., Joubert, A., Chardon, G., Dumas, V., Rech, J., Grosjean, C. and Dimkovski, Z. 2018. Topography of as built surfaces generated in metal additive manufacturing: A multi scale analysis from form to roughness. *Precis. Eng.* 52:249–265.

Carlson, R. 2001. Design of experiments, principles and applications. *J. Chemometrics* 15:495–496.

Charles, A., Bassan, P. M., Mueller, T., Elkaseer, A. and Scholz, S. G. 2019a. On the assessment of thermo-mechanical degradability of multi-recycled ABS polymer for 3D printing applications. In: Ball, P., Huaccho Huatuco, L., Howlett, R. J. and Setchi, R. (eds.), *Sustainable Design and Manufacturing 2019*. Springer Singapore.

Charles, A., Elkaseer, A., Salem, M., Thijs, L. and Scholz, S. 2019b. ANN-based modelling of dimensional accuracy in L-PBF. *Proc. Euspen/ASPE 2019*, Nantes, France, 39–42.

Charles, A., Elkaseer, A., Thijs, L., Hagenmeyer, V. and Scholz, S. 2019c. Effect of process parameters on the generated surface roughness of down-facing surfaces in selective laser melting. *Appl. Sci.* 9:1256.

Cheng, L., Liang, X., Bai, J., Chen, Q., Lemon, J. and To, A. 2019. On utilizing topology optimization to design support structure to prevent residual stress induced build failure in laser powder bed metal additive manufacturing. *Addit. Manuf.* 27:290–304.

Chowdhury, S. and Anand, S. 2016. Artificial neural network based geometric compensation for thermal deformation in additive manufacturing processes. *Proc. MSEC2016*, Blacksburg, Virginia, USA, T08A006.

Conner, B. P., Manogharan, G. P., Martof, A. N., Rodomsky, L. M., Rodomsky, C. M., Jordan, D. C. and Limperos, J. W. 2014. Making sense of 3-d printing: Creating a map of additive manufacturing products and services. *Addit. Manuf.* 1–4:64–76.

Cotteleer, M. and Joyce, J. 2014. 3d opportunity. In: Cotteleer, M. and Joyce, J. (eds.), *Additive Manufacturing Paths to Performance Innovation, and Growth*. Deloitte Review.

Cunningham, R., Narra, S. P., Montgomery, C., Beuth, J. and Rollett, A. D. 2017. Synchrotron-based X-ray microtomography characterization of the effect of processing variables on porosity formation in laser power-bed additive manufacturing of Ti-6Al-4V. *JOM* 69:479–484.

D'Addona, D. M. and Teti, R. 2013. Image data processing via neural networks for tool wear prediction. *Procedia CIRP* 12:252–257.

Dantan, J.-Y., Huang, Z., Goka, E., Homri, L., Etienne, A., Bonnet, N. and Rivette, M. 2017. Geometrical variations management for additive manufactured product. *CIRP Ann.* 66:161–164.

Davis, N. and Schwab, K. 2018. *Shaping the Future of the Fourth Industrial Revolution*. World Economic Forum.

De Baere, D., Valente, E. H., Mohanty, S. and Hattel, J. H. 2018. Modelling of the microstructural evolution of Ti6Al4V parts produced by selective laser melting during heat treatment. *Proc. euspen*, Venice, Italy, 249–250.

Dean, J. and Ghemawat, S. 2004. Mapreduce: Simplified data processing on large clusters. *J. ACM* 51:107–113.

DebRoy, T., Wei, H. L., Zuback, J. S., Mukherjee, T., Elmer, J. W., Milewski, J. O., Beese, A. M., Wilson-Heid, A., De, A. and Zhang, W. 2018. Additive manufacturing of metallic components – process, structure and properties. *Prog. Mater. Sci.* 92:112–224.

DebRoy, T., Zhang, W., Turner, J. and Babu, S. S. 2017. Building digital twins of 3D printing machines. *Scripta Mater*. 135:119–124.

DeCastro-García, N., Muñoz Castañeda, Á., Rodríguez, M. and Carriegos, M. V. 2018. On detecting and removing superficial redundancy in vector databases. *Math. Probl. Eng.* 2018:1–14.

Deris, A. M., Zain, A. M. and Sallehuddin, R. 2011. Overview of support vector machine in modeling machining performances. *Procedia Eng.* 24:308–312.

Dilberoglu, U. M., Gharehpapagh, B., Yaman, U. and Dolen, M. 2017. The role of additive manufacturing in the era of industry 4.0. *Procedia Manuf.* 11:545–554.

Duan, Y., Edwards, J. S. and Dwivedi, Y. K. 2019. Artificial intelligence for decision making in the era of big data – evolution, challenges and research agenda. *Int. J. Inf. Manage.* 48:63–71.

Elkaseer, A., Lambarri, J., Quintana, I. and Scholz, S. 2019. Laser ablation of cobalt-bound tungsten carbide and aluminium oxide ceramic: Experimental investigation with ANN modelling and GA optimisation. In: Dao, D., Howlett, R. J., Setchi, R. and Vlacic, L. (eds.), *Sustainable Design and Manufacturing 2018*. Springer International Publishing.

Elkaseer, A., Mueller, T., Azcarate, S., Philipp-Pichler, M., Wilfinger, T., Wittner, W., Prantl, M., Sampaio, D., Hagenmeyer, V. and Scholz, S. 2018a. Replication of overmolded orthopedic implants with a functionalized thin layer of biodegradable polymer. *Polymers* 10:707.

Elkaseer, A., Mueller, T., Charles, A. and Scholz, S. 2018b. Digital detection and correction of errors in as-built parts: A step towards automated quality control of additive manufacturing. *Proc. WCMNM*, Portorož, Slovenia, 389–392.

Elkaseer, A., Salem, M., Ali, H. and Scholz, S. 2018c. Approaches to a practical implementation of Industry 4.0. *Proc. ACHI*, Rome, Italy, 141–146.

Elkaseer, A. M., Dimov, S. S., Pham, D. T., Popov, K. P., Olejnik, L. and Rosochowski, A. 2016. Material microstructure effects in micro-endmilling of Cu99.9E. *Proc. Inst. Mech. Eng. Pt. B: J. Eng. Manuf.* 232:1143–1155.

Elkaseer, A. M., Dimov, S. S., Popov, K. B. and Minev, R. M. 2014. Tool wear in micro-endmilling: Material microstructure effects, modeling, and experimental validation. *J. Micro Nano-Manuf.* 2:044502.

Emon, M. O. F., Alkadi, F., Philip, D. G., Kim, D.-H., Lee, K.-C. and Choi, J.-W. 2019. Multi-material 3d printing of a soft pressure sensor. *Addit. Manuf.* 28:629–638.

Eyers, D. R. and Potter, A. T. 2017. Industrial additive manufacturing: A manufacturing systems perspective. *Comput. Ind.* 92–93:208–218.

Eylon, D. and Froes, F. H. 1990. Titanium powder metallurgy products. In: ASM, Handbook and Committee. (ed.), *Properties and Selection: Nonferrous Alloys and Special-Purpose Materials.* Cleveland, OH: ASM International.

Fang, K.-T., Liu, M.-Q., Qin, H. and Zhou, Y. 2018. Construction of uniform designs – deterministic methods. In: Fang, K.-T., Liu, M.-Q., Qin, H. and Zhou, Y. (eds.), *Theory and Application of Uniform Experimental Designs*. Springer Singapore.

Fauth, J., Elkaseer, A. and Scholz, S. G. 2019. Total cost of ownership for different state of the art FDM machines (3D printers). In: Ball, P., Huaccho Huatuco, L., Howlett, R. J. and Setchi, R. (eds.), *Sustainable Design and Manufacturing 2019*. Springer Singapore.

Fu, Y., Jha, D., Zhang, Z., Yuan, Z. and Ray, A. 2019. Neural network-based learning from demonstration of an autonomous ground robot. *Machines* 7:24.

Ghosh, A., Mal, P. and Majumdar, A. 2019. Simulated annealing. In: Ghosh, A., Mal, P. and Majumdar, A. (eds.), *Advanced Optimization and Decision-Making Techniques in Textile Manufacturing*. Taylor & Francis Group.

Grasso, M., Gallina, F. and Colosimo, B. M. 2018. Data fusion methods for statistical process monitoring and quality characterization in metal additive manufacturing. *Procedia CIRP* 75:103–107.

Gurrala, P. K. and Regalla, S. P. 2014. Multi-objective optimisation of strength and volumetric shrinkage of FDM parts. *Virtual Phys. Prototy.* 9:127–138.

Gyulai, D., Pfeiffer, A., Nick, G., Gallina, V., Sihn, W. and Monostori, L. 2018. Lead time prediction in a flow-shop environment with analytical and machine learning approaches. *Proc. IFAC,* 1029–1034.

Hällgren, S., Pejryd, L. and Ekengren, J. 2016. Additive manufacturing and high speed machining – cost comparison of short lead time manufacturing methods. *Procedia CIRP* 50:384–389.

Hallmann, M., Goetz, S., Schleich, B. and Wartzack, S. 2019. Optimization of build time and support material quantity for the additive manufacturing of non-assembly mechanisms. *Procedia CIRP* 84:271–276.

Han, J., Yang, J., Yu, H., Yin, J., Gao, M., Wang, Z. and Zeng, X. 2017. Microstructure and mechanical property of selective laser melted Ti6Al4V dependence on laser energy density. *Rapid Prototyping J.* 23:217–226.

Han, Q., Gu, H., Soe, S., Setchi, R., Lacan, F. and Hill, J. 2018. Manufacturability of ALSi10mg overhang structures fabricated by laser powder bed fusion. *Mater. Des.* 160:1080–1095.

Jurrens, K., Migler, K., Ricker, R., Pei, Z., Schmid, S., Love, L., Resnick, R. and Vorvolakos, K. 2013. *Measurement Science Roadmap for Metal-Based Additive Manufacturing*. National Institute of Standards and Technology.

Karpathy, A., Toderici, G., Shetty, S., Leung, T., Sukthankar, R. and Fei-Fei, L. 2014. Large-scale video classification with convolutional neural networks. *Proc. CVPR*, Columbus, OH, USA, 1725–1732.

Kasperovich, G., Haubrich, J., Gussone, J. and Requena, G. 2016. Correlation between porosity and processing parameters in Ti6Al4V produced by selective laser melting. *Mater. Des.* 105:160–170.

Kempen, K., Thijs, L., Yasa, E., Badrossamay, M., Verheecke, W. and Kruth, J. P. 2011. Process optimization and microstructural analysis for selective laser melting of ALSi10mg. *Proc. SFF*, Austin, Texas, USA, 484–495.

Khorasani, A., Gibson, I., Awan, U. S. and Ghaderi, A. 2019. The effect of SLM process parameters on density, hardness, tensile strength and surface quality of Ti-6Al-4V. *Addit. Manuf.* 25:176–186.

Kochenderfer, M. J. and Wheeler, T. A. 2019. *Algorithms for Optimization*. The MIT Press.

Koeppe, A., Hernandez Padilla, C. A., Voshage, M., Schleifenbaum, J. H. and Markert, B. 2018. Efficient numerical modeling of 3D-printed lattice-cell structures using neural networks. *Manuf. Lett.* 15:147–150.

Komineas, G., Foteinopoulos, P., Papacharalampopoulos, A. and Stavropoulos, P. 2018. Build time estimation models in thermal extrusion additive manufacturing processes. *Procedia Manuf.* 21:647–654.

Kruth, J.-P., Deckers, J., Yasa, E. and Wauthlé, R. 2012. Assessing and comparing influencing factors of residual stresses in selective laser melting using a novel analysis method. *Proc. Inst. Mech. Eng. Pt. B: J. Eng. Manuf.* 226:980–991.

Kuang, X., Roach, D. J., Wu, J., Hamel, C. M., Ding, Z., Wang, T., Dunn, M. L. and Qi, H. J. 2019. Advances in 4D printing: Materials and applications. *Adv. Funct. Mater.* 29:1805290.

Kwon, O., Kim, H. G., Ham, M., Kim, W., Kim, G.-H., Cho, J.-H., Kim, N. and Kim, K. 2020. A deep neural network for classification of melt-pool images in metal additive manufacturing. *J. Intell. Manuf.* 31:375–386.

Leach, R. and Smith, S. 2018. *Basics of Precision Engineering*. CRC Press.

LeCun, Y., Bengio, Y. and Hinton, G. 2015. Deep learning. *Nature* 521:436–444.

Lee, J., Bagheri, B. and Kao, H.-A. 2015. A cyber-physical systems architecture for Industry 4.0-based manufacturing systems. *Manuf. Lett.* 3:18–23.

Leirmo, T. S. and Martinsen, K. 2019. Evolutionary algorithms in additive manufacturing systems: Discussion of future prospects. *Procedia CIRP* 81:671–676.

Lingitz, L., Gallina, V., Ansari, F., Gyulai, D., Pfeiffer, A., Sihn, W. and Monostori, L. 2018. Lead time prediction using machine learning algorithms: A case study by a semiconductor manufacturer. *Procedia CIRP* 72:1051–1056.

Lipson, H. and Pollack, J. 2000. Automatic design and manufacture of robotic lifeforms. *Nature* 406:974–978.

Liu, L.-Y., Yang, Q.-S. and Zhang, Y. X. 2019. Plastic damage of additive manufactured aluminium with void defects. *Mech. Res. Commun.* 95:45–51.

Liu, S. and Shin, Y. C. 2019. Additive manufacturing of Ti6Al4V alloy: A review. *Mater. Des.* 164:107552.

Marrey, M., Malekipour, E., El-Mounayri, H. and Faierson, E. J. 2019. A framework for optimizing process parameters in powder bed fusion (PBF) process using artificial neural network (ANN). *Procedia Manuf.* 34:505–515.

Masci, J., Meier, U., Fricout, G. and Schmidhuber, J. 2013. Multi-scale pyramidal pooling network for generic steel defect classification. *Proc. IJCNN*, Dallas, Texas, USA, 1–8.

Mathieu, M., Couprie, C. and Lecun, Y. 2016. Deep multi-scale video prediction beyond mean square error. *CoRR* abs/1511.05440.

McCall, J. 2005. Genetic algorithms for modelling and optimisation. *JCoAM* 184:205–222.

Meireles, M. R. G., Almeida, P. E. M. and Simoes, M. G. 2003. A comprehensive review for industrial applicability of artificial neural networks. *ITIE* 50:585–601.

Mellor, S., Hao, L. and Zhang, D. 2014. Additive manufacturing: A framework for implementation. *Int. J. Product. Econ.* 149:194–201.

Mertens, R., Clijsters, S., Kempen, K. and Kruth, J.-P. 2014. Optimization of scan strategies in selective laser melting of aluminum parts with downfacing areas. *J. Manuf. Sci. Eng.* 136:061012.

Michopoulos, J. G., Iliopoulos, A. P., Steuben, J. C., Birnbaum, A. J. and Lambrakos, S. G. 2018. On the multiphysics modeling challenges for metal additive manufacturing processes. *Addit. Manuf.* 22:784–799.

Mies, D., Marsden, W. and Warde, S. 2016. Overview of additive manufacturing informatics: 'A digital thread'. *Integr. Mater. Manuf. Innov.* 5:114–142.

Mitra, A. 2011. The Taguchi method. *Wiley Interdiscip. Rev. Comput. Stat.* 3:472–480.

Mohanty, S. and Hattel, J. H. 2016. Reducing residual stresses and deformations in selective laser melting through multi-level multi-scale optimization of cellular scanning strategy. *Proc. SPIE*, San Francisco, CA, USA, 97380Z.

Möller, M., Baramsky, N., Ewald, A., Emmelmann, C. and Schlattmann, J. 2016. Evolutionary-based design and control of geometry aims for AMD-manufacturing of Ti-6Al-4V parts. *Physics Procedia* 83:733–742.

Montgomery, D. C. 2019. *Design and Analysis of Experiments*. John Wiley, Chichester.

Mourtzis, D., Vasilakopoulos, A., Zervas, E. and Boli, N. 2019. Manufacturing system design using simulation in metal industry towards Education 4.0. *Procedia Manuf.* 31:155–161.

Mukherjee, I. and Ray, P. 2006. A review of optimization techniques in metal cutting processes. *Comput. Ind. Eng.* 50:15–34.

Mukherjee, M. 2019. Effect of build geometry and orientation on microstructure and properties of additively manufactured 316L stainless steel by laser metal deposition. *Materialia* 7:100359.

Mukherjee, T. and DebRoy, T. 2019. A digital twin for rapid qualification of 3d printed metallic components. *Appl. Mater. Today* 14:59–65.

Nassef, A., Elkaseer, A., Abdelnasser, E. S., Negm, M. and Qudeiri, J. A. 2018. Abrasive jet drilling of glass sheets: Effect and optimisation of process parameters on kerf taper. *Adv. Mech. Eng.* 10:1–10.

Niaki, M. K., Torabi, S. A. and Nonino, F. 2019. Why manufacturers adopt additive manufacturing technologies: The role of sustainability. *J. Cleaner Prod.* 222:381–392.

Nielsen, M. A. 2015. How the backpropagation algorithm works. In: Nielsen, M. A. (ed.), *Neural Networks and Deep Learning*. San Francisco, CA: Determination Press.

Ning, Y., Wong, Y. S. and Fuh, J. Y. H. 2005. Effect and control of hatch length on material properties in the direct metal laser sintering process. *Proc. Inst. Mech. Eng. Pt. B: J. Eng. Manuf.* 219:15–25.

Noryani, M., Sapuan, S. M., Mastura, M. T., Zuhri, M. Y. M. and Zainudin, E. S. 2019. Material selection of natural fibre using a stepwise regression model with error analysis. *J. Mater. Res. Technol.* 8:2865–2879.

Pereyra, G., Tucker, G., Chorowski, J., Kaiser, Ł. and Hinton, G. 2017. Regularizing neural networks by penalizing confident output distributions. *Proc. ICLR 2017*, Toulon, France,

Pfrommer, J., Zimmerling, C., Liu, J., Kärger, L., Henning, F. and Beyerer, J. 2018. Optimisation of manufacturing process parameters using deep neural networks as surrogate models. *Procedia CIRP* 72:426–431.

Qin, J., Liu, Y. and Grosvenor, R. 2017. A framework of energy consumption modelling for additive manufacturing using internet of things. *Procedia CIRP* 63:307–312.

Rahm, E. and Do, H. 2000. Data cleaning: Problems and current approaches. *IEEE Data Eng. Bull.* 23:3–13.

Rehme, O. 2010. Cellular design for laser freeform fabrication. PhD diss., Hamburg University of Technology.

Ríos, S., Colegrove, P. A., Martina, F. and Williams, S. W. 2018. Analytical process model for wire + arc additive manufacturing. *Addit. Manuf.* 21:651–657.

Robinson, J., Ashton, I., Fox, P., Jones, E. and Sutcliffe, C. 2018. Determination of the effect of scan strategy on residual stress in laser powder bed fusion additive manufacturing. *Addit. Manuf.* 23:13–24.

Rong-Ji, W., Xin-hua, L., Qing-ding, W. and Lingling, W. 2008. Optimizing process parameters for selective laser sintering based on neural network and genetic algorithm. *Int. J. Adv. Manuf. Tech.* 42:1035–1042.

Salehi, M., Blum, M., Fath, B., Akyol, T., Haas, R. and Ovtcharova, J. 2016. Epicycloidal versus trochoidal milling-comparison of cutting force, tool tip vibration, and machining cycle time. *Procedia CIRP* 46:230–233.

Salem, M., Elkaseer, A., Saied, M., Ali, H. and Scholz, S. 2019. Industrial Internet of Things solution for real-time monitoring of the additive manufacturing process. *Proc. ISAT 2018*, Wroclaw, Poland, 355–365.

Sauerwein, M., Doubrovski, E., Balkenende, R. and Bakker, C. 2019. Exploring the potential of additive manufacturing for product design in a circular economy. *J. Cleaner Prod.* 226:1138–1149.

Schlaepfer, R. and Koch, M. 2015. What is Industry 4.0? In: Schlaepfer, R. and Koch, M. (eds.), *Industry 4.0 Challenges and Solutions for the Digital Transformation and Use of Exponential Technologies*. Deloitte AG.

Schmidt, M., Merklein, M., Bourell, D., Dimitrov, D., Hausotte, T., Wegener, K., Overmeyer, L., Vollertsen, F. and Levy, G. N. 2017. Laser based additive manufacturing in industry and academia. *CIRP Ann.* 66:561–583.

Scholz, S., Elkaseer, A., Salem, M. and Hagenmeyer, V. 2019. Software toolkit for visualization and process selection for modular scalable manufacturing of 3d micro-devices. *Proc. ISAT 2019*, Wroclaw, Poland, 160–172.

Shevchik, S. A., Kenel, C., Leinenbach, C. and Wasmer, K. 2018. Acoustic emission for in situ quality monitoring in additive manufacturing using spectral convolutional neural networks. *Addit. Manuf.* 21:598–604.

Shi, X., Ma, S., Liu, C. and Wu, Q. 2017. Parameter optimization for Ti-47Al-2Cr-2Nb in selective laser melting based on geometric characteristics of single scan tracks. *Opt. Laser Technol.* 90:71–79.

Shipley, H., McDonnell, D., Culleton, M., Coull, R., Lupoi, R., O'Donnell, G. and Trimble, D. 2018. Optimisation of process parameters to address fundamental challenges during selective laser melting of Ti-6Al-4V: A review. *Int. J. Mach. Tools Manuf.* 128:1–20.

Shukla, A., Cheema, G. and Anand, S. 2018. Semi-supervised clustering using neural networks. *CORR* 1806:01547.

Siebertz, K., Bebber, D. and Hochkirchen, T. 2010. Versuchspläne. In: Siebertz, K., Bebber, D. and Hochkirchen, T. (eds.), *Statistische Versuchsplanung – Design of Experiments (DOE)*. Springer-Verlag Berlin Heidelberg.

Sinico, M., Ametova, E., Witvrouw, A. and Dewulf, W. 2018. Characterization of AM metal powder with an industrial microfocus CT: Potential and limitations. *Proc. ASPE/euspen 2018*, Berkeley, CA, USA, 286–291.

Sivakumar, R. and Manivel, R. 2020. Analysis on overall equipment effectiveness of a PEMAMEK panel processing machine. *Mater. Today: Proc.* 21:367–370.

Smith, C. J., Derguti, F., Hernandez Nava, E., Thomas, M., Tammas-Williams, S., Gulizia, S., Fraser, D. and Todd, I. 2016. Dimensional accuracy of electron beam melting (EBM) additive manufacture with regard to weight optimized truss structures. *J. Mater. Process. Technol.* 229:128–138.

Srivastava, N., Hinton, G., Krizhevsky, A., Sutskever, I. and Salakhutdinov, R. 2014. Dropout: A simple way to prevent neural networks from overfitting. *J. Mach. Learn. Res.* 15:1929–1958.

Stulp, F. and Sigaud, O. 2015. Many regression algorithms, one unified model: A review. *NN* 69:60–79.

Sufiiarov, V. S., Popovich, A. A., Borisov, E. V., Polozov, I. A., Masaylo, D. V. and Orlov, A. V. 2017. The effect of layer thickness at selective laser melting. *Procedia Eng.* 174:126–134.

Sun, J., Yang, Y. and Wang, D. 2013. Parametric optimization of selective laser melting for forming Ti6Al4V samples by Taguchi method. *Opt. Laser Technol.* 49:118–124.

Sun, S., Brandt, M. and Easton, M. 2017. Powder bed fusion processes: An overview. In: Brandt, M. (ed.), *Laser Additive Manufacturing*. Cambridge, UK: Woodhead Publishing.

Tao, F., Qi, Q., Wang, L. and Nee, A. Y. C. 2019. Digital twins and cyber-physical systems toward smart manufacturing and Industry 4.0: Correlation and comparison. *Engineering* 5:653–661.

Tao, X., Zhang, D., Ma, W., Liu, X. and Xu, D. 2018. Automatic metallic surface defect detection and recognition with convolutional neural networks. *Appl. Sci.* 8:1575.

Tapia, G., Elwany, A. H. and Sang, H. 2016. Prediction of porosity in metal-based additive manufacturing using spatial gaussian process models. *Addit. Manuf.* 12:282–290.

Thompson, M. K., Moroni, G., Vaneker, T., Fadel, G., Campbell, R. I., Gibson, I., Bernard, A., Schulz, J., Graf, P., Ahuja, B. and Martina, F. 2016. Design for additive manufacturing: Trends, opportunities, considerations, and constraints. *CIRP Ann.* 65:737–760.

Tofail, S. A. M., Koumoulos, E. P., Bandyopadhyay, A., Bose, S., O'Donoghue, L. and Charitidis, C. 2018. Additive manufacturing: Scientific and technological challenges, market uptake and opportunities. *Mater. Today* 21:22–37.

Triantaphyllou, A., Giusca, C. L., Macaulay, G. D., Roerig, F., Hoebel, M., Leach, R. K., Tomita, B. and Milne, K. A. 2015. Surface texture measurement for additive manufacturing. *Surf. Topogr. Metrol. Prop.* 3:024002.

Uhlemann, T. H. J., Schock, C., Lehmann, C., Freiberger, S. and Steinhilper, R. 2017. The digital twin: Demonstrating the potential of real time data acquisition in production systems. *Procedia Manuf.* 9:113–120.

Vahabli, E. and Rahmati, S. 2016. Application of an RBF neural network for FDM parts' surface roughness prediction for enhancing surface quality. *Int. J. Precis. Eng. Manuf.* 17:1589–1603.

West, B. M., Capps, N. E., Urban, J. S., Pribe, J. D., Hartwig, T. J., Lunn, T. D., Brown, B., Bristow, D. A., Landers, R. G. and Kinzel, E. C. 2017. Modal analysis of metal additive manufactured parts. *Manuf. Lett.* 13:30–33.

Wilson, P. and Mantooth, H. A. 2013. Chapter 10 – Model-based optimization techniques. In: Wilson, P. and Mantooth, H. A. (eds.), *Model-Based Engineering for Complex Electronic Systems*. Oxford, UK: Newnes.

Witvrouw, A., Metelkova, J., Ranjan, R., Bayat, M., Charles, A., Moshiri, M., Dos Santos Solheid, J., Scholz, S., Baier, M., Carmignato, S., Stavroulakis, P., Shaheen, A., Thijs, L., De Baere, D. and Tosello, G. 2018. Precision additive metal manufacturing. *Proc. ASPE/euspen 2018*, Berkeley, CA, USA, 18–23.

Yeo, N. C. Y., Pepin, H. and Yang, S. S. 2017. Revolutionizing technology adoption for the remanufacturing industry. *Procedia CIRP* 61:17–21.

Zavala-Arredondo, M., London, T., Allen, M., Maccio, T., Ward, S., Griffiths, D., Allison, A., Goodwin, P. and Hauser, C. 2019. Use of power factor and specific point energy as design parameters in laser powder-bed-fusion (L-PBF) of ALSi10mg alloy. *Mater. Des.* 182:108018.

Zhang, B., Li, Y. and Bai, Q. 2017. Defect formation mechanisms in selective laser melting: A review. *Chin. J. Mech. Eng.* 30:515–527.

4

Modelling Techniques to Enhance Precision in Metal Additive Manufacturing

Sankhya Mohanty and Jesper H. Hattel

CONTENTS

4.1 Introduction

From a generic metrology perspective, while the actual additive manufacturing (AM) process influences the measured values, it has little effect on the methodology used for such measurement as well as the concepts used in the measurement terms and definitions. For instance, surface metrology terms such as *texture* (see Chapter 11) and coordinate metrology terms such as the *flatness deviation* (see Chapter 10) are agnostic to the metal AM technique used to produce the sample being measured. However, more recent methodologies have incorporated aspects of the particular AM processes to achieve better quantitative descriptions of the surfaces being measured. For example, in characterisation of AM metal surfaces, specific instruments and specific surface texture parameters can be chosen, depending on the actual surface features of interest resulting from the specific process – such as the presence of surface re-entrant features, which are not measurable using conventional methods, but become measurable using X-ray computed tomography and generalised surface texture parameters (Zanini et al. 2019a, Zanini et al. 2019b, Pagani et al. 2017, see Chapter 12). Similarly, from the design perspective, the actual metal AM technique plays a significant role, as differing design rules have been developed for the different metal AM processes, such as laser direct energy deposition, electron-beam melting, laser powder bed fusion, wire feed metal deposition and metal binder jetting. The incoherency in the design rules stems from the fact that the varied AM processes are fundamentally different in the manner in which they interact with and consolidate the material. Currently, metals and metallic alloys can be processed using all seven families of AM processes (see Chapter 1), each of which can involve different physical phenomena. Addressing each of the processes, the physics and associated numerical modelling techniques would require considerable space and are thus beyond the scope of the current book. Rather, in this chapter, the specific modelling techniques for precision metal AM are introduced and explained for the laser powder bed fusion (L-PBF) process, but the concepts are also broadly applicable to other metal AM techniques.

This chapter will introduce the multiple physical phenomena involved in metal AM in Section 4.2. Subsequently, modelling techniques at the part-scale, at the scale of the powder and at the scale of the microstructure will be addressed in greater detail in Sections 4.3–4.5. The remaining sections of this chapter will deal with more recent data-driven modelling techniques, uncertainty quantification and model-based optimisation methods, as applicable to metal AM.

4.2 Demystifying AM through Simulations

Chapters 2 and 3 introduced computational design and process development techniques for precision metal AM, respectively, and thus some of the physical aspects of the metal AM process have already been discussed – such as simplified AM models for distortion, stresses and cause-and-effect relationships between process parameters and AM performance indicators. However, prior to discussing the modelling techniques for precision metal AM, it is necessary to get a clear picture of the various physical phenomena that take place.

4.2.1 The Physics of Laser Powder Bed Fusion

The metal L-PBF process has arguably two distinct starting points – (a) the generation of the powder bed and (b) the production and delivery of the energy source (in this case the laser) – while the core of the process begins at the interaction of the energy source with the powder bed. For the current discussion, the process is first explored from the perspective of the powder bed.

The metal AM process begins with the deposition of the powder particles from a powder delivery system, for example, a hopper, onto the base plate. The powder particles fall down through a tube and get deposited as a heap on the base plate. The flow of the powder particles within the tube is often promoted through a low-pressure gas flow that ensures the powder particles do not agglomerate and block the powder delivery system. The phenomenon involved at this stage is termed *particulate flow* or *particle-laden flow* (see Maxey 2017 for further details on particle-laden flow). As the number of powder particles per cross-section transported within the tube increases, the effects of the powder rheology start to increasingly dominate the flow behaviour (see Vock et al. 2019 for details on powder rheology applicable to AM). The formation and behaviour of the resultant powder heap on the base plate is almost entirely governed by the same powder rheology characteristics.

At the next stage, the powder particles are spread over the base plate using a recoater blade. The blade is typically composed of a hard section (for example, metal or hard plastic) which comes into contact with most of the powder heap and may have a softer end section (silicone or rubber) which is able to glide over the deposited powder and/or any surface protrusions. A small gap equivalent to the intended layer thickness is maintained between the tip of the recoater blade and the base plate to allow the deposition of the required powder layer. Again, the rheological characteristics of the powder particles play a prominent role and govern the powder spreading. For example, if the particles have low flowability, the recoater would drag large lumps of powder across the base plate without spreading them to the required thickness. At subsequent recurring stages of the AM process, the powder is spread on top of the previous manufactured layers and thus the interaction of the recoater blade with the surface of the previous layer starts to become significant. The characteristically rough surface of the previous AM layer and the potential localised mechanical defects (such as warpage and/or delamination) directly influence the spreading of the powder and can even cause obstruction to the recoater blade, resulting in a 'recoater blade jump', as well as damage to the corresponding blade section. Thus, the mechanics of the recoater blade also becomes relevant from the second layer onwards.

The production and delivery of the energy source is the alternative starting point for the metal AM process. In the case of L-PBF, the energy source is a focused laser beam with high energy density. The high spatial and temporal coherence (see Zhou et al. 1985 for details) of the laser source allows it to be focused to a small spot (order of 10 μm–100 μm) and achieve a narrow spectrum (of the order of a few nanometres in terms of wavelength). The generated laser beam is then passed through several lenses that regulate (collimate, polarise and/or focus) the optical characteristics of the beam, and is then incident on a mirror that changes the incident location of the laser beam on the powder bed. The interaction of the laser beam with the lenses and the mirrors can typically be described by geometrical optics when the direction of propagation of the laser beam is the prime concern. However, when effects such as absorption of the laser beam in the optical elements such as mirrors and lenses (Lumeau et al. 2011), polarisation (Verschaffelt et al. 2001) or thermal lensing (Du et al. 1995) are under consideration, physical optics must be employed.

As discussed earlier in this section, the metal AM process begins when the laser beam is incident upon the powder bed. In broad terms, a portion of the laser beam is directly reflected at smooth powder/liquid surfaces (specular reflection) in a direction away from the processed area. A second portion undergoes diffuse reflection at the rough powder surfaces. A third portion is absorbed by the powder/bulk material, and a fourth portion is reflected into the powder bed and undergoes multiple cycles of reflection, absorption and scattering. The absorption of the laser beam depends on the characteristics of the laser (for example, central wavelength and spectrum), the properties of the material (albedo, refractive index, emissivity, etc.) as well as the characteristics of the powder bed (powder packing, powder morphology, powder composition, etc.). The actual laser–material interaction is often complex and can only accurately be defined through the radiative transfer equations combined with ray optics (see Section 4.4 for more details).

From the point that the laser is converted into a heat source, the metal AM process starts to locally resemble a generic welding process involving powder filler material. Neglecting the small region with the melt-pool, the metal AM process also globally resembles a generic post-solidification casting or non-homogenous, non-isothermal heat-treatment processes. Despite the similarities, the metal AM process remains considerably distinct, with unique process signatures and defects that cannot be adequately justified by treating metal AM as either of the aforementioned processes, thus bringing up the concepts of length and time scales discussed in the following section.

4.2.2 Challenges of Length and Time Scales

In metal AM, the distinct process signatures and defects arise due to various phenomena occurring across several length and time scales. These phenomena and their modelling techniques are addressed in greater detail in Section 4.3–Section 4.5, while a discussion on the length and time scales follows here.

The 'length scale' of a phenomenon refers to the particular length or distance at which the phenomenon becomes significant. The concept is especially important when understanding a complex process (such as metal AM) as a physical phenomenon occurring at different length scales can be decoupled, that is, it is possible to make a justifiable assumption that the phenomena do not directly affect each other. For example, heat transfer in large bodies (say a few centimetres in size) can be described adequately using Fourier's law and the heat conduction equations, while the heat transfer between a proton and an electron requires a different set of physics involving quantum statistical mechanics, particle dynamics and electromagnetic theory (see Marla et al. 2018 for different models relevant to laser processing). Nonetheless, the physical laws relevant at the lower length scale can always be used to describe the phenomena at the larger length scales, albeit often requiring rigorous handling from a mathematical/algorithmic perspective to ensure a solution can be found in a reasonable time. Analogously, 'time scale' refers to the shortest period of time over which the phenomena can be observed, and the property of decoupling is also applicable to phenomena at different time scales. Using the previous example of heat transfer in a large body, Fourier's law governing heat transfer is valid when considering time periods much larger than 10^{-12} s to 10^{-14} s, below which relativistic heat conduction equations become more applicable (Molina et al. 2014).

In the case of metal AM processes (Figure 4.1), from the process viewpoint, the smallest length scale typically corresponds to the wavelength of the laser beam being used (for example, 1.06 μm for a Nd YAG laser). However, from a material science perspective, even smaller features, such as nano-precipitates and micro-/nano-scale pores, become relevant for the metal AM process, thus making nanometre length scales significant. Warpage effects on metal AM components only become practically relevant in parts that are several centimetres in size (in each dimension). Thus, the range of relevant length scales for metal AM can vary from nanometre to centimetre scales, corresponding to at least seven orders of magnitude. Similarly, the relevant time scales for the process can vary from microseconds (the time for the laser beam to pass over a powder particle) to several hours (the time for a complete build), corresponding to between ten and eleven orders of magnitude.

When discussing the process length and time scales, it is prudent to have the perspective that qualification of metal AM parts might necessitate tracking the emergence of defects/features, such as melt-pool balling, porosity evolution, columnar dendritic microstructure

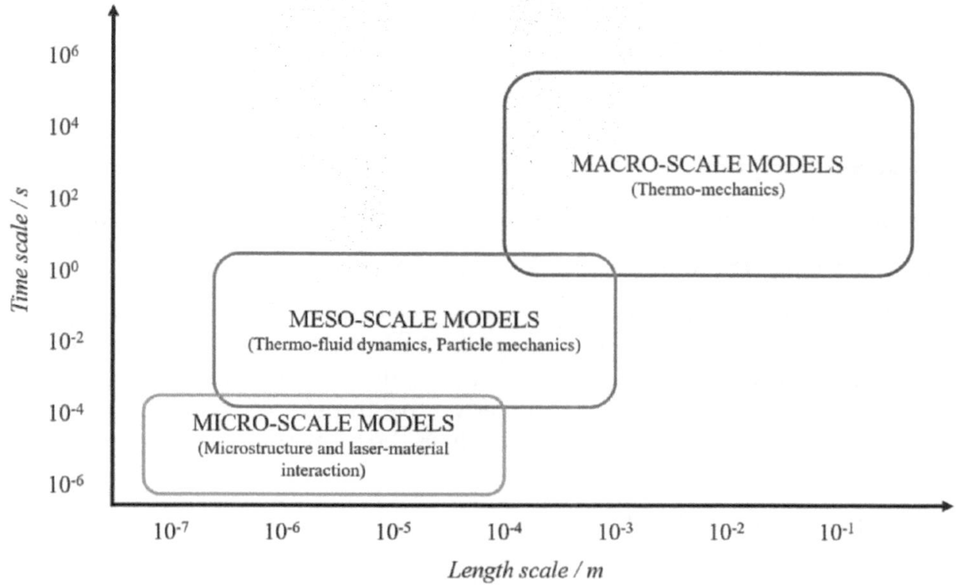

FIGURE 4.1
The length and time scales at which various models are applied for simulating the physical phenomena during metal AM.

growth, hot tearing, thermal stresses and warpage. Experimentally following the process through the entire range of length and time scales to which the aforementioned phenomena correspond would require significant resources and is practically infeasible. Computational modelling can be a potential approach to enable such qualification at a reasonable resource expenditure.

A key concept behind engineering applications of computational modelling is goal-oriented model selection, which resonates with the problem-solving principle of Occam's razor: that simpler solutions are more likely to be correct than complex ones. As physical phenomena occurring at different length scales can be decoupled, goal-oriented model selection first involves the identification of the most relevant length and time scales corresponding to the type of defect being tracked. The physics involved at the relevant length and time scales is handled with greater accuracy, while the contributions from phenomena occurring at larger or smaller scales are typically approximated. This approach has enabled grouping of the various computational models into broader categories of micro-, meso- and macro-scale models. In this chapter, macro-scale models (Section 4.3) will handle transient thermal stresses, plastic deformation, in-situ warpage and distortion, stress-induced damage and delamination. Meso-scale models (Section 4.4) will deal with powder-layer spreading, melt-pool formation and transient dynamics, balling effects, porosity evolution and surface-texture generation. Micro-scale models (Section 4.5) will tackle the metallurgical microstructures formed during metal AM processes, and the effect of material properties, such as strength, wear resistance and fatigue life. A robust computational approach will also include techniques for linking these models at different length and time scales. Several studies in recent years have summarised these micro-, meso- and macro-scale modelling approaches to metal AM (Moges et al. 2019, Kushan et al. 2018, Liu et al. 2018, Stavropoulos and Foteinopoulos 2018, Lindgren et al. 2016, Smith et al. 2016, Markl and Körner 2016, King et al. 2015a, King et al. 2015b, Schoinochoritis et al.

2014, Lavery et al. 2014), and subsequent sections of this chapter will draw upon concepts already described in these references.

4.3 Warpage and Distortion Predictions by Macro-Scale Modelling of AM

4.3.1 Understanding Thermal History, Residual Stresses and Distortions

In simple terms, the thermal history of any object can be calculated by adopting the assumption that the rate of change of energy in the body is equal to the sum of energy added to the body through an external heat source, the energy transferred into the body across its boundaries and the diffusive and advective transfer of energy inside the body. The thermal history is described mathematically in the form of a governing equation

$$\rho C_p \frac{\partial T}{\partial t} = \frac{\partial\left(k_{xx} \partial T / \partial x\right)}{\partial x} + \frac{\partial\left(k_{yy} \partial T / \partial y\right)}{\partial y} + \frac{\partial\left(k_{zz} \partial T / \partial z\right)}{\partial z} + u\rho C_p \frac{\partial T}{\partial x} + v\rho C_p \frac{\partial T}{\partial y} + w\rho C_p \frac{\partial T}{\partial z} + \Phi \tag{4.1}$$

where T is the temperature; t is the time; (x, y, z) are the spatial coordinates; (u, v, w) are the advection velocities along the x-, y- and z-axis, respectively; k_{xx}, k_{yy} and k_{zz} are the thermal conductivities in the different axes; ρ is the density; C_p is the specific heat; and Φ is the heat source term. The thermal interaction at the boundaries between the material and the surroundings can be represented as

$$-k\frac{\partial T}{\partial x} = -h\left(T_{amb} - T\right) + \sigma\varepsilon\left(T^4 - T_{amb}^4\right) \tag{4.2}$$

where h is the heat transfer coefficient between the material and the gaseous environment, T_{amb} is the temperature of the gaseous environment, ε is the emissivity of the material and σ is the Stefan–Boltzmann constant.

The governing equations for thermo-mechanical modelling are based on the principles of equilibrium and virtual work. The principle of equilibrium, based on Newton's second law of motion, states that in a body at equilibrium, the net force (external and inertial) on the body is zero as the linear momentum of the body is conserved. The virtual work statement has a simple physical interpretation: the rate of work done by the external forces subjected to any virtual velocity field is equal to the rate of work done by the equilibrating stresses on the rate of deformation of the same virtual velocity field. The principle of virtual work is thus the weak form of the equilibrium equations, wherein the total work done by different internal stresses and external forces needs to be zero, rather than the summation of the individual forces being zero. In the specific case of thermal processes, such as L-PBF, the primary source of these stresses in the body are the physical constraints (the support structures and base plate) that prohibit the thermal expansion of the material. In addition, due to the localised heating inherent to the L-PBF process, the cooler parts of the body can themselves act as constraints inducing localised tensile and compressive stresses, while the body as a whole remains in equilibrium. In certain scenarios, the stresses induced by the external or self-constraining forces can exceed the tensile/compressive yield stress of the material at the particular temperature and lead to plastic deformation (which is analogous to the material flowing in proportion to the applied stress). Inducing plastic deformation in such a manner is analogous to partially resetting the nominal location of the material within the body, albeit at a different temperature. Consequently, this resetting will lead to different thermal stresses when the local temperature changes back to the ambient, resulting in residual stresses in L-PBF components. Another possible scenario, especially in the case of overhanging structures, involves free thermal expansion and subsequent contraction of sections of the component, wherein the difference in the sequence of heating and cooling produces an overall deformed or distorted shape. The overhanging parts of the component in this case will continue to deform until the entire component is in a state of equilibrium (with or without localised residual stresses).

4.3.2 Goals and Challenges in Macro-Scale Modelling of AM Parts

The material experiences successive cycles of thermal expansion and shrinkage during L-PBF to produce localised compression and tension that induce residual stresses in the produced part. These thermally induced residual stresses partially relax when the support structures and surrounding powders are removed. Depending on part geometry, these stresses may cause fatigue crack growth and deformation in the final part, which may result in warping, loss of edge tolerance, loss of net shape and part failure. The residual stresses and distortion of the produced part are potentially affected by the build orientation, scanning strategy and preheating of the build platform (Hodge et al. 2016; Mercelis and Kruth 2006). In order to mitigate residual stresses, some of the known techniques are proper selection of scan strategies and build orientation, in-situ heating of the powder bed and ex-situ heat treatment before removing support structures. In addition, since larger residual stresses are observed at the top layer and substrate–part connection, filleting the edges can reduce stress concentration at these regions (Denlinger 2018). Thus, macro-scale process models should ideally be able to simulate each of these phenomena and behaviours.

4.3.3 Full-Scale, Reduced-Order and Effective Models

To simulate the L-PBF process at the macro-scale, the easiest approach is a simplified mathematical model that can be derived to predict the general profile of residual stresses using equilibria of forces and moments using general beam theory, albeit based on a number of assumptions. An alternative method is to use an inherent strain/eigenstrain approach (Seitien et al. 2019), wherein the residual stresses in any complex part can be calculated based on deformations measured in specific 'calibration' parts processed using the same process parameters. The core assumption of the inherent strain method is that the final residual stresses can be considered the cumulative effect of inelastic stresses generated during the production of each layer. Such models are useful when macro-simulations are necessary to quickly identify problem areas (for example, where support structures should be added) for optimisation of part orientation as well as for implementing an iterative geometry optimisation procedure to reduce deformation during actual production (Figure 4.2).

The more rigorous approach to determining the residual stresses in L-PBF processes is the thermo-mechanical finite element model (FEM) based analysis (Figure 4.3) that uses the temperature gradients determined from the melt pool model as thermal loading. From a thermo-mechanical perspective, the AM process resembles a multiple pass welding process, and thus different approaches have been employed based on the numerical methods

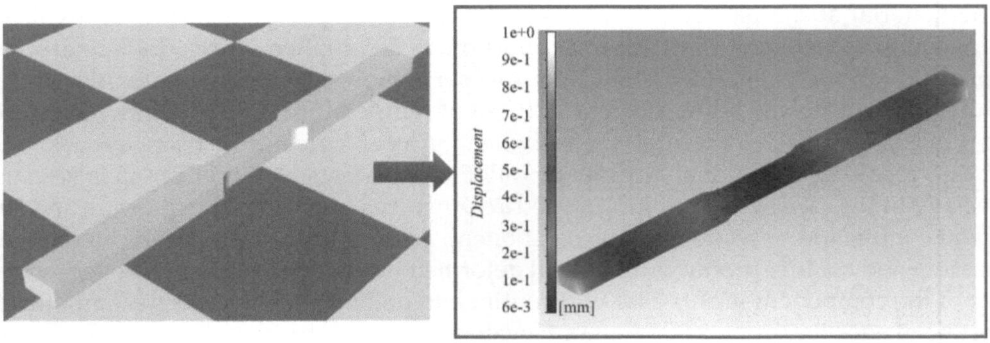

FIGURE 4.2
Deformations predicted in a component using the inherent strain method: (left) tensile bar geometry, (right) predicted deformation along the tensile bar.

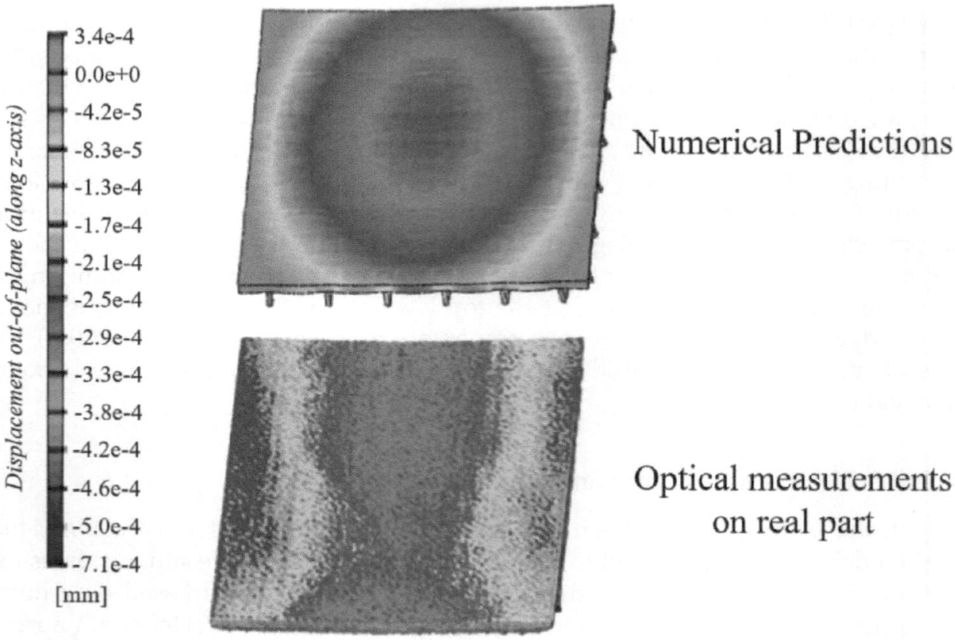

FIGURE 4.3
Deformation in a plate predicted using thermomechanical modelling compared to actual measurements. (Based on Bayat et al. 2019d.)

developed for welding simulations (including the inherent strain method discussed earlier) (Zhou et al. 2016; Francois et al. 2017). The thermo-mechanical modelling of AM can be divided into two main steps: (1) performing a pure thermal or heat-transfer analysis in order to obtain the nodal temperature in the FEM based model and (2) establishing a structural configuration to evaluate the mechanical response of the FEM-based model under applied nodal temperature gradients. The intended coupling between the thermal and mechanical models determines the manner in which these steps are carried out within the overall simulation. For the specific case of L-PBF, the decoupled approach is usually

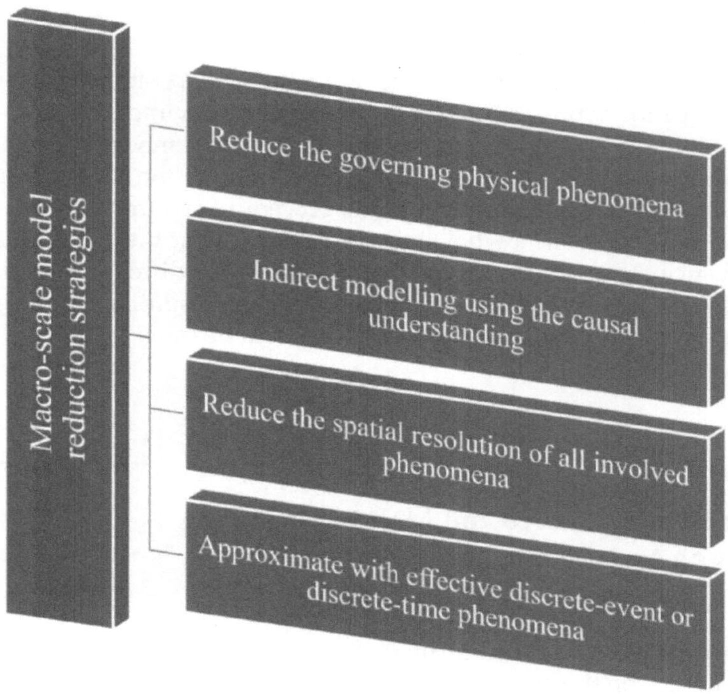

FIGURE 4.4
Four common model reduction strategies for macro-scale modelling of metal AM.

preferred since the analysis time is significantly lower compared to computation time for the coupled approach, with small differences in the prediction results.

A more computationally practical approach to the macro-scale modelling of L-PBF is to adopt the reduced order approach. The reduced order approach (Figure 4.4) uses simple assumptions to either (a) reduce the physical phenomena governing the process, (b) indirectly model the involved physical phenomena using their causal understanding, (c) reduce the spatial resolution of all involved phenomena and/or (d) approximate the continuous physical phenomena with effective discrete-event or discrete-time phenomena. A typical example of reducing the physical phenomena governing the process is observed when simulating the removal of the base plate from L-PBF components (Mercelis and Kruth 2006). While the actual removal process is localised and causes dynamic changes in the stress state of the component throughout the removal process (performed using either electro-discharge machining or mechanical processes), a majority of the studies use a steady-state approach to determine the resulting deformation in the component.

Indirect modelling of the governing physical phenomena during modelling is the rationale behind the inherent strain method discussed above. Here, the thermal modelling of the process is not used, and instead, assumed values of inelastic strains (based on experiment and process characterisation) are used to create an initial strain condition for each layer added to the component during mechanical simulation of L-PBF.

The method for reducing the spatial resolution of the involved phenomena has been applied for modelling the L-PBF process to remove the necessity of simulating the melt pool and direct laser material interaction. This is achieved by ensuring that the smallest modelled domain (corresponding to a single element in FEM or a single control volume in the finite volume technique) is much larger than the melt pool, thereby allowing for the domain to be deposited as a solid at a lower temperature, while still ensuring that the correct amount of energy has been added to the body during deposition (Bayat et al. 2019d).

The last technique for approximating a continuous process into a discrete-event or discrete-time process has been implemented within macro-scale modelling of L-PBF, by adding an entire layer (or multiple layers) within a single simulation time increment. Similar to the spatial resolution reduction method, this procedure ensures the correct amount of energy is added to the body as a whole but not necessarily at the correct locations at the correct time. In such cases, the predicted thermal fields can differ substantially from the actual thermal fields for more dense geometries but can be similar for more sparse/low-mass geometries, such as those composed of lattice structures.

4.4 Tracking Powders, Pores and Melt Pools during AM through Meso-Scale Models

4.4.1 Powder Bed Formation and Representation

The primary process of powder bed formation involves the re-coater moving and spreading a layer of particles on the build platform. The packing structure of the resultant L-PBF powder bed is dependent on parameters such as powder size and shape, particle size distribution, layer thickness, and re-coater shape. In addition to the powder bed parameters, the re-coater velocity also influences the surface structure and packing density of the powder bed (Mindt et al. 2016, Parteli and Pöschel 2016). During powder bed formation, a number of phenomena, such as friction, collision and adhesion, occur due to the interaction between micro-sized particles with elastic, frictional forces, gravity and van der Waal forces governing the final powder bed morphology (Dou et al. 2014, Xiang et al. 2016, Parteli 2013).

The simplest models for simulating the powder bed involve using empirical correlations between the packing structure and the powder characteristics (morphology, size distribution and layer thickness). Such correlations exist when assumptions of spherical powder particles and non-significant van der Waal forces are valid. Being based on geometrical considerations, these models inherently do not differentiate between powders of different materials. However, experimental results of packing density, powder bed porosity and packing structure can be used to extend and characterise these models with respect to the different materials of interest. These powder bed models can then cater to macro-scale simulations, such as those for residual stresses and deformations.

The thermal properties of conventional materials, based on experimental measurements, are typically available in the literature (Mills 2002). However, the values often correspond to bulk material properties and cannot be directly applied to powder beds, which are characterised by discrete spatial distributions of materials. Once the powder packing fraction and/or powder bed porosity is determined, other empirical methods can be used to estimate the effective material properties for the powder bed. The typical approach is to substitute the discrete powder bed by a continuum of material having equivalent material properties. The porosity of the powder bed does not affect the specific heat capacity of the material, but proportionately affects the volumetric enthalpy by influencing the equivalent density of the continuum media. Successively improving predictive models for emissivity and thermal conductivity of powder beds have been proposed (Sih and Barlow 1995a, Sih and Barlow 1995b). The powder bed has been considered as a network of thermal resistances connected at the contact points and via the gaseous medium, and a corresponding theoretical model accounting for the size-dependence of the conductivity has been proposed (Shapiro et al. 2004). The concept of the powder bed as a network of discrete thermal resistances with a dependency on effective thermal properties on the morphology of the powder bed is another alternative modelling approach (Gusarov and Kovalev 2009). In the case of a randomly packed powder bed formed of mono-sized spherical powder particles, the effective thermal conductivity can alternately be estimated by the modified Zehner-Schlünder-Damköhler equation (Sih and Barlow 2010), which takes into account the gas present in the spaces between the powder particles.

More rigorous and accurate models for powder beds correspond to the discrete element method (DEM) models that consider the actual size and morphology distributions for the powder. In DEM, interparticle forces, computed using non-linear Hertz theory, are explicitly considered (Herbold et al. 2015, Jia et al. 2011) to simulate the spreading of powder on top of the previously processed layers. The range of applicability of these models is limited by the large computational resource requirements, as well as accurate values of properties such as friction coefficients and van der Waal forces. These powder bed models (Figure 4.5) can cater more effectively to micro-/meso-scale simulations, such as those for melt-pool simulations and process-window identification.

4.4.2 Simulating Laser–Material Interactions

The laser–material interaction forms the core of the L-PBF process, essentially determining all the other emergent characteristics, such as porosity, residual stresses and metallurgical microstructures. From the material perspective, the models for simulation of the powder

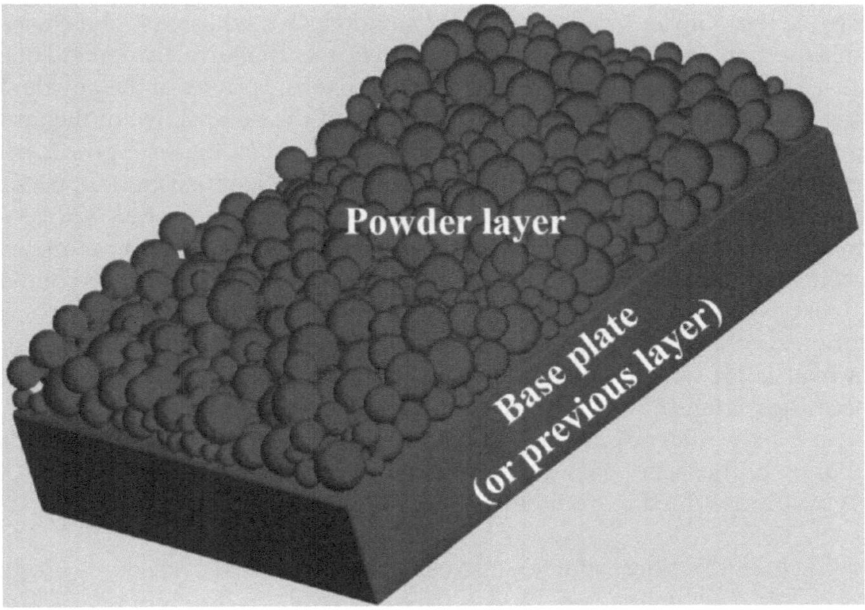

FIGURE 4.5
A typical powder bed generated using DEM simulation models.

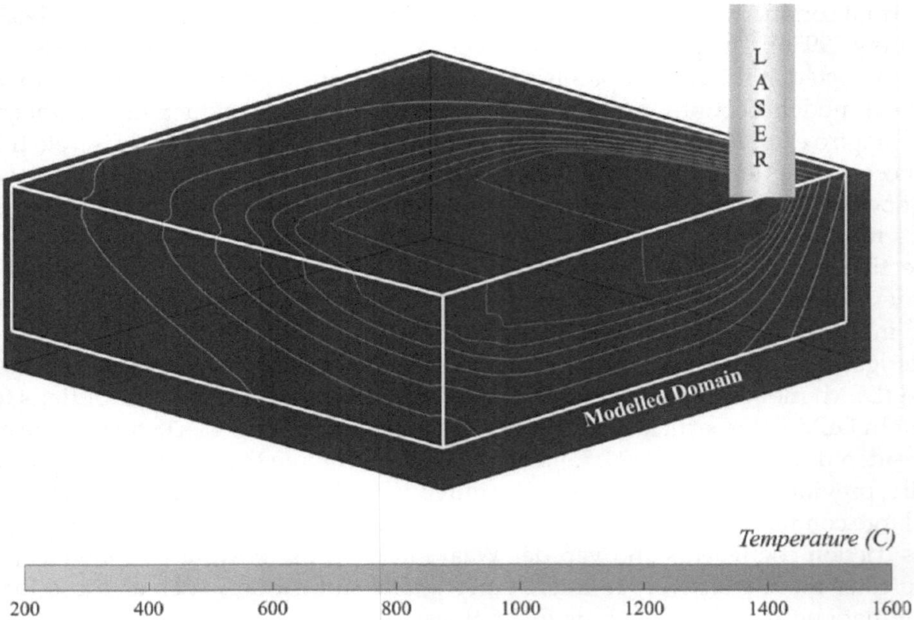

FIGURE 4.6
Thermal field predicted in an AM component based on indirect modelling of laser beam as a Gaussian heat source. (Based on Mohanty et al. 2014.)

bed have been described in Section 4.4.1. However, these models can be applied while considering the particle size distribution as Gaussian (most common), bimodal (provides higher powder packing density), uniform or mono-sized (ideal condition and easiest to simulate) (Moges et al. 2019). Each distribution type provides different packing densities and porosity of the powder bed, especially at small layer thicknesses. The choice of particle size distribution has a significant influence on factors that directly depend on packing density, such as the value and degree of fluctuations of absorptivity of the powder bed and the radiative transfer process (Moges et al. 2019). From the perspective of the laser beam, the amount of heat absorbed by the powder bed is governed by the laser power, beam spot size, absorptivity of the material, powder size and shape, size distribution, packing density, surface oxidation and contamination (King et al. 2015b, Boley et al. 2015). When the laser beam strikes the powder bed, multiple scattering of the laser beam occurs within the powder particles and in the melt pool, and hence, the penetration depth is comparable to the layer thickness (Gusarov and Kruth 2005). Multiple scattering of the laser beam causes the absorptivity of the powder bed to be higher than the absorptivity of laser on a flat surface (Boley et al. 2015), which becomes relevant when either the laser is first switched on or when processing regions have a downskin surface (i.e. overhanging surfaces).

The simplest laser–material interaction models are indirect models based on the intensity distribution of the laser beam incident on the powder bed (Figure 4.6). Here, the heat source is typically assumed to be either:

(a) a moving point/line heat source – also called the Rosenthal heat source (Rosenthal 1946);
(b) a moving uniform cylindrical heat source with radius equal to the laser spot size (Assouroko et al. 2016);

(c) a moving ellipsoidal heat source, commonly referred to as the Goldak heat source (Lindgren et al. 2016); or
(d) a moving Gaussian heat source with a normally distributed laser intensity (Mohanty et al. 2014).

Assuming the absorbed energy is to be constrained on the surface instead of the volume of the powder bed considerably reduces the predictive accuracy of these models, despite the reduced computational requirement. Nonetheless, these models are most applicable within macro-scale simulations where the modelled domain can be as large as the entire build chamber.

The more rigorous models for the laser–material interaction focus on determining the amount of absorbed energy in the powder bed via either:

(a) a homogeneous continuum radiative transfer equation (RTE) using effective powder porosity and surface areas (Gusarov and Kruth 2005);
(b) a ray tracing method that accounts for the effect of multiple reflections by tracking the trajectories of multiple laser rays (Bayat et al. 2019a); or
(c) a Beer-Lambert law-based approach that relates the attenuation of the irradiation intensity with penetration depth using an exponential decay assumption (Bayat et al. 2019b).

The continuum RTE approach and the Beer-Lambert approach are useful within medium fidelity models where the actual shape of the powder bed (and thus the direct laser interaction) is not of interest. The ray-tracing approach (Figure 4.7) is necessary for the high-fidelity micro-/meso-scale simulations where identification of defects and process windows are of interest.

4.4.3 Melt-Pool Dynamics in a Powder Bed

The melt-pool dynamics during the L-PBF process resemble other traditional processes, such as welding and cladding. Thus, numerical methods developed for the simulation of these processes have been further developed and employed to study the melt pool in L-PBF. The main driving forces in a melt pool are commonly considered to be derived from surface tension γ (a force in the normal direction at the liquid surface that is proportional to its total curvature) and Marangoni force (a shear force acting in the tangential plane caused by the tangential gradient of the normal force). The latter force emerges due to the temperature dependency of the surface tension and the tangential gradient of temperature. An additional force that comes into play is caused by the recoil pressure during vaporisation of the liquid metal (Bayat et al. 2019c).

The contributions of the surface tension and the recoil pressure towards the melt-pool dynamics are well understood from studies of welding processes. The contribution of the Marangoni force is more difficult to ascertain since it is strongly dependent on the nature of the temperature dependency of surface tension (i.e. whether the surface tension increases or decreases with increasing temperature). This $\partial\gamma/\partial T$ coefficient has been shown to have a significant influence on the length and width of the melt pool. A high negative $\partial\gamma/\partial T$ leads to a centrifugal convection flow with high velocity from the hot centre under the laser spot towards the colder boundary, while a high positive $\partial\gamma/\partial T$ will cause the flow to occur in the reverse direction. Consequently, having a negative $\partial\gamma/\partial T$ leads to a longer and wider

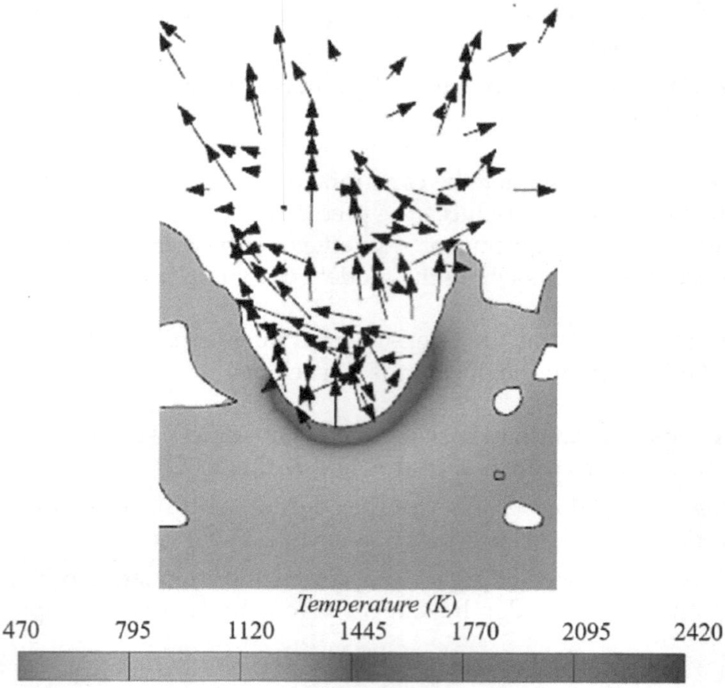

FIGURE 4.7
Ray tracing method implemented in a CFD model showing multiple reflections inside a melt pool. (Based on Bayat et al. 2019a.)

melt pool in the material as compared to a positive $\partial\gamma/\partial T$. The longer melt pools, however, correspond to larger melt surface areas, higher evaporation rates and stronger thermal gradients, which can lead to instability of the melt flow.

Instabilities in the melt pool can also be caused when the powder layer thickness increases, particularly due to the larger volume of gas within the powder particles that attempts to escape when the powder is molten. Thicker powder beds also promote keyhole formations leading to greater risk of instabilities arising from recoil pressure, spattering and increased melt flow velocities. The Kelvin Helmholtz instability criterion is a measure of the instability induced by the increased gas flow over the melt pool due to vaporisation and natural convection in the build chamber (or forced convections in the cases where a strong inert gas flow in the build chamber is ensured during metal AM) (Mukherjee et al. 2018). The criterion uses the Richardson number at the top of the melt pool, which is the ratio of buoyancy forces to shear forces, with values lower than 0.25 leading to Kelvin Helmholtz instability. The Plateau Rayleigh instability criterion is instead focussed on the geometrical aspects of the melt pool and suggests that instability occurs when the length of the melt pool exceeds its width by a factor of π. Figure 4.8 shows the velocity profile in a typical melt pool during metal AM and the different forces acting on it.

4.4.4 Evolution of Porosity during AM

The irregular melt pool dynamics often results in porosity throughout L-PBF components. Pores can be classified into three categories based on their origin, as follows.

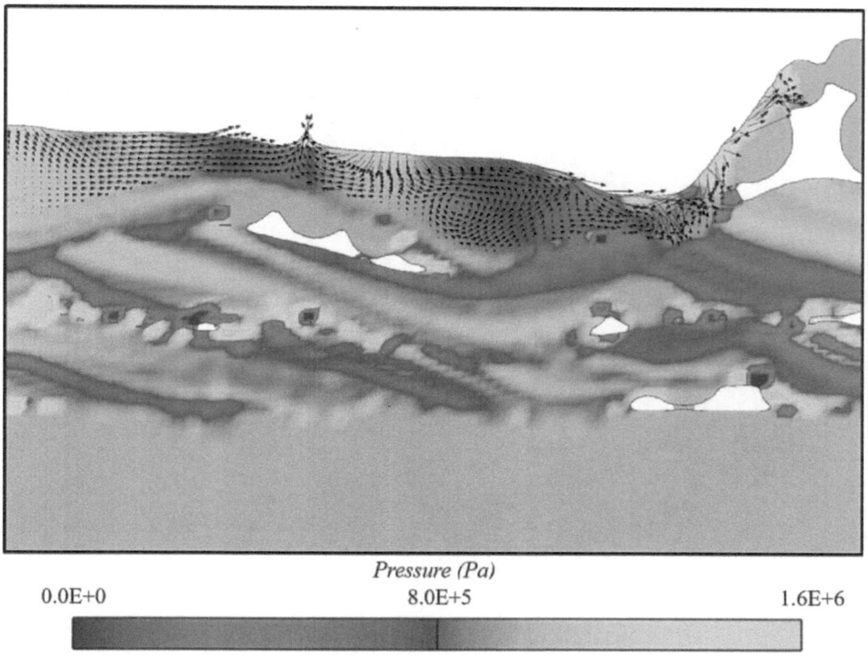

FIGURE 4.8
Melt-pool velocity vectors showing the direction of flow within the melt pool during AM. (Based on Bayat et al. 2019a.)

(a) Lack of fusion porosity

Lack of fusion porosity originates primarily due to either a lack of powder material in the scanned regions or to improper overlap of the molten regions arising from improper process parameter selection. Lack of material can arise from the improper spreading of the powder particles during the powder bed formation step (see Section 4.4.1 for corresponding models), or due to denudation occurring during the AM process (Figure 4.9). A third source of lack of fusion porosity (particularly at the surface) is the rapid solidification of the melt before it is able to spread and create a uniform surface. This often leads to open porosity on the intermediate surface of the components during processing, but this typically disappears as successive layers are built above the corresponding regions (the same is true for most lack-of-fusion porosity formed during the process).

(b) Keyhole porosity

Keyhole porosity originates from the collapse of keyholes during the AM process due to the entrapped metal vapour. These entrapped gas/vapour bubbles originate at the bottom of the front end of the melt pool and travel towards the tail of the melt pool along the melt-flow streamlines. Further, due to the lower density of the metal vapours, they can begin to ascend towards the surface of the melt pool. Simultaneously, the vapour bubbles also begin to cool down; they consequently shrink in size and can disappear when the size is sufficiently small. In certain cases, these bubbles can be trapped within the material, especially when the solidification rate is high, and create keyhole porosity (Figure 4.10).

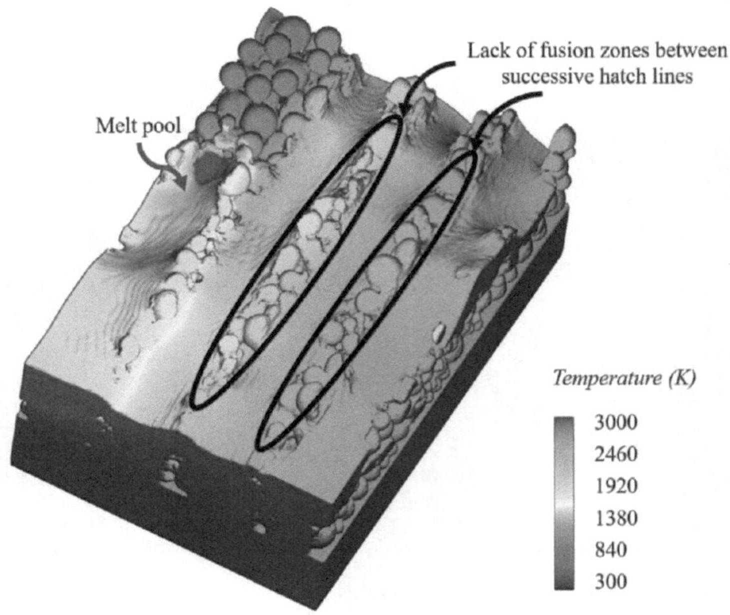

FIGURE 4.9
Lack of fusion defects predicted during metal AM. (Based on Bayat et al. 2019a.)

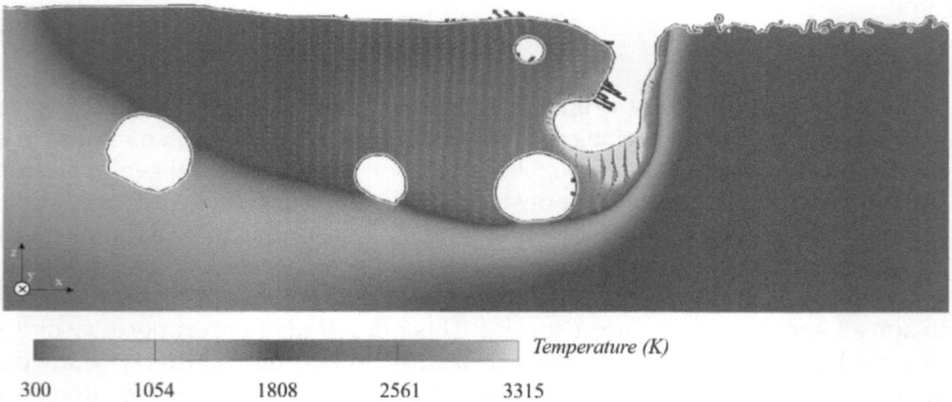

FIGURE 4.10
Keyhole porosity predicted during metal AM. (Based on Bayat et al. 2019c.)

(c) Gas porosity

Metallurgical pores are known to be small and spherical, originating from the dissolved gases in the melt pool or from pre-existing pores within the powder itself (Xia et al. 2017). Modelling has shown that the prevalence of these pores in the component depends on the longevity of the melt pool, i.e. if a particular region is within the melt pool for a longer time, the trapped gases could have enough time to rise to the surface and escape (Xia et al. 2017).

An interesting observation for porosity in AM components corroborated through modelling studies corresponds to the transition of the dominating type of porosity from keyhole porosity, to metallurgical/gas porosity, to open porosity, to improper overlap porosity as the laser scanning speed is progressively increased (or the laser power decreased).

4.4.5 Surfaces and Solidification during AM

The surface characteristics of L-PBF components are also governed by the melt-pool dynamics and the interaction with powder particles at the meso-scale. The instability of the melt pool (including the successive formation and collapse of keyholes) at high laser power or scanning speed dictates the surface topography emerging from the AM process (Figure 4.11), while the denudation and the spattering during the AM process often leads to surface discontinuities (see also Chapter 11). At the same time, the thickness of the powder layer also significantly influences the stability of the process, with thicker powder layers leading to progressively higher levels of surface texture. Thicker powder layers also promote dross formation by allowing the molten metal to spread through the adjacent powder domain before solidifying, thus trapping partially/un-molten powder particles at the periphery of the laser scanned track. Such effects are well known (forming the basis for laser sintering based AM technologies) and have led to different processing parameters being used in commercial L-PBF machines for near-surface regions of components as opposed to the interior bulk regions.

Modelling studies (King et al. 2015b, Bayat et al. 2019a) have indicated that the melt pool can behave in a dispersed and random way during the L-PBF process, leading to irregular track shapes. The driving force for this behaviour is the combined effects of the Marangoni

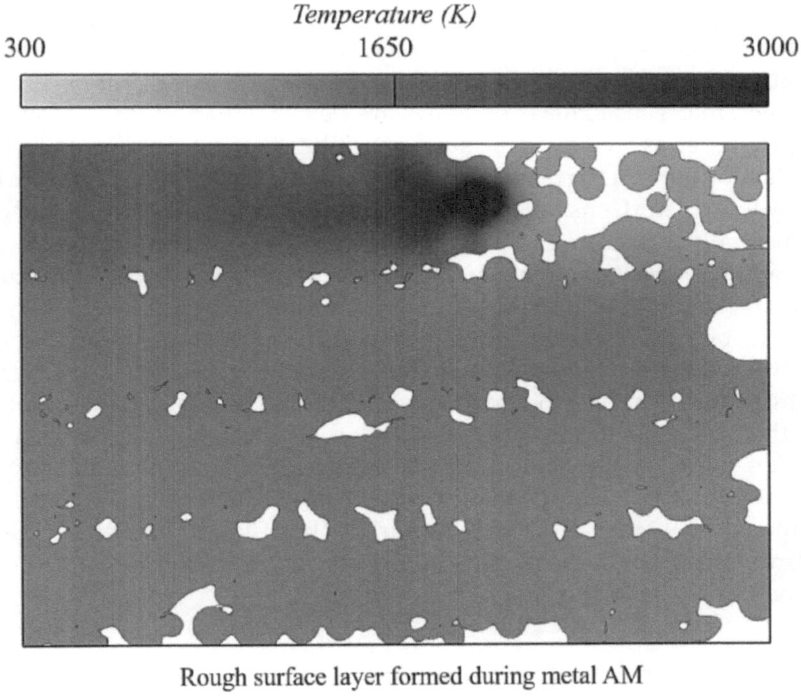

FIGURE 4.11
Rough surface emerging due to melt-pool fluctuations during metal AM. (Based on Bayat et al. 2019a.)

force developed at the surface and sub-surface regions of the melt pool, the recoil pressure due to material evaporation and gas expansion, and the discontinuous physical boundaries of the melt pool offered by the powder bed. Further, the metallic vapour often solidifies at the periphery of the melt pool leading to burr formation and it can lead to spatter formation when the recoil pressure is strong enough to displace small liquid metal droplets. Currently, modelling efforts at the meso-scale are capable of capturing these phenomena individually, but at the time of writing, to the authors' knowledge, a combined melt pool and vapour dynamic model is yet to be developed.

4.5 Microstructure Simulations in Precision AM

4.5.1 Understanding the Metallurgical Needs

The primary drive for advancement of metal AM processes in the initial decades since their invention has been the ability to produce complex shapes that cannot be produced easily by any other traditional manufacturing techniques. Hence, the growth of metal AM was promoted by industries such as aerospace, automotive and energy where the high cost of metal AM could be offset by the significantly larger gains. These industries, however, have stringent quality and certification procedures that take several years to create and implement (see Chapter 6). Consequently, while the metal AM process was attractive from a design perspective, there were considerable challenges in qualifying a metal AM component – especially from the materials perspective (see Chapter 9).

The initial industrial application of metal AM was to produce parts that could substitute more established components with known mechanical property requirements, the same materials as the original component had to be used (Ti6Al4V, 316L SS and IN718 are examples of such materials). However, alloys are developed and tuned to optimise their mechanical properties as well as their manufacturability using a preferred technique, i.e. certain alloys are optimised to create products by casting while other alloys are optimised for forming, welding, etc. Consequently, different additives had to be introduced into the powders of these alloys to ensure that they can be processed using metal AM.

One of the main assumptions when using such optimised alloys for producing parts is that the material properties within the part would be homogeneous (or at least predictably controlled). In metal AM, however, due to the repeated cycles of melting and solidification followed by cycles of in-situ heat treatment, the material is also created while creating the parts – leading to considerable inhomogeneity within the part. This material inhomogeneity is one of the major hurdles for the qualification of metal AM components across different industries. However, the same phenomena also lead to opportunities to create more exotic materials with unique localised properties – albeit when adequate control and optimisation of the metal AM process can be ensured. Thus, simulations of the metallurgical microstructure coming out of the metal AM process and determination of the consequent material properties is a growing field of research.

4.5.2 Metallurgical Modelling Techniques

The internal state variable approach is well suited to the development of models for non-isothermal microstructural evolution prevalent during the metal AM process. In general, a microstructure may be defined by different state variables such as grain size, volume fraction of grains and fraction of solid in solidification. The simplest method for modelling the microstructure thus entails calculating the thermal gradients and cooling rates at each location and comparing them with the compositional phase diagram for the material, as well as the corresponding time-temperature-transformation (TTT) diagram or the continuous-cooling-transformation (CCT) phase diagram.

Modelling of the microstructural transformation accurately may require the ability to track the local concentration of each element, the temperature at each location, the thermal and concentration gradients, the local geometrical configuration and so on, which are contributing factors to the overall thermodynamic state of the material at that location. The thermodynamic computations of these state variables are the basis for the phase field method (Lu et al. 2018), which is a rigorous, accurate method of predicting the local metallurgical changes. However, the large computational cost of solving several coupled phase field equations, in addition to the thermal equations, render the size of the domain that can be modelled within reasonable time relatively small.

For performing microstructural evolution simulations at larger scales, one of the common methods adopted is known as the *cellular automata* method (Jabbari et al. 2018, De Baere et al. 2020). This technique offers the advantage of requiring relatively low computational cost (as compared to full thermodynamic microstructural phase field simulations) but has different challenges, such as mesh anisotropy, where the shape of a growing grain also has a dependence on the shape of the mesh (rather than only on the modelled grain). Cellular automata has been used successfully to model the different types of grain morphologies, columnar and equiaxed, by coupling it with Kurz-Trivedi-Giovanola kinetics (Rappaz and Gandin 1993), and then successfully applied for metal AM simulations in two and three dimensions in recent years (Zinoviev et al. 2016, Zinovieva et al. 2018).

A more detailed overview of the different microstructural models relevant for metal AM can be found in a recent review (Markl and Körner 2016).

4.5.3 Revisiting Solidification during AM from a Metallurgical Perspective

A simple way to obtain quantitative information about the solidification microstructure from the metal AM process is by extracting the local thermal gradient and cooling rates from meso-scale melt-pool simulations. A low growth velocity is typically indicative of a low occurrence of nucleation in the liquid melt, while a high thermal gradient is typically required for directional growth. If the thermal gradient is high compared to the growth velocity, the entire microstructure is made up of columnar grains, which grow epitaxially on existing grains – this corresponds to the microstructure typically observed in the bulk of the metal AM components. A solidification front that grows faster, however, leads to excessive nucleation in the melt and an equiaxed microstructure – such as that observed on the downskin surfaces and outer edges of metal AM components. This behaviour can be captured quantitatively through the phenomenological morphology factor (equal to the quotient of the thermal gradient squared and the cooling rate), wherein a low morphology factor corresponds to an equiaxed microstructure, while a high morphology factor corresponds to a columnar microstructure (Bayat et al. 2019b). The morphology factor corresponds to the squared value of the Niyama function, which is known to be correlated with porosities in castings (Niyama et al. 1980, Tavakoli 2014).

The same distribution of columnar and equiaxed grains can also be predicted using the cellular automata method by directly simulating the nucleation and growth of the grains (Jabbari et al. 2018). In this case, the transient thermal fields can be calculated either simultaneously or prior to microstructure simulations, with the former allowing a coupling between the two models via the latent heat released during solidification, phase transformation and vaporisation. Since the microstructure forms by grains nucleating and growing independently, separate models for nucleation and grain growths need to be coupled to the cellular automata method. The Oldfield model (Oldfield 1966) is often used for solidification modelling, wherein a Gaussian distribution, with the mean undercooling and maximum possible nucleation density as driving parameters, describes the local nucleation density. Once nucleated, the grains start to grow outwards. The grain nucleation is typically random with respect to the preferred orientation of the material, and to distinguish between different grains, a characteristic misorientation angle is attributed to each grain with respect to one of the principal directions. The growth of the nucleated grain is promoted by four driving factors: thermal undercooling, curvature undercooling, kinetic undercooling and solute undercooling (more information can be found in Easterling 1992). The interaction of these driving forces of grain growth is governed by the minimisation of free energy in the system – a core thermodynamic principle.

The similar prediction capability of the morphology factor-based technique and the cellular automata technique can be seen in Figure 4.12, where the microstructure formed in a single track during metal AM is shown. Both methods predict that the microstructure is columnar on the boundary of the melt tack and becomes increasingly equiaxed towards the centre.

4.5.4 Need for Heat-Treatment as Post-Process

Although AM offers the possibility to produce complex parts, the material properties of a part produced by L-PBF are not optimal. For example, for Ti-6Al-4V, the microstructure resulting from the L-PBF process is entirely made up of martensitic α′ organised in columnar formal β grains, due to the high cooling rates. However, the bases for selection of such a material for a particular application are the optimal material properties designed around an equilibrium mix of α and β phases. Consequently, heat treatments are common post-processing steps in the L-PBF process chain in order to obtain the desired microstructure. A second reason for heat treatment after L-PBF is the ability to induce stress relaxation in components.

A simple heat-treatment procedure can involve two distinct phases: a high-temperature treatment phase and a cooling phase. A typical approach is to quickly ramp up the temperature in a heat-treatment furnace (Figure 4.13), thereby inducing a transient thermal field

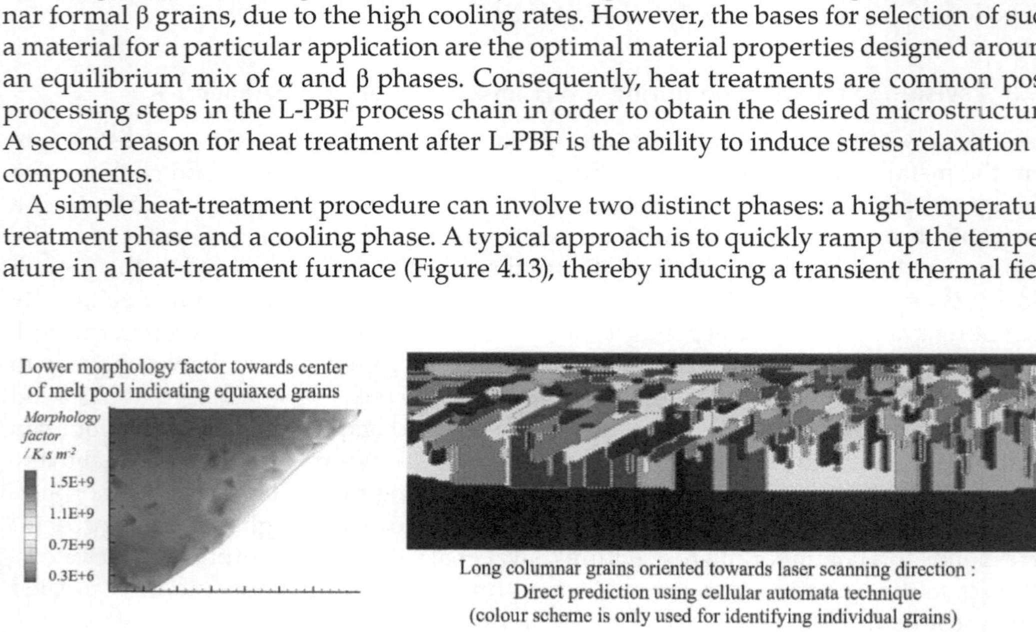

FIGURE 4.12
Morphology factor predictions within a single melt track and cellular automata prediction of the microstructure in the melt track. (Based on Bayat et al. 2019b [left] and Jabbari et al. 2018 [right].)

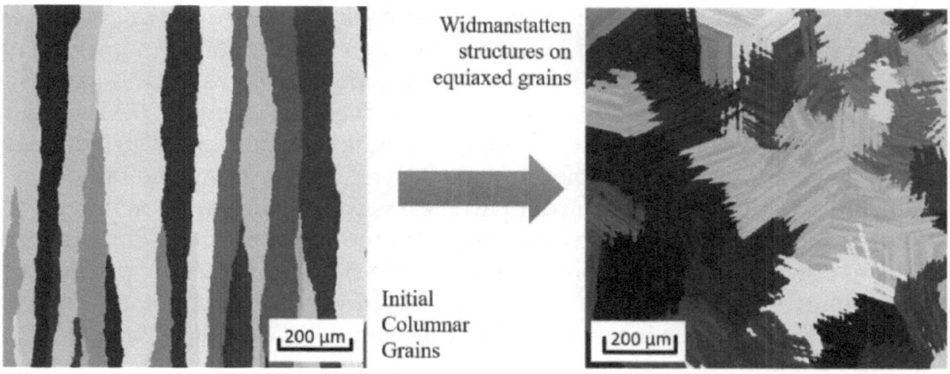

FIGURE 4.13
Influence of heat treatment on the microstructure of metal AM components. (Based on De Baere et al. 2019.)

with the component being heat-treated. Thus, for this phase, a temperature calculation is performed in tandem with the microstructural evolution calculation. However, during the cooling-down step, an isothermal assumption is adopted (except when cooling is achieved by quenching), with the temperature in the entire component decreasing continuously in time. More complex heat-treatment procedures can include successive heating and cooling phases as well as holding phases (where the component is allowed to remain at the elevated temperature, ensuring enough time for the transformation kinetics), where each step induces a specific microstructural or mechanical effect in the component.

In addition to the temperature history, the characteristics of the microstructure determined from the micro-scale phase transformation models can form valuable inputs to the macro-scale residual stress models discussed in Section 4.3. The residual stress and distortion in such cases can be predicted by utilising a thermo-elasto-plastic constitutive material model that includes the stress relaxation effect due to annealing (extracted from metallurgical/micro-mechanical simulations). The same constitutive material models are also relevant when simulating the L-PBF process, where in-situ heat treatment takes place throughout the process in regions away from the laser beam. Such modelling approaches become relevant when the residual stress states, and the consequent effect on component fatigue life, are of interest. The approach can also provide a more accurate deformation prediction and can be used as a final verification step for the optimisation studies conducted with simpler models.

4.6 Data-Driven Modelling for Process Windows

4.6.1 Data-Based Models

The various multiphysics models at different scales discussed in Sections 4.3–4.5 have an underlying commonality in their approach to simulating the metal AM process, i.e. they are all deterministic models. Thus, the coherency of the numerical predictions with the repeated experimental measurements, for any set of process parameters, will always be subject to the accuracy of the numerical replication of the particular experimental conditions.

For instance, the size of the powder particles used during metal AM often follows either unimodal or bimodal Gaussian distributions – the latter leading to better packing of the powder bed. The particle size also directly affects the powder flowability, as observed in high-speed X-ray imaging studies (Escano et al. 2018). The most complex DEM simulations for powder layer generation, therefore, attempt to include such nominal distribution information when generating the initial stack or virtual layer (for spreading and raindrop models respectively, see Section 4.3) of powder particles, as well as the variation in the powder distribution in subsequent layers. However, most micro- or macro-scale simulations use a representative powder layer based on average particle sizes, and thus the predicted thermal and melt-pool behaviour may differ from that observed with in-situ monitoring (Bayat et al. 2019c). The effect of such incoherency between simulation and experiments becomes significant when empirical models or constitutive equations have to be derived from the lower length/time scale models and applied within the macro-scale models.

In other words, the accuracy of the predictions of numerical models is subject to the uncertainty in the input parameters (both process parameters as well as numerical parameters, such as quantitative boundary conditions, material properties and numerical convergence/relaxation factors). In practical terms, these uncertainties can be categorised into epistemic uncertainties (uncertainty due to lack of knowledge) and aleatory uncertainties (uncertainty due to inherent stochasticity) of the models, which exist over and above the epistemic and aleatory uncertainties of the process and measurement techniques (see Chapters 10–12 for more details). The importance of uncertainty quantification (UQ) methods in computational modelling is highlighted by two ASME specification standards: ASME V&V 10 (2006) and ASME V&V 20 (2009). ASME V&V 10 contains the processes of verification, validation and UQ for computational solid mechanics models, and ASME V&V 10.1 provides detailed illustrative examples of the concepts. ASME V&V 20 provides a similar detailed methodology for verifying, validating and UQ of computational fluid dynamics (CFD) and heat-transfer models.

UQ for metal AM has received increasing attention in recent years with notable experimental studies (Delgado et al. 2012, Kamath 2016). However, the large number of process parameters render a detailed experimental UQ of the metal AM process prohibitively resource intensive. Consequently, simulation-based UQ of metal AM processes is emerging as a potential approach for process understanding and part qualification.

The earliest of such simulation-based UQ methods (Mohanty et al. 2014) had a focus on continuum thermal models for predicting temperatures and melt-track dimensions, followed by UQ investigations of arbitrary Lagrangian Eulerian 3D method (ALE3D)-based CFD models for tracking laser material interactions (Anderson 2015). A similar study with a more resource-efficient polynomial chaos expansion method to investigate the uncertainty in predictions of melt-track dimension with an analytical Eager-Tsai model and a FE-based thermal model (see Section 4.3 for more information on thermal models) has been conducted more recently (Tapia et al. 2018). Although the UQ activities for metal AM are more recent, the classical design of experiments approaches (including analyses of variance [ANOVA] and Taguchi methods) have been used since the early days of metal AM (see Chapter 3).

4.6.2 Digital and Physical Design of Experiments

A common strategy to improve a product's quality involves evaluating the robustness of the product's design against the loading conditions, that is, how well the product performs outside the recommended operating conditions. Similar procedures are applied in the field of manufacturing when evaluating the influence of process parameters on the desired properties – a procedure better known as design of experiments (DOE) (see Chapter 3 for more on this topic). Analogous to the physical world, the techniques of DOE can also be applied to numerical models (for example, fractional factorial DOE to identify the critical process parameters and material properties that significantly influence peak temperature in a FEM-based thermal model – Ma et al. 2015), where the different simulations would correspond to virtual experiments. The DOE can further be used to evaluate the robustness of the model's predictive capabilities by determining the effect of small changes in model inputs and model numerical parameters on the predicted model outputs (Figure 4.14). The former application of DOE (for small variations in model inputs) results in quantification of the uncertainty associated with the process, while the latter application (for small variations in model numerical parameters) results in quantification of the uncertainty associated with the model.

4.6.3 GIGO Approach to Model Calibration

Model calibration is an important step while setting up process simulations, yet it does not necessarily receive much attention. Note that the term *calibration* used in this context is not the same as that defined in Chapter 10, which explicitly requires measurement uncertainty to be estimated at each step. Here, calibration is the process of establishing the values for the parameters used and the relationships between variables in a model. In effect, the model calibration technique is the linking step between model verification (where it is

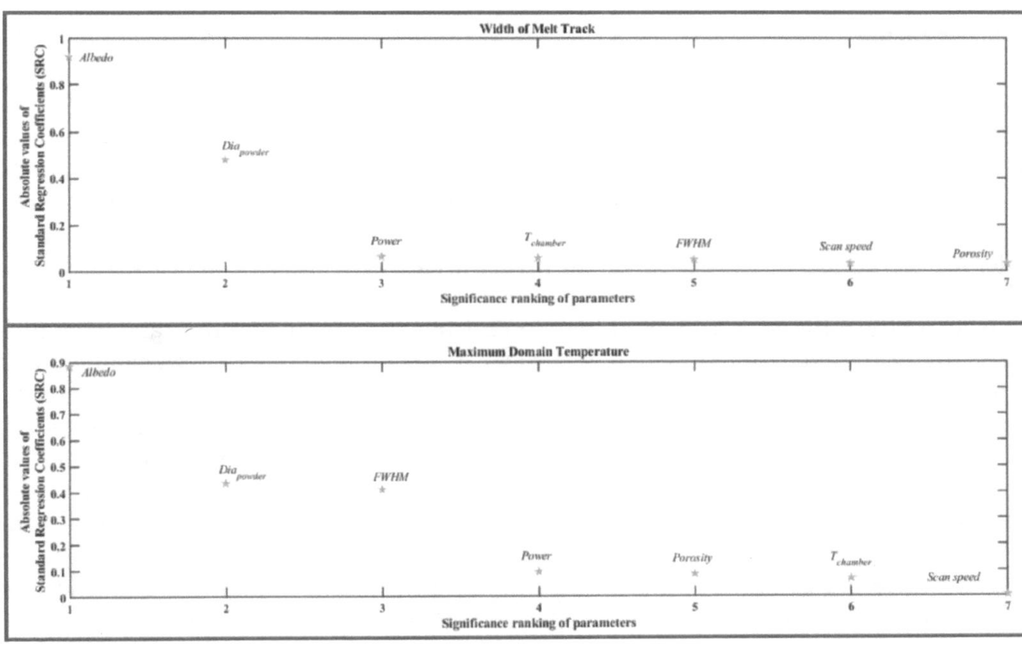

FIGURE 4.14
Relative parameter significance ranking for selected parameter with respect to width of melt track and maximum temperature during L-PBF. Dia_{powder} = mean diameter of powder particles, $T_{chamber}$ = average ambient temperature in the chamber, $FWHM$ = full width at half maximum of laser beam. Note: y-axis shows the absolute value of the standard regression coefficients (SRC) for a linear regression model extracted using Monte Carlo simulations based uncertainty quantification. (Based on Mohanty and Hattel 2016.)

ensured that the models are solved accurately) and model validation (where it is ensured that the correct models are solved to match experimental predictions). For multiphysics multi-scale models of metal AM, several non-measurable physical parameters that need to be estimated (for example, effective absorptivity, temperature-dependence of surface tension, high-temperature thermal expansion coefficient or phase/grain nucleation rate) exist along with several numerical parameters that should be chosen based on expert knowledge (for example, relaxation factors for iterations, accommodation factors for mushy zone models and exponents for hardening power laws). While rigorous experimentation can help determine the physical parameters, the selection of numerical parameters remains a challenge. Consequently, despite using the 'correct' values for all measurable and non-measurable parameters, as well as applying the appropriate models, the predictions of the simulation may not match the experimental results. This phenomenon (caused in this case by erroneous numerical parameter selection), has been referred to as 'garbage in, garbage out' (GIGO) in computer science terminology.

Due to the large number of unknown physical and numerical parameters and the resource limitations of experimental measurements, most research studies perform rudimentary calibration (for example, simple curve fitting) and consequently do not highlight the adopted model calibration approach. One of the approaches to resolving the associated challenge would be to perform UQ for the physical and numerical model parameters. The techniques for these sensitivity and uncertainty studies (Saisana et al. 2005, Saltelli et al. 2006) have been recently applied to metal AM (Tapia et al. 2018, Mohanty and Hattel 2016).

4.7 Concluding Remarks and Future Outlook

Simulation-based process qualification has become an active area of research, with metal AM being a prime candidate for application of such techniques. The increasing availability of computational resources and their incorporation in industrial/research environments further promotes the potential for tailoring properties and processes for metal AM using modelling techniques. Thus, several areas for continued development can easily be identified for modelling of metal AM.

The current techniques for model reduction for macro-scale modelling can be complemented by application of coupled global–local models, where thermo-mechanical and microstructural responses are calculated at different length and time scales within the same simulation. More optimally designed experimental results would then be necessary in the near future to better capture process parameter uncertainties and correlations for macro-, meso- and micro-scale models, eventually leading to better predictability of corresponding outputs. Such studies would pre-requisite application of appropriate sensitivity and identifiability analysis techniques that can indicate the adequacy of experiments towards calibrating the model parameters. Having such calibrated and well-characterised models of the metal AM process would open the possibilities for optimising the process/process-chain for newer materials and applications at significantly accelerated rates.

This chapter has highlighted the various models of interest for the metal AM process and discussed their applicability towards simulating known defects and behaviour of metal AM components. The high-fidelity models discussed in this chapter are relevant when critical components are to be produced in known materials as well as when specific defects and/or process windows need to be evaluated for new metallic alloys. On the other hand, the implementation of the highest fidelity models for daily continuous evaluation/control of the metal AM process is not feasible due to the long computation time at high computational resource requirements. Instead, simplified modelling approaches need to be adopted and calibrated according to the specific metal AM machine and process – the existing approach in industries venturing into metal AM for production of their specific products. The faster models, when supported with monitoring systems, can enable effective feed-forward control of the metal AM process. More specifically, these faster models can be used for quality control of the metal AM process within four major areas: (a) process window identification, (b) hotspot/porosity indicator, (c) local phase fraction/morphology prediction and (d) warpage/deformation estimation. The relevance of developing both approaches (highest fidelity modelling and faster modelling) remains high nonetheless due to the rapid progress in the field of industrial metal AM, which results in newer machines every year for the same core metal AM process. Leveraging the progress in either of the development pathways is the way to ensure that the various simulation possibilities are exploited with the highest efficiency for enabling precision in metal AM.

References

Anderson, A. 2015. Development of physics-based numerical models for uncertainty quantification of selective laser melting processes-*2015 Annual Progress Report*. Lawrence Livermore National Laboratory (LLNL), Livermore, CA, report number: LLNL-TR-678006.

ASME-V&V-10. 2006. An overview of the PTC 60/ASME V&V 10: Guide for verification and validation in computational solid mechanics. American Society of Mechanical Engineers.

ASME-V&V-20. 2009. An overview of ASME V&V 20: Standard for verification and validation in computational fluid dynamics and heat transfer. American Society of Mechanical Engineers.

Assouroko, I., Lopez, F., Witherell, P. 2016. A method for characterizing model fidelity in laser powder bed fusion additive manufacturing. *Proc. ASME 2016 IMechE Congress*, Phoenix, AZ. 1–13.

Bayat, M., Mohanty, S., Hattel, J. H. 2019a. Multiphysics modelling of lack of fusion voids formation and evolution in IN718 made by multi-track/multi-layer L-PBF. *Int. J. Heat Mass Trans.* 139:95–114.

Bayat, M., Mohanty, S., Hattel, J. H. 2019b. A systematic investigation of the effects of process parameters on heat and fluid flow and metallurgical conditions during laser-based powder bed fusion of Ti6Al4V alloy. *Int. J. Heat Mass Trans.* 139:213–230.

Bayat, M., Thanki, A., Mohanty, S., Witvrouw, A., Yang, S., Thorborg, J., Skat Tiedje, N., Hattel, J. H. 2019c. Keyhole-induced porosities in laser-based powder bed fusion (L-PBF) of Ti6Al4V: High-fidelity modelling and experimental validation. *Addit. Manuf.* 30:100835.

Bayat, M., Klingaa, C. G., Mohanty, S., De Baere, D., Thorborg, J., Tiedje, N. S., Hattel, J. H. 2019d. Part-scale thermo-mechanical modelling of distortions in L-PBF – Analysis of the sequential flash heating method with experimental validation. *Addit. Manuf.*

Boley, C. D., Khairallah, S. A., Rubenchik, A. M. 2015. Calculation of laser absorption by metal powders in additive manufacturing. *Appl. Opt.* 54(9):2477–2482.

De Baere, D., Mohanty, S., Hattel, J. H. 2020. Microstructural modelling of above β-transus heat treatment of additively manufactured Ti-6Al-4V using cellular automata. *Mater. Today Comm.* 24: 101031.

Delgado, J., Ciurana, J., Rodríguez, C. A. 2012. Influence of process parameters on part quality and mechanical properties for DMLS and SLM with iron-based materials. *Int. J. Adv. Manuf. Technol.* 60:601–610.

Denlinger, E. R. 2018. Residual stress and distortion modeling of electron beam direct manufacturing Ti-6Al-4 V. In: Gouge, M., Michaleris, P. (eds.), *Thermo-Mechanical Modelling of Additive Manufacturing*, Oxford, UK: Butterworth-Heinemann.

Dou, X., Mao, Y., Zhang, Y. 2014. Effects of contact force model and size distribution on microsized granular packing. *J. Manuf. Sci. Eng.* 136:021003.

Du, K., Biesenbach, J., Ehrlichmann, D., Habich, U., Jarosch, U., Klein, J., Loosen, P., Niehoff, J., Wester, R. 1995. Lasers for materials processing: Specifications and trends. *Opt. Quant. Elect.* 27:1089–1102.

Easterling, K. 1992. *Introduction to the Physical Metallurgy of Welding*, Oxford, UK: Butterworth-Heinemann.

Escano, L. I., Parab, N. D., Xiong, L., Guo, Q., Zhao, C., Fezzaa, K., Everhart, W., Sun, T., Chen, L. 2018. Revealing particle-scale powder spreading dynamics in powder-bed-based additive manufacturing process by high-speed X-ray imaging. *Sci Rep.* 8:15079.

Francois, M. M., Sun, A., King, W. E., Henson, N. J., Tourret, D., Bronkhorst, C. A., Carlson, N. N., Newman, C. K., Haut, T., Bakosi, J., Gibbs, J. W., Livescu, V., Vander Wiel, S. A., Clakre, A. J., Schraad, M. W., Blacker, T., Lim, H., Rodgers, T., Owen, S., Abdeljawad, F., Madison, J., Anderson, A. T., Fattebert, J. L., Ferencz, R. M., Hodge, N. E., Khairallah, S. A., Walton, O. 2017. Modeling of additive manufacturing processes for metals: Challenges and opportunities. *Curr. Op. Sol. State Mater. Sci.* 21:198–206.

Gusarov, A. V., Kruth, J. P. 2005. Modelling of radiation transfer in metallic powders at laser treatment. *Int. J. Heat Mass Transf.* 48:3423–3434.

Gusarov, A., Kovalev, E. 2009. Model of thermal conductivity in powder beds. *Phys. Rev. B.* 80:024202.

Herbold, E. B., Walton, O., Homel, M. A. 2015. Simulation of powder layer deposition in additive manufacturing processes using the discrete element method. Report No. LLNL-TR-678550.

Hodge, N. E., Ferencz, R. M., Vignes, R. M. 2016. Experimental comparison of residual stresses for a thermomechanical model for the simulation of selective laser melting. *Addit. Manuf.* 12:159–168.

Jabbari, M., Baran, I., Mohanty, S., Comminal, R., Sonne, M. R., Nielsen, M. W., Spangenberg, J., Hattel, J. H. 2018. Multiphysics modelling of manufacturing processes: A review. *Adv. Mech. Eng.* 10:1–31.

Jia, T., Zhang, Y., Chen, J. K. 2011. Dynamic simulation of particle packing with different size distributions. *J. Manuf. Sci. Eng.* 133:021011.

Kamath, C. 2016. Data mining and statistical inference in selective laser melting. *Int. J. Adv. Manuf. Technol.* 86:1659–1677.

King, W., Anderson, A. T., Ferencz, R. M., Hodge, N. E., Kamath, C., Khairallah, S. A. 2015a. Overview of modelling and simulation of metal powder bed fusion process at Lawrence Livermore National Laboratory. *Mater. Sci. Technol.* 31:957–968.

King, W. E., Anderson, A. T., Ferencz, R. M., Hodge, N. E., Kamath, C., Khairallah, S. A., Rubenchik, A. M. 2015b. Laser powder bed fusion additive manufacturing of metals; Physics, computational, and materials challenges. *Appl. Phys. Rev.* 2:041304.

Kushan, M. C., Poyraz, O., Uzunonat, Y., Orak, S. 2018. Systematical review of the numerical simulations of laser powder bed additive manufacturing. *Sigma J. Eng. Nat Sci.* 36:1193–1210.

Lavery, N. P., Brown, G. R. S., Sienz, J., Cherry, J., Belblidia, F. 2014. A review of computational modelling of additive layer manufacturing – multiscale and multiphysics. *Sustainable Des. Manuf.* Sdm14-091:651–673.

Lindgren, L. E., Lundbäck, A., Fisk, M., Pederson, R., Andersson, J. 2016. Simulation of additive manufacturing using coupled constitutive and microstructure models. *Addit Manuf.* 12-B:144–158.

Liu, J., Jalalahmadi, B., Guo, Y. B., Sealy, M. P., Bolander, N. 2018. A review of computational modeling in powder-based additive manufacturing for metallic part qualification. *Rapid Prototyping J.* 24:1245–1264.

Lu, L. X., Sridhar, N., Zhang, Y. W. 2018 Phase field simulation of powder bed based additive manufacturing. *Acta Mater.* 144:801–809.

Lumeau, J., Glebova, L., Glebov, L. B. 2011. Near-IR absorption in high-purity photothermorefractive glass and holographic optical elements: Measurement and application for high-energy lasers. *Appl. Opt.* 50:5905–5911.

Ma, L., Fong, J., Lane, B., Moylan, S., Filliben, J., Heckert, A., Levine, L. 2015. Using design of experiments in finite element modeling to identify critical variables for laser powder bed fusion. *Proc. Solid Free. Fabr. Symp 2015*, Austin, TX, 219–228.

Markl, M., Körner, C. 2016. Multiscale modeling of powder bed–based additive manufacturing. *Annu. Rev. Mater. Res.* 46:93–123.

Marla, D., Zhang, Y., Hattel, J.H., Spangenberg, J. 2018. Modeling of nanosecond pulsed laser processing of polymers in air and water. *Modell. Simul. Mater. Sci. Eng.* 26(5).

Maxey, M. 2017. Simulation methods for particulate flows and concentrated suspensions. *Annu. Rev. Fluid Mech.* 49:171–193.

Mercelis, P., Kruth, J. 2006. Residual stresses in selective laser sintering and selective laser melting. *Rapid Prototyp. J.* 12:254–265.

Mills, K.C. 2002. *Recommended Values of Thermophysical Properties for Selected Commercial Alloys*, Woodhead Publishing Ltd., Abington, Cambridge, 211–217.

Mindt, H. W., Megahed, M., Lavery, N. P., Holmes, M. A., Brown, S. G. R. 2016. Powder bed layer characteristics: The overseen first-order process input. *Metall. Mater. Trans. A Phys. Metall. Mater. Sci.* 47:3811–3822.

Moges, T., Ameta, G., Witherell, P. 2019. A review of model inaccuracy and parameter uncertainty in laser powder bed fusion models and simulations. *J. Manuf. Sci. Eng.* 141:040801-1.

Mohanty, S., Hattel, J. H. 2014. Numerical model based reliability estimation of selective laser melting process. *Physics Procedia*, 56:379–389.

Mohanty S., Hattel J. H. 2016. Improving accuracy of overhanging structures for selective laser melting through reliability characterization of single track formation on thick powder beds. *Proc. SPIE* 9738:97381B.

Molina, J. A. L, Maria, J. R., Berjano, E. 2014. Fourier, hyperbolic and relativistic heat transfer equations: A comparative analytical study. *Proc. R. Soc. A.* 470:20140547.

Mukherjee, T., Wei, H. L., De, A., DebRoy, T. 2018. Heat and fluid flow in additive manufacturing – Part I: Modeling of powder bed fusion. *Comput. Mater. Sci.*150:304–313.

Niyama, E., Uchida, T., Morikawa, M., Saito, S. 1980. Predicting shrinkage in large steel castings. *J. Japan Found. Soc.* 53:635–640.

Oldfield, W. 1966. A quantitative approach to casting solidification: Freezing of cast iron. *ASM Trans.* 59:945–961.

Pagani, L., Qi, Q., Jiang, X., Scott, P. J. 2017. Towards a new definition of areal surface texture parameters on freeform surface. *Measurement*. 109:281–291.

Parteli, E. J. R. 2013. DEM simulation of particles of complex shapes using the multisphere method: Application for additive manufacturing. *AIP Conf. Proc.*, 1542:185–188.

Parteli, E. J. R., Pöschel, T. 2016. Particle-based simulation of powder application in additive manufacturing. *Powder Technol.* 288:96–102.

Rappaz M., Gandin C. A. 1993. Probabilistic modelling of microstructure formation in solidification processes. *Acta Metall. Mater.* 41:345–60.

Rosenthal, D. 1946. The theory of moving sources of heat and its application to metal treatments. *ASME*. 43:849–866.

Saisana, M., Saltelli, A., Tarantola, S. 2005. Uncertainty analysis and sensitivity analysis techniques as tools for the quality assessment of composite indicators. *J. R. Stat. Soc. -A.* 168:307–323.

Saltelli, A., Ratto, M., Tarantola, S., Camppolongo, F. 2006 Sensitivity analysis practices: Strategies for model based inference. *Reliab. Eng. Syst. Saf.* 91:1109–1125.

Seitien, I., Chiumenti, M., van der Veen, S., Sebastian, M. S., Garciandia, F., Echeverria, A. 2019. Empirical methododlogy to determine inherent strains in additive manufacturing. *Comput. Math. Appl.* 78:2282–2295.

Schoinochoritis, B., Chantzis, D., Salonitis, K. 2014. Simulation of metallic powder bed additive manufacturing processes with the finite element method: A critical review. *Proc. Inst. Mech. Eng. Part B J. Eng. Manuf.* 231:96–117.

Shapiro, M., Dudko, V., Royzen, V., Krichevets, Y., Lekhtmaker, S., Grozubinsky, S., Brill, M. 2004. Characterization of powder beds by thermal conductivity: Effect of gas pressure on the thermal resistance of particle contact points. *Part. Syst. Charact.* 21:268–275.

Sih, S., Barlow, J. 1995a. The prediction of the thermal conductivity of powders. *Proc. Solid Free. Fabr. Symp.*, Austin, TX, 1995.

Sih, S., Barlow, J. 1995b. Emissivity of powder beds. *Proc. Solid Free. Fabr. Symp.*, Austin, TX, 1995.

Sih, S., Barlow, J. 2010. The prediction of the emissivity and thermal conductivity of powder beds. *Part. Sci Tech.* 22:427–440.

Smith, J., Xiong, W., Yan, W., Lin, S., Cheng, P., Kafka, O. L., Wagner, G. J., Cao, J., Liu, W. K. 2016. Linking process, structure, property, and performance for metal-based additive manufacturing: Computational approaches with experimental support. *Comput. Mech.* 57:583–610.

Stavropoulos, P., Foteinopoulos, P. 2018. Modelling of additive manufacturing processes: A review and classification. *Manuf. Rev.* 5(2)

Tapia, G., Khairallah, S., Matthews, M., Kng, W.K., Elwany, A. 2018. Gaussian process-based surrogate modeling framework for process planning in laser powder-bed fusion additive manufacturing of 316L stainless steel. *Int. J. Adv. Manuf. Technol.* 94: 3591–3603.

Tavakoli, R. 2014. On the prediction of shrinkage defects by thermal criterion functions. *Int. J. Adv. Manuf. Tech.* 74:569–579.

Verschaffelt, G., Panajotov, K., Albert, J., Nagler, B., Peeters, M., Danckaert, J., Veretennicoff, I., Thienpont, H. 2001. Polarisation switching in vertical-cavity surface-emitting lasers: From experimental observations to applications. *Opto-Electron. Rev.* 9:257–268.

Vock, S., Kloden, B., Kirchner, A., Weissgarber, T., Kieback, B. 2019. Powders for powder bed fusion: A review. *Prog. Addit. Manuf.* 1–15.

Xia, M., Gu, D., Yu, G., Dai, D., Chen, H., Shi, Q. 2017. Porosity evolution and its thermodynamic mechanism of randomly packed powder-bed during selective laser melting of Inconel 718 alloy. *Int. J. Mach. Tools. Manuf.* 116:96–106.

Xiang, Z., Yin, M., Deng, Z., Mei, X., Yin, G. 2016. Simulation of forming process of powder bed for additive manufacturing. *J. Manuf. Sci. Eng.* 138:081002.

Zanini, F., Pagani, L., Savio, E., Carmignato, S. 2019a. Characterisation of additively manufactured metal surfaces by means of X-ray computed tomography and generalised surface texture parameters. *CIRP Annal.* 68(1).

Zanini, F., Sbettega, E., Sorgato, M., Carmignato, S. 2019b. New approach for verifying the accuracy of X-ray computed tomography measurements of surface topographies in additively manufactured metal parts. *J. Nondestr. Eval.* 38(1):12.

Zhou, B., Kane, T. J., Dixon, G. J., Byer, R. L. 1985. Efficient, frequency-stable laser-diode-pumped Nd:YAG laser. *Opt. Lett.* 10:62–64.

Zhou, X., Zhang, H., Wang, G., Bai, X. 2016. Three-dimensional numerical simulation of arc and metal transport in arc welding based additive manufacturing. *Int. J. Heat Mass Trans.* 103:521–537.

Zinoviev, A., Zinovieva, O., Ploshikhin, V., Romanova, V., Balokhonov, R. 2016. Evolution of grain structure during laser additive manufacturing: Simulation by a cellular automata method. *Mater. Des.* 106:321–329.

Zinovieva, O., Zinoviev, A., Ploshikhin, V. 2018. Three-dimensional modeling of the microstructure evolution during metal additive manufacturing. *Comput. Mater. Sci.*141:207–220.

5

Secondary Finishing Operations

Bethan Smith, David Gilbert and Lewis Newton

CONTENTS

5.1 Introduction

This chapter focuses on the secondary finishing operations (as defined in Section 5.2) applicable for metal additive manufactured (AM) products. Due to the benefits of AM parts in high-value manufacturing sectors, such as aerospace and medical (for example, mass saving and part complexity), there is an increasing need for AM parts to be suited for applications such as fluid flow, tribology, fatigue loading and aesthetics. In order to meet the needs of these applications, AM parts must, in many cases, have their as-built surface texture reduced via post-build operations in order to, for example, reduce crack propagation, improve flow properties, allow better part tolerance control and improve assembly surface contact (outlined in more detail in Section 5.3). The challenges that metal AM brings to these secondary finishing operations, as compared to traditional manufacturing, can be grouped into those caused by the specific characteristics of the metal AM part, for example, high as-built surface texture, extreme geometric complexity (including no line-of-sight surfaces) and varying surface texture characteristics across different part surfaces. There are also challenges caused by the maturity of the technology, for example, part variation across different machines and cases of part distortion after build, which the finishing process must address. In addition, there is currently limited available guidance for the selection of secondary finishing operations appropriate for AM parts, with a gap in the published standards available for both selection and application of finishing operations specific to AM; this is discussed in Section 5.4.

The secondary operations that can be selected for AM parts are wide-ranging, numerous and have been used for traditional manufactured parts for a number of years (see Section 5.6). However, the selection of secondary finishing operations for AM parts from the processes available is dictated, in most cases, by the complexity of access to the surface to be finished (Section 5.7). As such, the most commonly used techniques for AM parts are those with no fixed tool, such as mass finishing (where loose media is energised within the part), chemical techniques and electropolishing-based processes (where in both cases, a fluid is used to access the part). Other considerations, however, can eliminate these fluid-based techniques from selection, as the final functional and surface properties required and the part tolerances may not be achievable without a fixed tool; this is addressed in Section 5.5. As AM parts are used more widely, new secondary finishing techniques have been developed specifically for them. The combination of robotic and new sensing developments has had an impact on improving control for a number of techniques as well as the introduction of software which can support process chain selection at the design stage; this is discussed in Section 5.10.

5.2 Basic Definition of Secondary Finishing

5.2.1 What Is Considered to Be Secondary Finishing in This Chapter?

Within this chapter, the following activities will be considered as secondary finishing techniques in an AM context:

- The act of removing, re-shaping, or re-distributing surface material to add to, improve upon, or achieve a certain property of the component at the surface. Specifically, in the context of AM, secondary finishing is concerned with processing of surfaces that result in a reduction in certain measures of a surface, such as surface texture or form parameters.
- Processes by which changing the number, direction, and/or magnitude of significant high-spatial frequency surface texture components results in an increase in one or more surface functionalities, often identified by a change in a statistical characterisation of the surface.

5.2.2 Not Included in the Scope of This Chapter

Processes where the improvement of surface finish by modification of surface texture is not the predominant mechanism for improving component performance (for example, peening), despite these processes having side effects that may improve surface finish, are not considered in this chapter.

Since a key application for AM is light-weighting and topology optimisation (see Chapter 2) of load bearing structures, particularly in the aerospace sector (Wohlers Associates 2018), it is often the surface texture of the AM part and properties of the build material itself that significantly impact part performance. As such, processes whereby material is added to the component to achieve a surface texture (for example, coating (Sahoo et al. 2017]) are not considered secondary finishing but rather further downstream processes in the AM process chain.

5.3 Why Do AM Surfaces Need to Be Finished?

AM shows promise as an enabling technology for a step-change in component design for a number of high-value and/or low-volume manufacturing industries. Applications ranging from topology optimisation for fatigue performance and light-weighting in aerospace to bespoke one-off components for patient implants and surgery guides in the medical industry are common candidates for the use of AM. However, these applications have a number of functional requirements that components must meet, which may be affected by surface topographies. Note that surface topography/texture measurement and characterisation are discussed in detail in Chapter 11.

5.3.1 Impact of Surface Topography on Function

5.3.1.1 Fatigue Applications

Products manufactured by aerospace, power generation and automotive industries frequently employ components that undergo both low- and high-cycle fatigue loading. Concentration of cyclic stresses on sharp micro-scale features such as cracks and indentations can, over time, enlarge the pre-existing crack or indentation, which may propagate through the component, leading to mechanical failure.

There are many factors that affect the resistance to fatigue failure of components, including material properties, loading stresses and environmental conditions. Different loading conditions will affect how the crack propagates, such as out-of-plane shearing (mode I), crack 'opening' (mode II) and in-plane shearing (mode III), shown in Figure 5.1.

In addition to these factors, however, the geometry of the crack or indentation itself strongly affects how a component may perform under cyclic stress loading. Both the size and shape of the crack directly influence the extent to which stresses are locally concentrated. In general, larger cracks with sharper tips result in higher stress concentrations, leading to poorer performance in fatigue applications. As such, surfaces with a large number of pits (negative skew) are undesirable for fatigue performance. Finishing processes that only remove surface peaks and leave surface valleys largely untouched may not be capable of significantly improving the fatigue performance of components. Instead, processes for material removal or redistribution should be employed in which the size, sharpness and number of surface valleys is reduced.

Secondary finishing operations where mechanical cutting or compression is involved can impart compressive residual stresses (Jomaa et al. 2014), which can improve the mechanical performance of parts.

- *Fluid flow applications* – A wide range of industries, including power-generation, aerospace, automotive and medical sectors, manufacture components which have features and geometries designed for the direction and delivery of fluids. In these applications, where fluid flow must be controlled – such as gas flow over a compressor component – surface topography can significantly affect flow properties (Gloss and Herwig 2010). In both liquid and gas fluid flows, breakdown of laminar flows into turbulence results in loss of control over the direction of flow, and flow instabilities may cause significant performance discontinuities (such as turbulence around an aircraft wing during flight).

  As with fatigue applications, a number of factors affect how a fluid flows over a solid surface, such as fluid viscosity, temperature and velocity. Further to these factors, surface topography may have a significant effect on how a fluid flows at the boundary layer between the fluid body and component surface. Rougher surfaces

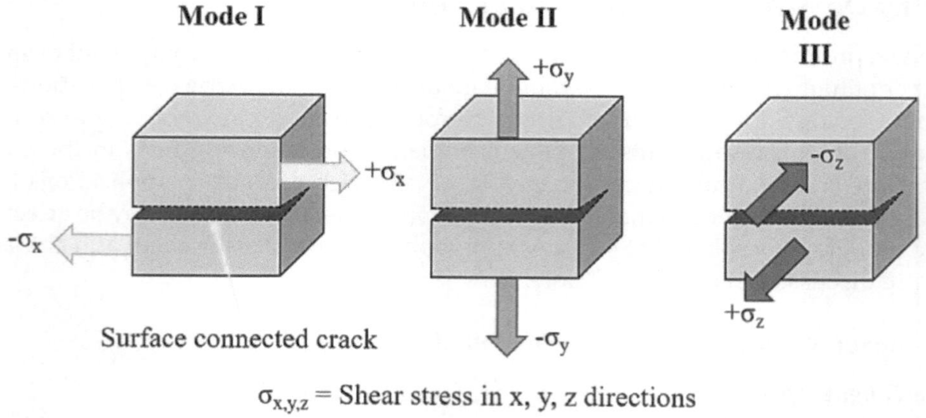

FIGURE 5.1
Mode I, II and III crack loading due to stresses σ in x, y and z directions.

with a significant number of surface peaks can cause microturbulences, such as shown in Figure 5.2, resulting in friction between the fluid and the component.

- *Bearing surface applications* – Many industries manufacture components that may come into sliding contact with other components or surfaces. The surfaces which bear against other components or surfaces are typically manufactured with specific surface texture parameter targets, such as *Ra* or *Rp* (see Chapter 11). Surface asperities will be the first point of contact between two bearing surfaces, and loading will be concentrated on these points of contact. This may result in scouring and scratching on the softer of the two surfaces in the case of linear sliding, or vibration and significant run-out in rotating applications. Surfaces with fewer peaks, even if there are a significant number of pits and valleys (shown in Figure 5.3), may perform better in bearing applications.
- *Aesthetic applications* – Surface finishing is often used to alter the aesthetic appearance of parts, where the surface finish from the manufacturing process is not

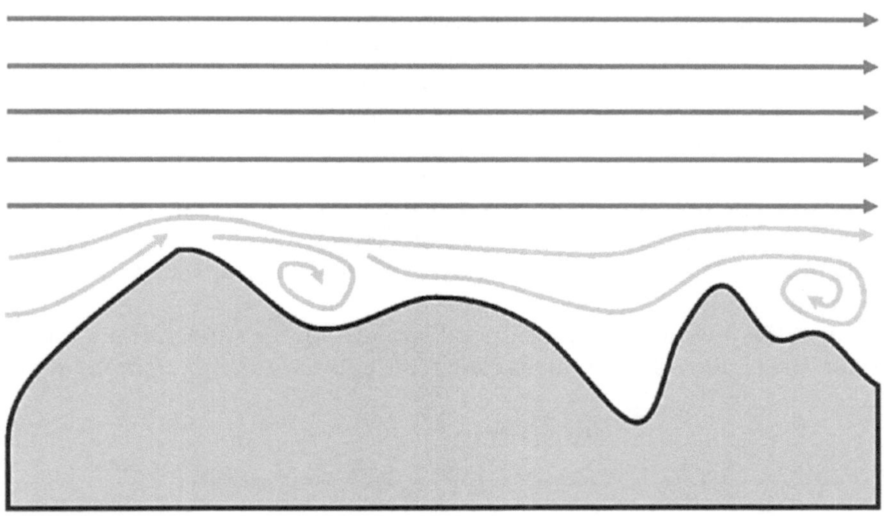

FIGURE 5.2
Surface peaks causing micro-turbulence in fluid flows.

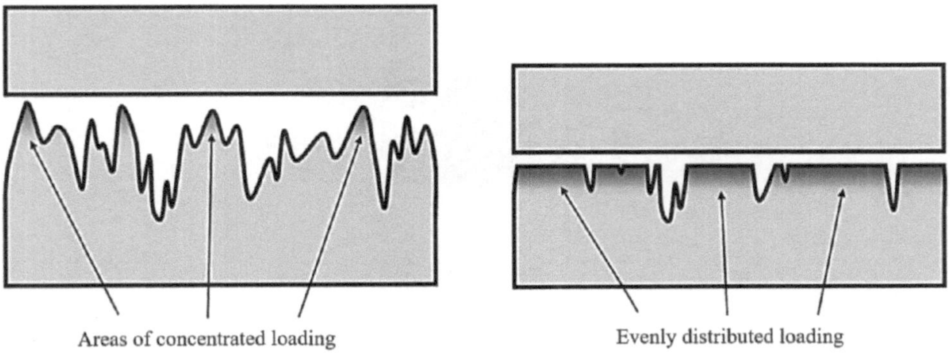

FIGUR E 5.3
Surface peak loading in bearing applications.

suitable for the end-use application. Whilst surface finishing for aesthetic purposes does not necessarily improve part function, it can add value to the product for consumers or customers (Ashby and Johnson 2003, Fleming 2014).

- *Corrosion resistance applications* – Rates of material corrosion can be accelerated where rough surfaces textures lead to higher interfacial surface areas between the component material and the corrosive environment. Biomedical applications can present these challenges to AM components (Bertolini et al. 2017).

5.3.2 Examples of AM Surfaces

The additive as-built surfaces described in this section often exhibit features that are significantly different to the topographies required for the applications described in Section 5.4.1. The focus here is primarily on powder bed AM parts as the most commonly used AM process in precision applications today (Wohlers Associates 2018); where features are significantly different from powder bed AM, this is highlighted and commented upon.

Powder bed fusion (PBF) metal parts, and typically other metal AM parts, tend to have higher levels of surface texture when compared to more traditionally manufactured parts, such as those from casting, forging or machining (Gibson et al. 2015). Metal PBF parts are created either through electron beam powder bed fusion (EB-PBF) or laser powder bed fusion (L-PBF), where the source is either an electron beam or a laser source (typically infrared, but wavelength may vary with build material) respectively (Kruth et al. 2005, Gibson et al. 2015, Körner 2016). Each manufacturing process, its principles and its respective process parameter settings have an effect on the properties of the surface of the final part, often investigated through characterisation using surface texture parameters (Townsend et al. 2016, Senin et al. 2018, Leach et al. 2019), as illustrated in Figure 5.4 and discussed in Chapter 11.

For PBF parts, the presence of semi-sintered particles on the surface, the size of the layer thickness and the pattern of weld tracks contribute to the surface topography significantly

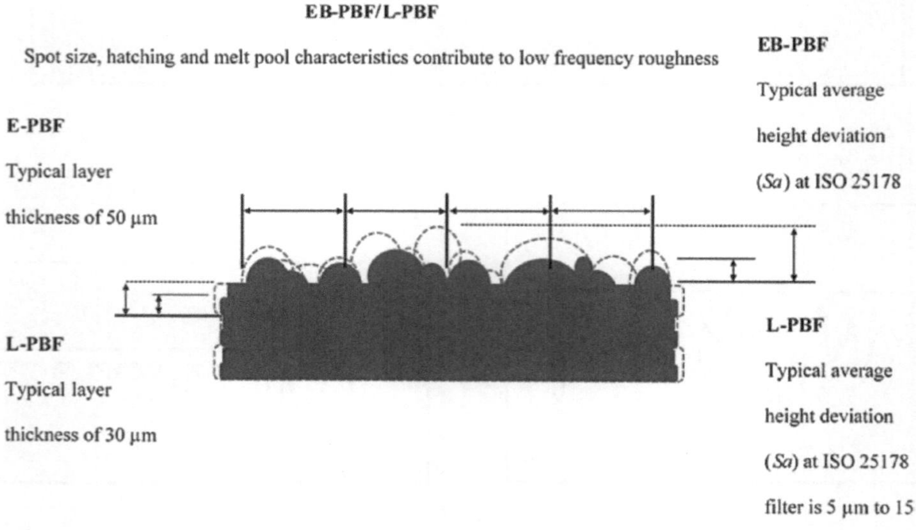

FIGURE 5.4
Example of PBF part top surface made up of semi-sintered particles. L-PBF and EB-PBF surface texture values and layer thickness are compared.

(Calignano et al. 2013, Triantaphyllou et al. 2015, Sidambe 2017, Senin et al. 2018). As manufacturers move towards faster methods, particle sizes and layer thicknesses have increased, with some recent EB-PBF machines now working at 90 μm layer thicknesses (GE Additive 2018).

Partial sintering of build powder can lead to much rougher surfaces, especially on the 90° (parallel to build direction) and sloped surfaces due to the 'stair-stepping' effect (Strano et al. 2013, Boschetto et al. 2017). Even with smaller powder particles and layers, stair stepping on non-horizontal surfaces still give rise to rough surfaces on PBF parts that need to be finished (Figure 5.5).

Supporting material is also a key feature of PBF AM parts, with L-PBF parts tending to have solid anchoring supports to the build platform. EB-PBF parts, on the contrary, have fewer supports, not necessarily anchored to the build platform but instead to the neighbouring semi-sintered powder cake. These supports tend to be removed manually which can leave witness marks of the support removal procedure (Gibson et al. 2015). Alongside these witness marks, the EB-PBF process leaves more semi-sintered powder around internal geometries, which can be difficult to remove.

The support structures in both cases must be removed, most commonly using band sawing, wire-electro-discharge machining (EDM) or through manual means. These removal processes often leave significant material remaining which the final finishing operation must consider, or alternatively they will leave deterministic marks (i.e. machining marks) on the surface, which are significantly different to the AM as-built surface present on other areas of the part. In non-PBF AM techniques, such as direct energy deposition or the more recently developed metal binder jetting, the built components are self-supporting by virtue of the build process operating principles. As such, components produced by these processes do not require support removal, resulting in surfaces typically free of witness marks brought about by support removal processes (Gisario et al. 2019).

Non-supported, overhanging surfaces in both PBF techniques also demonstrate particularly rough characteristics (though not as rough as support removal witness marks) due to the angle of the build and the risk of material collapse during the build process (Morgan et al. 2016, Zwier and Wits 2016).

It can be seen that metal AM components have a number of features that may impact the final application for which the component can be used. In most cases, a final surface finishing operation is required before the AM part can be successfully implemented in the final functional application.

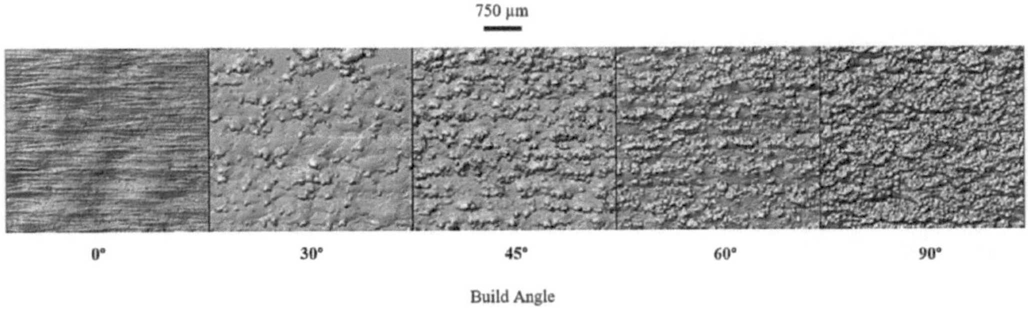

FIGURE 5.5
Example of the effect of build angle on surface texture in PBF parts. (Courtesy of FlexiFinish Project/Sandwell.)

5.4 Specification Standards in Secondary Finishing

This section provides a brief overview of the current specification standards landscape that will directly impact the application, selection and development of finishing processes for AM parts (see Chapter 6 for a wider overview of the current status with specification standards for AM). Gaps in the current standards are highlighted and the implications for the use of AM in precision applications are described.

In Chapter 6, Chapter 10 and Chapter 11, the specific measurement standards for AM are discussed and, therefore, although these standards are referenced here, they will not be covered in detail. Tables 5.1– 5.4 give an overview and summary of the relevant standards in the secondary finishing processes area. This section has been split into four topics: texture assessment, mechanical testing, design and quality and specific process applications. The list is not exhaustive but gives an overview of the most commonly referenced standards in the areas relevant to this chapter.

- *Texture assessment* – There has been significant work undertaken to date to update and modify standards to reflect current measurement technology, and these are discussed in more detail in Chapter 11. As such, the current standards shown in Table 5.1 give a good basis for assessing the impact of secondary finishing processes. Further work remains to develop methods to fully characterise the as-built AM surfaces (as outlined in Chapter 11), but, for the purposes of this chapter, the current standards are adequate for assessment of finishing techniques.
- *Mechanical testing* – Table 5.2 demonstrates that many of the standards in the testing area have been updated in order to give better clarity for the AM processes and for comparison of baseline properties. For example, ASTM F3122 (ASTM International 2014c) highlights the areas where emerging AM technology may struggle to conform to existing standards in the generation of thin and small diameter test pieces. In relation to surface finishing, however, there is no new guidance for how to meet some of the stringent requirements imposed by the standards for test specimens.

TABLE 5.1

Summary of Relevant Standards in the Texture Assessment Area

Texture Assessment	
The standards in this section focus on the assessment of texture following processing or build of the component. This is important to note for finishing processes, as in most cases, it is the surface texture specified that is used to assess the success of the secondary finishing operation.	
Relevant Standard	***Summary***
ASME B46.1 Surface Texture (Surface Roughness, Waviness and Lay) **ISO 4287:1997** Surface Texture: Profile method (ISO 4287 1997, ASME Standard 2002)	Surface texture guidance for traditional materials, defining roughness, waviness and lay and associated parameters. ISO 4287 in particular focusses on procedures for generating and defining profile parameters.
ISO 25178 Geometric Product Specification (GPS) – Surface Texture: Areal (ISO 25178 2010)	Covers in particular surface texture/areal surface texture measurements, with reference to areal parameters and applicable measurement technologies. This is particularly relevant to AM as many surfaces need these areal parameters to describe the surface due to the complexity (see Chapter 11). The ISO 25178 series also covers contact and optical measurement methods.

TABLE 5.2

Summary of Relevant Standards in the Mechanical Testing Area

Relevant Standard	Summary
Mechanical Testing Assessment *The standards in this section focus on how the components should be tested and reported for a range of mechanical properties. Embedded in a number of these standards is reference to the surface finish and tolerance required of these components, or the method of finishing that may need to be performed in order to meet the specification of the testing procedures.*	
ASTM F3122-14 Standard Guide for Evaluating Mechanical Properties of Metal Materials Made via Additive Manufacturing Processes (ASTM International 2014c)	This overarching standard references many common standards for mechanical testing of metals that have been published for many years. A few are highlighted below. For the majority of tests, the current standards are applicable. However, it is acknowledged in the case of rod-shaped specimens with small diameters and sheet metal samples that AM may not be able to produce these samples. This is relevant to tensile testing (ASTM E8, E21, E1450 and ISO 6892, ISO 15579 and ISO 19819), compression testing (ASTM E9 and E209), and tensile rupture and yield (ASTM E292, E740). For bearing testing (ASTM E238), it is noted that the surface finish requirements and thickness of specimens may be problematic for AM. The standard also references a large range of tests for fatigue (ASTM E466 and ISO 1099), hardness, fracture toughness and crack growth that are applicable to AM processes.
ASTM E466 Load Controlled Constant Amplitude Fatigue Tests of Metals **ISO 1099** Axial Force-Controlled Method of Fatigue Testing of Metallic Materials (ISO 1099:2006 2006, ASTM International 2015)	Both standards specify standard methods of fatigue testing metals at room temperature where strains are predominately elastic. Specimen geometries are specified and, in some cases, can be notched or un-notched depending on the standard or method. Due to the requirement on the surface finish, AM parts are often machined to meet this standard, as any minor surface cracking will influence the fatigue result for the bulk material.
ASTM E8/M Standard Test Methods for Tension Testing of Metallic Materials **ASTM E9-09 (2018)** Standard Test Methods of Compression Testing of Metallic Materials at Room Temperature (ASTM Standard E9-09 2012, ASTM International 2013)	This standard outlines the test specimen geometry and tolerances for tensile testing. Specific mention of powder metallurgy manufactured parts is made, which has most relevance to AM. The geometry for the test specimen is specifically outlined. It is noted that an un-machined sample will have a reduced ultimate tensile strength (UTS) compared to a machined sample. In general, specimen machining is highlighted as a common reason for variation in results, and the specimen should be free of chatter marks, rough surfaces, grooves, etc. ASTM E9-09 outlines the procedure and specimen geometry for compression testing. This requires a machined surface finish of 1.6 μm or better.

It is understood that machining is the preferred method to reduce the surface finish for tensile and fatigue testing, but there remains a gap in understanding about whether the test specimens should, or can, change in geometry or tolerance when testing different surface finishing methods. It is critical to understand this point for precision applications – as for many AM parts, machining is not possible due to the part complexity and, therefore, part mechanical properties should be tested in the state that most represents the final part. In this case, the current standards cannot be followed without deviation when testing alternative finishing operations for AM.

- *Design and quality* – There are a number of recent standards in the design area focussed on AM technology, as shown in Table 5.3. With reference to design considerations, there is clear guidance for considering the secondary finishing operation at the design stage, and the implication on cost, part accuracy and final part

TABLE 5.3

Summary of Relevant Standards in the Design Area

Design and Quality *The standards in this section are included so as to view the surface finishing processes as part of a larger process chain. Surface finish processing is referenced here both as a design consideration for AM, as well as part of a quality process.*	
Relevant Standard	***Summary***
ASTM F3318-18 Additive Manufacturing – Finished Part Properties Specification for AlSi10Mg with Powder Bed Fusion **ASTM F3055-14a** Additive Manufacturing Nickel Alloy with Powder Bed Fusion **ASTM F2924-14** Standard Specification for Additive Manufacturing Ti64Alv with Powder Bed Fusion (ASTM International 2014b, 2014a)	These standards outline the steps required to produce a good quality AlSi10Mg, Ti6Al4v and nickel alloy powder-bed part and the data behind this. It is stated that parts manufactured to this specification are often, but not necessarily, post-processed via machining, grinding, EDM, polishing, etc. to achieve desired surface finish and critical dimensions. It is stated that in ordering the parts, the surface finishing operations should be stated, as well as a manufacturing plan with process sequences and specifications. The final surface finish and operation should be agreed with the supplier and purchaser. Parts should be tested as per the mechanical and texture standards referenced in Table 5.1 and Table 5.2.
ASTM/ISO 52910: 2017 Standard Guidelines for Design for AM (ASTM International 2017)	This standard covers key areas for consideration when designing an AM part. In particular, surface finish is referenced as a limitation, specifically around support removal and high, variable surface texture. Removal of powder is considered for internal cavities, and reference is made that, depending on accuracy and finish requirements, the part may require finish machining, polishing, grinding, bead blasting or shot peening, as well as coatings such as painting or electroplating. Designing for the whole process chain is also a consideration as well as the cost of the post-processing steps – with attention on whether manual post-processing is needed or if it can be automated.

function. However, there is no guidance on the selection of the correct surface finishing method, and currently no standard that addresses this.

There is only brief guidance for ensuring supplier finish quality when purchasing AM parts, for example, in standard ASTM F3055 (ASTM International 2014b) for nickel and similarly in ASTM F2924-14 (ASTM International 2014a) for Ti6Al4V. These standards include agreement on the finishing operation to be used and the texture required on the final product. The most common finishing operations noted in the ASTM F3055 (ASTM International 2014b) and ASTM F2924 (ASTM International 2014a) standards for AM are machining, grinding, EDM and polishing. However, there is a far broader range of processes currently being used in the AM field, which are not covered by these standards.

- *Processes and applications* – There are limited standards available for specific surface finishing techniques, and none that are specific to the challenges provided by AM parts, as can be seen in Table 5.4. This means that users can specify the final finish and tolerance required to meet a certain standard (such as outlined in ASME B46.1 [ASME Standard 2002]). The user can also agree the method that the supplier uses to reach this finish. However, the reasoning on why this method has been chosen cannot be specified under any current standards for AM. This chapter will provide some insight into the current good practice for how secondary finishing operations should be selected, and the considerations relevant to precision AM products.

TABLE 5.4

Summary of Relevant Standards in the Finishing Process Area

Specific Process and Application

The standards in this section relate to specific processes and their process setup. Unlike many other manufacturing methods, AM almost always requires surface finishing or polishing following build. A variety of techniques can be used, and a wide variety is used due to the range in design complexity that this technology allows. This is therefore not a comprehensive list, as many standards are reliant on knowing the specific application.*

*Relevant Standard***	*Summary*
ASTM B912 Standard Specification for Passivation of Stainless Steels using Electro polishing (ASTM International 2010)	A commonly referenced standard for electro polishing of steels (Harrison Electropolishing 2012), covers processing and steps, such as that basis materials are to be free of visible defects and undergo preparatory cleaning. Evaluation of specimens during passivation is also covered.
SSPC-SP6 Joint Surface Preparation Standard: Commercial Blast Cleaning	This standard covers the use of blast cleaning abrasives for cleaning of steel prior to the application of coatings. Other standards in this series cover a range of cleaning requirements, with special reference to preparing the surface for coating or painting.
BS EN 10088/2 Stainless Steels – Technical delivery conditions for sheet/plate for general purposes **ASTM A480/A480M-18a** Standard Specification for General Requirements for Flat-Rolled Stainless and Heat-Resisting Steel Plate, Sheet and Strip **NiDi No.9012** Finishes for Stainless Steel (BS EN 10088-5:2009 2014)	All of these standards incorporate surface finishing for stainless steel. The standards define process routes and surface finish specifications, including 'special finishes' such as mechanical polishing. Definitions from these standards appear to have been rolled out to other metals as well as stainless steel. The most commonly used terms are #4 (general purpose with 120–150 mesh abrasive), #2B (a bright cold-rolled finish) and #8 (a highly polished highly reflective 'mirror like surface' generated through finer abrasives) (Williams Metalfinishing Inc. 2013). These terms have been used by a range of industries for specifying abrasives relevant to these finishes on a wide range of sheet metal.

* For machining and grinding, many of the standards are very application-specific and therefore not relevant within this context, and they have not been included. For barrel and vibratory finishing techniques, no finishing standards for process setup could be found.

** Commonly used AM finishing processes have been considered here, such as bead blasting, electro polishing, barrel and vibratory mass finishing, machining and grinding.

5.5 Challenges for Finishing Operations for AM Parts

5.5.1 Typical Operational Challenges for Metal AM Components Due to Surface Morphologies and Topographies

5.5.1.1 Challenges of Surface Topography

For applications, such as medical implants, and functions, such as fluid flow, fatigue resistance and more, there are a variety of features on the surface that need to be removed or altered to ensure suitable functionality (Section 5.3.1). Large asperities, such as particles, particle clusters and spatter across top, side and overhanging surfaces, may need to be removed in order to improve mating surfaces, reduce wear with other components and reduce risk of contamination or damage, especially when used in medical applications. The size and shape of these asperities can differ across different orientations of the surface with respect to the build (Grimm et al. 2015, Triantaphyllou et al. 2015, Sidambe 2017), which will have an influence on the surface texture parameters for the surface, surface function and type of finishing required.

Porosity on the surface, or sub-surface porosity that can emerge through removal of the upper surface, also needs to be considered as it is known to influence fatigue life and act as crack-propagation zones (Alrbaey et al. 2016, Bagehorn et al. 2017, Brandão et al. 2017). By altering the surface to relieve it of any porosity, by re-melting or by finishing mechanisms, the fatigue life can be improved (Bagehorn et al. 2017, Brandão et al. 2017).

5.5.1.2 Supporting Material and Witness Marks

For aesthetic, bearing surface and medical applications, a high level of surface finish and surface uniformity may be required. After supports have been removed, there may remain some remnants of the support structure. As a part of the surface topography, these remnants are seen as large plateau regions that sit higher than the bulk surface and as a result may require more aggressive surface finishing to remove. Support removal techniques may also leave process-specific features on the AM surface that do not match with the surrounding surface topography, and these may require removal to ensure that a consistent surface topography can be achieved.

5.5.1.3 Distortion

Distortion of the part may have a significant impact on assembly and functional operations. Although distortion does not directly impact or change the final part surface finish, it does impact the choice of the technique that can perform the material removal, as the part in some areas is different to the geometry expected and planned. If the technique selected cannot account for a change in shape caused by distortion (through either quick process re-planning or high process conformability, such as adaptive programming or flexible tooling), this can lead to high scrap or re-work rates.

5.5.2 Geometrical Challenges for Finishing Operations

AM processes allow unprecedented design freedom in the build phase of component manufacture (see Chapter 2). This freedom can lead to geometries and part shapes that it has previously not been possible to manufacture. In contrast, the large majority of traditional finishing techniques are not able to access these types of complex geometries. This can often lead to a situation where the finishing process (which is required to meet the final product requirements) limits the design freedom within the additive platform. For example, Figure 5.6 demonstrates geometries often seen in three common case-study geometries (as noted elsewhere Wohlers Associates 2018]), which each highlight a challenge for traditional finishing technology with AM products.

A key benefit of AM technology is the ability to optimise the topology of a part such that its load-bearing requirements are met, whilst the amount of material required to make the component is minimised. This can result in significant mass savings for applications such as aerospace, for example in load-bearing brackets. This topology optimisation, however, often results in complex freeform geometries (struts) within a part, which leads to the presence of tight corners and small gaps in the part that restrict access for methods using hard tools (blue features in Figure 5.6). However, there may be a need for these geometries to be finished for fatigue applications (Section 5.3.1). It can be common practice for these kinds of geometries to be processed manually, where operator dexterity aids with finishing these complex freeform topologies; however, there still may be challenges with tool access. Loose media finishing processes may also face challenges with these features, as

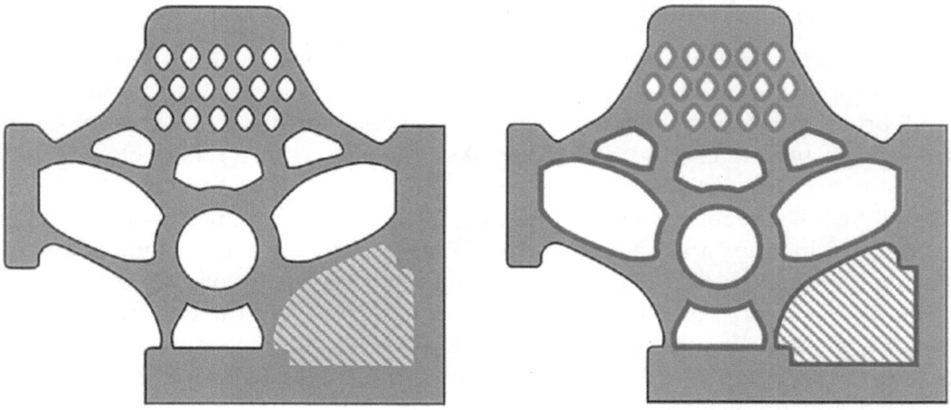

FIGURE 5.6
Example of common features seen in AM parts. Blue highlights those common in a topology-optimised structures, red an example of lattice structures and purple an area of no line-of-sight.

the broad range of feature radii (both convex and concave curvature) result in significant variation on the finished surface topographies and localised material removal rates.

The use of lattice structures within AM is another common way to reduce component mass (Hanzl et al. 2017). The internal structures within a lattice may be impossible to access with a hard tool or small media due to the convoluted tool path required to reach the internal struts, leading to no line-of-sight access (red features on Figure 5.6). Chemical methods are often applied, but these are dependent on the material properties (Dong et al. 2019). Difficulties in accessing surfaces for finishing leads to challenges in the exploitation of lattices, especially in applications such as space or medicine, where the final surface finish is required to demonstrate cleanliness from semi-sintered AM particles. The lattice geometry also presents a fixturing and datum feature challenge, which is applicable to a number of complex geometries, especially in precision applications. The complexity of the parts often causes specialist fixtures to be required to both hold the part in order to withstand machining forces and to provide datum features for the planned toolpath. This is often overcome by additional features being added to the part at the design stage.

Another common way to exploit the benefit of AM is to combine several individual parts from one assembly into one fully functional part. This can result in a number of challenges for the secondary operations. In some cases, part consolidation can result in enclosed features or passages with no line-of-sight access (purple features on Figure 5.6). These no-line-of-sight passage features are common in fluid channels or cooling applications and, therefore, commonly have surface requirements much improved from the as-built AM surfaces (Section 5.3.1). Depending on how much material needs to be removed and the size of the passage, there may be a limitation on the number of appropriate finishing techniques from tens of processes to just one or two. Within these internal passages, it may then be difficult to control the material removal, especially where significant variations on the internal geometry are concerned (such as changes in internal channel cross-sectional area).

As the AM process has combined a number of individually functioning parts into just one, this can also lead to a number of different surface requirements on the same part. The requirement on the secondary operation, therefore, is to remove different levels of material from different areas. This can be a challenge for certain finishing techniques, in particular for loose media-based processes where masking will be required in order to control surface finishes in specific, targeted areas.

5.5.3 AM Process Chain Challenges for Finishing Operations

It is clear that surface finishing operations must be a key consideration when assessing the AM process for any application. However, when assessing the finishing operation within an overall process plan, it is not only the specific geometric or surface features of AM parts that are significant, but also these features taken together within the whole AM process chain.

As metal AM is still relatively new and complex, there are numerous potential build parameter combinations and sources of variability. As such, the likelihood of two parts made using the same model of AM machine, but at different sites or even different times of day, being identical is much lower than from traditional production methods, at the current time. The result of this potential variation between supposedly identical builds is that surface finishing operations for AM processes must be developed and used to gain standardised and predictable outputs from variable inputs within the production process chain (Figure 5.7). This is in comparison to the traditional use of secondary finishing operations to improve further on a standard input, as is usually the case in traditional manufacturing methods such as forging, casting or machining from billet.

Therefore, the secondary finishing operation or operations applied to AM processes today must be able to both manage varying material removal requirements in order to meet the specification of the final application (Section 5.3), as well as being compliant enough to accommodate part variation outside of tolerance and create a standard output for the AM production process chain.

5.5.4 Finishing Challenges for AM in Precision Applications

Sections 5.5.1–5.5.3 demonstrate that there are a number of challenges that AM presents for secondary operations in particular. These challenges are summarised below.

- High initial (as-built) texture.
- Surface texture variation across part after build.
- Geometric complexity.

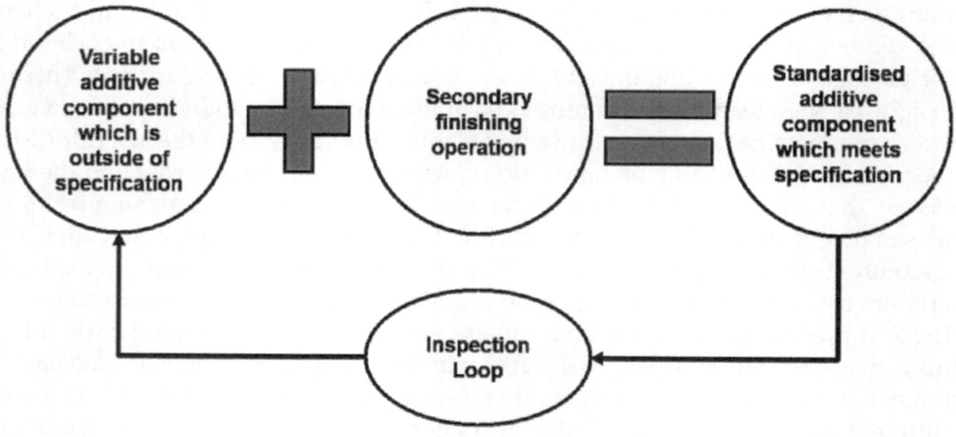

FIGURE 5.7
Example of a typical surface finishing operation loop for additively manufactured parts.

- Part distortion.
- Geometric variation from build to build on a specific machine.
- Machine-to-machine variation in build output.

Although geometric complexity is by far the most significant challenge, the impact of the texture variation across the build is significant for precision applications, as it causes a need for precise control of material removal in the secondary finishing stage to create a repeatable and within-tolerance final part. Conversely, the machine-to-machine build output variation and distortion still present in many AM processes mean that secondary operations often need to be easily adaptable and flexible to varying inputs in order to reduce scrap and re-work (hence a significant amount of manual finishing takes place). This often cannot be achieved whilst also satisfying the requirement for a highly controlled and repeatable operation. Furthermore, the increasing geometric complexity of parts reduces the range of technologies available that can achieve the finish required for the final function. This combination of factors leads to significant challenges in applying precision engineering principles within secondary operations for AM. However, by understanding the functional requirements of the part, using comprehensive process selection and feeding back into the design cycle, these challenges can be minimised or overcome.

5.6 Available Secondary Finishing Processes

Secondary finishing processes, which perform one or more of the activities described in Section 5.2.1, typically fall into one of the following categories.

- *Mechanical* – Use of physical contact between a solid or liquid tool and/or media, whereby material removal, re-shaping and re-distribution is achieved by deformation, cutting or shear mechanisms. Finishes achieved by these processes are often called *mechanical* finishes, which have designations related to abrasive grit mesh sizes and the visual appearance of the finish (Simmons 2011).
- *Chemical* – Use of chemical agents, such as acids, which attack the parent material of the AM component, resulting in dissolution of surface material for removal or re-deposition (Tamaki et al. 1989, Keping and Jingli 1998).
- *Electrical* – Application of electrical current to the component surface, resulting in localised heating and material sublimation and/or melting. An electrical potential difference may also be applied between the tool and the component to assist chemical finishing processes (Zhang et al. 2018).
- *Thermal* – Application of thermal energy to the component surface, used to sublimate and/or melt the parent material resulting in material removal and/or re-distribution (Gora et al. 2016).
- The processes described in this chapter fall into one of two sub-categories: conventional and non-conventional machining methods. Within these categories, there may be processes which fall into one (or more) of the mechanical, chemical, electric or thermal process categories, as shown in Figure 5.8.

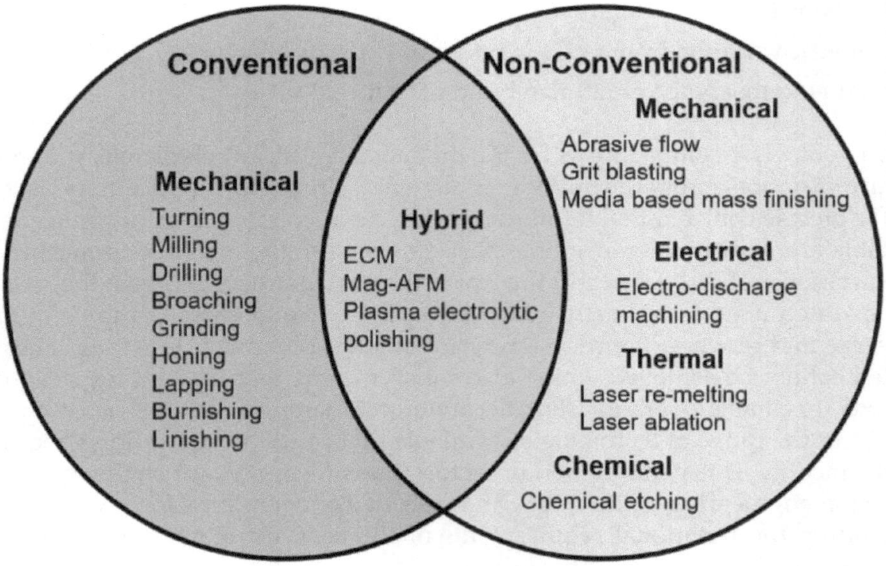

FIGURE 5.8
Conventional and non-conventional finishing methods falling under mechanical, thermal, chemical, electrical and hybrid sub-categories.

5.6.1 Conventional Machining Methods

Conventional machining methods are typically those whose primary method of material removal or displacement is through mechanical interactions between a solid tool and the workpiece (Degarmo et al. 2005). These processes are suitable for use on a large range of materials and geometries; however, to some extent, they are limited in their application to geometries found on parts made through AM due to the need for a rigid connection between the machinery, the tool and the workpiece.

- *Turning* – Turning involves rotation of the workpiece, typically at high speeds, and using a hard, typically single-point cutting tool to remove material (Ratnam 2017). Turning is best suited to finishing of features with co-axial circular cross sections, as it is limited to rotationally symmetric parts (such as tubes or cylinders).
- *Milling* – Milling involves fixturing of the workpiece and using a range of cutting tools from three- to six-axis configurations, removing material to generate features (Karayel 2009).
- *Drilling* – Drilling involves axial translation of a hard, helical cutting edge (or edges) into a workpiece to remove material, typically to create holes (Balaji et al. 2018).
- *Broaching* – Broaching involves passing a multi-faceted cutting tool, which typically removes material closer and closer to the final dimension with each successive cutting face brought into contact with the workpiece. There are generally three sections to a broaching tool, including roughing, semi-finishing and final finishing cutting edges, which achieve the final finish in a single pass of the broaching tool (Degarmo et al. 2005).

- *Grinding* – Grinding involves abrasion of surface material from a workpiece by dragging hard abrasive grains across the material surface. Abrasive grains will fracture and reveal new, sharp cutting edges as the tool progressively wears. Grinding is capable of finishing complex geometries when performed manually, robotically (Xi et al. 2017) or with a five- or six-axis machine centre using in-process monitoring for process optimisation (Maggioni and Marzorati 2013).
- *Honing* – Honing is similar to grinding in its material removal mechanism; however, the tools differ in that a honing tool must have a degree of compliance, such that the path of the cutting edges follows the pre-existing geometry. The effect of compliance causes honing processes to have good form retention and can achieve fine finishes with no significant change to the component form. Honing is best suited to flat and cylindrical parts (Choudhury 2016).
- *Lapping* – Lapping involves rubbing abrasive grains, either loose or embedded in a soft 'lap' material, such as ceramic slurry, between the workpiece and a tool, creating micro-scale scoring – similar to grinding and honing – to polish a surface (Irvin 2011, Doi et al. 2015, Choudhury 2016). Lapping is best performed manually or with sophisticated multi-axis machine centres due the complex, 'randomised' toolpaths required to create a good quality lapped surface.
- *Burnishing* – Burnishing processes improve surface finish by plastic deformation of material through rub (Cubberly and Bakerjian 1989, Choudhury 2016). Burnished surfaces typically exhibit compressive residual stresses, which may also enhance the mechanical performance of the component.
- *Linishing* – Linishing is similar to grinding, with the primary aim of improving the flatness of a surface. Linishing is less effective than grinding for bulk material removal, and the formats of tools intended for linishing (including tool shape and abrasive grain size) are able to produce finer surface finishes than typical grinding tools. Compliant tooling can produce even finer finishes, where tool geometries conform to component surfaces to remove sub-micrometre-scale asperities, such as in the shape-adaptive grinding process (Beaucamp et al. 2015a).

5.6.2 Non-Conventional Machining Methods

Non-conventional machining methods are those methods that do not fall under the conventional machining category. Some of the non-conventional machining methods' primary mechanisms of material displacement or removal are mechanical, just as with conventional methods. However, the tooling and machinery required to apply these processes to a workpiece, such as robotics and special purpose machines, are significantly different to those required for the application of conventional machining methods, such as computer numerical control (CNC) machine centres.

- *Abrasive flow* – Abrasive flow, also known as *abrasive flow machining* (AFM), involves passing a transport medium (typically a slurry of viscous fluid) which contains abrasive particles through internal channels of components (Williams and Melton 1998, Williams et al. 2007). As the fluid is forced through these internal channels, the abrasive grains act upon the walls of the channels to remove material and improve the surface finish.

- *Chemical etching* – A wide range of surface finishing processes can be termed *chemical etching*, where material is removed by chemical (typically acid) dissolution to reduce the size of prominent surface asperities. Specific formulations are required to achieve efficient surface finishing, dependent on the component material.
- *Electro-discharge machining (EDM)* – EDM achieves material removal through the discharge of electric current between two electrodes (one of which is the tool, the other is the component), separated by a dielectric fluid. The mechanism of material removal is thermal melting and evaporation caused by high temperatures generated by sparks between the tool and the workpiece. Particularly useful for machining very hard materials, EDM can be performed using dies or wires.
- *Grit blasting* – Grit blasting involves the acceleration of hard particles towards a component surface, resulting in a multitude of mechanical impact interactions on the component surface (Chohan and Singh 2017). Grit blasting is used commonly to dislodge and remove semi-sintered particles present on the surface of AM components as a primary post-processing step. The impact of the grit particles also acts to deform prominent surface asperities on the component surface, improving the surface finish.
- *Magnetic finishing* – Magnetic finishing involves the forcing of magnetic abrasive particles against a component surface using a controlled, localised electric field. This process offers access to surfaces that may not be accessible by conventional processes, as the magnetic field permeates the component material.
- *Media-based mass finishing* – Media-based mass finishing covers a wide range of processes whereby abrasive media is used to remove material from component surfaces (Jamal and Morgan 2017). Processes such as barrel finishing, tumbling, drag finishing and centrifugal high-energy finishing all fall within this category. The abrasive media may be made from a wide range of materials including plastics, metals, ceramics or even nut shells. The way in which media is passed over component surfaces varies within this category, but there are fundamental similarities between the processes, one of which offers a significant cost benefit to component post-processing: these processes can finish multiple components with a range of geometries at one time (Crisbasan 2018).
- *Ultrasonic polishing* – Ultrasonic polishing achieves material removal in a similar way to AFM, where hard abrasive particles in a relatively soft tool are forced upon a component surface. Abrasive particle motion is achieved by ultrasonic oscillation of the probe. The soft probe ensures good form retention whilst achieving material removal for surface finishing.
- *Laser re-melting* – Laser re-melting, also known as *laser polishing*, uses a continuous wave or relatively long pulse duration pulsed laser (milliseconds to microseconds) to achieve melting of the component surface to a depth of a few tens of micrometres, up to a few hundred micrometres, dependent on process parameters (Laserage 2016). The molten material flows by surface tension, resulting in a flatter surface; however, the surrounding texture will influence the final surface topography after material re-solidification.
- *Laser ablation* – Laser ablation, also known as *laser micro polishing*, achieves surface finish improvements by material removal as opposed to material redistribution in laser re-melting. Short pulses of laser radiation, typically in the nanosecond to

femtosecond regimes, remove material by melting and evaporation. This process typically generates a smaller heat affected zone than laser re-melting but is an overall slower process (dos Santos Solheid et al. 2018).

- *Electro-chemical machining (ECM)* – ECM is a hybrid process wherein a chemical agent is used to reduce the effects of wear on the tool, which can be found in EDM. ECM is particularly well suited to finish hard materials and complex geometries such as turbine blades (Hirtenberger 2017, South West Metal Finishing 2018). There are many variants within the ECM category, which cover the ranges of tooling geometries and machinery used for application of the material removal process.
- *Plasma electrolytic polishing* – Plasma electrolytic polishing is another hybrid process very similar to ECM, with the addition of a thin plasma layer generated over the component surface. The layer acts to reduce the surface temperature of the component, leading to fewer post-processing stages required to obtain desirable surface material properties.
- *Magnetic abrasive flow machining (Mag-AFM)* – Mag-AFM is a hybrid process utilising the principles of AFM and enhancing the level of control available to the process operator by the application of localised magnetic fields. By using magnetic abrasive particles, a magnetic field can be used to generate regions of increased contact force between the abrasive particles and the component surface (Kiani et al. 2016).

5.6.3 Emerging Technologies Developed for AM

There have been many recent developments of the above-described established finishing techniques in response to AM requirements in particular. The techniques outlined and described below have all come to the market in the last five years and are advertised specifically for the AM finishing market. All of these processes are based on the technologies described in Sections 5.6–5.6.2, with specific developments for the AM markets. This list is not exhaustive, but aims to outline the current state of the art. Future developments and a view on how these technologies may develop will be outlined in Section 5.10.

5.6.3.1 Chemical Processes

Chemical-based finishing processes offer the advantage of allowing access to complex features. This is due to the finishing mechanism; the chemical agents are delivered as a fluid designed to erode and dissolve the specific metal on which they are acting (South West Metal Finishing 2018). Therefore, wherever the fluid can reach, some erosion and smoothing of the surface can occur. Recently, companies have been offering chemical processes specifically tuned for common AM alloys and designed to remove the large amounts of material needed. Hirtization (Hirtenberger 2017) and Almbrite (South West Metal Finishing 2018) are two examples. As with any chemical process, there are significant health and safety concerns, especially around disposal processes, which must be considered.

5.6.3.2 Hybrid Mass Finishing and Chemical

In order to address some of the limitations of both mass finishing and chemical techniques, a new range of hybrids have emerged targeting AM parts. The isotropic super finishing (ISF) process from REM is one such example (REM Surface Engineering 2016). These processes combine micrometre-scale abrasive media and a targeted chemical solution that aims to improve the material removal rate of each process individually (Brar et al. 2015). The use of free-flowing media also allows the techniques to reach areas often not accessible via machine tools. As in traditional mass finishing, the media must have sufficient energy to remove material via impacting the surface; this energy can be reduced when the media is forced through long, complex channels, as can be common in topologically optimised AM parts (Jamal and Morgan 2017) and consolidated assemblies.

5.6.3.3 Hybrid Mass Finishing and Electropolishing

Electropolishing is used frequently for finishing of complex AM parts; however, the AM surfaces accessible to the technique can become limited when the distance to the cathodes increases, and the electrolyte fluid can be hazardous (Landolt 1987). A recent technology has addressed some of these challenges by using solid media as the electrolyte, effectively using the media to carry ions away from the part rather than using an electrolytic fluid (DLyte 2019). This technology claims to remove only material from the peaks of the part and, therefore, improve tolerances, retain the part shape and reduce polishing time (DLyte 2019). Access to surfaces smaller than the media size remains a challenge, however, as in any mass finishing technique.

5.6.3.4 Electropolishing Developments

Electropolishing is an electro chemical technique where the part sits in an electrolyte fluid which, as an electrical current is passed through the part, leads to corrosion of the anode (the part) and plating of the cathode via ion diffusion through the electrolyte. As it removes material preferentially at burrs or peaks (where current density is highest), the technique is often used for deburring or micro polishing of surfaces (Landolt 1987). In order to develop the technique for complex AM parts, a number of researchers have investigated the process, with many focussing on parameter development as well as cathode distance/location to the AM surface (Kim and Park 2019, Wu et al. 2019). Researchers have also investigated advanced modelling of the electrolyte bath to improve material removal in the areas required, with subsequent changes including the placement of the cathodes and agitation of the electrolyte fluid specific to the AM geometry (Han and Fang 2019). Companies are also offering the possibility to build the cathode in a reverse image of the part, possibly in the same build as the AM part. This cathode statically mirrors the AM part, allowing a small distance for the electrolyte to flow between the cathode and the part, increasing material removal and allowing internal and external machining of geometries (Extrude Hone 2018a). Developments in filtration and the type of electrolyte fluid in order to reduce environmental impact and increase material removal are ongoing (Han and Fang 2019).

5.6.3.5 Mass Finishing Targeted at AM

Mass finishing is a commonly used and an established technique for AM parts due to its application to complex geometries. Common to most types of mass finishing is the use of free-flowing abrasive media of varying sizes and shapes. This media is energised in different ways depending on the technique; for example, in its simplest form, a barrel is vibrated, thus agitating the media and part within (Jamal et al. 2017). More recently, there have been advances in the amount of energy that can be imparted to the media and how this is controlled. An example of this development is stream finishing, whereby the part is fixed in a specific orientation and moved through the media barrel, whilst the media is simultaneously moved in the opposite direction. Stream finishing can give good results on

AM parts, and (as the parts are fixtured) the angles and material removal can be controlled (Fintek 2019). A range of companies are also offering varying high-energy systems, where the part remains loose in the barrel to give the material removal needed on AM parts but keeps the flexibility for a range of geometries (Crisbasan 2018). Alongside these developments, companies are also now targeting mass finishing processes in the aerospace sector, where AM is increasingly being used, through improving the automation and control of these systems (MMP Technology 2013, AM Solutions 3D Post Processing 2019)

5.7 What Processes Are Appropriate for AM?

Secondary finishing processes are, in general, not specific in terms of their applicability to AM or non-AM geometries. The majority of secondary finishing processes discussed in this chapter have been used in general manufacturing for many years and have been developed for application with components made by conventional methods. Novel applications of these well-established techniques have been identified for use in secondary finishing of AM components, as well as driving the development of new and hybrid techniques. There are a number of considerations to make when selecting a secondary finishing process, relating to the intended application for the part, the levels of process control required, and how a part has been built. Processes that are suitable for secondary finishing of geometries which can be made conventionally are likely suitable for application to similar geometries on AM parts. For example, a flat external surface, or simple bore, is easily finished using many of the techniques discussed in this chapter. These types of geometries can be finished with a range of techniques, regardless of how they have been manufactured. It should be noted that some of these techniques may be limited in terms of the initial surface textures for which they can be applied, such as shape-adaptive grinding (Beaucamp et al. 2015b).

Application-driven surface texture requirements may also dictate which secondary finishing techniques are appropriate for use on an AM geometry. There may be cases where an AM part has application-driven requirements that require specific processes, such as a mirror finish on an aesthetic feature, but the AM geometry itself restricts which processes can be used. Complex geometries enabled by AM build processes can offer significant challenges in terms of tool access or physical/optical line-of-sight. Therefore, a feature requiring a specific finish that may only be achievable using a limited range of secondary finishing techniques may only be suitable for finishing by another subset of techniques, which may not overlap with the techniques that can meet the surface finish requirements.

Furthermore, some of the techniques discussed in this chapter may have metallurgical and/or material effects or impacts. This is a similar consideration to application-driven surface texture requirements, where the intended end-use of the part dictates what the material properties and behaviours need to be. These requirements may also then impose restrictions or limitations on which processes can be selected for use in secondary finishing operations.

Table 5.5 shows some of the key benefits and limitations of the processes discussed in this chapter. Their applicability to typical AM geometries must be assessed on a case-by-case basis; however, there are some general rules that may help to narrow down the selection to a smaller range of processes. Table 5.6 gives additional detail to some of the generalised limitations of the processes discussed in this chapter.

Figure 5.9 shows some of the more common AM geometries that may offer specific challenges for secondary finishing processes.

TABLE 5.5
Secondary Finishing Techniques and Their Strengths and Limitations with Respect to Common AM Feature Types and Geometries

			Suitable features				
Process	**Strengths**	**Limitations**	**Simple external**	**Complex external**	**External details**	**Simple internal**	**Complex internal**
Turning	High dimensional tolerances	Only rotational features	✓	①	X	✓	X
Milling	Complex external features	May require multiple operations	✓	✓	✓	✓	X
Drilling	Relatively cheap equipment	Primarily for through-holes	X	X	✓	✓	X
Broaching	Roughing to final finishing in a single pass	Expensive tools	✓	②		✓	✓
Grinding	Can be performed on a machine centre, manually or robotically	Tool access limited for small features	✓	✓	③	④	X
Honing	Good form retention	Expensive equipment	✓	⑤	X		X
Lapping	Achieves superfinishes	Requires a highly skilled operator	✓		X	⑥	X
Burnishing	Creates superior surface mechanical properties by material displacement	Requires relatively smooth surface to start	✓		X	✓	X
Linishing	Similar to grinding	Similar to grinding	✓	✓	③	④	X
Abrasive flow	Can access complex internal features	Expensive equipment and relatively long setup times	⑦		X	✓	✓
Chemical etching	Relatively fast process	Requires use of hazardous chemicals	✓	✓	⑧	✓	✓
Electro-discharge	Very high tolerances	Requires shaped electrodes for complex features	✓	✓		✓	⑨
Grit blasting	Low consumable cost	Low process control	✓	✓	⑩	⑪	X
Magnetic abrasive			✓	✓	⑫	✓	⑬

(Continued)

TABLE 5.5 (CONTINUED)

Secondary Finishing Techniques and Their Strengths and Limitations with Respect to Common AM Feature Types and Geometries

			Suitable features				
Process	**Strengths**	**Limitations**	**Simple external**	**Complex external**	**External details**	**Simple internal**	**Complex internal**
Media-based mass finishing	Can finish multiple parts at once	Feature rounding, may require masking	✓	✓	③		X
Ultrasonic polishing	No loss of detail through tool contact		✓	✓	✓	✓	X
Laser re-melting	Fine detail and high process control; preserves overall geometry	May introduce thermal damage and oxidation, especially on thin features	✓	✓	✓	⑪	X
Laser ablation	Fine detail and high process control	Relatively slow; thermal effects can be limited at the cost of processing speed	✓	✓	✓		X
Plasma electrolytic polishing	Relatively short processing time	Requires special electrolyte solutions for each different metal	✓	✓	✓	✓	X
Electrochemical machining	Achieves superfinishes; suitable for hard-to-machine materials	Expensive equipment; requires use of hazardous chemicals	✓	✓	⑧	✓	✓

TABLE 5.6

Common Limitations of Secondary Finishing Techniques for AM Features and Geometries

①	Largely limited to rotational features (those with circular cross sections).
②	Only suitable for 2.5D features and details.
③	Minimum feature size limited by media size.
④	Limited to relatively large (10s of millimetres in diameter or more) cylindrical bores.
⑤	Difficult to apply to freeform surfaces.
⑥	Difficult to generate required toolpaths on internal features.
⑦	Difficult to create a seal between an external part surface and equipment.
⑧	Process may round edges and remove fine details.
⑨	Shaped electrode required for complex surfaces, which may limit complexity of internal features that can be accessed.
⑩	Difficult to process small features where grits can become stuck.
⑪	Limited effectiveness with shallow incidence angles.
⑫	Difficult to generate targeted magnetic field over small areas.
⑬	Difficult to manipulate magnetic field accurately within a component.

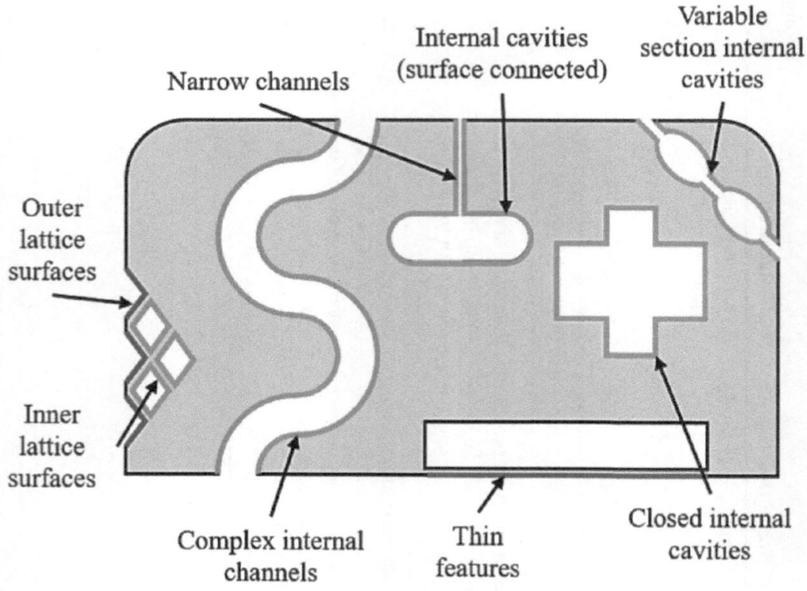

FIGURE 5.9
Typical geometrical features seen on AM components.

5.7.1 Narrow Channels

Narrow channels may be difficult to finish due to the limitations they place on tool access. Straight channels may be suitable for drilling (mechanical or EDM) or electrochemical processes, provided that a thin enough tool can be obtained. Longer, narrow channels (channels with a high aspect ratio between length and width) may only be appropriate for non-conventional methods, such as EDM drilling.

5.7.2 Complex Internal Channels

Internal channels which have non-straight paths (for example, those which could not be drilled) have line-of-sight restrictions on which processes may be applicable. Neither physical delivery of a tool nor optical processes which require line-of-sight are applicable to these features. Flexible tooling or liquid-based techniques such as abrasive-flow machining can be used instead.

5.7.3 Internal Cavities (Surface Connected)

Surface connected internal cavities (i.e. features where the major dimensions are significantly greater than those of access channels) have similar restrictions in terms of tool access to narrow channels, as mechanical tooling cannot be delivered to the salient surface. As such, chemical methods are likely to be most appropriate for finishing features of this type. However, closed-ended features may also cause stagnation of finishing media in mass finishing techniques or inhibit process control in chemical or electrochemical techniques.

5.7.4 Variable Cross-Section Internal Channels

Processes that can finish internal channels often require localised control to achieve consistent finishes throughout the length of the channel. Electrochemical or abrasive flow techniques may require localised process control, for example by use of a shaped electrode for electrochemical processes, to maintain surface texture consistency.

5.7.5 Outer Lattice Surfaces

Outer surfaces of a lattice may be suitable for secondary finishing by a range of techniques, but care must be taken not to cause damage to the feature when using mechanical techniques due to contact forces in material removal, or thermal damage when using laser-based processes. Low energy mass-finishing techniques or chemical processes may be most appropriate for finishing both outer and inner lattice surfaces.

5.7.6 Inner Lattice Surfaces

Secondary finishing processes that are suitable for use on outer lattice surfaces may not be appropriate for inner surfaces, such as mechanical techniques and their tool access requirements. Therefore, if an entire lattice surface needs to be finished, then different processes may be required. Localised control of chemical and electrical processes may also cause issues for finishing surfaces deep into a lattice feature.

5.7.7 Thin Features

As with lattice features, thin features will pose challenges to mechanical processes where contact forces could cause part deformation or loading. Thermal loading can also cause issues in terms of localised metallurgical changes, part deformation or inconsistent surface texture. Chemical and electrochemical processes, where there is no mechanical and no bulk thermal loading (there may be thin layers of high thermal loads in electric processes), can be most appropriate for these features.

5.7.8 Closed Internal Cavities

Surfaces that are part of completely closed internal cavities, where there is no access for physical, optical or chemical tooling, in general cannot be finished through a secondary operation. If a closed internal cavity is required as a feature of the part for its functionality, which must also be finished to a certain texture, then the part can be manufactured in multiple parts and then assembled into the final geometry. Alternatively a hybrid AM machine combining a machining tool with powder-bed laser technology could be considered, in this case the part is machined every few layers, allowing internal cavities to be machined as they are built.

5.8 Other Considerations for Finishing Operations in AM

Aside from the challenges arising from component geometry that can drive the selection for AM secondary finishing operations, many other aspects must be considered. As a large proportion of metal AM processes are powder bed (Wohlers Associates 2018), there are risks associated with trapped powder being released further down the process chain, especially as parts are often handed over to other sites for processing after the build. For example, if parts are having supports removed via an EDM process, significant amounts of powder can be released into the machine from areas that were previously inaccessible. This can cause maintenance issues as well as potential health and safety hazards associated with handling or breathing the powder particles (Graff et al. 2017, Du Preez et al. 2018). Full assessment of these issues should be made at the design stage, and considerations for all of the required secondary operations to deliver a functional part for the intended application must be made.

Due to the geometric and part variation challenges, AM parts are often manually finished using hand tools. Aside from the repeatability and control challenges associated with manual finishing, there are also health and safety considerations. For example, there are inherent risks to the operators from dust inhalation (Graff et al. 2017) and concerns about potential repetitive strain injuries. These challenges are not new to the industry but must be considered in relation to the additive material and product and the high number of secondary finishing processes that the part undergoes. Lastly, as AM parts become more complex, there is an increasing move towards processes that use chemical agents to remove material; this is at odds with current legislation (Regulation EC No 1907/2006 on the Registration, Evaluation, Authorisation and Restriction of Chemicals – REACH) aiming to reduce industrial chemical use. Thorough assessments of the chemicals required and total environmental impact of the part must be made when justifying chemical processing methods (Thiery and Wojczykowski 2012).

5.9 How to Impact AM Design for Finishing

The biggest challenge for the selection of current available finishing technologies is the application to additional geometric complexities as commonly seen in AM part designs. Therefore, it is critical to design not only for the AM build, but also for the post-processes;

otherwise it may be impossible, or prohibitively time consuming and costly, to finish the part in order to meet the specifications needed. This design process should involve first defining the requirements based on the part function and not, for example, from historical data based on previous manufacturing methods. Once the requirements are set, the designer should identify the limitations of any finishing technique that the part requires based on the final part function and utilise this knowledge in the design of the part. Design rules are available for AM parts in a range of literature (Leutenecker-Twelsiek et al. 2016) (see Chapter 2), and it is clear that the limitations and properties of the secondary finishing operations should have a significant impact on these rules as they develop.

5.10 Future Work

5.10.1 New Processes and Technologies in Development

As the uptake of additive technologies spreads, there are an increasing number of finishing techniques that are coming onto the market designed specifically for AM parts (Section 5.6.3) (Hirtenberger 2017, Extrude Hone 2018b, DLyte 2019). However, the majority, if not all, of these techniques are based on technology that has been available for a number of years.

It can be seen in Figure 5.10 that the latest significant development in component finishing was the advent of closed-loop systems first available in the 1980s. In contrast, component 'build' techniques have accelerated and improved since the advent of AM techniques in the 1990s, and a like-for-like development in finishing has not kept up – leading to the challenges outlined in this chapter. One of the most common techniques for finishing of AM parts is the use of abrasives, the basics for which were invented in the pre-industrial age.

There are, however, a number of processes being researched in academia to try to address the AM finishing challenges; these are outlined below.

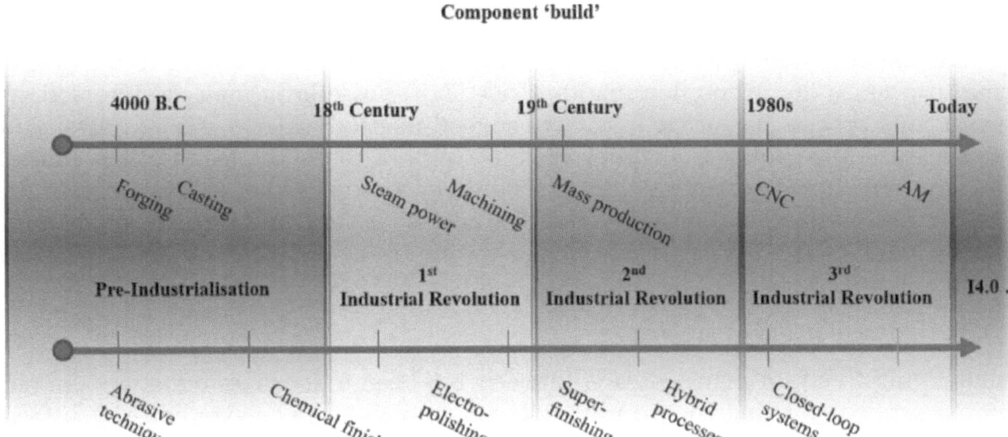

FIGURE 5.10
Development of component build techniques compared to secondary finishing developments.

5.10.1.1 Hybrid AFM

Abrasive flow machining is defined in Section 5.6.2. Hybrid AFM, in contrast, covers the use of different external technologies to improve the traditional AFM method, for example to improve material removal or control. The hybrid processes still use a media-laden fluid, but this media or delivery method is often changed or an external force applied in addition to the traditional hydraulic pressure. AFM is used for additive parts, but limitations can exist when the surface is very rough, or complex passages mean a uniform surface is difficult to achieve (Extrude Hone 2018a) due to carrier-fluid flow discontinuities and instabilities. A number of academic groups are researching improvements to this technology via the use of hybrid techniques. Research to date has focussed on improving control of the media whilst in the part, for example by using ferromagnetic media within the fluid and applying a magnetic field from outside the part (Kiani et al. 2016) to achieve localised process control inside the part. Increasing material removal rates have also been investigated by combining different processes with AFM to create a hybrid AFM technique, for example applying rotational movement to the part, applying ultrasonic movement to the fluid, using magneto-rheological polishing fluids and combining AFM with electrochemical machining (Brar et al. 2015). Improved tooling configurations and process parameters via modelling and simulation have also been investigated to improve AFM as a technique for finishing AM parts. Furthermore, Peng et al. suggest that alongside surface reduction, AFM may also result in improved compressive stress at the AM surface (Peng et al. 2018).

5.10.1.2 Laser Polishing

As the industrial use of lasers has increased, the availability and quality of laser sources continues to improve (Laserage 2016). This has led to increased research on the use of lasers for removing and/or altering surface texture. Current research focuses on two areas. The first is using continuous wave or long-pulse (typically microsecond or longer) infrared laser sources to re-melt the surfaces. This area can reduce the AM surface by 90% in a relatively short time but can cause significant recast layers and tensile stresses at the surface (Gora et al. 2016, Li et al. 2019). The second area of research is in using short-pulse (typically nanosecond or shorter) lasers to ablate the surface peaks. Laser ablation does not typically create recast layers to the same extent as re-melting, as thermal processes do not predominantly drive short-pulse laser-material interactions. However, this process has a much slower removal rate for surfaces with a high initial level of texture (dos Santos Solheid et al. 2018). In both cases, the technique is limited by the requirement for near-normal line-of-sight but could have benefits in targeted finishing for AM and automation, as well as for accessing very fine features on external surfaces (Obeidi et al. 2019).

5.10.1.3 Automation and Modelling

An underlying trend for many research projects has been to focus on removing the reliance on manual de-burring and grinding techniques often used on external AM surfaces (Symplexity 2017, Vialva 2018). The use of sensing and robotics has allowed technologies such as laser polishing and grinding to be applied to complex AM parts, without the need for manual intervention (The Manufacturing Technology Centre 2019b). Automation of these processes also allows for a reduction in manual handling during and between physical processes; build plates, AM parts and tooling, can all be delivered and deployed using automated systems. As AM grows, it is likely that these solutions will be taken up more widely by industry, despite the initial high costs.

Similarly, there is also a trend for improved modelling of processes in order to better control and predict process outcomes. In the mass finishing area, where the motion and wear of loose media is difficult to model and, therefore, the output of processes using loose media is difficult to predict, recent research has focussed on solving these modelling challenges (Jamal et al. 2017). This research could have a positive impact on industrial access to, and control of, these techniques in the future for AM parts.

5.10.2 Future of This Field

The outlook in this field for the next years is difficult to predict, with this area being heavily linked to the continued and rapid development of additive machines, technologies and processes. However, based on current trends, the following areas can expect to see significant development:

5.10.2.1 Internal Targeted Finishing

As additive parts become increasingly complex, with a trend towards integrating not just two or three parts, but tens or hundreds of parts (for example, blisk components for aerospace) into just one additive part, the requirement for controlled surface finishes of complex internal passageways and difficult-to-access surfaces will grow (Wohlers Associates 2018). The current inability to reach these surfaces will become a significant barrier as additive technologies find uses in a wider range of aerospace parts with critical air flow and fatigue requirements. As AM becomes common in a wider range of sectors, with new applications such as electronics and consumer products, complexity is also likely to increase (Wohlers Associates 2018). The need for smoother parts with aesthetic requirements for different sectors will also drive innovation in secondary finishing operations. Current developments and emerging finishing technologies have shown a move towards an attempt to address more complex parts through hybridisation and new media (as discussed in Section 5.6.3), due to the increasing part complexities being produced by AM, and the current high cost required to finish these parts. However, it is unlikely that without a significant new development in this area that existing technologies, even with further improvements, will be able to address all the challenges that increasingly complex internal additive surfaces will present. New developments in AM machines, for example the recently developed Matsuura Hybrid Lumex, which can perform subtractive machining on the part as it builds (Matsuura UK 2019), may mean that the requirement for finishing of features, such as complex internal passages, is reduced – with finishing possible during the build stage, allowing access to otherwise impossible-to-finish areas. Developments in AM design, process capability, and modelling may also reduce the likelihood for finishing requirements across the whole part and, therefore, lessen the need for internal finishing of components.

5.10.2.2 Hybrid Technologies

Significant improvements for the final AM product have already been seen by combining techniques (Section 5.6.3). This practice is likely to become more commonplace as the increased geometric complexities enabled by AM bring with them a wide range of finishing challenges. It is predicted that it will become common to see processes such as chemical or electropolishing combined with media-based techniques (for example, the Dlyte and REM techniques outlined in Section 5.6.3) (REM Surface Engineering 2016, DLyte 2019).

Similarly, improved control will be combined with those techniques that provide access advantages in order to enable additive technologies in precision applications. For example, the uptake of hybrid abrasive flow techniques may become common, with additional control achieved through techniques such as magnetic fields controlling the abrasive flow (Kiani et al. 2016). Additionally, the surface finishing techniques may see significant developments in modelling and simulation, which may allow accurate predictions to be made, especially for the difficult-to-predict loose media methods, enabling the use of these previously difficult-to-control techniques in more precision applications.

5.10.2.3 Design Processes

In recent years, the AM field has seen a rapid expansion of design and modelling packages targeted at supporting the additive build process (see Chapter 2). Software packages can now automatically generate an optimised support structure methodology, topological optimisation and orientation of the part based on load conditions and can highlight areas where distortion will be an issue in the build. It is likely that this trend will also influence the design packages available for 'design for finishing in AM'. This topic has already been addressed in various research projects in the area (Encompass EWF 2016, The Manufacturing Technology Centre 2019a) and, therefore, it is predicted that industrial software packages supporting AM design will soon include finishing capability for a range of techniques and tools which flag difficult-to-finish areas. These intelligent software packages should have a direct impact on increasing the knowledge of users in the AM field, reducing some of the complex requirements that can cause significant challenges in post-build operations.

5.10.2.4 Specification Standards

It can be seen from the review in Section 5.4 that there is a lack of specification standards for finishing processes specific to AM and methods to select the most appropriate technique based on the final part function and geometry. It is predicted that as AM moves into precision applications, more standards will become necessary to ensure quality across suppliers and to give guidance on process selection (see also Chapter 6).

5.10.2.5 Automation and Targeted Finishing

It is clear that the developments in sensing, control and data informatics will have a significant impact on AM as a whole, and in particular the secondary finishing operations. The increasing use of robotics and automated systems will see age-old techniques, such as manual finishing and barrel finishing, being developed and integrated into automated systems with improved control. This is already occurring, as demonstrated in Section 5.6.3. For example, in-process control, enabled by new sensing technologies and artificial intelligence software, could lead to automated finishing cells utilising existing finishing technologies but improving, learning and updating automatically as they process parts, building on the work seen in recent research projects (Symplexity 2017, Vialva 2018). This could form a step-change for finishing of AM parts, allowing techniques to adapt process parameters based on each individual part (and individual areas/features on any given part) and learn from the history of AM parts processed previously.

Based on improvements in designing for the AM process chain, it is also likely that requirements will be more strongly linked to, and described by, the final part's intended function. This will lead to an increased requirement for targeted finishing, with only critical surfaces required to be finished rather than blanket requirements over the complete part. This will play to the strengths of automated systems, with the focus on improved control and reduced re-work over the ability to finish highly complex internal surfaces. However it must be noted that as targeted finishing based on function becomes more common this will cause validation challenges, especially within industries such as aerospace where significant amounts of data are required to qualify new methods to meet the historical specifications previously followed.

References

Alrbaey, K., Wimpenny, D. I., Al-Barzinjy, A. A. and Moroz, A. 2016. Electropolishing of re-melted SLM stainless steel 316L parts using deep eutectic solvents: 3 × 3 full factorial design. *J. Mater. Eng. Perform.* 25:2836–2846.

AM Solutions 3D Post Processing 2019. *Products: Machines for Post Processing of 3D Printed Parts.* Untermerzbach, Germany: Rösler Oberflächentechnik GmbH.

Ashby, M. and Johnson, K. 2003. The art of materials selection. *Mater. Today* 6:24–35.

ASME B46.1 2002. Surface Texture, Surface Roughness, Waviness and Lay. ASME Standard.

ASTM A967/A967M part 17 2010. Standard Specification for Chemical Passivation Treatments for Stainless Steel Parts. ASTM International.

ASTM E466 part 15 2015. Standard Practice for Conducting Force Controlled Constant Amplitude Axial Fatigue Tests of Metallic Materials. ASTM International.

ASTM E8/E8M 2013. Standard Test Methods for Tension Testing of Metallic Materials. ASTM International.

ASTM E9 part 9 2012. Standard Test Methods of Compression Testing of Metallic Materials at Room Temperature. ASTM International.

ASTM F2924 part 14 2014a. Standard Specification for Additive Manufacturing Titanium-6 Aluminum-4 Vanadium with Powder Bed Fusion. ASTM International.

ASTM F3055 2014b. Standard Specification for Additive Manufacturing Nickel Alloy (UNS N07718) with Powder Bed Fusion. ASTM International.

ASTM F3122 part 14 2014c. Standard Guide for Evaluating Mechanical Properties of Metal Materials Made via Additive Manufacturing Processes. ASTM International.

Bagehorn, S., Wehr, J. and Maier, H. J. 2017. Application of mechanical surface finishing processes for roughness reduction and fatigue improvement of additively manufactured Ti-6Al-4V parts. *Int. J. Fatigue* 102:135–142.

Balaji, M., Venkata Rao, K., Mohan Rao, N. and Murthy, B. S. N. 2018. Optimization of drilling parameters for drilling of TI-6Al-4V based on surface roughness, flank wear and drill vibration. *Measurement* 114:332–339.

Beaucamp, A., Namba, Y. and Charlton, P. 2015a. Process mechanism in shape adaptive grinding (SAG). *CIRP Ann.* 64:305–308.

Beaucamp, A. T., Namba, Y., Charlton, P., Jain, S. and Graziano, A. A. 2015b. Finishing of additively manufactured titanium alloy by shape adaptive grinding (SAG). *Surf. Topogr. Metrol. Prop.* 3:024001.

Bertolini, R., Bruschi, S., Bordin, A., Ghiotti, A., Pezzato, L. and Dabalà, M. 2017. Fretting corrosion behavior of additive manufactured and cryogenic-machined Ti6Al4V for biomedical applications. *Adv. Eng. Mater.* 19:1500629.

Boschetto, A., Bottini, L. and Veniali, F. 2017. Roughness modeling of AlSi10Mg parts fabricated by selective laser melting. *J. Mater. Process. Technol.* 241:154–163.

Brandão, A. D., Gumpinger, J., Gschweitl, M., Seyfert, C., Hofbauer, P. and Ghidini, T. 2017. Fatigue properties of additively manufactured AlSi10Mg-surface treatment effect. *Procedia Struct. Integr.* 7:58–66.

Brar, B. S., Walia, R. S. and Singh, V. P. 2015. Electrochemical-aided abrasive flow machining (ECA2FM) process: A hybrid machining process. *Int. J. Adv. Manuf. Technol.* 79:329–342.

BS EN 10088-5:2009 2014. Stainless Steel – Part 5: Technical Delivery Conditions for Bars, Rod, Wire Sections and Bright Products of Corrosion Resisting Steels for Construction Purposes. British Standards Institute.

Calignano, F., Manfredi, D., Ambrosio, E. P., Iuliano, L. and Fino, P. 2013. Influence of process parameters on surface roughness of aluminum parts produced by DMLS. *Int. J. Adv. Manuf. Technol.* 67:2743–2751.

Chohan, J. S. and Singh, R. 2017. Pre and post processing techniques to improve surface characteristics of FDM parts: A state of art review and future applications. *Rapid Prototyp. J.* 23:495–513.

Choudhury, I. A. 2016. Introduction to finish machining and net-shape forming. In: Hashmi, M. S. J. (ed.) *Comprehensive Materials Finishing*, Oxford: Elsevier.

Crisbasan, A. 2018. *CHE Finishing Process Developed for Additive Manufacturing Company*. Coventry, UK: ActOn Finishing Ltd.

Cubberly, W. and Bakerjian, R. 1989. *Tool and Manufacturing Engineers Handbook (Desk Edition)*. Dearborn, MD: Society of Manufacturing Engineers.

Degarmo, E. P., Kohser, R. A. and Klamecki, B. E. 2005. *Materials and Processes in Manufacturing*. 10th edn. Singapore: John Wiley & Sons.

DLyte 2019. *The First Dry Electropolishing System*. Barcelona, Spain: GPA INNOVA.

Doi, T. K., Ohnishi, O., Uhlmann, E. and Dethlefs, A. 2015. Lapping and polishing. In: Marinescu, I. D., Doi, T. K. and Uhlmann, E. (eds.) *Handbook Of Ceramics Grinding and Polishing*. Boston, MA: William Andrew Publishing.

Dong, G., Marleau-Finley, J. and Zhao, Y. F. 2019. Investigation of electrochemical post-processing procedure for Ti-6Al-4V lattice structure manufactured by direct metal laser sintering (DMLS). *Int. J. Adv. Manuf. Technol.* 104:3401–3417.

ENCOMPASS EWF 2016. Integrated design decision support to cover the whole manufacturing chain for a laser powder bed fusion (L-PBF) process end-to-end. ENCOMPASS EWF.

Extrude Hone 2018a. Abrasive flow machining. Bavaria, Germany: Extrude Hone.

Extrude Hone 2018b. *Extrude Hone Coolpulse*™ process provides solutions to unlock 3D printed parts with demanding surface finishing. Bavaria, Germany: Extrude Hone.

Fintek 2019. Stream finishing turbine blades for maximum surface performance. Bury, UK: Fintek.

Fleming, R. W. 2014. Visual perception of materials and their properties. *Vision Res.* 94:62–75.

GE Additive 2018. Arcam Q20plus. Arcam EBM.

Gibson, I., Rosen, D. and Stucker, B. 2015. *Additive Manufacturing Technologies: 3D Printing, Rapid Prototyping, and Direct Digital Manufacturing*. 2nd edn. New York, NY: Springer New York.

Gisario, A., Kazarian, M., Martina, F. and Mehrpouya, M. 2019. Metal additive manufacturing in the commercial aviation industry: A review. *J. Manuf. Syst.* 53:124–149.

Gloss, D. and Herwig, H. 2010. Wall roughness effects in laminar flows: An often ignored though significant issue. *Exp. Fluids* 49:461–470.

Gora, W. S., Tian, Y., Cabo, A. P., Ardron, M., Maier, R. R. J., Prangnell, P., Weston, N. J. and Hand, D. P. 2016. Enhancing surface finish of additively manufactured titanium and cobalt chrome elements using laser based finishing. *Phys. Procedia* 83:258–263.

Graff, P., Ståhlbom, B., Nordenberg, E., Graichen, A., Johansson, P. and Karlsson, H. 2017. Evaluating measuring techniques for occupational exposure during additive manufacturing of metals: A pilot study. *J. Ind. Ecol.* 21:S120–S129.

Grimm, T., Wiora, G. and Witt, G. 2015. Characterization of typical surface effects in additive manufacturing with confocal microscopy. *Surf. Topogr. Metrol. Prop.* 3:014001.

Han, W. and Fang, F. 2019. Fundamental aspects and recent developments in electropolishing. *Int. J. Mach. Tools Manuf.* 139:1–23

Hanzl, P., Zetkova, I. and Dana, M. 2017. A comparison of lattice structures in metal additive manufacturing. *Proc. 28th Int. DAAAM Symp. 2017*:0481–0485.

Harrison Electropolishing 2012. Electropolishing industry standards. Houston, TX: Harrison Electropolishing.

Hirtenberger 2017. Hirtisation; Surface treatment of 3D printed metal parts. Hirtenberg, Austria: Hirtenberger Engineered Surfaces GmbH.

Irvin, M. 2011. Diamond lapping and lapping plate control. *Production Machining Magazine*. Available at: https://www.productionmachining.com/articles/diamond-lapping-and-lapping-plate-control.

ISO/ASTM 52910 2017. Standard Practices – Guidelines for Design for Additive Manufacturing. ASTM International.

ISO 1099 2006. Metallic Materials: Fatigue Testing. Axial Force-Controlled Method. International Organisation for Standardisation.

ISO 25178 Series 2010. Geometric Product Specifications (GPS) – Surface Texture: Areal. International Organisation for Standardisation.

ISO 4287 2009. Geometrical Product Specification (GPS) – Surface Texture : Profile Method – Terms, Definitions and Surface Texture Parameters. International Organisation for Standardisation.

Jamal, M. and Morgan, M. 2017. Design process control for improved surface finish of metal additive manufactured parts of complex build geometry. *Inventions* 2:36.

Jamal, M., Morgan, M. N. and Peavoy, D. 2017. A digital process optimization, process design and process informatics system for high-energy abrasive mass finishing. *Int. J. Adv. Manuf. Technol.* 92:303–319.

Jomaa, W., Songmene, V. and Bocher, P. 2014. Surface finish and residual stresses induced by orthogonal dry machining of AA7075-T651. *Materials (Basel)* 7:1603–1624.

Karayel, D. 2009. Prediction and control of surface roughness in CNC lathe using artificial neural network. *J. Mater. Process. Technol.* 209:3125–3137.

Keping, H. and Jingli, F. 1998. Study on chemical polishing for stainless steel. *Trans. IMF* 76:24–25.

Kiani, A., Esmailian, M. and Amirabadi, H. 2016. Abrasive flow machining: A review on new developed hybrid AFM process. *Int. J. Adv. Des. Manuf. Technol.* 9:61–71.

Kim, U. S. and Park, J. W. 2019. High-quality surface finishing of industrial three-dimensional metal additive manufacturing using electrochemical polishing. *Int. J. Precis. Eng. Manuf. – Green Technol.* 6:11–29.

Körner, C. 2016. Additive manufacturing of metallic components by selective electron beam melting – a review. *Int. Mater. Rev.* 61:361–377.

Kruth, J. P., Mercelis, P., Van Vaerenbergh, J., Froyen, L. and Rombouts, M. 2005. Binding mechanisms in selective laser sintering and selective laser melting. *Rapid Prototyp. J.* 11:26–36.

Landolt, D. 1987. Fundamental aspects of electropolishing. *Electrochim. Acta* 32:1–11.

Laserage 2016. Growing demand for advanced technology drives the laser industry. *Laserage*. Available at: http://www.laserage.com/growing-demand-advanced-technology-drives-laser-industry/.

Leach, R. K., Bourell, D., Carmignato, S., Donmez, A., Senin, N. and Dewulf, W. 2019. Geometrical metrology for metal additive manufacturing. *CIRP Ann.* 68:677–700.

Leutenecker-Twelsiek, B., Klahn, C. and Meboldt, M. 2016. Considering part orientation in design for additive manufacturing. *Procedia CIRP* 50:408–413.

Li, Y.-H., Wang, B., Ma, C.-P., Fang, Z.-H., Chen, L.-F., Guan, Y.-C. and Yang, S.-F. 2019. Material characterization, thermal analysis, and mechanical performance of a laser-polished Ti alloy prepared by selective laser melting. *Metals (Basel)* 9:112.

Maggioni, M. F. and Marzorati, E. 2013. *In-process quality monitoring via sensor data fusion: Chatter control in grinding*. Milan, Italy: Politecnico di Milano.

The Manufacturing Technology Centre 2019a. *AMAZE - Building confidence in additive manufacturing*. Coventry, UK: MTC.

The Manufacturing Technology Centre 2019b. *Flexible and automated finishing and post-processing cell for high value AM components – FlexiFinish*. Coventry, UK: MTC.

Matsuura UK 2019. *Unique metal sintering CNC milling hybrid technology*. Coalville, UK: Matsuura Machining.

MMP Technology 2013. *First surface high precision surface finishes*, Munich, Germany. First Surface.

Morgan, H. D., Cherry, J. A., Jonnalagadda, S., Ewing, D. and Sienz, J. 2016. Part orientation optimisation for the additive layer manufacture of metal components, *Int. J. Adv. Manuf. Technol.* 86:1679–1687.

Obeidi, M., McCarthy, E., O'Connell, B., Ul Ahad, I. and Brabazon, D. 2019. Laser polishing of additive manufactured 316L stainless steel synthesized by selective laser melting. *Materials (Basel)* 12:991.

Peng, C., Fu, Y., Wei, H., Li, S., Wang, X. and Gao, H. 2018. Study on improvement of surface roughness and induced residual stress for additively manufactured metal parts by abrasive flow machining. *Procedia CIRP* 71:386–389.

Du Preez, S., Johan de Beer, D. and Lodewykus du Plessis, J. 2018. Titanium powders used in powder bed fusion: Their relevance to respiratory health. *South African J. Ind. Eng.* 29.

Ratnam, M. M. 2017. Factors affecting surface roughness in finish turning. In: Hashmi, M. S. J. (ed.) *Comprehensive Materials Finishing*. Oxford, UK: Elsevier.

REM Surface Engineering 2016. *Isotropic superfinishing process*, St Neots, UK: REM Surface Engineering.

Sahoo, P., Das, S. K. and Davim, J. P. 2017. Surface finish coatings. In: Hashmi, M. S. J. (ed.) *Comprehensive Materials Finishing*. Oxford: Elsevier.

dos Santos Solheid, J., Jürgen Seifert, H. and Pfleging, W. 2018. Laser surface modification and polishing of additive manufactured metallic parts. *Procedia CIRP* 74:280–284.

Senin, N., Thompson, A. and Leach, R. 2018. Feature-based characterisation of signature topography in laser powder bed fusion of metals. *Meas. Sci. Technol.* 29:045009.

Sidambe, A. T. 2017. Three dimensional surface topography characterization of the electron beam melted Ti6Al4V. *Met. Powder Rep.* 72:200–205.

Simmons, H. L. 2011. *Olin's Construction: Principles, Materials, and Methods*. 11th edn. Hoboken, NJ: John Wiley & Sons.

South West Metal Finishing 2018. *Almbrite additive layer surface technology*, EIC Grinding.

Strano, G., Hao, L., Everson, R. M. and Evans, K. E. 2013. Surface roughness analysis, modelling and prediction in selective laser melting. *J. Mater. Process. Technol.* 213:589–597.

SYMPLEXITY 2017. *Symbiotic human-robot solutions for complex surface finishing operations*, SYMPLEXITY.

Tamaki, Y., Miyazaki, T., Suzuki, E. and Miyaji, T. 1989. Polishing of titanium prosthetics (Part 6). The chemical polishing baths containing hydrofluoric acid and nitric acid. *J. Japanese Soc. Dent. Mater. Devices* 8:103—109.

Thiery, L. and Wojczykowski, K. 2012. *Impact of REACH Regulation on the Global Finishing Market, NASF SUR/FIN 2012*. Las Vegas: National Association for Surface Finishing.

Townsend, A., Senin, N., Blunt, L., Leach, R. K. and Taylor, J. S. 2016. Surface texture metrology for metal additive manufacturing: A review. *Precis. Eng.* 46:34–47.

Triantaphyllou, A., Giusca, C. L., Macaulay, G. D., Roerig, F., Hoebel, M., Leach, R. K., Tomita, B. and Milne, K. A. 2015. Surface texture measurement for additive manufacturing. *Surf. Topogr. Metrol. Prop.* 3:024002.

Vialva, T. 2018. MTC presents FlexiFinish project – An automated post-processing solution. *3D Print. Ind. Auth. 3D Print.* July.

Williams Metalfinishing Inc. 2013. *Williams Metalfinishing types of finishes*. Sinking Spring, UK: Williams Metalfinishing.
Williams, R. E. and Melton, V. L. 1998. Abrasive flow finishing of stereolithography prototypes. *Rapid Prototyp. J.* 4:56–67.
Williams, R. E., Walczyk, D. F. and Dang, H. T. 2007. Using abrasive flow machining to seal and finish conformal channels in laminated tooling. *Rapid Prototyp. J.* 13:64–75.
Wohlers Associates 2018. *Wohlers Report* 2018: *3D Printing And Additive Manufacturing State of the Industry*. Fort Collins, Colorado: Wohlers Associates.
Wu, Y.-C., Kuo, C.-N., Chung, Y.-C., Ng, C.-H. and Huang, J. C. 2019. Effects of electropolishing on mechanical properties and bio-corrosion of Ti6Al4V fabricated by electron beam melting additive manufacturing. *Materials (Basel).* 12:1466.
Xi, F. J., Chen, T. and Guo, S. 2017. Robotic polishing and deburring. In: Hashmi, M. S. J. (ed.) *Comprehensive Materials Finishing*. Oxford: Elsevier.
Zhang, Y., Li, J. and Che, S. 2018. Electropolishing mechanism of Ti-6Al-4V alloy fabricated by selective laser melting. *Int. J. Electrochem. Sci.* 13:4792–4807.
Zwier, M. P. and Wits, W. W. 2016. Design for additive manufacturing: Automated build orientation selection and optimization. *Procedia CIRP* 55:128–133.

6

Standards in Additive Manufacturing

David Butler and Peter Woolliams

CONTENTS

6.1 Introduction

As metal-based additive manufacturing (AM) adoption increases, there is a rush to catch up by various standards and industry bodies to develop relevant specification standards to ensure conformity across various platforms. Among the key players are the American Standards and Testing of Materials (ASTM), who have been leading the drive for specification standards through the F42 subcommittee and, more recently, with the establishment of International Centres of Excellence located in the USA, UK and Singapore, to lead the research and validation aspects. The International Organization for Standardization (ISO) has also published a number of standards in conjunction with ASTM ensuring greater adoption and reducing the amount of duplication. Specifically, ISO Technical Committee 261 has been tasked with developing these standards. Table 6.1 provides a list of current ISO standards while Table 6.2 lists those under development. A recent overview of metal AM standards development is given elsewhere (Leach et al. 2019).

In addition to ASTM and ISO, other organisations involved in developing standards include the National Aerospace and Defense Contractors Accreditation Program (NADCAP), the National Aeronautics and Space Administration (NASA), the American Welding Society (AWS) and the Society of Automotive Engineers (SAE) International. Table 6.3 and Table 6.4 provide an overview of the current and draft SAE AM standards.

NADCAP is an industry-managed programme that is focused on improving quality and reducing costs of special process accreditations throughout the aerospace and defence industries. In 2013, its Welding Task Group was assigned responsibility to assess the industry needs and develop audit criteria capable of assessing suppliers utilising AM technology. In early 2017, they released for industry usage the checklist AC7110/14 *Nadcap Audit Criteria for Laser and Electron Beam Metallic Powder Bed Additive Manufacturing* (2017).

NASA has been one of the early adopters of AM in relation to design flexibility, cost and schedule challenges of system development, and manufacture. Each of NASA's current

TABLE 6.1

Current ISO AM Standards

Standard	Title
ISO 17296-2:2015	Additive manufacturing – General principles – Part 2: Overview of process categories and feedstock
ISO 17296-3:2014	Additive manufacturing – General principles – Part 3: Main characteristics and corresponding test methods
ISO 17296-4:2014	Additive manufacturing – General principles – Part 4: Overview of data processing
ISO/ASTM 52900:2015	Additive manufacturing – General principles – Terminology
ISO/ASTM 52901:2017	Additive manufacturing – General principles – Requirements for purchased AM parts
ISO/ASTM 52910:2018	Additive manufacturing – Design – Requirements, guidelines and recommendations
ISO/ASTM 52915:2016	Specification for additive manufacturing file format (AMF) Version 1.2
ISO/ASTM 52921:2013	Standard terminology for additive manufacturing – Coordinate systems and test methodologies
ISO/ASTM 52902:2019	Additive manufacturing – Test artefacts – Standard guideline for geometric capability assessment of additive manufacturing systems

TABLE 6.2

ISO AM Standards under Development

Standard	Title
ISO/DIS 14649-17	Industrial automation systems and integration – Physical device control – Data model for computerized numerical controllers – Part 17: Process data for additive manufacturing
ISO/ASTM DIS 52900	Additive manufacturing – General principles – Fundamentals and vocabulary
ISO/ASTM DIS 52903-1	Additive manufacturing – Standard specification for material Extrusion based additive manufacturing of plastic materials – Part 1: Feedstock materials
ISO/ASTM DIS 52903-2	Additive manufacturing – Standard specification for material extrusion based additive manufacturing of plastic materials – Part 2: Process – Equipment
ISO/ASTM FDIS 52904	Additive manufacturing – Process characteristics and performance – Practice for metal powder bed fusion process to meet critical applications
ISO/ASTM DTR 52905	Additive manufacturing – General principles – Non-destructive testing of additive manufactured products
ISO/ASTM CD TR 52906	Additive manufacturing – Non-destructive testing and evaluation – Standard guideline for intentionally seeding flaws in additively manufactured (AM) parts
ISO/ASTM DIS 52907	Additive manufacturing – Technical specifications on metal powders
ISO/ASTM AWI 52908	Additive manufacturing – Post-processing methods – Standard specification for quality assurance and post processing of powder bed fusion metallic parts
ISO/ASTM AWI 52909	Additive manufacturing – Finished part properties – Orientation and location dependence of mechanical properties for metal powder bed fusion
ISO/ASTM DIS 52911-1	Additive manufacturing – Technical design guideline for powder bed fusion – Part 1: Laser-based powder bed fusion of metals
ISO/ASTM DIS 52911-2	Additive manufacturing – Technical design guideline for powder bed fusion – Part 2: Laser-based powder bed fusion of polymers
ISO/ASTM PWI 52911-3	Additive manufacturing – Technical design guideline for powder bed fusion – Part 3: Standard guideline for electron-based powder bed fusion of metals
ISO/ASTM CD TR 52912	Additive manufacturing – Design – Functionally graded additive manufacturing
ISO/ASTM PWI 52913	Additive manufacturing – Process characteristics and performance – Standard test methods for characterization of powder flow properties
ISO/ASTM PWI 52914	Additive manufacturing – Design – Standard guide for material extrusion processes
ISO/ASTM WD 52916	Additive manufacturing – Data formats – Standard specification for optimized medical image data
ISO/ASTM NP 52917	Additive manufacturing – Round Robin Testing – Guidance for conducting Round Robin studies
ISO/ASTM CD TR 52918	Additive manufacturing – Data formats – File format support, ecosystem and evolutions
ISO/ASTM NP 52919-1	Additive manufacturing – Test method of sand mold for metalcasting – Part 1: Mechanical properties
ISO/ASTM NP 52919-2	Additive manufacturing – Test method of sand mold for metalcasting – Part 2: Physical properties
ISO/ASTM PWI 52920-1	Additive manufacturing – Qualification principles – Part 1: Conformity assessment for AM System in industrial use
ISO/ASTM PWI 52920-2	Additive manufacturing – Qualification principles – Part 2: Conformity assessment at Industrial additive manufacturing centers
ISO/ASTM DIS 52921	Additive manufacturing – General principles – Standard practice for part positioning, coordinates and orientation
ISO/ASTM PWI 52922	Additive manufacturing – Design – Directed energy deposition

(Continued)

TABLE 6.2 (CONTINUED)

ISO AM Standards under Development

Standard	Title
ISO/ASTM PWI 52923	Additive manufacturing – Design decision support
ISO/ASTM CD 52924	Additive manufacturing – Qualification principles – Quality grades for additive manufacturing of polymer parts
ISO/ASTM WD 52925	Additive manufacturing – Qualification principles – Qualification of polymer materials for powder bed fusion using a laser
ISO/ASTM PWI 52926-1	Additive manufacturing – Qualification principles – Part 1: Qualification of machine operators for metallic parts production
ISO/ASTM PWI 52926-2	Additive manufacturing – Qualification principles – Part 2: Qualification of machine operators for metallic parts production for PBF-LB
ISO/ASTM PWI 52926-3	Additive manufacturing – Qualification principles – Part 3: Qualification of machine operators for metallic parts production for PBF-EB
ISO/ASTM PWI 52926-4	Additive manufacturing – Qualification principles – Part 4: Qualification of machine operators for metallic parts production for DED-LB
ISO/ASTM PWI 52926-5	Additive manufacturing – Qualification principles – Part 5: Qualification of machine operators for metallic parts production for DED-Arc
ISO/ASTM PWI 52927	Additive manufacturing – Process characteristics and performance – Test methods
ISO/ASTM PWI 52928	Powder life cycle management
ISO/ASTM PWI TR 52929	Guideline for installation/operation/performance qualification (IQ/OQ/PQ) of laser-beam powder bed fusion equipment for production manufacturing
ISO/ASTM AWI 52931	Additive manufacturing – Environmental health and safety – Standard guideline for use of metallic materials
ISO/ASTM WD 52932	Additive manufacturing – Environmental health and safety – Standard test method for determination of particle emission rates from desktop 3D printers using material extrusion
ISO/ASTM PWI 52933	Additive manufacturing – Environment, health and safety – Consideration for the reduction of hazardous substances emitted during the operation of the non-industrial ME type 3D printer in workplaces, and corresponding test method
ISO/ASTM PWI 52934	Additive manufacturing – Environmental health and safety – Standard guideline for hazard risk ranking and safety defence
ISO/ASTM CD 52941	Additive manufacturing – System performance and reliability – Standard test method for acceptance of powder-bed fusion machines for metallic materials for aerospace application
ISO/ASTM CD 52942	Additive manufacturing – Qualification principles – Qualifying machine operators of metal powder bed fusion machines and equipment used in aerospace applications
ISO/ASTM CD 52950	Additive manufacturing – General principles – Overview of data processing

human spaceflight programmes – the Space Launch System, Orion Spacecraft and the Commercial Crew Programme – is developing AM hardware and establishing a significant future role for AM in these systems. In many cases, the timeline for qualification of this early AM hardware and certification of its associated systems has been condensed, compared to that for the typical introduction of new manufacturing technologies. Two documents developed by the NASA Marshall Space Flight Centre (MSFC) were published in 2017 and provide an overarching framework of methodologies to meet the intent of existing requirements for spaceflight hardware. The two documents consist of MSFC-STD-3716, *Standard for Additively Manufactured Spaceflight Hardware by Laser Powder Bed Fusion in Metals* (2017), and an associated specification MSFC-SPEC-3717, *Specification for Control and Qualification of Laser Powder Bed Fusion Metallurgical Processes* (2017).

TABLE 6.3

Current SAE International AM Standards

Standard	Title
AMS7000 (2018)	Laser Powder Bed Fusion (L-PBF) Produced Parts, Nickel Alloy, Corrosion and Heat-Resistant, 62Ni -21.5Cr - 9.0Mo - 3.65Nb Stress Relieved, Hot Isostatic Pressed and Solution Annealed
AMS7001 (2018)	Nickel Alloy, Corrosion and Heat-Resistant, Powder for Additive Manufacturing, 62Ni - 21.5Cr - 9.0Mo - 3.65 Nb
AMS7002 (2018)	Process Requirements for Production of Powder Feedstock for Use in Laser Powder Bed Additive Manufacturing of Aerospace Parts
AMS7003 (2018)	Laser Powder Bed Fusion Process
AMS7004 (2019)	Titanium Alloy Preforms from Plasma Arc Directed Energy Deposition Additive Manufacturing on Substrate – Ti6Al4V – Stress Relieved
AMS7005 (2019)	Plasma Arc Directed Energy Deposition Additive Manufacturing Process

TABLE 6.4

SAE International AM Standards under Development

Standard	Title
AMS7006	Alloy 718 Powder
AMS7007	Electron Beam Powder Bed Fusion Process
AMS7008	Nickel Alloy, Corrosion and Heat-Resistant, Powder for Additive Manufacturing, Ni-Cr22-Fe18-Mo9–Co
AMS7009	Additive Manufacturing of Titanium 6Al4V with Laser-Wire Deposition – Annealed and Aged
AMS7010	Laser-Wire Directed Energy Deposition Additive Manufacturing Process
AMS7011	Additive Manufacture of Aerospace Parts from T-6Al-4V using the Electron Beam Powder Bed Fusion (EB-PBF) Process
AMS7012	17-4PH Powder for Additive Manufacturing
AMS7015	Ti-6Al-4V, Powder for Additive Manufacturing
AMS7016	Laser-Powder Bed Fusion (L-PBF) Produced Parts, 17-4PH H1025 Alloy
AMS7017	Titanium 6-Aluminum 4-Vanadium Powder for Additive Manufacturing, ELI Grade
AMS7018	Aluminium Alloy Powder 10.0Si – 0.35Mg (Compositions similar to UNS A03600)
AMS7020	Aluminium Alloy Powder, F357 Alloy
AMS7021	Stainless Steel Powder, 15-5PH Alloy
AMS7022	Binder Jetting Process
AMS7023	Gamma Titanium Aluminide Powder for Additive Manufacturing, Ti-48Al-2Nb-2Cr
AMS7024	Inconel 718 L-PBF Material specification
AMS7025	Metal Powder Feedstock Size Classifications for Additive Manufacturing

SAE International is a global association of engineers and technical experts focusing on mobility engineering. In 2015, the SAE AMS-AM, Additive Manufacturing, a technical committee in SAE International's Aerospace Materials Systems Group, was established with responsibility for developing and maintaining aerospace material and process specifications. The committee is also responsible for other SAE technical reports for AM, including precursor materials, additive processes, system requirements and post-build materials, pre-processing and post-processing, non-destructive testing and quality assurance. Over 300 global participants from more than fifteen countries representing aircraft, spacecraft, engine original equipment manufacturers (OEMs), material suppliers, operators, equipment/system suppliers, service providers, regulatory authorities and defence agencies are active in the committee. There are currently six subcommittees: Metals, Polymers, Non-destructive Inspection, General, Data Management and Regulatory Coordination. In addition to standards development, SAE's AMS-AM Data Management subcommittee is preparing data submission guidelines describing the minimum data requirements necessary to generate specification minimum values for both metals and polymers.

In 2019, the AWS released its D20.1:2019, *Specification for Fabrication of Metal Components using Additive Manufacturing*. This is an all-encompassing standard covering process qualification, validation, operator training and part traceability for both powder bed and directed energy deposition (DED) processes. D20.1 provides detail on the setup and acceptance of machines, operator training and the determination of material data. It covers geometries and numbers of material test specimens, machine key performance variables, includes testing and acceptance criteria for different criticality classes of components being made and feedstock material being used. D20.1 provides extensive checklists to enable auditing to the standard and additional guidance information for validation, key performance variable checking, design and production process control as well as a number of suggested recording forms for many aspects of the process.

6.2 AM Standards Roadmaps

6.2.1 America Makes

Aiming to provide a comprehensive and holistic approach to the development of AM standards, America Makes, a public–private consortium, partnered with the American National Standards Institute (ANSI) to establish the Additive Manufacturing Standardisation Collaborative (AMSC). While not chartered to draft standards, the AMSC has engaged all the major stakeholders in AM, including end users, equipment manufacturers, consumable suppliers and research institutions, to develop a *Standardization Roadmap for Additive Manufacturing* (America Makes and ANSI 2018). Originally published in 2017, an updated version was published a year later.

6.2.2 Identified Gaps in the Roadmaps

The roadmaps identified a total of ninety-three open gaps and corresponding recommendations across five topical areas: 1) design; 2) process and materials (precursor materials, process control, post-processing and finished material properties); 3) qualification and certification; 4) non-destructive evaluation; and 5) maintenance. Of that total, eighteen gaps/recommendations were identified as high priority, fifty-one as medium priority and twenty-four as low priority. A 'gap' means that no published standard or specification exists that covers the particular issue in question. In sixty-five cases, additional research and development was needed.

6.3 AM Powder Feedstock Characterisation Standards

AM metal powder characteristics have a major influence on both how the powder behaves in the AM process and how it affects the quality and performance of the part. A number

TABLE 6.5

Powder Features

Powder property	Category	Parameter
Intrinsic	Morphology	Particle size distribution
		Particle shape
		Particle density
Extrinsic	Microstructure	Internal porosity
		Crystal structure
	Chemistry	

of the current standards are based on what has been carried out in the pharmaceutical sector, which has a long history of dealing with powders in the micrometre range. For any particular powder, a number of properties can be measured, which are listed in Table 6.5.

Since 2018, ASTM has run an Additive Manufacturing Powder Proficiency testing programme to establish competence in testing metallic AM powders using a range of existing testing standards, which is shown in Table 6.6. The work is needed to ensure that the methods can be relied upon to give repeatable and accurate results when used by competent parties, as many of these standards were not developed for free-flowing metal powders.

6.3.1 Feedstock Sampling Strategy

Prior to any powder analysis, the question needs to be answered as to whether any sample selected is representative of the bulk powder. Powder sampling is an area of study that has been well covered by the plastics and pharmaceutical industries (Allen 1990). Allen identified four stages of sample size reduction from bulk size to measurement sample, as shown in Table 6.7.

There are a large number of ways in which a gross powder can be sampled, with the sampling method based upon whether the powder is free flowing or cohesive and whether the powder is stationary (static) or moving (dynamic).

- *Static non-flowing materials* – This category is comprised of fine cohesive powders, sticky materials, moist material or fibrous solids. In AM, factors such as moisture content could alter flowability results and are, therefore, something that should be kept in mind (Allen 1990). Provided the powder has been passed through a mixer, surface sampling using a scoop is generally considered to be an adequately representative measure of powder particle size distribution (PSD).
- *Static free-flowing materials* – Free-flowing powders have a greater tendency than non-flowing powders to segregate when poured into a heap or into a container. Fine powders tend to move toward the centre of a powder mass, whilst the larger, coarser particles tend to roll down to the outside of a pile when being poured. Furthermore, particle movement can be exacerbated when powder storage containers are subject to vibration, causing coarse material to migrate towards the surface. Interestingly, this phenomenon happens even if the larger particles are denser than their smaller counterparts (Allen 1990). These tendencies make sampling of a free-flowing powder a complex task.
- *Dynamic (moving) powders* – Dynamic powder sampling refers to sampling which is carried out when the powder is in motion and can be achieved through continuous

TABLE 6.6

ASTM Proficiency Testing Programme of AM Metal Powders, Tests Covered

ASTM B212-13	Test Method for Apparent Density of Free-Flowing Metal Powders using the Hall Flowmeter Funnel
ASTM B213-17	Test Methods for Flow Rate of Metal Powders Using the Hall Flowmeter Funnel
ASTM B214 -16 (equivalent to ISO 4497)	Test Method for Sieve Analysis of Metal Powders
ASTM B215-15	Practices for Sampling Metal Powders
ASTM B417-18	Test Method for Apparent Density of Non-Free-Flowing Metal Powders Using the Carney Funnel
ASTM B527-16	Test Method for Tap Density of Metal Powders and Compounds
ASTM B822-17 (equivalent to ISO 13320)	Test Method for Particle Size Distribution of Metal Powders and Related Compounds by Light Scattering
ASTM B855-17	Test Method for Volumetric Flow Rate of Metal Powders Using the Arnold Meter and Hall Flowmeter Funnel
ASTM B923-16	Test Method for Metal Powder Skeletal Density by Helium or Nitrogen Pycnometry
ASTM B964-16	Test Methods for Flow Rate of Metal Powders Using the Carney Funnel
ASTM E1019-18	Test Methods for Determination of Carbon, Sulfur, Nitrogen, and Oxygen in Steel, Iron, Nickel, and Cobalt Alloys by Various Combustion and Inert Gas Fusion Techniques
ISO 3923-1	Metallic powders – Determination of apparent Density – Funnel method NOTE: This method is equivalent to ASTM B212 and ASTM B417
ISO 13322-1	Particle size analysis – Image analysis methods – Static image analysis methods
ISO 13322-2	Particle size analysis – Image analysis methods – Dynamic image analysis methods
Composition can be determined by suitable wet chemical, wavelength dispersive X-ray, atomic emission techniques such as:	
ASTM E539-19	Standard Test Method for Analysis of Titanium Alloys by X-Ray Fluorescence Spectrometry
ASTM E572-13	Standard Test Method for Analysis of Stainless and Alloy Steels by Wavelength Dispersive X-ray Fluorescence Spectrometry
ASTM E1834-18	Test Method for Analysis of Nickel Alloys by Graphite Furnace Atomic Absorption Spectrometry
ASTM E2465-19	Standard Test Method for Analysis of Ni-Base Alloys by Wavelength Dispersive X-Ray Fluorescence Spectrometry
ASTM E2594-09	Standard Test Method for Analysis of Nickel Alloys by Inductively Coupled Plasma Atomic Emission Spectrometry (Performance-Based Method)
ASTM E2823-17	Standard Test Method for Analysis of Nickel Alloys by Inductively Coupled Plasma Mass Spectrometry (Performance-Based)
ASTM E3047-16	Standard Test Method for Analysis of Nickel Alloys by Spark Atomic Emission Spectrometry

or intermittent methods. While the preferred sampling method is to extract while the powder is in motion (i.e. being poured between two containers), it may not always be possible or feasible to do so. For static measurement, a device called a 'powder thief' is employed, which consists of a long rod with a sampling chamber along its length. Using an open-and-shut mechanism, powder can be extracted which is representative of the bulk powder lot.

In order to make the sample size more manageable, the material needs to be further sub-divided, and this can be achieved through a variety of methods including cone and quartering, scoop sampling, table sampling, chute splitting and spin riffling, which are

TABLE 6.7
Terms Describing Powder Shapes in ASTM B243-19

Term	Definition
acicular	needle-shaped
flake	flat or scale-like; thickness is small compared with other dimensions
granular	approximately equidimensional; non-spherical shapes
irregular	lacking symmetry
needles	elongated and rod-like
nodular	irregular, having knotted, rounded or similar shaped
platelet	composed of flat particles having considerable thickness
plates	flat particles of metal powder having considerable thickness
spherical	globular-shaped

described in ASTM B215-15 (2015). Cone and quartering involve mounding a pile of powder into a cone-shaped heap, flattening it with a spatula, dividing it into four sections, and repeating the process on one of the sections, so that the final sample is one sixteenth the size of the original sample. Scoop sampling simply involves using a scoop to select a portion of the bulk sample. Table sampling involves pouring the bulk sample of powder down an inclined plane that has a series of structures and holes used to divide it. Chute splitting is a process in which samples are divided into two lots, via dispersion, through a series of chutes. Spin riffling involves pouring the bulk sample into a hopper that empties onto a vibratory chute that leads to a series of sample containers contained in a rotating ring. According to Allen (1990), the spin riffler provided statistically significant results which were superior to other methods and should be the method of choice.

ASTM B215-15 (2015) describes the procedures for sampling metal powders. The standard focuses on bulk powder sampling from storage tanks or blenders as well as pre-packaged powders in containers such as bags. The standard also provides specific instructions for the collecting of metal powder from a moving stream.

6.3.2 Particle Size Determination and Distribution

In AM, powder particle size determines the minimum part layer thickness, as well as the minimum buildable feature sizes on a part. A number of commonly used methods for particle size determination include sieving, gravitational sedimentation, microscopy-based techniques and laser light diffraction.

Particle size determination and distribution are addressed by several ASTM standards. ASTM B214-16 (2016) provides specifications for the sieving process, recommending the arrangement of sieves in consecutive order by the size of their opening. The coarsest sieve should be placed at the top with a collection pan placed at the bottom below the finest sieve. Sieve meshes made from brass, bronze or stainless-steel wires and ranging from 5 μm to 1 mm are used. The sieves and collecting pan are placed and fastened, along with the powder, into a sieve shaker and agitated for fifteen minutes. While the standard provides the sieve dimensions to be 200 mm in diameter and either 25.4 mm or 50.8 mm, ASTM E161-17 (2017) states that a smaller sieve diameter of 76.2 mm can be used, albeit with the same depth range.

A commonly used technique for measuring the PSD is the use of light scattering, which is described in ASTM B822-17 (2017). The method can either be used with the powder dispersed in a liquid or in a carrier gas. The method is suitable for particles in the range of 0.4 μm to 2 mm. By passing the particles through the path of a light beam, light scattering occurs. Photodetector arrays collect the scattered light and convert it into electrical signals, which are then analysed. The analysed signal is converted to size distribution through the theories of Fraunhofer diffraction, Mie scattering or a combination of both (ASTM B822-17 2017, Simmons and Potter 2010).

6.3.3 Morphology Characterisation Methods

The morphology of powder particles determines how well the particles lay or pack together and thus, in AM powder bed systems, is an important factor in determining the required part density. Prior to any characterisation technique being employed, it is necessary to ensure that each powder particle has an equal chance of being selected for analysis, and this has been discussed in Section 6.3.1.

Particles can be characterised either qualitatively or quantitatively. One method is single-number classification, in which a shape is defined by applying one number associated with a feature of a particle (Hawkins 1993). Single-number classification suffers from two disadvantages: ambiguity, in that more than one outline shape can have the same resulting value; and the difficulty in reproducing the shape from a single number. Figure 6.1 illustrates the ambiguity with the classification with all the objects (a–h) having the same projected area but clearly different shapes. Each shape would behave differently with respect to its flowability and packing, resulting in parts being produced with varying densities.

Qualitative assessment involves imaging the particles against a bright background to provide a silhouette of each particle (Cox 1927). Each particle shape has, over the years, been assigned an adjective to describe it, and this has been enshrined in ASTM B243-19 (2019). Table 6.6 lists some of the more common powder shapes. It should be noted that, as some definitions are similar, the standard does lack some scientific rigour.

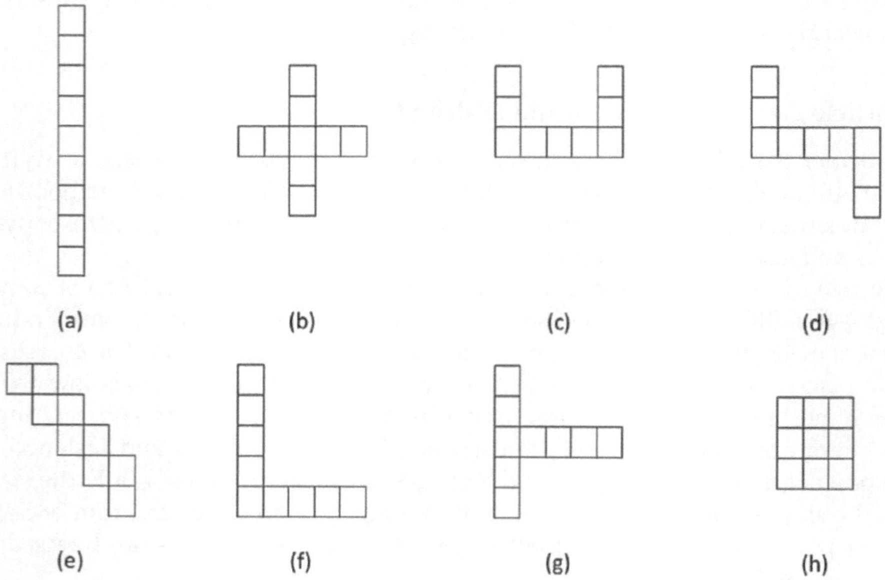

FIGURE 6.1
Shapes illustrating the ambiguity of a single number classification system.

6.3.4 Flow Characteristics

How well a powder flows is an important criterion for both powder bed and powder blown AM processes. Methods of determining the flow rate of powders use two types of flowmeters, the Carney funnel (ASTM B964-16 2016) and the Hall flowmeter funnel (ASTM B213-17 2017).

There are two types of methods for determining the flow rate: a static flow method and a dynamic flow method. The static flow method involves the user covering the funnel opening with their finger and pouring in a pre-determined mass of powder. Once the funnel is loaded with powder, the finger is removed from the orifice and the timing taken for all the powder to exit the funnel. The dynamic flow method does not require the covering of the orifice, and the timing begins as soon as the powder begins flowing out of the funnel and stops when all the powder has flowed through. The flow rate value, reported in units of seconds per gram, is determined by dividing the measured time taken for all of the powder to exit the funnel by the mass of the powder sample. The difference between the Hall flowmeter funnel and the Carney funnel is the size of their orifice, with the Hall flowmeter having a 2.54 mm opening, and the Carney flowmeter having a 5.08 mm opening. The Carney flowmeter is used only for powder that will not flow through the Hall flowmeter.

The process of timing powder as it flows through a Hall flowmeter is described in ASTM B855-17 (2017). However, this standard specifies measuring the flow by volume instead of by mass, thereby eliminating the variable of powder density. The ability to flow and pack is a function of interparticle friction, so as the surface area increases, the friction increases, which gives less efficient flow and packing. The acquisition of the correct volume of powder to use requires the use of an Arnold meter (see Section 6.3.6). Once the volume to be tested is attained, the process follows that of ASTM B213-17 (2017). This flow rate value is determined by dividing the measured time taken for all of the powder to exit the funnel by the volume of the powder sample, and is reported in units of seconds per centimetre cubed. Care is required to reduce the impact of user repeatability and environmental conditions (for example, moisture) on the final result. Other techniques, such powder rheometry or angle of repose, can provide useful dynamic powder metrics, but they are yet to be standardised for use with metal AM powders.

6.3.5 Thermal Characterisation

The melting of powder by a high energy source is the essence of many AM processes. Thus, a good understanding of the thermal properties of powders is highly significant to AM. A comprehensive summary of techniques used to determine the conductivity of materials was undertaken by Sih and Barlow (1992) who divided the techniques into two categories: steady-state and transient. While the techniques focused on solid specimens, the authors claimed that the methods were also applicable to powders.

A number of steady-state techniques are centred on heating up the specimen in an insulated environment and measuring various parameters, including numerous temperature readings and input power. The guarded-hot-plate method is the only method described in an ASTM standard (ASTM C177-19 2019) for steady-state measurements. Other methods include the cylindrical method and the spherical and ellipsoidal methods, which use a heater positioned in the centre of a cylindrical, spherical or ellipsoidal specimen such that heat radiates out from the centre of the sample, with temperatures being measured at different radii and heat transfer principles being applied to determine the thermal conductivity.

In all of the steady-state methods, the calculation of the thermal conductivity involves simple heat-transfer equations, and the heat losses by radiative heat transfer are assumed to be negligible. However, there is often difficulty in obtaining the required specimen shape. On the other hand, with transient methods, obtaining the thermal conductivity is much faster, although the calculations are more involved.

Transient heat techniques typically measure the temperature as a function of time, allowing for heat transfer principles to be applied to calculate the thermal conductivity. Two techniques that follow the approach are the transient hot wire method and the thermal probe method (Sih and Barlow 1992). An alternative method is based on measuring electrical resistance in a metallic strip placed within the specimen. By knowing the input temperature and output voltage, the thermal conductivity of the surrounding material can be determined. A fourth method which lends itself to determining the thermal conductivity of a powder is the flash method (Parker et al. 1961), in which a high-intensity light pulse is focused onto the surface of the specimen. The temperature of the surface on the opposite side of the specimen is then measured. The thermal diffusivity is determined by the shape of the curve of temperature versus time at the opposing surface. The heat capacity is determined by the maximum temperature reached by the opposing surface. Finally, the thermal conductivity is determined by multiplying the heat capacity, thermal diffusivity and density.

6.3.6 Density Determination

A number of methods for determining the apparent density of metal powder are contained in ASTM standards. Defined as the ratio of mass to a given volume of powder, the apparent density can be simply determined using the Hall flowmeter and the Carney funnel. ASTM B212-13 (2013) describes using the Hall flowmeter, where the powder flows through the flowmeter into a cup of definite value. The powder should mound over the cup, and a non-magnetic spatula is used to level off the top surface of the powder flush with the sides of the cup. The mass is determined using a balance to measure the cup and powder, and the apparent density is calculated from the measured mass divided by the volume. The process of determining the apparent density of metal powder through the use of a Carney funnel is outlined in ASTM B417-18 (2018). This process mirrors that of the apparent density determination process using the Hall flowmeter funnel, only the Carney flowmeter funnel is used instead.

The technique of determining the apparent density using the Arnold meter is described in ASTM B703-17 (2017). An Arnold meter is a steel block with a cavity in the middle and a powder delivery sleeve. A powder delivery sleeve is placed on either side of the die cavity and powder is poured into the sleeve, which is slid across the cavity, allowing the powder to fill the die cavity. The sleeve is then slid back across the cavity to level the amount of powder flush with the steel block. The amount of powder in the die cavity is placed on a balance to obtain the mass. The apparent density is the mass divided by the volume of the die cavity.

ASTM B527-16 (2016) describes the method for determining the tap density of metallic compounds and powders. Tap density is defined as the density of a powder that has been tapped to settle the contents in a container. A known mass of powder is poured into a graduated cylinder. The cylinder is loaded into a tapping apparatus, which taps against the base of the apparatus at a rate between 100 taps per minute and 300 taps per minute. When there is no further decrease in volume due to tapping, that volume is used in the calculation of the tap density, which is the mass divided by the volume.

One test for internal porosity (as some powder production methods can lead to varying degrees of trapped gas porosity) is helium pycnometry (ASTM B923-16 2016); this provides the skeletal density of the powder, which should be equivalent to the expected bulk density for the alloy.

6.3.7 Chemical Composition

In practice, the metallic feedstock used in AM is not totally pure and contains other materials. These added materials consist of different elements that, when combined and used in AM, produce material properties specific to the bulk material type. However, the overall chemical composition of the metal powder can have an influence on the final part's mechanical properties.

Powder properties can also change over time, or in the case of recycled powder, due to repeated exposure to the AM build chamber environment. Techniques that are commonly applied to determine the chemical composition of powder include microanalysis, surface analysis and bulk analysis methods.

Only bulk analysis methods and, specifically, the inert gas fusion technique, are described in standards. The inert gas fusion technique is used to determine the oxygen, nitrogen and hydrogen content in metals, as described by ASTM E1409-13 (2013), ASTM E1447-09 (2016), and ASTM E2792-16 (2016). In this technique, a sample is held in chamber directly above a graphite crucible, which is brought up to a temperature of around 3000 °C. An inert gas flows over the crucible to remove any contaminants before the crucible temperature is lowered and the sample is added. As the sample melts, the oxygen present reacts with the carbon in the graphite crucible to form carbon monoxide and carbon dioxide. The nitrogen present is released as molecular nitrogen, and the hydrogen is released as hydrogen gas. The gases are carried out of the chamber onto a detector by the inert gas.

6.4 Processes

While the main focus has been on developing standards around the definitions, data formats and design guidelines, there has been interest in developing process standards. In the ASTM/ISO standards roadmap, standards have been categorised as general top-level category AM and specialised AM standards. Both categories, AM and specialised AM standards, have identified processes as an area of interest to develop new standards.

6.5 Part Verification

Part verification can cover a wide range of measurements in order to provide confidence in the part's ability to undertake the function for which it is intended. These measurements can cover areas such as mechanical, dimensional, tribological and thermal properties. EN ISO 17296-3 (2014) and AWS D20.1 provide lists of recommended material validation tests and test artefacts.

6.5.1 Tensile Properties

Tensile tests are used to determine the response of a material to withstand tensile loads until their failure point is reached. ASTM E8/8M-16a (2016) is the standard for tensile testing of metallic materials under standard laboratory conditions, while ASTM E21 (2017) covers testing at elevated temperatures. In addition to the ASTM standards, ISO 6892 (2019) covers tensile testing of metallic materials. These tensile testing standards specify different coupon types, including flat and cylindrical coupons, as well as loading conditions, such as strain rates. Work is ongoing to ensure that the coupon geometries can provide consistent results for AM materials, with proposals for new micro-scale coupons being developed so that the actual properties can be measured by cutting test specimens from actual test parts (for instance, to evaluate thin wall or graded property parts).

6.5.2 Compressive Properties

The compressive strength of metals can be measured by following these two ASTM standards that cover testing at room temperature (ASTM E9-19 2019) and at elevated temperatures (ASTM E209-18 2018). The stress-strain curve, compressive strength and elastic modulus can be found. Table 6.8 shows typical results for two metals which have been subjected to compression testing.

6.5.3 Hardness Measurement

Hardness can be defined as a measure of how resistant the metal is to various kinds of permanent change in shape when subjected to compressive force. A number of different hardness tests exist which vary according to the type and geometry of the indenter and the number of indentations.

The Brinell hardness test uses a sphere to indent a hole on the test coupon surface. The diameter of the indentation is used to calculate the hardness value. ASTM E10-18 (2018) and ISO 6506 (2014) describe the standard test method for Brinell hardness for metallic materials.

The Knoop and Vickers hardness test methods are similar to the Brinell hardness test but employ a pyramid-shaped indenter. The diagonal length of the indentation is used to determine the hardness. The face angle of the pyramid is different between the Knoop and Vickers tests. ASTM E384-17 (2017) describes the test methods for Knoop and Vickers and ISO 4545 (2017) provides the standard for the Knoop hardness test.

The Rockwell hardness of metallic materials is described in ASTM E18-19 (2019). The standard covers both pyramid and spherical indenters. The test performs indentations multiple times at a single spot with increasing force. The depth of the indenter is measured by the machine directly providing the indentation depth. Table 6.9 shows the hardness results for the same materials deposited using different AM processes.

TABLE 6.8

Compression Test Results for Metallic Samples

Process	Material	Porosity	E/GPa	Compressive strength/ MPa	References
Laser Powder Bed Fusion	Titanium	55	0.687	-	Sing, et al. (2016a)
	Ti6Al4V	70	5.1 ± 0.3	155 ± 7	Wang, et al. (2016)

TABLE 6.9
Hardness Test Results for Metallic Samples

Process	Material	Microhardness/HV	References
Laser Powder Bed Fusion	Ti6Al4V	479-613	Sing, et al. (2016b)
Electron Beam Powder Bed Fusion	Ti6Al4V	358-387	

6.5.4 Fatigue Measurement Methods

Fatigue can be described as the weakening of a material due to repeated applied loads or cyclic loading. While a number of standards exist to measure the fatigue properties of conventionally produced metal, little has been published for AM metals.

ASTM E466-15 (2015) and ISO 1099 (2017) are the standards for axial-force fatigue testing. The test involves a sample being pulled axially with a periodic force function, which typically takes the form of a sine wave. For notched samples, where information is required about the material resistance to crack growth, ASTM E647-15e1 (2015) and ISO 12108 (2018) describe the standard test method. The surface condition of AM-made parts can have a significant impact on the results obtained, so it should be carefully considered.

The creep-fatigue test method is similar to the fatigue test but is carried out at elevated temperatures. The creep-fatigue test method is able to provide both the strain/stress over time curve and the stress/strain hysteresis curve. ASTM E2714-13 (2013) describes the standard test method.

6.5.5 Fracture Toughness

Fracture toughness can be described as a material's ability to contain a crack to resist fracture. Several standards exist, with ASTM E399-19 (2019) measuring the K_{Ic} (plane-strain fracture toughness) while ASTM E1820-18ae1 (2018) gives the fracture toughness from R-curves (crack resistance). ISO 12737 (2010) describes how to determine the plane-strain fracture toughness, while ISO 12135 (2016) provides a unified method of test to determine quasistatic fracture toughness.

6.5.6 Other Properties

In addition to those described above, other standards exist to determine properties, such as rupture strength (ASTM E292-18 2018), Young's modulus (ASTM E111-17 2017), Poisson's ratio (ASTM E132-17 2017) and shear modulus (ASTM E143-13 2013).

6.6 Surface Standards

A part's surface can be considered as a fingerprint back to the actual manufacturing process and the respective input parameters, while at the same time it can provide information about the functional performance of the part (see Chapter 11 for a thorough treatment on AM surface measurement and characterisation).

6.6.1 Profile and Areal Surfaces

Traditionally, the surface has been characterised by taking a profile measurement to be representative of the whole surface. While this has been sufficient for many manufacturing processes that generate isotropic surfaces, the same cannot be said for those that have anisotropic properties. ISO 4287 (1997) is the standard for defining the profile surface texture parameters. With the interest in providing more representative areal measurements, which can provide more information regarding surface structure, direction and the relationship between features, a set of standards have been developed as part of the Geometrical Product Specification (GPS) family. More details of these standards are given in Chapter 11.

6.7 Dimensional Standards

6.7.1 Performance Verification of Coordinate Measuring Machines

In order to provide a basis of comparison for performance between various coordinate measuring machine (CMM) manufacturers, a family of standards was established. ISO 10360 (many sub-parts for specific measurement tools) defines the procedure for performance verification of CMMs. The standards cover a wide range of configurations including contact and non-contact probing systems as well as additional axial stages. While most parts have been issued, part 11 and part 13 are still under development. These standards are discussed in detail in Chapter 10.

6.8 Non-Destructive Evaluation Standards

6.8.1 Current Standards

Non-destructive evaluation (NDE) is primarily used in AM to detect and characterise defects (for more information on defects, their sources and control, see Chapters 9 and 13). Defects found in direct energy deposition (DED) systems are similar to those detected in welding. Common defects for DED are:

1. Poor surface finish.
2. Porosity.
3. Incomplete fusion.
4. Lack of geometrical accuracy.
5. Undercuts at the toe of welds between adjoining weld beads.
6. Non-uniform well beads.
7. Holes or voids.
8. Non-metallic inclusions.
9. Cracking.

For powder bed fusion (PBF), the common defects are:

1. Unconsolidated powder.
2. Lack of geometrical accuracy.
3. Reduced mechanical properties.
4. Inclusions.
5. Voids.
6. Layer voids or porosity.
7. Cross-layer.
8. Porosity.
9. Poor surface finish.
10. Trapped powder.

In DED, material is fused together by melting as it is being deposited. DED processes are primarily used to either add features to an existing structure or to repair worn parts. In PBF, powder is deposited onto a build platform and, using a localised energy source, is fused to form a section through the component. PBF, unlike DED, does not exhibit similarities to welding; however, common defects, such as porosity and voids, are similar to welding defects.

In addition to welding, some common casting defects – gas porosity, cracking and inclusions – are similar to DED and PBF defects. Hence, NDE standards for casting will also be described and their suitability to AM defects discussed.

6.8.2 Welding Standards

A number of NDE standards exist which cover various aspects of inspection in welding. ISO 5817 (2014) and ISO 10042 (2018) specify the welding quality requirements; both these standards feed into ISO 17635 (2016), which acts as the bridge between quality levels and the acceptance levels for indications. ISO 17635 also describes the method selection process, which is split into six method-specific standards:

1. Radiographic.
2. Magnetic particle.
3. Eddy current.
4. Penetrant.
5. Ultrasonic.
6. Visual examination.

The respective method-specific standard describes the test procedure and characterisation acceptance levels (see also Chapter 9). As each method has its own inherent errors based on material properties, component or target defect, a combination of methods is normally applied. In the case of radiography and ultrasonic, a number of sub-method standards exist as shown in Figure 6.2 and Figure 6.3. Currently, there a number of advanced NDE methods which lack any standard; these include ultrasonic phased array, thermography and X-ray computed tomography. A number of these methods offer great potential for AM applications (see, for example, Chapters 9 and 12).

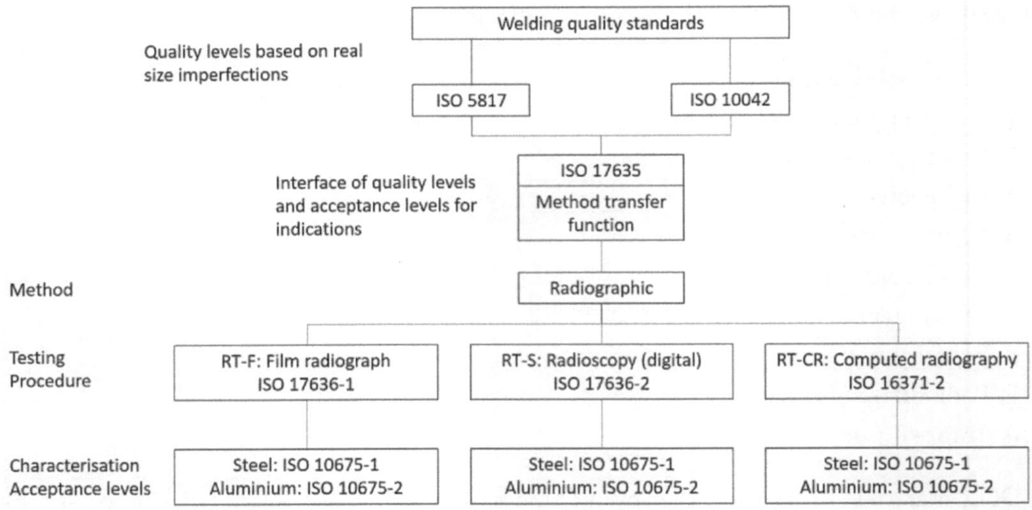

FIGURE 6.2
Diagram showing the current structure of NDE standards (radiography only) for welding. (Adapted from ISO17635 [2016].)

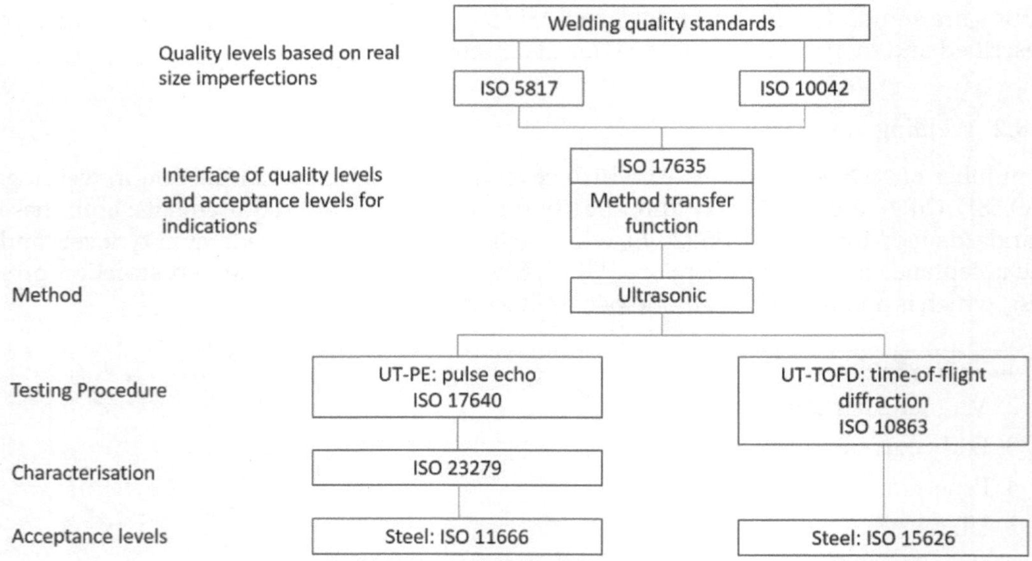

FIGURE 6.3
Diagram showing the current structure of NDE standards (ultrasonic only) for welding. (Adapted from ISO17635 [2016].)

6.8.3 Casting Standards

ISO 4990 (2015) provides an overarching guide for casting NDE standards. It categorises casting flaws into surface discontinuity and internal discontinuity. Each of the five main conventional NDE methods has its own standard as follows:

1. Visual examination ISO 11971 (2008) (surface).
2. Magnetic particle inspection ISO 4986 (2010) (surface).
3. Liquid particle inspection ISO 4987 (2010) (surface).
4. Ultrasonic examination ISO 4992 (2006) (internal).
5. Radiographic testing ISO 4993 (2015) (internal).

The case for advanced NDE methods for castings is the same as that for welding, with no standards for X-ray computed tomography, phased array ultrasonic and thermography.

6.9 Future and Planned Standards Activities

The development of AM specific standards and those that provide guidance for the most appropriate use of other existing standards is still in the early stages (Leach et al. 2019). Table 6.10 is a list of standards currently in draft. It is envisaged that, in addition to the activities listed in the table, other areas of focus for the AM community include the following:

- Continual updating of the terminology as new AM processes (for example, print and sinter or material jetting) become more industrially relevant.

TABLE 6.10

Areas for Potential New AM Standards

Area	Topic
AM machines	Facility Requirements for Metal Powder Bed Fusion
	Initial Qualification, Operational Qualification and Part Qualification of Metal Powder Bed Fusion Machines
	Acceptance Testing of Powder-Bed Fusion AM Machines for Metallic Materials for Aerospace Application
	Specifying Gases and Nitrogen Generators Used with Metal Powder Bed Fusion Machines
	Cleaning Metal Powder Bed Fusion Machines
	Creating Maintenance Schedules and Maintaining Metal Powder Bed Fusion Machines
	Calibration of Metal Powder Bed Fusion Machines and Subsystems
	Establishing a Personnel Training Program for Metal Powder Bed Fusion Part Production
	Qualification of operators for additive manufacturing equipment used in aerospace applications
Feedstock	Creating Feedstock Specifications for Metal Powder Bed Fusion
	Receiving and Storing of Metal Powders Used in Powder Bed Fusion
	Metal Powder Reuse in the Powder Bed Fusion Process
	Disposal of Metal Powders Used for Powder Bed Fusion
Process	Digital Workflow Control for the Metal Powder Bed Fusion Process
	Establishing Manufacturing Plan and Sequence of Operation Work Flow for Metal Powder Bed Fusion Part production
	Storage of Technical Build Cycle Data
Others	Post Thermal Processing of Metal Powder Bed Fusion Parts
	Metallographic Evaluation of Metal Powder Bed Fusion Test Specimens and Parts
	Process Specification for Directed Energy Deposition (Wire+Arc, Wire+Beam, Laser Blown Powder) Used in Aerospace Application

- More detailed guidance of NDE to enable faster, more cost-effective quality assurance of parts (see Chapter 9).
- Specific guidance for testing, specification, safety and recycling of powder feedstock for AM.
- Standards for control and specification of the process and feedstock for various DED and polymer extrusion processes.
- The training and assessment of trained personnel for specific roles, for example, machine operator and AM designer, so that end-users know that staff have the right level of knowledge and understanding to undertake their work.
- More detailed standards outlining the installation, commissioning, maintenance and ongoing validation of machines.
- Guidance on the assessment and use of in-process sensing and feedback sensing, which are getting added to new AM commercial platforms.
- Round robin testing – to help provide assured data that can be re-used and added to the required materials databases.
- Health and safety – powder (and other feedstock) hazards, for personal and environmental impact.
- Standards looking at data transfer, medical imaging and specifically for high-technology sectors such as aerospace.

References

AC7110/14. 2017. Nadcap Audit Criteria for Laser and Electron Beam Metallic Powder Bed Additive Manufacturing, NADCAP.

Allen, T. 1990. *Particle Size Measurement*. 4th ed. London: Chapman and Hall.

America Makes and ANSI. 2018 Standardization Roadmap for Additive Manufacturing (Version 2.0) Available at https://www.ansi.org/standards_activities/standards_boards_panels/amsc/America-Makes-and-ANSI-AMSC-Overview.

ASTM B212-13. 2013. Standard Test Method for Apparent Density of Free-Flowing Metal Powders Using the Hall Flowmeter Funnel, ASTM International.

ASTM B213-17. 2017. Standard Test Methods for Flow Rate of Metal Powders Using the Hall Flowmeter Funnel, ASTM International.

ASTM B214-16. 2016. Standard Test Method for Sieve Analysis of Metal Powders, ASTM International.

ASTM B215-15. 2015. Standard Practices for Sampling Metal Powders, ASTM International.

ASTM B243-19. 2019. Standard Terminology of Powder Metallurgy, ASTM International.

ASTM B417-18. 2018. Standard Test Method for Apparent Density of Non-Free-Flowing Metal Powders Using the Carney Funnel, ASTM International.

ASTM B527-16. 2016. Standard Test Method for Determination of Tap Density of Metallic Powders and Compounds, ASTM International.

ASTM B703-17. 2017. Standard Test Method for Apparent Density of Metal Powders and Related Compounds Using the Arnold Meter, ASTM International.

ASTM B822-17. 2017. Standard Test Method for Particle Size Distribution of Metal Powders and Related Compounds by Light Scattering, ASTM International.

ASTM B855-17. 2017. Standard Test Method for Volumetric Flow Rate of Metal Powders Using the Arnold Meter and Hall Flowmeter Funnel, ASTM International.

ASTM B923-16. 2016. Standard Test Method for Metal Powder Skeletal Density by Helium or Nitrogen Pycnometry, ASTM International.

ASTM B964-016. 2016. Standard Test Methods for Flow Rate of Metal Powders Using the Carney Funnel, ASTM International.

ASTM C177-19. 2019. Standard Test Method for Steady-State Heat Flux Measurements and Thermal Transmission Properties by Means of the Guarded-Hot-Plate Apparatus, ASTM International.

ASTM E8/E8M-16a. 2016. Standard Test Methods for Tension Testing of Metallic Materials, ASTM International.

ASTM E9–19. 2019. Standard Test Methods of Compression Testing of Metallic Materials at Room Temperature, ASTM International.

ASTM E10-18. 2018. Standard Test Method for Brinell Hardness of Metallic Materials, ASTM International.

ASTM E18–19. 2019. Standard Test Methods for Rockwell Hardness of Metallic Materials, ASTM International.

ASTM E21-17. 2017. Standard Test Methods for Elevated Temperature Tension Tests of Metallic Materials, ASTM International.

ASTM E111–17. 2017. Standard Test Method for Young's Modulus, Tangent Modulus, and Chord Modulus, ASTM International.

ASTM E132–17. 2017. Standard Test Method for Poisson's Ratio at Room Temperature, ASTM International.

ASTM E143–13. 2013. Standard Test Method for Shear Modulus at Room Temperature, ASTM International.

ASTM E161-17. 2017. Standard Specification for Precision Electroformed Sieves, ASTM International.

ASTM E209-18. 2018. Standard Practice for Compression Tests of Metallic Materials at Elevated Temperatures with Conventional or Rapid Heating Rates and Strain Rates, ASTM International.

ASTM E292-18. 2018. Standard Test Methods for Conducting Time-for-Rupture Notch Tension Tests of Materials, ASTM International.

ASTM E384-17. 2017. Standard Test Method for Microindentation Hardness of Materials, ASTM International.

ASTM E399-19. 2019. Standard Test Method for Linear-Elastic Plane-Strain Fracture Toughness, ASTM International.

ASTM E466-15. 2015. Standard Practice for Conducting Force Controlled Constant Amplitude Axial Fatigue Tests of Metallic Materials, ASTM International.

ASTM E647-15e1. 2015. Standard Test Method for Measurement of Fatigue Crack Growth Rates, ASTM International.

ASTM E1409-13. 2013. Standard Test Method for Determination of Oxygen and Nitrogen in Titanium and Titanium Alloys by Inert Gas Fusion, ASTM International.

ASTM E1447-09. 2016. Standard Test Method for Determination of Hydrogen in Titanium and Titanium Alloys by Inert Gas Fusion Thermal Conductivity/Infrared Detection Method, ASTM International.

ASTM E1820–18ae1. 2018. Standard Test Method for Measurement of Fracture Toughness, ASTM International.

ASTM E2714–13. 2013. Standard Test Method for Creep-Fatigue Testing KIc of Metallic Materials, ASTM International.

ASTM E2792-16. 2016. Standard Test Method for Determination of Hydrogen in Aluminum and Aluminum Alloys by Inert Gas Fusion, ASTM International.

AWS D20.1. 2019. Specification for Fabrication of Metal Components using Additive Manufacturing, American Welding Society.

Carson, J. W., Pittenger, B. H. 1998. *Bulk Properties of Powders*. ASM International.

Cox, P. 1927. A method of assigning numerical and percentage values to the degree of roundness of sand grains. *Journal of Paleontology* 1:179–183.

Hawkins, A. E. 1993. *The Shape of Powder-Particle Outlines*. Baldock, UK, Research Studies Press Ltd.

ISO 1099. 2017. Metallic Materials – Fatigue Testing – Axial Force-Controlled Method, International Organization for Standardization.
ISO 4287. 1997. Geometrical Product Specifications (GPS) – Surface Texture: Profile Method – Terms, Definitions and Surface Texture Parameters, International Organization for Standardization.
ISO 4545-1. 2017. Metallic Materials – Knoop Hardness Test – Part 1: Test Method, International Organization for Standardization.
ISO 4986. 2010. Steel Castings – Magnetic Particle Inspection, International Organization for Standardization.
ISO 4987. 2010. Steel Castings – Liquid Penetrant Inspection, International Organization for Standardization.
ISO 4990 (2015), Steel Castings – General Technical Delivery Requirements, International Organization for Standardization.
ISO 4992-1. 2006. Steel Castings – Ultrasonic Examination – Part 1: Steel Castings for General Purposes, International Organization for Standardization.
ISO 4993. 2015. Steel and Iron Castings – Radiographic Testing, International Organization for Standardization.
ISO 5817. 2014. Welding – Fusion-Welded Joints in Steel, Nickel, Titanium and their Alloys (Beam Welding Excluded) – Quality Levels for Imperfections, International Organization for Standardization.
ISO 6506-1. 2014. Metallic Materials – Brinell Hardness Test – Part 1: Test Method, International Organization for Standardization.
ISO 6892-2. 2018; Metallic Materials – Tensile Testing – Part 2. Method of Test at Elevated Temperature, International Organization for Standardization.
ISO 10042. 2018. Welding – Arc-Welded Joints in Aluminium and Its Alloys – Quality Levels for Imperfections, International Organization for Standardization.
ISO 10360 part 1. 2000. Geometrical Product Specifications (GPS) – Acceptance and Reverification Tests for Coordinate Measuring Machines (CMM) – Part 1: Vocabulary, International Organization for Standardization.
ISO 10360 part 2. 2009. Geometrical Product Specifications (GPS) – Acceptance and Reverification Tests for Coordinate Measuring Machines (CMM) – Part 2: CMMs Used for Measuring Size, International Organization for Standardization.
ISO 10360 part 3. 2000. Geometrical Product Specifications (GPS) – Acceptance and Reverification Tests for Coordinate Measuring Machines (CMM) – Part 3: CMMs with the Axis of a Rotary Table as the Fourth Axis, International Organization for Standardization.
ISO 10360 part 4. 2000. Geometrical Product Specifications (GPS) – Acceptance and Reverification Tests for Coordinate Measuring Machines (CMM) – Part 4: CMMs Used in Scanning Measuring Mode, International Organization for Standardization.
ISO 10360 part 5. 2010. Geometrical Product Specifications (GPS) – Acceptance and Reverification Tests for Coordinate Measuring Machines (CMM) – Part 5: CMMs Using Single and Multiple-Stylus Contacting Probing Systems, International Organization for Standardization.
ISO 10360 part 6. 2001. Geometrical Product Specifications (GPS) – Acceptance and Reverification Tests for Coordinate Measuring Machines (CMM) – Part 6: Estimation of Errors in Computing Gaussian Associated Features, International Organization for Standardization.
ISO 10360 part 7. 2011. Geometrical Product Specifications (GPS) – Acceptance and Reverification Tests for Coordinate Measuring Machines (CMM) – Part 7: CMMs Equipped with Imaging Probing Systems, International Organization for Standardization.
ISO 10360 part 8. 2013. Geometrical Product Specifications (GPS) – Acceptance and Reverification Tests for Coordinate Measuring Machines (CMM) – Part 8: CMMs with Optical Distance Sensors, International Organization for Standardization.
ISO 10360 part 9. 2013. Geometrical Product Specifications (GPS) – Acceptance and Reverification Tests for Coordinate Measuring Machines (CMM) – Part 9: CMMs with Multiple Probing Systems, International Organization for Standardization.

ISO 10360 part 10. 2016. Geometrical Product Specifications (GPS) – Acceptance and Reverification Tests for Coordinate Measuring Machines (CMM) – Part 10: Laser Trackers for Measuring Point-To-Point Distances, International Organization for Standardization.

ISO/CD 10360 part 11. 2019. Geometrical Product Specifications (GPS) – Acceptance and Reverification Tests for Coordinate Measuring Machines (CMM) – Part 11: CMMs Using the Principle of Computed Tomography (CT), International Organization for Standardization.

ISO 10360 part 12. 2016. Geometrical Product Specifications (GPS) – Acceptance and Reverification Tests for Coordinate Measuring Machines (CMM) – Part 12: Articulated Arm Coordinate Measurement Machines (CMM), International Organization for Standardization.

ISO/CD 10360 part 13. 2019. Geometrical Product Specifications (GPS) – Acceptance and Reverification Tests for Coordinate Measuring Machines (CMM) – Part 13: Optical 3D CMS, International Organization for Standardization.

ISO 10675-1. 2016. Non-Destructive Testing of Welds. Acceptance Levels for Radiographic Testing. Steel, Nickel, Titanium and Their Alloys, International Organization for Standardization.

ISO 10675-2. 2017. Non-Destructive Testing of Welds. Acceptance Levels for Radiographic Testing. Aluminium and Its Alloys, International Organization for Standardization.

ISO 10863. 2011. Non-Destructive Testing of Welds. Ultrasonic Testing. Use of Time-of-Flight Diffraction Technique (TOFD), International Organization for Standardization.

ISO 11666. 2018. Non-Destructive Testing of Welds. Ultrasonic Testing. Acceptance Levels, International Organization for Standardization.

ISO 11971. 2008. Steel and Iron Castings – Visual Examination of Surface Quality, International Organization for Standardization.

ISO 12108. 2018. Metallic Materials – Fatigue Testing – Fatigue Crack Growth Method, International Organization for Standardization.

ISO 12135. 2016. Metallic Materials – Unified Method of Test for the Determination of Quasistatic Fracture Toughness, International Organization for Standardization.

ISO 12737. 2010. Metallic Materials – Determination of Plane-Strain Fracture Toughness, International Organization for Standardization.

ISO 15626. 2018, Non-Destructive Testing of Welds. Time-of-Flight Diffraction Technique (TOFD). Acceptance Levels, International Organization for Standardization.

ISO 16371-2. 2017. Non-Destructive Testing. Industrial Computed Radiography with Storage Phosphor Imaging Plates. General Principles for Testing of Metallic Materials Using X-Rays and Gamma Rays, International Organization for Standardization.

ISO 17296-3. 2014, Additive Manufacturing – General Principles – Part 3: Main Characteristics and Corresponding Test Methods, International Organization for Standardization.

ISO 17635. 2016. Non-Destructive Testing of Welds. General Rules for Metallic Materials, International Organization for Standardization.

ISO 17636-1. 2013. Non-Destructive Testing of Welds. Radiographic Testing. X- and Gamma-Ray Techniques with Film, International Organization for Standardization.

ISO 17636-2. 2013. Non-Destructive Testing of Welds. Radiographic Testing. X- and Gamma-Ray Techniques with Digital Detectors, International Organization for Standardization.

ISO 17640. 2018. Non-Destructive Testing of Welds. Ultrasonic Testing. Techniques, Testing Levels, and Assessment, International Organization for Standardization.

ISO 23279. 2017. Non-Destructive Testing of Welds. Ultrasonic Testing. Characterization of Discontinuities in Welds, International Organization for Standardization.

Leach, R. K., Bourell, D., Carmignato, S., Donmez, A., Senin, N., Dewulf, W. 2019. Geometrical metrology for metal additive manufacturing. *CIRP Annals* 68:677–700.

NASA – MSFC-STD-3716. 2017. Standard for Additively Manufactured Spaceflight Hardware by Laser Powder Bed Fusion in Metals, NASA.

NASA – MSFC-SPEC-3717. 2017. Specification for Control and Qualification of Laser Powder Bed Fusion Metallurgical Processes, NASA.

Parker W. J., Jenkins R. J., Butler C. P., Abbott G. L. 1961. Flash method of determining thermal diffusivity, heat capacity, and thermal conductivity. *Journal of Applied Physics* 32:1679–1684.

Sih, S. S., Barlow, J. W. 1992. The measurement of the thermal properties and absorptances of powders near their melting temperatures. *Proceedings SFF*, Austin, TX, 131–140.
Simmons, J. H., Potter, K. S. 2010. *Optical Properties*. San Diego, CA: Academic Press.
Sing, S. L., Yeong, W. Y., Wiria, F. E., Tay, B. Y. 2016b. Characterisation of titanium lattice structures fabricated by selective laser melting using an adapted compressive test method. *Experimental Mechanics* 56:735–748.
Sing, S. L., An, J., Yeong, W. Y., Wiria, F. E. 2016a. Laser and electron-beam powder-bed additive manufacturing of metallic implants: a review on processes, materials and designs. *The Journal of Orthopaedic Research* 34:369–385.
Wang, J., Xie, H., Weng, Z., Senthil, T., Wu, L. 2016. A novel approach to improve mechanical properties of parts fabricated by fused deposition modelling. *Journal Materials and Design* 5:152–159.

7

Cost Implications of Precision Additive Manufacturing

Martin Baumers

CONTENTS

7.1 Introduction

On a fundamental level, activities or courses of action are normally evaluated though an assessment of the benefits and costs involved, with the difference between both manifesting as value arising to some agent or owner. Any systematic analysis of the value of an activity or process is thus likely to involve a treatment of said benefits and costs, which may originate from core processes but also from secondary aspects, such as quality assurance and metrology in the manufacturing setting (see, for example, Savio 2012 or Savio et al. 2016). This chapter outlines basic methods used for the assessment of the costs of additive manufacturing (AM). It further explores how these methods can be extended to form a state-of-the-art cost model suitable for the characterisation of precision AM.

The label *AM* describes a group of technologies capable of combining material layer by layer to manufacture geometrically complex products in a single digitally controlled process step, entirely without moulds, dies or other tooling. Additionally, AM is characterised as a parallel manufacturing approach allowing the contemporaneous production of multiple, potentially unrelated, components or products. A simple intuition associated with AM is that the technology gives the user the ability to effectively 'print' objects based on digital 3D design data. Following this popular narrative, anyone who possesses 3D design data or can create these data is able to manufacture the objects desired. This possibility has captivated the imagination of many engineers, scientists and journalists – and has led to a considerable amount of 'hype' with respect to the impact that the technology might have on the manufacturing environment and wider society. Rather unhelpfully, AM

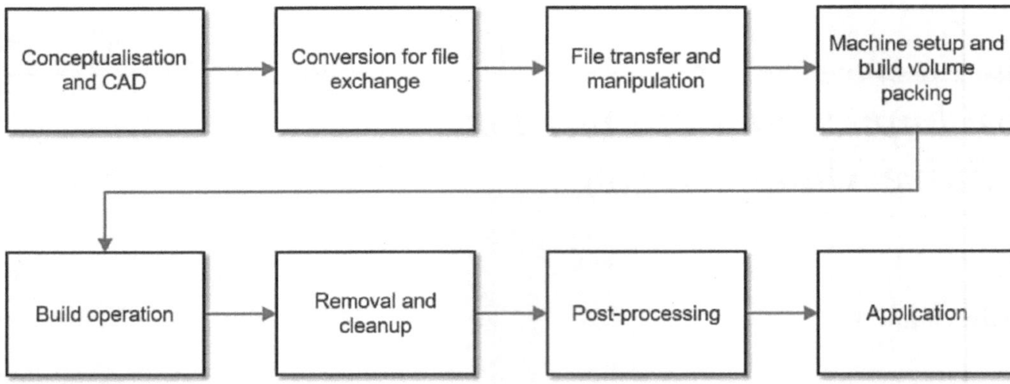

FIGURE 7.1
The generic AM process. (From Gibson et al. 2014. With permission.)

is also frequently portrayed as a single-step manufacturing process in which the finished product emerges from the AM build process instantaneously, ready for shipping or use. Reassuringly, however, many companies' knowledge of AM has now matured to a level that allows them to make realistic decisions on the use of AM (D'Aveni 2018).

An assessment of the AM process flow found in industry reveals that it is in fact a multi-step process in which the core material deposition step is preceded and followed by a number of additional processes. In this vein, Gibson et al. (2014) characterise AM as an eight-step generic process, as shown in Figure 7.1.

As necessary elements of the AM process, it is clear that all steps shown in Figure 7.1 impart some form of value and will likewise impose costs. This aspect is particularly relevant where a high degree of precision and accuracy is required in manufacturing execution. Since each operation can result in unpredictable and often undesirable variation, possibly leading to process failure, product rejection or, in the worst case, product failure during use, the need to identify and rectify variation in the process will create further costs. If these costs are prohibitive, the rationale for the adoption of AM techniques requiring a high degree of precision or consistency will be undermined.

Besides facing a number of technological limitations and organisational challenges (Khorram Niaki and Nonino 2017), AM processes are associated with three major advantages over conventional manufacturing techniques (see, for example, Tuck et al. 2008, Gardan 2016, Rayna and Striukova 2016, Baumers et al. 2017, Beltrametti and Gasparre 2018). The first advantage is that AM is a source of operational flexibility in manufacturing execution. This characteristic results in a potential for responsive 'on-demand' production, elimination of the requirement to hold inventories and accelerated cycles of new product development. In a more general sense, the extended operational flexibility afforded by AM yields novel operational practices (Holmström et al. 2017).

The second advantage relates to the freedom of geometry offered by AM. This freedom implies that the adoption of AM can be used to access an entirely different and enlarged solution space in terms of the designs and functionalities that can be realised. From a technical perspective, this freedom stems from the absence of physical tooling such as moulds, tools or dies, which, in turn, affords AM the ability to manufacture complex geometries (Hague et al. 2003), consolidate multiple components into single parts (Stevenson et al. 2017) and also customise or differentiate individual units to end-user requirements (Tuck et al. 2008, Eyers et al. 2008, Berman 2012).

The third advantage reflects AM's commercial impact by posing a threat to conventional manufacturing business models. By constituting a relatively automated manufacturing approach in which large amounts of value can be created within the material deposition operation, the adoption of AM may render some kinds of existing manufacturing competences obsolete, thereby changing the nature of value creation and value delivery in manufacturing (Rayna and Striukova 2016). In applications in which AM has achieved the status of the default manufacturing technology, such as in the hearing aid industry, the commercial validity of business models based on conventional manufacturing processes is severely diminished (see, for example, D'Aveni 2015).

While the commercial opportunities of AM mentioned above have been analysed in the relevant literature with an emphasis on applications featuring medium part sizes, and without placing additional requirements on precision or achieving exacting tolerances (see, for example, Hopkinson and Dickens 2001 or Baumers et al. 2019), it is likely that the outlined opportunities will also drive the adoption of precision AM. In order to set out the cost implications of precision AM, this chapter first provides a brief primer on cost estimation (Section 7.2) followed by a summary of how these methods can be applied to AM cost estimation in general (Section 7.3). Section 7.4 proposes an extension to existing AM cost models to accommodate the increased need for inspection processes in precision AM. Section 7.5 briefly discusses salient features of this cost model. This chapter concludes with a summary and an outline of additional perspectives of relevance (Section 7.6).

7.2 A Primer in Manufacturing Cost Modelling

The aim of cost estimation is to put the leadership of an organisation in a position to make decisions relating to the control of activities and processes in the present and the planning of future ones. This is especially relevant in the manufacturing context, where cost estimation is used to determine and control existing physical processes and machinery, modify existing products, shape future designs and decide on the acquisition of new equipment. The accuracy and consistency of cost estimates can be vital for the organisation since it determines the effectiveness of overall decision-making, thus affecting overall performance in the business context.

If costs are overestimated, it is likely that the organisation will forego sales and lose goodwill in the market. If costs are underestimated, the organisation is likely to suffer financial losses and instability. Owing to the importance of cost modelling, the existing literature on cost modelling is well-established and extensive (Niazi et al. 2006). A range of textbooks on product cost estimation approaches used in manufacturing are available, addressing a wide range of issues, including production cost estimation for standard parts, cost assessment of complex products, cost optimisation, approximate and detailed costing techniques to support design activities, assessment of overhead costs and life cycle costing approaches (see, for example, Brimson 1991, Ostwald 1992). For an overview of costing methods suitable for advanced manufacturing technologies, see Niazi et al. (2006).

A definition of the scope of any cost analysis requires a decision on which types of costs to include in the assessment. A useful starting point for such a decision is to employ a general categorisation in terms of 'well-structured costs' and 'ill-structured costs' (Son 1991). The term 'well-structured costs' denotes costs that are reasonably well understood by accountants in an organisation. Such costs include, for example, expenses incurred for raw materials or labour. 'Ill-structured costs', in contrast, constitute expenses that are not

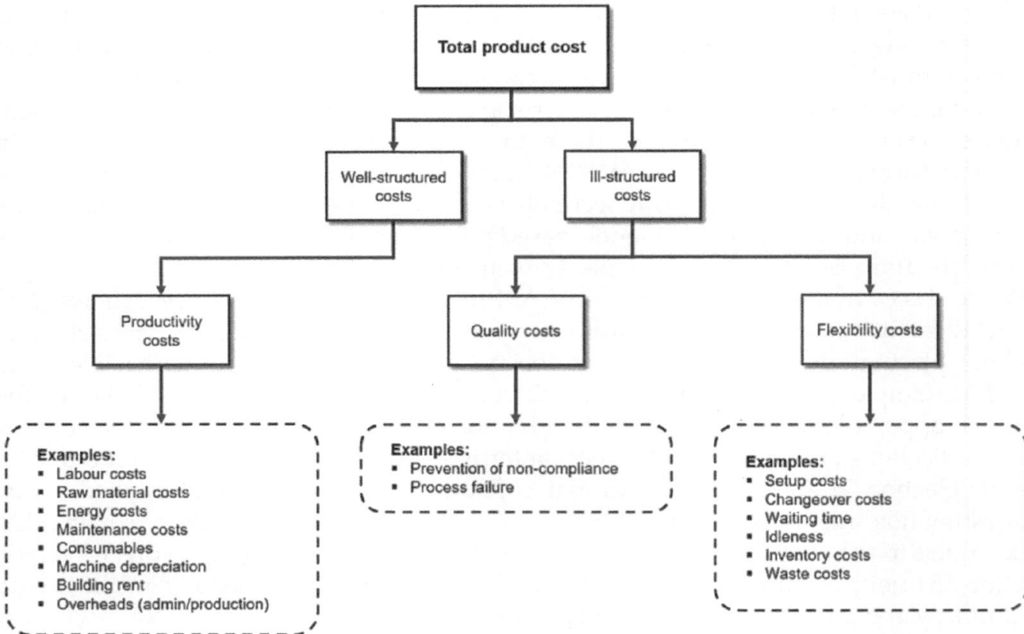

FIGURE 7.2
Types of cost in product cost estimation. (Adapted from Baumers and Tuck 2019.)

well understood. This may be due to a lack of knowledge, information or limitations in accounting practise. As a further and more detailed way to distinguish between costs arising in manufacturing systems, it is possible to delineate between productivity costs, quality costs and flexibility costs. Figure 7.2 summarises the relationship between the above-described categories and provides examples for each type of cost.

Investigations of production costs are carried out with one of two broad objectives in mind. The first objective is to predict costs. In this case, the outcome will form what is known as a cost estimator. Cost estimators give insight into the absolute cost performance of a manufacturing configuration and their validity is established, usually in hindsight, through how accurately and consistently they have predicted the costs. The second objective is to understand or explain costs. Cost investigations of the second type are known more generally as cost models and are designed to represent the relationships between different cost variables and parameters. Cost models are judged on their performance in capturing and representing important aspects in appropriate ways. This lends cost models an explanatory character that is not normally present in cost estimators. It is important to note that the two objectives are not mutually exclusive, such that cost models are often used to predict costs and, vice versa, cost estimators can be used to explain cost relationships.

7.3 Developing an AM Costing Framework

Assessments of the costs of AM are required by a variety of stakeholders and decision makers. These include researchers, current technology users, potential technology adopters, technology developers, software providers and the investment community. While several of these groups specify their own cost analyses, the best-documented and most accessible cost frameworks are contained in the academic literature (see, for example, Alexander et al. 1998, Hopkinson and Dickens 2001, Ruffo et al 2006, Atzeni and Salmi 2012, Rickenbacher et al. 2013, Baumers and Holweg 2019). In terms of the available work, a general distinction can be drawn between parametric costing techniques and activity-based costing techniques (Di Angelo and Di Stefano 2011).

Parametric costing captures costs as abstract mathematical relationships between different variables and parameters that are estimated via statistical methods. Due to this abstract nature, parametric costing techniques carry the advantage of not requiring a fundamental understanding of the underlying technologies and are usually specified quite broadly.

Activity-based costing frameworks, on the other hand, are constructed on the basis of a thorough technical understanding of the modelled processes. This includes an identification of the elementary operations, components and activities, and the relevant relationships between these. On this basis, a mapping of cost characteristics to particular elements can be constructed. Usually following the logic of summing up different cost elements, activity-based costing frameworks combine these characteristics to form an overall cost estimate.

The seminal AM cost model was proposed by Alexander et al. (1998) and belongs to the activity-based category. Since then, several extensions to this framework have been made, mainly aiming to provide a more useful scope and to increase realism. Figure 7.3 summarises the components of a state-of-the-art activity-based cost model, originally devised for the investigation of the unit cost of a metallic component used in the manufacture of an industrial food-packaging machine (Baumers et al. 2017).

As for many manufacturing applications, the most significant direct costs in AM are likely to arise through the consumption of raw materials, including one or more build materials and, where required, sacrificial support materials. Since three-dimensional digital design files are employed to control the process (normally in the STL format), it is routinely possible to estimate the total volume of the products and support structures contained in the build volume through specialised software. In combination with data on the price of raw materials, a direct cost estimate can be specified quite easily in the majority of cases. It should further be noted that raw material waste streams form a significant source of cost in some AM technology variants; therefore, direct costs are likely to include a parameter for material wastage, often entering as a material 'refresh ratio'.

In activity-based costing methods, a second major type of cost is indirect cost. Normally established as a function of build time, indirect costs contain aspects such as maintenance, consumables, depreciation and various overheads. Examples for elements of indirect costs are provided in Figure 7.3.

In the recent AM costing literature, a range of additional aspects have been incorporated in activity-based costing frameworks (see, for example, Baumers et al. 2017). Such aspects include the following:

1. The cost effect of capacity utilisation. Acknowledging that the degree to which the available build volume is used will have an effect on unit costs, the number of units contained or the geometric build volume fraction can be used to break down the total cost of the build to the unit cost (Ruffo et al. 2006).
2. Inclusion of the cost effect of build failure. Entering as a probability, for example as a function of the number of layers deposited, cost models can incorporate the likely effects of outright build failure, material failure and product rejection.

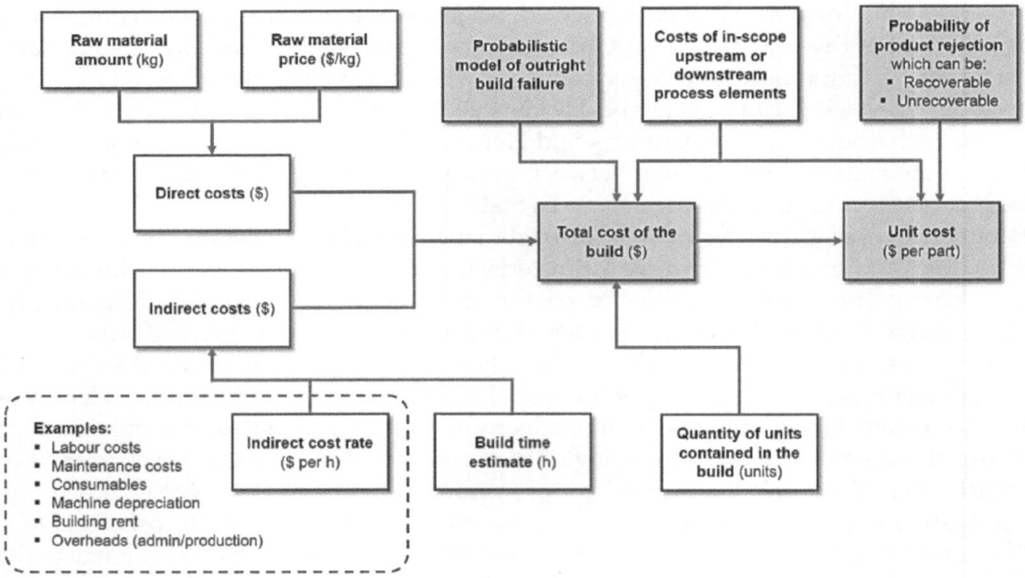

FIGURE 7.3
An AM cost estimation framework, including ill-structured elements (shaded).

It should be noted, however, that the inclusion of probabilistic terms changes the nature of such specifications into expected cost models (Baumers and Holweg 2019).

3. Extension of the scope of the cost model to include relevant process elements outside of the core build process, such as those encompassed by the generic AM process described in Section 7.1 (see also Gibson et al. 2014). Importantly, additional process elements can relate to post-process metrology, inspection and quality assurance processes that are central in precision manufacturing.

After the construction of the overall cost model, reflecting the costs of the build operation in its entirety, a frequently required additional step is to break down the total cost of the build to the unit cost level or to the cost of an individual part. As in the preceding steps, relevant well-structured and ill-structured aspects pertaining to the individual unit of output can be incorporated at this point. These could reflect the costs of ancillary process elements if they relate to the individual unit or additional risks of product rejection, either correctable or non-correctable.

7.4 Specifying a Simple Cost Model for Precision AM

This section constructs a model for precision AM by extending the activity-based costing framework proposed by Baumers et al. (2017). Prior to cost modelling, a necessary preliminary step is to define an appropriate scope for the model. In this section, it is assumed that the distinguishing characteristic of the cost model for precision AM is its emphasis on the detection, avoidance or elimination of undesirable variation in the AM process. Thus, additional expenses involving quality-related activities are accommodated in the cost model.

Initially, however, the cost of the AM process itself is modelled. The total cost C_{Build} of the build operation, including preparation and ancillary post-processing but excluding quality-related activities, can be specified as follows:

$$C_{\text{Build}} = \frac{\text{DC}_{\text{Build}} + \text{PLC} + T_{\text{Build}} \times \text{IC}_{\text{Build}}}{P(l)} + \text{FLC} + q \times \text{VLC} \tag{7.1}$$

where DC_{Build} forms the total direct cost of materials (including wastage of unprocessed material due to degradation and other losses) and process energy consumption, PLC denotes the fixed process labour cost increment required to execute the process (subject to the risk of build failure), T_{Build} is a build time estimate covering the duration of the full AM process including machine warm-up and cool-down (if applicable), and IC_{Build} is the indirect cost rate containing all overheads and machine costs.

The cost effect of build failure and the possible requirement for build repetition can be incorporated into the cost model by dividing the affected cost elements over an expression of the probability of successful build completion, following the scrappage cost model proposed by Ashby (2005). A simple yet effective model of this kind was developed by Baumers and Holweg (2019) as a variant of the mean time between failure concept used in reliability engineering (see, for example, Hopp and Spearman 2011). Here, the probability of successful build completion enters as a survival function P of the overall number of layers l deposited in the build process and is affected by a fixed and process-specific failure parameter λ. Since this parameter is treated as constant, the model implies that the process is stable in that no systematic or time-dependent failure mechanisms occur. Thus, the failure model $P(l)$ can be specified as

$$P(l) = (1 - \lambda)^{l}. \tag{7.2}$$

The remaining cost elements included in the total cost model C_{Build}, which are unaffected by the risk of build failure, are a fixed labour cost increment, FLC, associated with post-build operations occurring on a per-build basis, such as product retrieval and machine turnaround; and a variable labour cost increment, VLC, occurring on a per-part basis for all q parts contained in the build, such as removal of excess build material and initial post-processing. In this simple specification, all non-labour cost elements outside of the core build operation are ignored for simplicity.

The next step is to form a unit cost model, UC, on the basis of C_{Build}. If the analysed build operation contains multiple different product geometries, it is possible to proceed by associating a cost share with the ith part in the build through multiplication by its volume fraction, which is defined as the volume V of part i divided by the sum of the volumes of all j parts contained in the build. Alternatively, if the q units contained in the build volume all share the same geometry, meaning that they are instances of the same design, it is possible to further simplify the unit cost model UC_i by dividing C_{Build} by q. This model proceeds with the more complex and realistic case of the contained geometries being different.

The unit cost level is also the appropriate locus for the estimation of precision manufacturing-specific costs associated with post-build inspection and quality diagnostic activities. Analogous to the model used to assess the core AM process, the total costs of inspection C_{Inspect} are specified through the sum of their direct costs $\text{DC}_{\text{Inspect}}$, arising through inspection labour and consumables; and the indirect costs of inspection, which are obtained by multiplying the estimated duration of the inspection $\text{T}_{\text{Inspect}}$ by an indirect cost rate of inspection $\text{IC}_{\text{Inspect}}$, reflecting overheads and the capital costs of inspection equipment, such that

$$C_{\text{Inspect}} = \left(\text{DC}_{\text{Inspect}} + T_{\text{Inspect}} \times \dot{\text{IC}}_{\text{Inspect}} \right) \times q. \tag{7.3}$$

Thus, the unit cost model for the ith product contained in the build, incorporating both the costs of the AM process (from Equation 7.1) as well as the precision-specific costs (from Equation 7.3), can be expressed as

$$\text{UC}_i = C_{\text{build}} \frac{V_i}{\sum_j V_j} + \frac{C_{\text{Inspect}}}{q}. \tag{7.4}$$

At this point, the cost effect of the probability of product rejection can be attached to the model as done in the above. Again, seeking a very simple specification, it is assumed that the probability of product rejection is simply a constant P_{Reject}. Thus, the total unit cost model TUC_i is obtained by multiplying UC_i by the inverse of the complement of P_{Reject}, again following Ashby's (2005) scrappage model:

$$\text{TUC}_i = \text{UC}_i \times \left(1 - P_{\text{Reject}}\right)^{-1}. \tag{7.5}$$

7.5 A Brief Discussion of the Cost Model for Precision AM

Since the cost model developed in Section 7.4 combines a wide range of elements, spanning technological and commercial aspects, a range of relevant aspects arising from the model for further discussion is chosen. For each aspect discussed, issues of practical relevance, important conceptual connections to other elements in the process model and the simplifications made in our approach are highlighted.

7.5.1 Indirect Cost Rates

As an activity-based costing approach, several cost elements in the presented model are attributed through time. This applies to the build activity itself (T_{Build}) and to the inspection activity (T_{Inspect}). Evaluating the cost impact of these time-dependent cost elements requires the estimation of reasonably accurate indirect cost rates, IC_{Build} and $\text{IC}_{\text{Inspect}}$, measured as costs arising per unit of operating time (for example, \$/hour or €/hour). Since the available information on indirect costs is normally obtained on an annualised basis, indirect costs are broken down to an hourly rate by dividing the annual cost through the number of operating hours per year. It is instructive to briefly discuss three relevant aspects contained in these rates.

1. Indirect cost rates of the kind discussed above reflect the purchase price of the hardware and secondary systems. This will include not only the AM machine and inspection equipment but also ancillary systems, such as raw material handling stations. As capital equipment, which carries a residual value at any given point of time, the purchase cost of such machinery enters as a depreciation cost over time. In most available AM models, equipment is subject to a straight-line depreciation method with a depreciation period of five to ten years. In case of other ownership models (for example, hire purchase or lease), appropriate alternative arrangements must be made in place of depreciation. Additional costs entering as indirect costs are maintenance costs, the expenses for system calibration and the costs of consumables, such as protective gases, filters, seals and components with a replacement interval.
2. Indirect cost rates may contain labour costs incurred during the operation of a process. This approach is useful if a technician operates an AM system on a one-to-one basis, implying that the machine requires constant supervision and technician activity. This is not the case in the AM cost model specified in the above. Here, a fixed process labour cost element, PLC; a fixed post-process labour cost, FLC; and a variable labour cost, VLC, are included. It is important to note that only PLC is subject to the risk of build failure. Further, the inspection cost model contains no explicit labour cost element – so if there is a labour cost associated with inspection, this will form part of the direct costs $DC_{Inspect}$ or the indirect cost element $IC_{Inspect}$. Whenever specifying labour costs, it should also be noted that employer contributions need to be taken into account.
3. Indirect cost rates incorporate the overheads allocated to the operation of the investigated systems. Such overheads may result from production itself, for example in the form of building space required to house the equipment or other infrastructure costs. Additional overheads are administrative and relate to computer equipment, communications and software licences. Some cost models include energy costs in the production overheads, but this may also be included as a direct cost.

7.5.2 Capacity Utilisation

Since AM constitutes a parallel manufacturing technology allowing the contemporaneous processing of multiple components, any fixed costs incurred can be amortised across multiple components. This implies that there is potentially a strong efficiency-related incentive to maximise the utilisation of the available build volume capacity. However, when considering cost models of the kind developed in this chapter, it is important to realise that capacity utilisation enters in a second way, through the degree of machine utilisation over time, as part of the indirect cost rate, which relies on an estimation of the share of operating hours of overall time. Metrics such as overall equipment effectiveness are useful for the measurement of such aspects of capacity utilisation (see, for example, Bicheno and Holweg 2016).

Further complexity is introduced to the costing framework if the inspection process exhibits capacity or throughput bottlenecks, which is likely to be the case in precision AM if the inspection process is overly time consuming (see Chapter 13 for state-of-the-art inspection times). However, it is assumed that assessing capacity utilisation for inspection processes will carry less complexity than for AM processes due to their serial nature, which obviates the build volume utilisation problem.

7.5.3 Integration with Other Operational Processes

An initial overview of the scope of the cost modelling project can be obtained through a process mapping exercise, as done for example by Baumers and Holweg (2019). Importantly, such an exercise will help define the appropriate boundaries of the cost investigation. As described in the generic AM process shown in Figure 7.1, a number of pre- and post-processing steps are typically included in a useful AM cost model. In the model presented in Section 7.4, these costs are reflected purely as labour costs (PLC, FLC and VLC). Depending on the nature and characteristics of these secondary processes, significant additional expenses may be incurred which, in turn, warrant inclusion of specific extra items in the cost model.

If the AM process and the inspection process, or other ancillary processes for that matter, are discrete, workflow organisation and scheduling techniques should be used to determine efficient patterns of operation. Well-known scheduling environments include single machine, identical machines in parallel, machines in parallel with different speeds, unrelated machines in parallel, flow shop, flexible flow shop, job shop, flexible job shop and open shop. An exhaustive theoretical treatment of scheduling theories is provided by Pinedo (2012).

A further aspect that has received considerable attention over the past decade is the requirement to adapt designs to the characteristics of the AM process. In particular, the emergence of AM has led to a wave of research on computational design tools capable of exploiting the available design space, thereby leading to products with optimised geometries (for an overview, see Liu et al. 2018 and Chapter 2). In this context, the general reasoning is that optimised product configurations need to take into account all the elements of the product life to avoid island solutions that may sacrifice efficiency. This is likely to be of high relevance for precision AM, where stringent requirements are placed on product features and significant additional expenses may be incurred for inspection.

7.5.4 Relationship between Failure Parameters and Costs of Inspection

The total unit cost model TUC presented in Section 7.4 proposes relationships between overall unit cost and a number of cost parameters, including the instantaneous constant failure rate λ, the probability of product rejection P_{Reject} and the cost of inspection $C_{Inspect}$. In the case of a closed-loop system, it would be expected that the occurrence of process variation – detected during the AM process itself or during inspection – leads to an automatic correction mechanism. However, this is not currently the case in most applications of AM. Nevertheless, since a process implementation in the commercial manufacturing context demands an acceptable level of product quality and conformance to design parameters (which are of course highly application-specific), it is reasonable to expect a negative relationship between $C_{Inspect}$ and the failure and rejection parameters in the model, λ and P_{Reject}.

As illustrated in Figure 7.4, it is possible to posit a mechanism in which a higher expenditure $C_{Inspect}$ leads to better or increased feedback of failure and rejection information to the design and production planning activities, for example in the form of an information loop, perhaps as a product of a failure modes and effect analysis (see, for example, Stamatis 2003). The resulting reduction of λ and P_{Reject} is expected, in turn, to lead to a lower TUC. This suggests a mechanism in which the cost of additional inspection activities is offset over time by decreases in manufacturing cost.

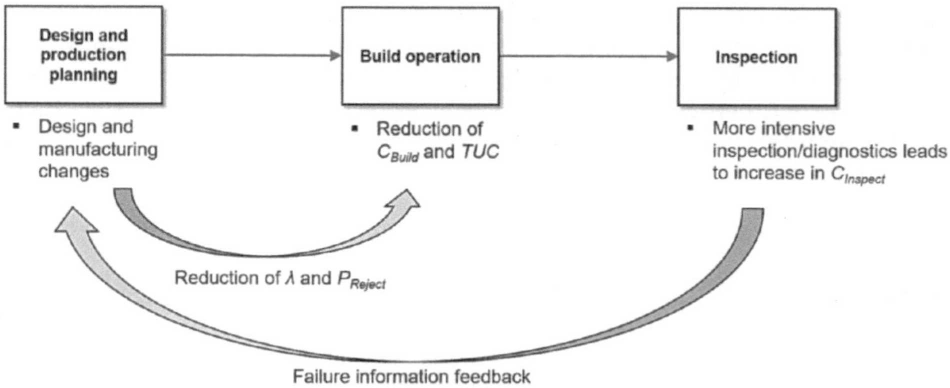

FIGURE 7.4
Total cost effect of increases in inspection effort.

7.6 Summary and Additional Perspectives

After providing a brief primer in activity-based cost modelling, this chapter has provided an outline of how AM-specific manufacturing cost models can be constructed. On this basis, this chapter has proposed a specification for a robust cost model featuring a dedicated inspection element targeting precision AM. A number of additional explanations were provided in terms of the model structure and its uses within the commercial manufacturing setting. This chapter has also proposed a feedback mechanism showing how an increased level of inspection, associated with higher cost, can be offset through lower probabilities of process failure and product rejection. It should be noted that it is not possible to construct such arguments without incorporating ill-structured aspects relating to process failure and product rejection in cost models. It has also been stressed that AM cost models tend to be highly process- and geometry-specific. As noted above, the estimated TUC levels are shaped by the scope of the cost analysis. It is, therefore, important to provide these criteria alongside cost estimates when drawing comparisons, ideally together with a summary of cost metrics and a summary of the investigated build configuration.

Offsetting inspection costs through the reduction of process failure or product rejection, as suggested above, forms a highly relevant topic in AM. In general, AM is marked by process failure rates far in excess of what can be observed in conventional manufacturing (see, for example, Baumers et al. 2017). As has been shown in this chapter, investigating such aspects hinges on the successful collection and use of digital information across various manufacturing processes (and ideally beyond this). Naturally, such initiatives are related to timely engineering concepts such as the 'digital twin', denoting a rich virtual model of any (actual or potential) physical entity, including objects, systems, humans and locations that can be employed for a multitude of purposes and exist simultaneously with the entity (see, for example, El Saddik 2018).

From a manufacturing control perspective, which is of the highest significance in the precision manufacturing context, it needs to be stressed that AM is characterised by a freedom of geometry (Hague et al. 2003). This freedom arises from the absence of physical tooling and an inherent 'fungibility' in the process, which describes the possibility of treating the available units of build space as almost perfect mutual substitutes (Baumers et al. 2017). While ultimately being responsible for the attraction of AM, this freedom introduces a host of challenges, perhaps captured best by what is known as Ashby's Law of Requisite Variety (Ashby 1961): for a system to be under control, the number of possible states of its control function must be at least as great as the number of possible states the controlled system can take. Of course, harnessing this 'variety to destroy variety' will be costly, especially in AM.

References

Alexander, P., Allen, S., Dutta, D. 1998. Part orientation and build cost determination in layered manufacturing. *Computer-Aided Des*. 30:343–356.

Ashby, W. R. 1961. *An Introduction to Cybernetics*. Chapman & Hall Ltd, London, UK.

Ashby, M. F. 2005. *Materials Selection in Mechanical Design*. 3rd ed. Butterworth-Heinemann, Burlington, MA.

Atzeni, E., Salmi, A. 2012. Economics of additive manufacturing for end-usable metal parts. *Int. J. Adv. Manuf. Technol*. 62:1147–1155.

Baumers, M., Beltrametti, L., Gasparre, A., Hague, R. 2017. Informing additive manufacturing technology adoption: Total cost and the impact of capacity utilisation. *Int. J. Prod. Res*. 55:6957–6970.

Baumers, M., Holweg, M. 2019. On the economics of additive manufacturing: Experimental findings. *J. Oper. Manag*. 65:794–809.

Baumers, M., Tuck, C. 2019. Developing an understanding of the cost of additive manufacturing. In *Additive Manufacturing – Developments in Training and Education*. Springer, Cham, Switzerland.

Baumers, M., Wildman, R., Wallace, M., Yoo, J., Blackwell, B., Farr, P., Roberts, C. J. 2019. Using total specific cost indices to compare the cost performance of additive manufacturing for the medical devices domain. *Proc. Inst. Mech. Eng. Part B: J. Eng. Manuf*. 233:1235–1249.

Beltrametti, L., Gasparre, A. 2018. Industrial 3D printing in Italy. *Int. J. Manuf. Technol. Manag*. 32:43–64.

Berman, B. 2012. 3-D printing: The new industrial revolution. *Bus. Horiz*. 55:155–162.

Bicheno, J., Holweg, M. 2016. *The Lean Toolbox: A Handbook for Lean Transformation*. Picsie Books, Buckingham, UK.

Brimson, J. A. 1991. *Activity Accounting: An Activity-based Costing Approach*. John Wiley & Sons, New York.

D'Aveni, R. 2015. The 3-D printing revolution. *Harvard Bus. Rev*. 93:40–48.

D'Aveni, R. 2018. *The Pan-industrial Revolution: How New Manufacturing Titans Will Transform the World*. Houghton Mifflin, Boston, MA.

Di Angelo, L., Di Stefano, P. 2011. A neural network-based build time estimator for layer manufactured objects. *Int. J. Adv. Manuf. Technol*. 57(1–4):215–224.

El Saddik, A. 2018. Digital twins: The convergence of multimedia technologies. *IEEE MultiMedia*. 25:87–92.

Eyers, D. R., Won, H., Y. Wang, Dotchev, K. 2008. Rapid manufactured enabled mass customisation: Untapped research opportunities in supply chain management. In *Proceedings of the Logistics Research Network Annual Conference*, edited by A. Lyons. Liverpool, UK.

Gardan, J. 2016. Additive manufacturing technologies: State of the art and trends. *Int. J. Prod. Res*. 54:3118–3132.

Gibson, I., Rosen, D.W., Stucker, B. 2014. *Additive Manufacturing Technologies*. Springer, New York.

Hague, R., Campbell, I., Dickens, P. 2003. Implications on design of rapid manufacturing. *Proc. Inst. Mech. Eng. Part C: J. Mech. Eng. Sci*. 217:25–30.

Holmström, J., Liotta, G., Chaudhuri, A. 2017. Sustainability outcomes through direct digital manufacturing-based operational practices: A design theory approach. *J. Cleaner Prod*. 167:951–961.

Hopkinson, N., Dickens, P. 2001. Rapid prototyping for direct manufacture. *Rapid Proto. J.* 7:197–202.

Hopp, W. J., Spearman, M. L. 2011. *Factory Physics*. Waveland Press, Long Grove, IL.

Khorram Niaki, M., Nonino, F. 2017. Additive manufacturing management: A review and future research agenda. *Int. J. Prod. Res.* 55:1419–1439.

Liu, J., Gaynor, A. T., Chen, S., Kang, Z., Suresh, K., Takezawa, A., Li, L., Kato, J., Tang, J., Wang, C. C., Cheng, L. 2018. Current and future trends in topology optimization for additive manufacturing. *Struct. Multidiscip. Optim.* 57:2457–2483.

Niazi, A., Dai, J. S., Balabani, S., Seneviratne, L. 2006. Product cost estimation: Technique classification and methodology review. *J. Manuf. Sci. Eng.* 128:563–575.

Ostwald, P. F. 1992. *Engineering Cost Estimating*. 3rd ed. Prentice Hall, Englewood Cliffs, NJ.

Pinedo, M. 2012. *Scheduling*. 29th ed. Springer, New York.

Rayna, T., Striukova, L. 2016. From rapid prototyping to home fabrication: How 3D printing is changing business model innovation. *Technol. Forecasting Social Change* 102:214–224.

Rickenbacher, L., Spierings, A., Wegener, K. 2013. An integrated cost-model for selective laser melting (SLM). *Rapid Proto. J.* 19:208–214.

Ruffo, M., Tuck, C., Hague, R. 2006. Cost estimation for rapid manufacturing-laser sintering production for low to medium volumes. *Proc. Inst. Mech. Eng., Part B: J. Eng. Manuf.* 220:1417–1427.

Savio, E. 2012. A methodology for the quantification of value-adding by manufacturing metrology. *Ann. CIRP* 61:503–506.

Savio, E., De Chiffre, L., Carmignato, S., Meinertz, J. 2016. Economic benefits of metrology in manufacturing. *Ann. CIRP* 65:495–498.

Son, Y. K. 1991. A cost estimation model for advanced manufacturing systems. *Int. J. Prod. Res.* 29:441–452.

Stamatis, D. H. 2003. *Failure Mode and Effect Analysis: From Theory to Execution*. ASQ Quality Press, Milwaukee, WI.

Stevenson, A., Baumers, M., Segal, J., Macdonell, S. 2017. How significant is the cost impact of part consolidation within AM adoption? *Solid Freeform Fabrication 2017: Proc. SFF*, Austin, USA.

Tuck, C. J., Hague, R. J., Ruffo, M., Ransley, M., Adams, P. 2008. Rapid manufacturing facilitated customization. *Int. J. Comp. Int. Manuf.* 21:245–258.

8

Machine Performance Evaluation

Shawn Moylan

CONTENTS

8.1 Introduction

This chapter focuses on evaluating the performance of metal additive manufacturing (AM) machines. Given the focus of this book on the production of precision AM parts and the introduction of precision engineering principles to AM, much of the attention in this chapter will be paid to quantifying the ability of AM machines to repeatably and predictably produce parts with the desired geometry. Characterising a machine to achieve other outcomes, for example to produce parts with a desired microstructure or mechanical properties, will also be discussed, and this distinction will be noted where appropriate. Many of the concepts of machine performance evaluation are consistent despite the desired outcome, but certain priorities and requirements may be different for varied applications.

Discussion in this chapter is limited to machines used to directly produce metallic AM parts. This essentially limits the discussion to powder bed fusion (PBF) machines and directed energy deposition (DED) machines. Some binder jetting machines are capable of producing metal parts, but these are often with a multi-step process, including production of a green part with AM technology and post-process heat treatment for consolidation (or sintering) or infiltration of the part with other materials to create a fully dense part (Gokuldoss et al. 2017). Regardless, the similarities between binder jetting machines and PBF machines mean that many of the same concepts apply. Some sheet lamination systems are also capable of producing metal parts, but parts from these machines are usually machined in a post-process step to achieve the desired geometry. Again, similarities in the machine design between these systems and DED systems mean that many of the same concepts still apply. Hybrid machines – where traditional processes, such as machining, and additive processes are found on the same platform – are outside the scope of this chapter, but once again, some general principles will still apply. Machines built upon a machining centre or turning centre frame and using DED processes will likely be characterised in a similar manner to a traditional machine tool (ISO 230 series, ASME B5.54, ASME B5.57). Machines that are built upon a PBF frame will likely benefit from many of the methods outlined in this chapter.

The test methods detailed in this chapter will be mostly applicable to users of AM systems. Some of the methods may require a level of control over the machine that is not typically available to a user, and these methods may be more appropriate for machine builders. Such methods are discussed nonetheless because there may come a time in the near future when various aspects of the machines become more available to users.

8.1.1 Definitions

- *Part geometry* – As covered in detail in Chapter 10, surface topography is essentially everything that makes up the geometry of an object's surface. In addition to the surface topography, a part's shape is determined by its geometric dimensions. The dimensions, surface form and surface texture combine to describe the *part geometry*. Most of the discussion in this chapter will focus on realising accurate dimensions and achieving desired form, with only cursory discussion on measuring a machine's ability to create a particular surface texture.
- *Machine coordinate system* – The coordinate systems for AM machines, that is, the definitions of the *X*-axis, *Y*-axis, *Z*-axis and origin, are governed by ISO/ASTM 52921 (2013). This definition is also consistent with ISO 841 (2001). The discussion of axes and coordinate systems in this chapter will generally follow the conventions detailed in these specification standards. In general, the machine coordinate system is a right-hand rectangular system with three principal axes labelled as *X*, *Y* and *Z*, with rotary axes about each of these labelled *A*, *B* and *C*, respectively. However, it should be noted that axes in machines are normally defined either by a specific motion direction or associated with a physical artefact (for example, the axis of a cylinder) (see, for example, ISO 230-1 2012). ISO/ASTM 52921 (2013) does

not follow this convention, defining the Z-axis as perpendicular to the layers and the X-axis as perpendicular to the Z-axis and parallel to the front of the machine. These definitions make the machine axes and the machine coordinate system very difficult to physically realise, which can present ambiguities. Many PBF machines that use a linear motion actuator in the recoating system refer to the direction of the motion as along the X-axis. However, by the definition in ISO/ASTM 52921, the Z-axis is perpendicular to the layers and the X-axis is perpendicular to the Z-axis and parallel to the front of the machine. The recoating axis is independent of these and, therefore, will be referred to as the recoating axis, or R-axis, in this chapter.

- *Functional point* – Most often, the functional point in a machine is the point where the workpiece or part is being formed (see, for example, ISO 230-1 2012). For example, in a PBF machine, the functional point is where the energy beam meets the top surface of the powder bed. The functional point is a single point that can move within the machine's working volume. The methods of moving the functional point are different for PBF and DED processes. For example, the movement of DED machines is generally accomplished by motors driving linear axes, whereas movement in a PBF machine is accomplished by electromagnetic scanning or galvanometer systems.
- *Machine error motions* – Each linear axis of a machine has six error motions (ISO 230-1), i.e. unwanted linear and angular motions of a commanded move along a nominally straight line (see Leach and Smith 2018 for a detailed general discussion on error motions). Linear positioning error motion is the unwanted motion in the direction of motion of the axis. Straightness error motions are the unwanted motions in the two directions orthogonal to the direction of travel. For example, if the direction of travel is along the X-axis, there will be straightness errors in the Y-direction and the Z-direction. Finally, there are three angular error motions: the unwanted rotations around each axis of the coordinate system. These rotation errors are commonly referred to as *roll* (around the axis of motion), *pitch* (around the horizontal orthogonal axis) and *yaw* (around the vertical orthogonal axis). Simple nomenclature can denote errors in any primary axis with an *E* (for error) followed by a subscript letter for the direction of the error motion, followed by a subscript letter for the axis of interest. For example, a rotation error about the Y-direction of the Z-axis is denoted as E_{BZ}.
- *System, machine and process* – The convention in this chapter is to refer to the AM 'system' as the combination of the AM machine and the AM process. The AM machine is the actual equipment and components, including control software, used to build the AM part. The AM process is the physics of melting and re-solidifying metal to form the AM part. This chapter will focus primarily on the machine. This is an important distinction, because phenomena such as residual stresses have a major impact on part geometry and mechanical properties but are considered outside the scope of this chapter since they are the results of the process more than the machine (see Chapter 3).
- *Artefact* – an artefact is a physical reference part or object that is measured for the purpose of determining errors. A test artefact, as referred to in this chapter, is a specific part that is built by the machine of interest. The test artefact is measured to determine the errors in the artefact, which can inform a user about the performance of the machine of interest. This differs from a reference artefact, which is a part of known (or independently determinable) geometry that is measured to assess a specific aspect of a machine's performance. For example, when testing the

straightness error motion of a machine's axis, the measurement may be the linear displacement of the axis relative to a reference artefact, such as a straightedge.

8.1.2 Motivation

The idea behind precision engineering for metal AM is not meant to imply that AM parts will soon have the same levels of accuracy or achieve the same levels of tolerance as precision parts, or even traditionally manufactured parts. However, the principles of precision engineering, such as error budgeting, determinism, etc. (Slocum 1992, Leach and Smith 2018), are still relevant, and their adoption by the AM community can lead to significant improvement for AM parts. For example, the principles of error budgeting for part geometry permeate this chapter, with the concept that errors contributed by the machine can be separated from errors contributed by the process, and understanding the relative contributions of each will help users concentrate on the areas where they can make the largest and/or quickest improvements. The general idea here is that AM, a relatively new technology, can benefit from, and build upon, knowledge developed over many years in other fields. AM culture might lean toward 'trial and error' or 'guess and check' approaches, because one of the advantages of the process is that changes can be made and implemented quickly, with little or no need for changes to the tooling or setup. Balancing these tendencies with the more deterministic approaches found in precision engineering should lead to better AM parts.

The need for machine performance evaluation is two-fold:

1. There is a need for quantitative criteria to judge system performance. The information gained in measuring machine performance will be vital in demonstrating conformance to specifications, standards or quality management systems. Furthermore, this information will be used extensively in communications about machine performance, whether those are between a vendor and customer or within an individual supply chain.
2. Understanding machine performance allows a user to make more informed decisions. These decisions might be anything from where in the build volume to place a part with particularly tight tolerances to when to schedule preventative maintenance or how to compensate for measured errors.

The Standardization Roadmap for Additive Manufacturing published by ANSI and America Makes (ANSI 2018) identified 'machine calibration and preventative maintenance' as a high priority, stating that there is 'an urgent need to develop guidelines on day-to-day machine calibration checks'. The Roadmap also identified a medium priority for machine qualification. Both of these areas still have required research and development needs before specification standards can be developed. However, much of what has been learned to date in these areas will be discussed in this chapter.

8.1.3 Background

Perhaps counterintuitively, AM has much in common with many traditional processes, especially machining. For example, raw material or feedstock is input into the machine and the process is performed by the system to create the part. This is true whether the material is a cast ingot or metal powder, or whether the process is milling or PBF. There is a significant divergence when examining the material or mechanical properties of additive versus traditional parts, but there is less breakdown if the examination is focused on part geometry. As such, much of the background for AM machine performance evaluation is actually the history of machine tool metrology.

Part geometry errors are the result of error motions of the machine combined with the physics of the process. For a traditional process, the process physics usually play a small role compared to the error motions of the machine, and measuring the well-known parametric and geometric errors of each machine axis goes a long way to predicting part geometry. For AM, the process physics play a relatively larger role (Cooke and Soons 2010) due to the relatively large size of the melt pool (compared to typical error motions) as well as the dynamic nature of the melt pool. As a result, attention needs to be given to separating these sources of error to allow efficient compensation or adjustment for systematic errors. For example, it is often desirable to quantify the systematic and non-systematic portions of the error so that adjustments can be made for the systematic errors and expected performance limitations can be drawn from the non-systematic errors.

Methods for machine tool performance evaluation have been standardised for some time now, and most of those standards have been through several revisions (Donmez et al. 1986, Slocum 1992, ISO 230-1 2012, ISO 230-2 2014, ASME B5.54 2005, ASME B5.57 2012). As a result, the methods have been rigorously tested and validated over the years, and consensus has led to good practice and expected results from a variety of machines and measurement processes. Much of the knowledge in machine tool metrology is well summarised in the ISO 230 series of standards (there are ten individual standards in the series) as well as in ASME B5.54 (2005) for machining centres and ASME B5.57 (2012) for turning centres. Most of the methods detailed in these standards involve individual, independent measurements of the components of a machine tool. However, ASME B5.54 (2005) and ISO 10791-7 (2014) describe test artefacts, notably the circle-diamond-square artefact, that can be used to characterise machine tool performance. It is worth mentioning here how the circle-diamond-square artefact was designed with each feature intended to highlight a well-known error motion of machine tools, similar to how the test methods for individual machine components isolate a specific type of error motion.

Within the field of AM, standards exist requiring the need for machine performance evaluation, verification and compensation (these terms are defined in Chapter 10), but the methods of performing the measurements are not detailed. SAE AMS7003 (2018) includes these requirements and adds an appendix listing the minimum measurement elements. Similarly, a NASA standard, MSFC-SPEC 3717 (2017) requires that 'calibration' and machine qualification be part of a *Qualified Metallurgical Process.* (Note that 'calibration' is used differently in the standard compared to this book. Typically, 'calibration' refers to a measurement [BIPM 2012] rather than compensation of manufacturing machines. It is used here for consistency with the standard.) Notably, both of these standards prioritise material and mechanical properties over part geometry. Furthermore, the standards do not require specific quality metrics to be met. Part of the reason for this is that different applications may have slightly different requirements, but another part of the reason is that the expected performance and limitations of the machines are constantly changing as new machines with new capabilities become available and are not fully known. To wit, the *Standardization Roadmap for Additive Manufacturing* (ANSI 2018) states that research and development are needed 'to determine how errors in machine components affect output quality so that tolerances can be developed for machine calibration'. In contrast, many of the process-specific machine tool metrology standards (for example, ISO 10791-7) list specific tolerances for machine performance.

8.1.4 Organisation of This Chapter

Although it may be appropriate for standard specifications, such as AMS7003 (2018) and MSFC-SPEC 7317 (2017), not to require certain methods to meet their requirements, instead allowing vendors to meet requirements in innovative and cost competitive ways, it is difficult for new users and for widespread adoption when the learning curve is long. The benefit of highly descriptive test methods in standards such as ISO 230-1 (2012) is that it is clear for everyone how the measurements are done and how the information is gathered, allowing for more open communication. This chapter attempts to address some of these details. Included in the discussion will be what needs to be measured and often multiple methods for measuring it. Composite measurement methods, such as when using test artefacts (see Section 8.2), are discussed in detail with some good practice for users designing their own artefacts. Component methods are evaluated (see Section 8.3), often with comparisons to machine tool metrology, with suggestions on what changes might be appropriate when adapting the methods for AM machines. A middle ground, using a two-dimensional (2D) artefact (see Section 8.4) as a test that isolates beam scanning but is a composite for the entire scanning system, is also discussed in detail. Finally, gaps in the knowledge base and potential areas for research are discussed (see Section 8.5).

8.2 Three-Dimensional Test Artefacts

Manufacturing and measuring a three-dimensional (3D) test artefact results in a composite test of machine performance. This means that most, if not all, of the errors in the system combine to create errors in the test artefact. The major advantage of this approach is that building a test artefact is directly aligned with the intended purpose of the machine – to make parts. It may be tempting for users to build and test a functional part that is typical of their application, or even the actual part itself, then measure that part as a sort of test artefact. This might be acceptable if the user expects to only make this one part with their machine. However, a standardised test artefact that is optimised to the intent of evaluating machine performance is usually a superior approach. The test artefact can, and should, be designed to highlight specific expected errors in the machine or specific characteristics in actual parts that need to be achieved. Furthermore, the test artefact can be designed to accommodate the best measurement equipment available to the user, whereas actual 3D parts tend to be difficult to measure with low uncertainty. A standard test artefact can lead to easy comparisons across machine platforms or over the course of time.

8.2.1 Key Contributions to 3D Test Artefacts

Thousands of AM test artefact designs likely exist in practice. Indeed, it is common for teachers of AM courses to assign their students the task of designing a part to highlight the capabilities of an AM machine or of AM processes generally. Furthermore, machine vendors and users often have their own designs for test artefacts that highlight their machines' capabilities or allow for adjustments to machine settings or process parameters. The vast majority of these artefacts are not discussed in the literature, so little is known about their specific characteristics or the specific intents of the designs.

Many test artefacts are discussed in the literature, and review articles have captured many of the different designs (Moylan et al. 2012, Rebaioli and Fassi 2017, Toguem et al. 2018, Leach et al. 2019). A couple of these designs have been highly influential. In an early book about rapid prototyping with the stereolithography process, Richter and Jacobs (1992) laid out rules to consider when designing a test artefact for stereolithography processes. The National Institute of Standards and Technology (NIST) provides an AM test artefact geometry that is freely available by download from their website (NIST Additive Manufacturing Test Artifact) (see Figure 8.1). Moylan et al. (2014b) describe the criteria used in designing the NIST artefact and details how it was used to measure and adjust the performance of a PBF machine. Notably, these rules, and the vast majority of artefacts proposed in the literature, lead to variations on the theme of one relatively large part with many features intended to highlight different aspects of machine performance.

The recently published ISO/ASTM 52902 (2019) takes a different approach, demonstrating the evolved thinking of the practicality of a test artefact. Instead of proposing one part that attempts to fit all needs, ISO/ASTM 52902 describes several parts intended to highlight different aspects of machine performance. Many users have different requirements, and many additive machines have very different performance expectations. This approach recognises those facts and allows the user to configure the test artefacts in the build in a way that best addresses their needs (see Figure 8.2). Furthermore, rather than having one part that spans the entire build volume, ISO/ASTM 52902 (2019) describes several relatively smaller artefacts. This allows a user to position the artefacts in different places in the machine volume, such as at the centre and at the outer edge of the build platform, to characterise performance in different locations of the machine, either quantitatively or qualitatively or

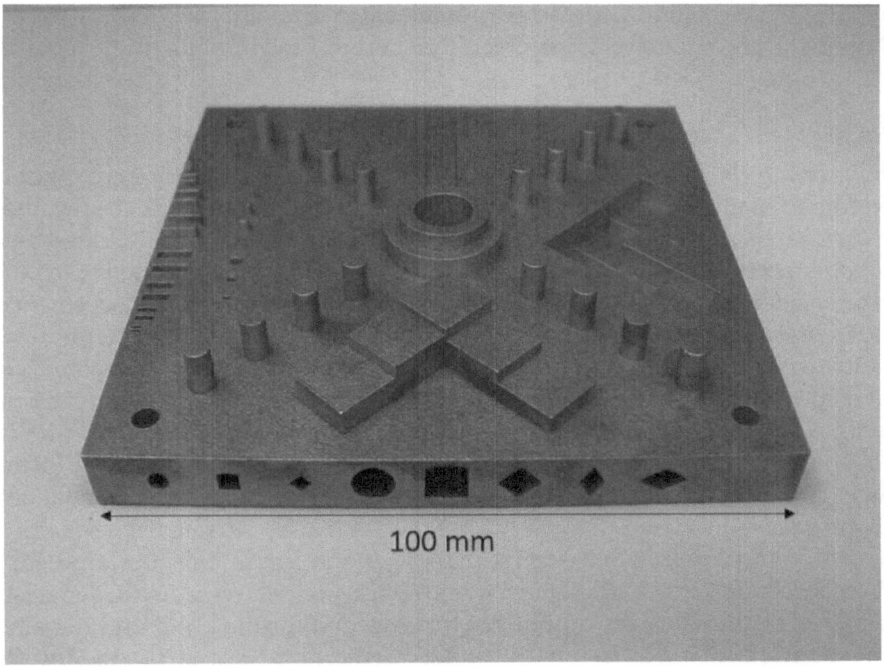

FIGURE 8.1
The NIST AM Test Artefact is designed to highlight potential errors seen in a variety of AM systems. The main structure of the artefact is 100 mm × 100 mm.

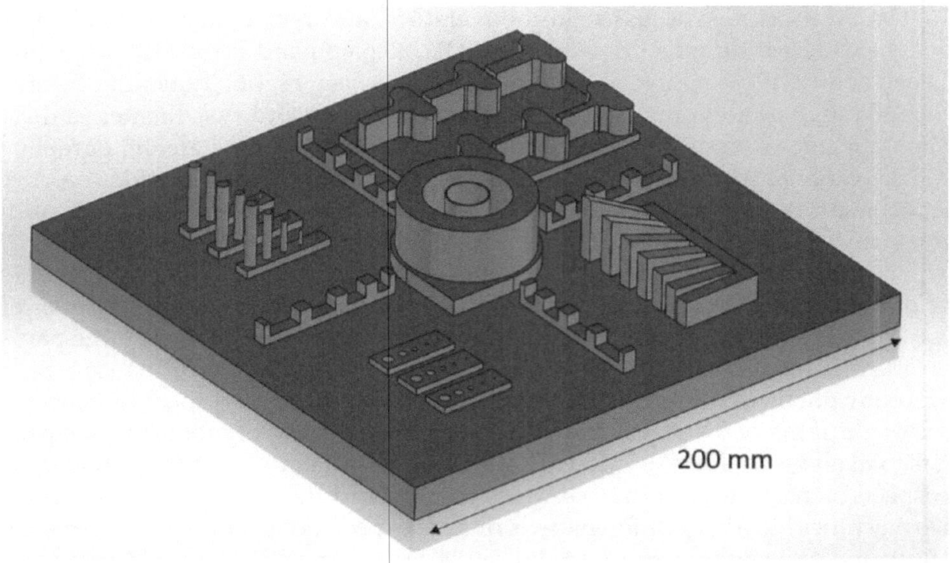

FIGURE 8.2
The test artefacts in ISO/ASTM 52902 (2019) can be configured by the user to best fit their need. In the example shown here, the configuration allows a quick test of *X*- and *Y*-axis performance of a machine with expected coarse resolution. Note that an artefact for evaluating surface texture is also available, but not pictured here.

both. The standard describes a linear accuracy artefact, a circular accuracy artefact, four sets of artefacts intended to test resolution or minimum feature size and an artefact to test surface texture. The resolution artefacts are available in different sizes and different aspect ratios, allowing users to choose the artefacts that best test the limitations of their machine.

8.2.2 Strengths and Challenges of 3D Test Artefacts

The main strength of 3D test artefacts was mentioned earlier, but there are other reasons why this method of machine performance evaluation may be preferred. The fact that many different aspects of machine performance, such as its ability to create a desired geometry, desired microstructure, desired density, etc. all in one part is often attractive to users. It is also worth considering what a physical test artefact can do that direct measurement of components (discussed in more detail in Section 8.3) cannot. It is rather difficult for direct measurement to efficiently and effectively measure the minimum feature size achievable by a system. It is also difficult to predict the achievable surface texture by measuring machine components. The MSFC-STD 3716 (2017) standard recognises that the quality of surfaces and level of detail resolved are often representative of the overall 'health' of a PBF machine, and changes in these characteristics can be among the earliest indicators of changes in the system. Fortunately, these two characteristics are easy to assess using test artefacts.

Unfortunately, test artefacts have some challenges and limitations. Building parts costs time and resources. For this reason, most test artefacts are relatively small, and a test build is rather sparse, minimising the amount of time needed for the build and the amount of feedstock material consumed in the test build. Furthermore, it may be difficult to determine the cause or reason for the observed error in the test artefact. Related to this, it may be difficult to separate machine errors from process errors; systematic errors from non-systematic errors; and errors resulting from poor performance of one component of the machine from that of another component of the machine (for example, errors resulting from poor performance of the energy beam from errors resulting from poor performance of the positioning system). There is a trade-off between capturing a lot in one relatively simple test and being able to easily determine the cause(s) of errors and acting to correct.

8.2.3 Considerations for 3D Test Artefact Design

Careful design of a test artefact will allow a user to maximise the benefits and minimise or overcome the challenges. While great care has gone into all of the designs discussed in the literature and in standards, the challenge with these artefact designs is that they are meant for the general user. Designs intended to be used broadly will get most of the users most of the way there for assessing the performance of their machines. However, given the breadth of the AM industry, even limiting only to metal parts, it is unlikely that these broad-based solutions will address every aspect of machine performance evaluation needed for an individual user, let alone all individual users. Standard specifications, such as MSFC-SPEC 3717, acknowledge that there will be cases where users need to design their own test artefacts: 'the design of the reference part(s) is not specified and may vary with the design needs and priorities of the organization'.

The first consideration for designing the artefact is to assess the needs and priorities of the organisation. Some organisations may need to demonstrate the ability of the system to achieve certain tolerances described in geometrical product specification standards such as ISO 1101 (2017) or ASME Y14.5 (2018). Other organisations may need an artefact to build at regular time periods as quick checks on performance. Some organisations will need to test the entire work volume of their systems; others may only be interested in one or a few particular areas of the build. The possibilities are endless. Some organisations will need to assess only one specific type of PBF machine; others may need to assess only DED machines; and others may need to assess multiple machine platforms with the same design. Each of these considerations could lead to a unique configuration of standard artefacts or a unique test artefact design. If the needs and priorities are less clear, a standard artefact design is probably the best approach.

The next consideration for the design of the artefact should be how the artefact will be measured. The primary objective for any test artefact should be establishing quantitative metrics for machine performance. Quantitative metrics require measurement. This is not a major limitation for this type of method, since most users interested in making parts need to be able to somehow measure those parts. The user should assess whether or not the test artefact, including each of its individual features, is easily accessed by the measurement techniques on hand. One of the most impressive aspects of AM is its ability to produce freeform surfaces and internal geometries. It is tempting to include these features in a test artefact to demonstrate capabilities, but these types of features tend to be difficult to measure with low uncertainty without high-end metrology equipment (for example, a metrology X-ray computed tomography system) (Carmignato et al. 2017). Low measurement uncertainty is paramount. Ideally, the task-specific measurement uncertainty would be ten times smaller than the expected geometric error being measured on the test artefact, but this is often just a 'rule of thumb'; it becomes difficult to draw unambiguous conclusions if the measurement uncertainty is less than four times smaller than the measured geometric error (Khanam and Morse 2009; ANSI Z540.3). The upshot of all of this is that simple shapes and features are often preferred because they are easier to access, and measurement procedures (including fitting reference geometry and number of samples) are rigorous. Rivas Santos et al. (2019) discuss the 'design for metrology' for test artefacts and present measurement results for a specific artefact with different measurement methods.

The priority for any remaining considerations may depend on the needs and priorities of the user or may be considered of equal importance with each other. These considerations are the features to be selected, the sizes of the features and artefact(s), the placement of the test artefact(s) and the machine configuration being tested. Regarding the features to be tested, the MSFC-SPEC-3717 (2017) calls for features examining detail resolution (i.e. minimum feature size) and surface texture on prominent surfaces, such as horizontal, vertical, inclined and free-standing (i.e. overhang) surfaces. Other guidance (see Rebaioli 2017) calls for evaluating the fourteen geometric dimensioning and tolerancing call outs, such as flatness, straightness, circularity, profile, position, etc. Again, these may vary depending on the specific needs and priorities. In general, it is likely beneficial to have both protruding (for example, posts) and recessed (for example, holes) features. Regarding the size, the original thinking in the community (Richter and Jacobs 1992, Byun and Lee 2003, Campanelli et al. 2007) was that a test artefact needed to be large enough to test the entire machine volume. While testing throughout the volume, especially at the outer extremities, is still considered good practice, thinking has evolved to favour smaller geometries that can be replicated in various positions in the build volume (ISO/ASTM 52902:2019). Small artefacts have the benefit of building quicker, consuming less feedstock material and being less prone to distortion from residual stress. Placement and orientation of the test artefacts may be best considered in coordination with the machine configuration. For example, when evaluating a DED machine with stacked linear axes, it might be beneficial to align the artefact(s) or features with the individual axes, allowing the errors in the aligned features to be more easily linked with the individual axes during evaluation. Such a configuration may be more difficult on a PBF machine where the energy beam is actuated by a scanning system that does not necessarily align with the machine coordinate system.

Note that some sources in the literature suggest using multiples of the same feature or test artefact in a build to assess repeatability (Scaravetti et al. 2008). It is often beneficial to include multiple similar features in different locations in the build volume, but this does not test the machine's repeatability. Repeatability is generally considered to be the ability to independently repeat performance over a short period of time under similar circumstances. Multiple parts or features in one build are not independent; the presence of one part influences the properties of other parts in the build, for example, spatter particles, local thermal history or vibration in the recoating system caused by the presence of other parts. Many errors are position-dependent, including many systematic errors (for example, beam shape in laser PBF machines or angular error motion in a DED axis), whereas repeatability (or lack thereof) is more a measure of non-systematic errors (BIPM 2012). Furthermore, when positioning multiple parts or features throughout the build volume, it is likely good practice to use a non-uniform – perhaps even a pseudo-random or stratified-random target position – approach to avoid masking any periodic errors in the machine.

8.3 Component Testing

Breaking down the potential sources of error in a part and building an error budget from the bottom up usually requires isolating the components in the error budget and quantifying them individually. This approach is often better for making fine adjustments to reach precision tolerances, because it is often easier to tweak an individual component than to tweak the entire interconnected system. Furthermore, this approach often highlights that certain components perform much more poorly than others in the machine, allowing a user to concentrate their time on adjusting the most sensitive components. For these reasons, some users may wish to pursue an evaluation and qualification scheme that uses more measurements of individual components than building of test artefacts. This may be especially true of machine builders who have more access to the individual components than machine users.

8.3.1 Key Contributions to Component Testing

The topic of measuring individual components of a machine to derive a better understanding of machine performance and expected AM part tolerances has seen little mention in the literature (Lu et al. 2018) but has been a consistent topic of discussion at precision-engineering conferences focused on AM (see Moylan et al. 2014, McGlauflin and Moylan 2016, Jared et al. 2018). Much of this discussion is based on the approach of building upon standard measurement methods in machine tool metrology. In this way, the ASME B5.54 and ISO 230 series of standards (especially ISO 230-1 on geometric errors, ISO 230-2 on linear positioning error and ISO 230-3 on thermal effects) are key contributions themselves. The key differences between error motion tests for AM machines and tests for machine tools will be discussed in more detail later in this section.

Of course, error motions are not the only sources of geometric error in AM parts, with the energy source also playing a major role. However, there is a dearth of literature in this area. For electron beam systems, the focus is on the scanning system more than the electron beam itself (Guo et al. 2015). For laser beam systems, researchers may look at beam diagnostics as a solved problem, because standards such as the ISO 11146 series (for example, ISO 11146-1:2005, ISO/TR 11146-3:2004) for beam shape and ISO 11554 (2017) for power have existed for some time and have gone through several reviews. Alternatively, the view among AM researchers may be that measuring the energy beam is further along in the technology readiness level (TRL) scale and, therefore, is the domain of industry development, not basic or applied research. Indeed, many commercial providers of beam diagnostic equipment have offerings targeting AM users and applications (for example, Koglbauer 2018, Bergman 2016, Kirkham 2017, Dini 2018). The topic more popular in research is intentional beam shaping (Metel et al. 2019, Faidel et al. 2016).

Other components likely contribute as well, for example build platform heating or gas handling, but they are not discussed as much in the literature.

8.3.2 Strengths and Challenges of Component Testing

In addition to the strengths related to error budgeting, evaluation and adjustment mentioned earlier, direct measurement of components provides some other benefits to the user. Direct measurement can often be done with much lower measurement uncertainty, which may be necessary to achieve precision tolerances. Also, direct measurements do not consume raw materials.

However, the hurdles and challenges of direct measurement of components are often significant. One of the reasons that measurement uncertainty can be much lower is that direct measurement of components often requires specialised equipment. A larger organisation with many AM machines may benefit from purchasing some of this equipment, but a small or medium manufacturer with only one AM machine may find the cost of the equipment too high compared to the benefit. Another major challenge is that not all of the components may be accessible to the users. This is especially true in PBF systems. The access may be limited by proximity, space or control. For example, gaining access to individual mirrors in a laser PBF (L-PBF) machine is problematic; machine builders often buy these systems off the shelf and do not even have that level of access themselves. Furthermore, build chambers tend to be smaller in size, and measurement equipment larger than a build platform may not fit into the machine to perform the desired measurement. Another example is that machine builders and their maintenance staff may have access to a level of control that allows the laser beam to be fired and translated across a beam profile measuring device, but a user might not have access to the same level of control and may have to go through a complicated work-around to complete such a test.

One other consideration when performing these tests is the impact of not operating the machine in the same state as when it builds. Making discrete, targeted movements or individual commands to individual machine components is something that is usually best accomplished in the 'stand-by' or 'jog' mode of the machine, but some machines may exhibit different performance in these modes, especially if velocity or positioning direction are different between jog mode and building. Furthermore, some measurement equipment may need to reside outside the build chamber, which may require machine operation with a door open or window removed. Both of these cases mean that the machine will not be at the operating temperature and will not have the same environment (whether that be vacuum, circulating gas or shield gas) as when the machine is building. Users need to take care that the measured deviations during testing will actually reflect the deviations when the machine is operating.

8.3.3 General Principles of Component Testing

The approaches discussed in this section stem from machine tool metrology and laser beam metrology standards, methods and good practice. These tests tend to be a progression of measurements starting with an environmental test for machine drift, while very little is changed in the machine. This initial test helps to set the baseline and contribute to the measurement uncertainty estimation for tests that follow. Next are tests to determine the error motions of the individual motion components of the machine. Once the individual error motions are determined, the alignment errors between the machine axes are determined. The error motions and alignment errors can be used to create an error map of the machine, leading to an empirical model, often using homogeneous transformation matrices (Leach and Smith 2018), that can predict geometric errors at any location within the work volume (or within the ranges of the error measurements) (Donmez et al. 1986). The likely next progression is to perform additional component measurements (for example, spindle performance on a machine tool or laser performance on an AM system). Alternatively, additional motion measurements may be desired, such as thermal drift when repeatedly exercising axes or coordinated motion of pairs of axes.

A principle that permeates all the measurements of error motions is the concept of the functional point. In machine tool metrology, the functional point is where the cutting tool would contact the part. Good practice is that tests of geometrical characteristics apply setups that represent the relative point between the tool and the workpiece. While a similar principle applies to measurement of AM systems, some slight differences deserve attention. For a DED machine, the functional point is relatively simple: the workpiece is the same as for machine tool metrology, and the tool is represented by the point where the feedstock material intersects the centre of the energy beam. For PBF, the functional point is where the energy beam intersects with the top of the powder layer. Note that the top surface of the powder bed is created by movement of the recoating system, not movement of the Z-axis. Furthermore, the tool analogy here is more difficult because the energy beam does not have the same physical presence. Best application of the 'tool-to-workpiece' measurement for PBF systems likely involves measuring the workpiece movement (Z-axis) relative to the recoating system that creates the top surface of the powder, and measuring the laser system also relative to the recoating system. However, more research is needed to verify this practice (Moylan et al. 2014).

8.3.4 Z-Axis

AM, by definition, involves building a part layer upon layer, which means that nearly every AM machine has a linear actuator for a Z-axis. Slight differences may exist between machines, such as a moving tool versus a moving workpiece, but measurement of the Z-axis error motion is common across nearly all machines. Additionally, the Z-axis is very similar to a machine tool's Z-axis and, therefore, can be measured in a similar manner.

Tests for linear positioning error (ISO 230-2 2014), straightness error (ISO 230-1 2012) and angular errors (ISO 230-1 2012), many of which can be conducted simultaneously, can follow setups nearly identical to the setups described in machine tool metrology standards. For PBF machines, the only difference in the setup is the previously mentioned measurement of the Z-axis relative to the recoating system. For DED systems, the setup is identical to that in the standards. The equipment is also applicable, with laser interferometers, gauge block/dial indicator, straightedge/dial indicator and differential level setups having been described in the literature (McGlauflin and Moylan 2016, Lu et al. 2018, Moylan et al. 2014) (see Figure 8.3).

The differences in the measurements come in the procedure. First, AM machines typically only position in one direction during a build. As such, bi-directional tests of the Z-axis are not needed, only uni-directional tests. Another difference is that the error motions must be measured at different scales. The geometry of large parts might be affected by error motions over the entire range of the Z-axis. This calls for measurement of the entire Z-axis, likely hundreds of millimetres. However, the geometry of an individual layer has impact on the part's surface texture (see Chapter 11) and on the process stability and consistency (and, therefore, the material or mechanical properties). This calls for testing on the scale of individual layers, likely tens or hundreds of micrometres. Combining these two into one test is impractical. A solution to this is to perform two slightly different tests. One test is of the entire range of the Z-axis, almost exactly as described in machine tool metrology standards. These standards provide guidelines on the number of intermediate targets, locations of the targets, etc. It is likely that users will want to conduct multiple uni-directional runs, using random or pseudo-random spacing of the targets on the order of several millimetres. A second test is of positioning at the layer scale. With this test, the target spacing should be the same as a typical layer thickness, and the range of the entire test should be on the order of a millimetre or two. Users will likely want to perform layer-level tests at multiple (probably more than two) positions along the Z-axis range. The positions for layer level tests may be chosen randomly (or pseudo-randomly), or the location closest to the top of build platform (where a regular build would begin) may be given preference, since it will certainly factor into nearly every build. Furthermore, locations furthest from the platform in PBF machines may be tested under loads similar to a full bed of powder. More research is needed to verify that these tests are good practice, but early research is promising (McGlauflin and Moylan 2016). Furthermore, more research is needed to test

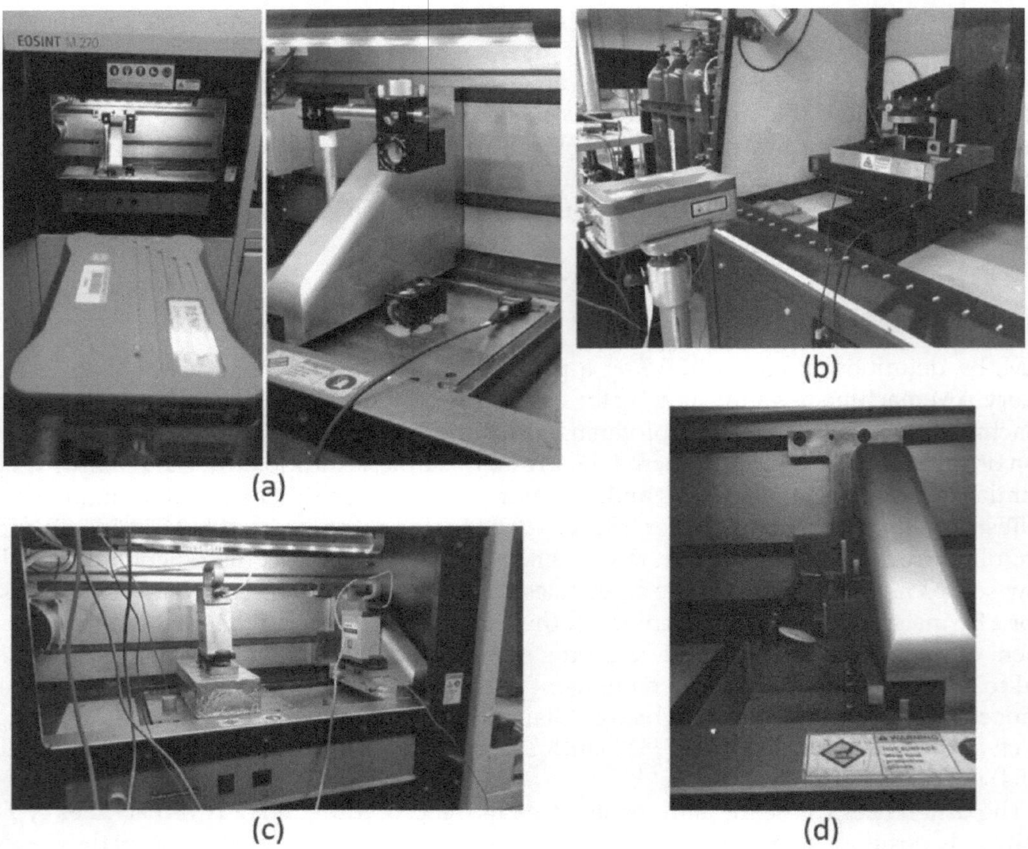

FIGURE 8.3
Typical setups to measure linear axis error motions described in ISO 230-1 (2012) and ASME B5.54 (2005) can also be used for a *Z*-axis in PBF and DED machines: (a) a laser interferometer setup for pitch error of the *Z*-axis, E_{AZ}, in a PBF machine; (b) a laser interferometer setup for yaw error of the Z-axis, E_{BZ}, in a DED machine; (c) an electronic level setup in a PBF machine for E_{AZ}; and (d) an indicator and gauge block setup for measuring linear displacement error of the Z-axis, E_{ZZ}, at individual layer height targets.

whether or not the two tests can be combined, spacing some targets by millimetres and other targets by layer thicknesses in the same run.

8.3.5 Directed Energy Deposition Machine Error Motions

DED machines are essentially vertical machining centres with energy beam and material feed in place of a spindle. Beyond the parametric tests outlined in ISO 230-1 (2012) and IOS 230-2 (2014), which apply to the *X*- and *Y*-axes completely, multi-axis motion tests, such as the circular tests described in ISO 230-4 (2005) and diagonal tests described in ISO 230-6 (2002), are also applicable (see Figure 8.4). The main benefit of these multi-axis tests is the speed of evaluation. Multi-axis tests can often give users a sense of the overall geometric performance of their machine in a matter of minutes or a few hours with only three or four setups, compared to multi-day tests of each individual error motion. However, the trend in machine tool performance evaluation toward a single metric for volumetric accuracy (Weikert 2004, Schwenke et al. 2008, Muralikrishnan et al. 2016) is likely not applicable to

FIGURE 8.4
An overall view (left) and close-up (right) of a circular test using a telescoping ball-bar on a DED machine.

DED because the process contributes more to part errors than in machining, and the layer-upon-layer building makes complex motion evaluation less important.

8.3.6 Powder Bed Fusion Machine Error Motions

Movement of the energy beam in the machine *X*- and *Y*-axes in a PBF machine is usually carried out by a galvanometer (a combination of two rotating mirrors that deflect the laser beam in L-PBF) or by an electromagnetic scanner (in the case of electron beam PBF). Either way, the motions of components are usually not measurable individually. If the rotating mirrors were accessible, a simple axis of rotation measurement could capture error motions similar to a machine tool's rotary stage. Regardless, an alternative approach is a 2D artefact, which is described in Section 8.4.

The overall part geometry is mostly insensitive to error motions of the recoating axis, but certain error motions directly impact layer thickness and, therefore, process stability and consistency. The recoating system of a PBF machine is often an actuator that resembles actuators on a machine tool, either a linear actuator or a rotary stage. Again, tests similar to those conducted on machine tools are likely warranted. For a linear motion recoating system, only two error motions impact the geometry of the top layer of powder: straightness in the *Z*-direction (E_{ZR}) and roll (E_{AR}) (see Figure 8.5).

8.3.7 Energy Beam Diagnostics

Given that a central premise of PBF and DED processes is delivery of energy to the work surface, understanding the characteristics of the energy beam should be of critical importance. The size, shape, power, power density, position (centroid and focal plane) and beam

FIGURE 8.5
A close-up of the sensor nest mounted to the recoating arm and the underlying reference artefact (a flat plate). This setup was used to measure straightness (E_{ZR}) and roll (E_{AR}) of the recoating axis simultaneously.

quality, among others, will all have significant impact on the process physics and thereby the final parts being produced. Since the overall focus of this chapter is on geometry, this subsection will exclude discrete power measurement in favour of other characteristics that impact geometric performance more directly.

As mentioned in Section 8.3.1, the main concepts of laser beam diagnostics are well established in standards (ISO 11146-1 2005, ISO/TR 11146-3 2004). In fact, the laser included in any commercially available AM system likely already conforms to specifications related to these standards in benchtop tests when the laser system was provided to the AM machine builder (since most laser systems are sourced from reputable laser providers, not built in-house by AM machine builders). However, in the AM machine, the energy beam usually passes through a scanning system and focusing optics, so it is good practice for the user to perform beam diagnostics at the point of application (i.e. in the AM machine itself) where all potential influences on the beam are captured, such as distortion or thermal lensing (Bergman 2016).

Measuring the energy beam in the AM machine offers some advantages but more challenges. The main advantage is that the motion system is integrated into the AM machine, so evaluation of beam shape by scanning over a slit or of beam propagation (which is generally one of the more difficult characteristics to measure) by testing at multiple *Z*-axis locations is simplified. However, for L-PBF machines, where the beam is scanned using a galvanometer and often focused using an f-theta lens, there are a range of incidence angles, sometimes as large as 20° (Koglbauer et al. 2018). This is a challenge because most beam diagnostic systems are only able to measure when they are aligned with the laser (either in-line or perpendicular), so measurement may be limited to only one position (i.e.

0,0). Furthermore, a lack of control of the beam position may prevent users from accomplishing some tests. Moving the beam to specific positions often involves programming the system to build a 'test part' that has a geometry that will ensure the beam will move to a specific point along a specific path. However, a typical user may not be able to program the machine to position the laser at a certain point and remain at that point to conduct a test. PBF build chambers tend to be rather confined and impenetrable, so bulky equipment, or some that requires cooling, may not fit in an AM machine and will certainly make testing at multiple locations difficult.

Good practice for energy beam diagnostics is a little notional at this point in time because there is little literature on the topic or standards that are specific to AM applications. However, following existing standards that govern beam diagnostics generally, such as ISO 11146, is certainly a good starting point. Testing the beam in multiple locations in the build envelope, especially at the extremes, will likely provide helpful information, but there may be large uncertainty for these measurements if alignment of the sensor and the beam is difficult. Users should consider the interactions between the beam, scanner, optics, and all other components between the beam output and the workpiece. Measurements of the beam characteristics in the machine can be compared to benchtop test results (if available) and repeated over time to diagnose causes of any problems.

With the current trends toward high-speed compound measurements or sensor fusion (Bergman 2016, Kirkham 2017, Koglbauer 2018), users must take care to understand tradeoffs. In these applications, speed or additional measurands may be favoured over precision or completeness of measurement. If precision is a priority, it is often preferable to perform separate measurements for individual characteristics. For example, Koglbauer et al. (2018) present a method to measure beam characteristics as well as scan speed and scanning position. However, beam waist with this method is confined to two predefined perpendicular orientations, not a complete representation of power density distribution. It may be preferable to have separate beam diagnostics and complement those with a 2D artefact, as discussed in Section 8.4.

8.3.8 Non-Geometric Measurements

Scanning speed is often cited as one of the most critical process parameters that affect part quality. However, little attention is paid to actually measuring the speed and quantifying any error compared to the programmed speed. Koglbauer et al. (2018) discuss the potential for their recently developed method to measure scan speed. Also, a fixed-field-of-view high-speed camera can be used to quantify scan speed (Criales et al. 2017). For DED machines, tests for machine tools likely apply (ISO 10791-6 2014). According to anecdotal evidence, the difference between actual speed and programmed speed at steady state is small compared to other error motions in the machines, but deviations during the transient stage of motion (i.e. while accelerating or decelerating) may be comparatively large.

Another aspect related to scanning speed is the timing of switching the energy beam on and off. Because most machines attempt to use short scan lengths, the energy beam is switched off and on very frequently. Any error in this timing could result in geometric errors, poor surface texture and sub-surface porosity. The most common method of evaluating this characteristic is usually as part of a 2D artefact, discussed in more detail in Section 8.4.

Other aspects of machine performance should likely be part of a larger evaluation or qualification scheme (AMS 7003), but their influence may be more on mechanical properties, microstructure or defect formation than on part geometry. For example, the gas flow inside an AM build chamber can affect cooling rates, which directly affect microstructures (Heigel et al. 2015). However, gas flow can also impact powder denudation, which can have a second-order effect on geometry or surface texture (Matthews et al. 2016). One challenge in measuring gas flow is that, although the measurement method may be rather simple (Heigel et al. 2015), there is large uncertainty in how much the method itself (i.e. the size, shape and positions of the measurement devices) affects the flow.

8.4 Two-Dimensional Test Artefacts

A middle ground between quantifying machine performance using 3D test artefacts exclusively or testing individual components of the machine exclusively is using a 2D test artefact to quantify some aspects of machine performance. This approach is most beneficial when the individual components of the lateral positioning (i.e. in the machine *X*- and *Y*-axes) and the energy beam may be inaccessible to the user. This is most often the case in PBF machines, so users of these machines will likely find the most benefit, whereas this is rarely the case in DED machines, and those users will have little or no use for this approach.

Creation of a 2D artefact is mentioned in MSFC-SPEC-3717 as part of the optical system evaluation. Although this might seem to validate approaches that combine beam diagnostics with beam positioning (discussed in Section 8.3), the standard seems to call for a separate test. The standard states in a note, 'lasing purposeful markings into a flat, solid plate, and evaluating the marking against metrics (based on past performance) may provide sufficient evidence of scanner head health' (MSFC-SPEC-3717 2017).

8.4.1 Strengths and Challenges of 2D Test Artefacts

The general approach for creating 2D test artefacts is to use the energy beam and positioning system to mark (sometimes referred to as etching) a solid plate. Similar to a 3D test artefact, the 2D artefact is then measured, and the deviations in the marked pattern from the designed pattern can inform a user on error motions of the positioning system or poor performance of the focusing optics. The benefit of this approach is that it isolates the beam scanning and the beam performance from other aspects of the system, such as feedstock delivery, gas circulation and error motions of the *Z*-axis. This test method would likely be part of a larger evaluation or qualification approach that also involves direct measurement of other aspects of the machine.

Beyond the fact that the 2D artefact method has not been as rigorously tested as other methods discussed in this chapter, it suffers from two main challenges. The bigger challenge is the measurement of the 2D test artefact. Optical coordinate measuring machines or machine vision systems with high-resolution cameras (see Chapter 10) are likely needed to quantitatively measure the marked pattern. A scanner with a calibrated scale may be useful for certain patterns, but these provide quantification only relative to the scale, and the uncertainty in the scale position may be large. Many machine users may not have access to such equipment. Furthermore, measurement by optical instruments may show increased uncertainty when little contrast exists between the marked pattern and the overall texture of the plate. This is discussed in more detail in Section 8.4.3. Furthermore, the determination of the laser position using the actual scan pattern may not be simple, especially if the melt-pool boundary is difficult to determine or the melt pool is not symmetrical. The second challenge is that programming the pattern may be challenging, or even inaccessible, for some users. Many of the metal AM machines on the market, especially PBF machines, have closed or 'black-box' controllers, meaning that the user may not be able to easily program simple lines or curves. This is usually overcome by designing an actual part where one layer of that part will create the desired pattern. Single lines and curves can be generated by designing very thin-walled parts or by designing solid parts but turning off the infill pattern.

8.4.2 Key Contributions to 2D Test Artefacts

Unfortunately, there is not much discussion on the use of 2D artefacts to evaluate AM machines in the literature. However, the use of 2D targets or patterns for evaluating performance of office printers (i.e. 2D printers) or imagers is well established. For example, Rochester Institute of Technology (RIT) has produced eight books of test targets for this purpose (Chung et al. 2008), and the United States Air Force (USAF) resolution test chart, originally standardised in 1951, is still readily available from many major optics providers (MIL-STD-150A 1951). Some of these patterns, such as the Siemens star (ISO 15775 1999), have provided inspiration for 3D test artefact designs (Jared et al. 2014, Chang et al. 2015). However, the focus of these artefacts and associated analyses tends to be on resolution more than positioning accuracy or geometric error motions (Li et al. 2018).

Analysis of 2D patterns is limited to only a few works in the literature. Tang et al. (2004) describe a method for evaluating and compensating distortion and scaling by placing a piece of white paper on the build platform, using the laser and scanner to 'draw' a 300 mm × 300 mm square and measuring deviations. Land (2014) detailed the use of a grid pattern to create an error map and adjustment scheme. Land's work found + or × markings on the plate to be most effective for lower uncertainty measurement. This work also discussed some of the challenges encountered with a lack of contrast between the marking and the plate and suggested that using a chrome-on-glass plate may be good practice. Lu et al. (2018) describe a pattern that combines a grid of circular shapes, rectangles, lines, concentric circles and a quadrifolium (see Figure 8.6). Preliminary experiments in this work established an optimum density of the circular shapes in the grid, settling on 2 mm diameter and 6 mm pitch. The shapes included in the pattern were chosen to produce an error map (using the grid) and test the machine's ability to satisfy the geometrical tolerance call outs in ASME Y14.5 (2018).

8.4.3 Considerations for Designing a 2D Test Artefact

Because there are no standards and little literature detailing patterns or 2D artefact designs for AM, users are likely to design their own. When doing so, much of the considerations for designing 3D artefacts (see Section 8.2.3), such as individual needs and priorities, apply here as well. However, some slight differences do exist, where the guidance below should be helpful. It should be noted that this guidance is more targeted towards PBF machines because the need for 2D artefacts to test beam scanning is more applicable to users of these machines.

Measurement is again a critical consideration in the design and use of 2D artefacts, but for different reasons than with 3D artefacts. Because of the 2D nature of these artefacts, they will almost certainly be measured optically. As such, a high contrast between the marked portion and un-marked portion of the test plate is beneficial. This may be

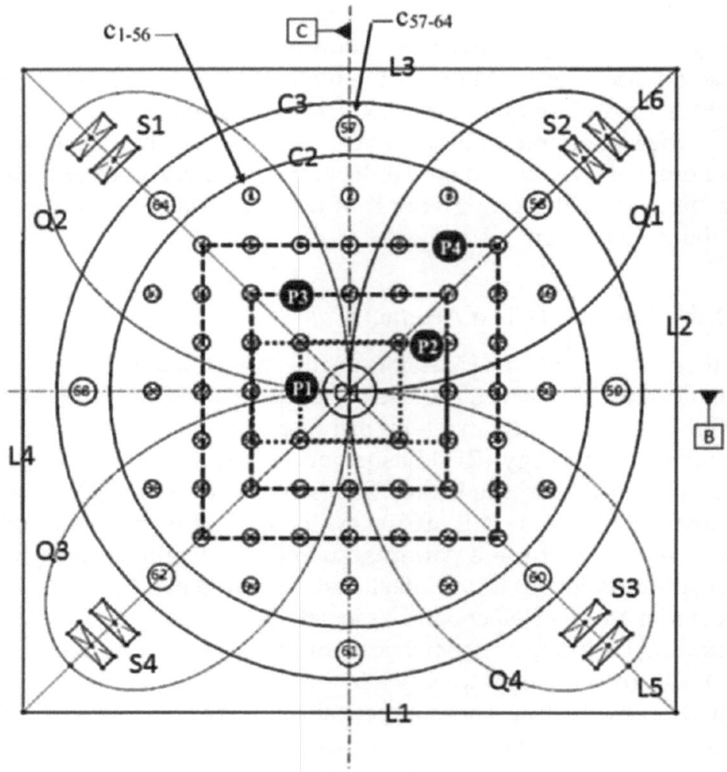

FIGURE 8.6
An example 2D artefact combining a grid pattern, concentric circles, a large square, and a quadrifolium. (From Lu et al., 2018. With permission.)

accomplished with a coating on a plate (for example, chrome on glass) or a surface treated plate (for example, anodised aluminium) where the energy beam removes the surface layer, exposing the underlying material with high contrast. It should be noted that surface treated plates may have larger flatness error, which may negatively impact measurement uncertainty.

Although a grid pattern is a simple and logical 2D artefact, it might not be the best choice for machine users. The benefit of the grid pattern is that it can lead directly to an error map and compensation of the scanner (Halme et al. 2010). As such, builders of scanners and AM machines likely use these patterns when setting up machines. A user looking to quantify residual error after the builder has tuned their positioning system might inadvertently replicate the pattern used for initial compensation, which would mask or minimise residual errors. It is likely advantageous for a machine user to program a different pattern, with various radii, angles and spacings between lines, arcs and other shapes. This should be done on a bare plate (i.e. no powder), exposing only one layer of this part.

It is good practice to include elements or features in a 2D artefact to test different aspects of machine performance. For example, a user might want to include a Siemens star or other diminishing artefact to test resolution as well as concentric circles to test coordinated motion and scaling errors, and a large square to test for scanner system distortions. Furthermore, it is good practice to have features or elements at the extreme positions (i.e. the edges) of the build envelope as well as at the centre. Again, machine configuration should be considered so that certain components can be isolated; users of L-PBF machines should be aware that lines parallel to the machine *X*- and *Y*-axes may still require coordinated motion between the two rotating mirrors (Tang et al. 2004).

Analogous to all of this would be to scan the beam directly onto a position-sensitive device (for example, a photosensitive diode focal plane array). This would allow a direct measurement of the pattern, removing the post-process measurement as a source of uncertainty. However, to do this would need the beam to be attenuated to a level that would not destroy the measurement device, and to do so in a way that ensures the performance of the positioning is the same as during a regular build. Furthermore, uncertainty in these measurements would depend on the resolution of the photo sensitive diode focal plane array. Preliminary research has started in this area but is still inconclusive.

8.5 Areas for Future Research

The fact that many of the methods discussed in this chapter have been fine-tuned over the years in other applications gives confidence that they will continue to be relevant for AM machines in the years to come. As such, the research is likely to be less focused on developing brand-new methods and more on improving the applicability of the test methods to AM machines and processes. More research is needed in determining how sensitive AM part geometry is to errors in various machine characteristics, to better inform the required level of measurement uncertainty for each test. For example, if it is determined that DED part errors are relatively insensitive to common error motions, quicker tests, such as circular tests with a telescoping ball bar, may be more appropriate than time-consuming laser interferometer tests.

The trend in AM machine design seems to be toward bigger and faster machines. For L-PBF machines, this often means multiple lasers acting in the same build envelope. One of the major machine performance challenges posed by multiple lasers is measuring their relative positions and alignments. This will likely be a topic for research in the coming years. In a similar direction, new machine designs that allow for continuous building (versus iterating exposure and recoating) are emerging. Again, the concepts of measuring these machines will not change, but the specifics of the measurements will need to change to match the new designs and applications.

Disclaimer

References

AMS7003. 2018. Laser powder bed fusion process. SAE International.

ANSI Z540.3. 2006. Requirements for the calibration of measuring and test equipment. American National Standards Institute.

ANSI, America Makes. 2018. Standardization roadmap for additive manufacturing—version 2.0. American National Standards Institute/National Center for Defense Manufacturing and Machining.

ASME B5.54-2005. Methods for performance evaluation of computer numerically controlled machining centers. American Society of Mechanical Engineers.

ASME B5.57-2012. Methods for performance evaluation of computer numerically controlled lathes and turning centers. American Society of Mechanical Engineers.

ASME Y14.5-2018. Dimensioning and tolerancing. American Society of Mechanical Engineers.

Bergman, S. 2016. Beam diagnostics improve laser additive manufacturing. Industrial Laser Solutions online (November). https://www.industrial-lasers.com/additive-manufacturing/article/16485613/beam-diagnostics-improve-laser-additive-manufacturing (Accessed 30-8-2019).

BIPM, IEC, IFCC, ILAC, ISO, IUPAC, IUPAP and OIML. 2012. International vocabulary of metrology –basic and general concepts and associated terms. Bureau International des Poids et Mesures, JCGM 200.

Byun, H.-S., Lee, K. 2003. Design of a new test part for benchmarking the accuracy and surface finish of rapid prototyping processes. *Computational Science and its Applications. ICCSA: 989.*

Campanelli, S. L., Cardano, G., Giannoccaro, R., Ludovico, A.D., Bohez, E. L. J. 2007. Statistical analysis of the stereolithographic process to improve the accuracy. *Comput.-Aided Des.* 39(1):80–86.

Carmignato, S., Dewulf, W., Leach, R.K. 2017. *Industrial X-Ray Computed Tomography*. Springer.

Chang, S., Li, S., Ostrout, N., Jhuria, M., Mottal, S.A., Sigg, F. 2015. Geometric element test targets for visual inference of a printer's dimension limitations. *Proc. SFF Symp.* Austin, TX: 1491–1503.

Chung, R., Hsu, F., Riordan, M., Sigg, F. 2008. Test Targets 8.0: A collaborative effort exploring the use of scientific methods for color imaging and process control. RIT Scholar Works.

Cooke, A., Soons, J. 2010. Variability in the geometric accuracy of additively manufactured test parts. *Proc. SFF*, Austin, TX, August: 1–12.

Criales, L. E., Arisoy, Y. M., Lane, B., Moylan, S., Donmez, M. A., Ozel, T. 2017. Laser powder bed fusion of nickel alloy 625: Experimental investigations of effects of process parameters on melt pool size and shape with spatter analysis. *Int. J. Mach. Tool and Mfg.* 121:22–36.

Dini, C. 2018. Why beam analysis is crucial for additive manufacturing. *Laser Tech. J.* 1/2018:35–37.

Donmez, M. A., Blomquist, D. S., Hocken, R. J., Liu, C. R., Barash, M. M. 1986. A general methodology for machine tool accuracy enhancement by error compensation. *Precis. Eng.* 8(4):187–196.

Faidel, D., Laskin, A., Behr, W., Natour, G. 2016. Improvement of selective laser melting by beam shaping and minimized thermally induced effects in optical systems. *Appl. Opt.* 56(26):7413–7418.

Gokuldoss, P. K., Kolla, S., Eckert, J. 2017. Additive manufacturing processes: Selective laser melting, electron beam melting and binder jetting – selection guidelines. *Materials* 10(6):672–684.

Guo, C., Zhang, J., Zhang, J., Wenjun, G., Yao, B., Lin, F. 2015. Scanning system development and digital beam control method for electron beam selective melting. *Rapid Prototyping J.* 21(3):313–321.

Halme, R.-J., Kumpulainen, T., Tuokko, R. 2010. Enhancing laser scanner accuracy by grid correction. *Proc. SPIE.* 7590:1–9.

Heigel, J. C., Michelaris, P., Reutzel, E. 2015. Thermo-mechanical model development and validation of directed energy deposition additive manufacturing of Ti-6Al-4V. *Addit. Manuf.* 5:9–19.

ISO 230-1:2012. Test code for machine tools – Part 1: Geometric accuracy of machines operating under no-load or quasi-static conditions. International Organization for Standardization.

ISO 230-2:2014. Test code for machine tools – Part 2: Determination of accuracy and repeatability of positioning of numerically controlled axes. International Organization for Standardization.

ISO 230-3:2007. Test code for machine tools – Part 3: Determination of thermal effects. International Organization for Standardization.

ISO 230-4:2005. Test code for machine tools – Part 4: Circular tests for numerically controlled machine tools. International Organization for Standardization.

ISO 230-6:2002. Test code for machine tools – Part 6: Determination of positioning accuracy on body and face diagonals (diagonal displacement tests). International Organization for Standardization.

ISO 841:2001. Industrial automation systems and integration – numerical control of machines – coordinate system and motion nomenclature. International Organization for Standardization.

ISO 1101:2017. Geometrical product specifications (GPS) – geometrical tolerancing – tolerances of form, orientation, location, and run-out. International Organization for Standardization.

ISO 10791-6: 2014. Test conditions for machining centres – Part 6: Accuracy of speeds and interpolations. International Organization for Standardization.

ISO 10791-7: 2014. Test conditions for machining centres – Part 7: Accuracy of finished test pieces. International Organization for Standardization.

ISO 11146-1:2005. Laser and laser-related equipment – test methods for laser beam widths, divergence, angles and beam propagation ratios – Part 1: Stigmatic and simple astigmatic beams. International Organization for Standardization.

ISO 11554:2017. Optics and photonics – lasers and laser-related equipment – test methods for laser beam power, energy and temporal characteristics. International Organization for Standardization.

ISO 15775:1999. Information technology – office machines – method of specifying image reproduction of colour copying machines by analog test charts – realisation and application. International Organization for Standardization.

ISO/ASTM 52902:2019. Additive manufacturing – test artifacts – geometric capability assessment of additive manufacturing systems. International Organization for Standardization / ASTM International.

ISO/ASTM 52921-13. 2013. Standard terminology for additive manufacturing – coordinate systems and test methodologies. International Organization for Standardization / ASTM International.

ISO/TR 11146-3:2004. Laser and laser-related equipment – test methods for laser beam widths, divergence, angles and beam propagation ratios – Part 3: Intrinsic and geometrical laser beam classification, propagation and details of test methods. International Organization for Standardization.

Jared, B. H., Saiz, D., Schwaller, E., Koepke, J., Lopez Martinez, M. 2018. *Proc. ASPE and EUSPEN*, Raleigh, July, vol. 69:74–77.

Jared, B. H., Tran, H. D., Saiz, D., Boucher, C., Dinardo, J. 2014. Metrology for additive manufacturing parts and processes. *Proc. ASPE*, Berkeley, April, vol. 57:131–134.

Khanam, S. A., Morse, E. 2009. Test uncertainty & test uncertainty ratio (TUR). *Proc. ASPE*, Albuquerque, April.

Kirkham, K. 2017. Using sensor fusion to analyze laser processing in additive manufacturing. *Tech Briefs* online (May). https://www.techbriefs.com/component/content/article/tb/features/articles/26897 (Accessed 30-8-2019).

Koglbauer, A. 2018. More than beam profiling. *Laser Tech. J.* 3/2018:40-44.

Koglbauer, K., Stefan, W., Märten, O., Reinhard, K. 2018. A compact beam diagnostic device for 3D additive manufacturing systems. *Proc. SPIE* 10523:1052316-1-1052316-7.

Land, W. S. 2014. Effective calibration and implementation of galvanometer scanners as applied to direct metal laser sintering. *Proc. ASPE*, Berkeley, April, vol. 57:151–156.

Leach, R. K., Smith, S.T. 2018. *Basics of Precision Engineering*. Boca Raton, FL: CRC Press.

Leach, R. K., Bourell, D., Carmignato, S., Dewulf, W., Donmez, A., Senin, N. 2019. Geometrical metrology for metal additive manufacturing. *Ann. CIRP* 68:677–700.

Li, H., Verdi, L., Chang, S. 2018. Benchmarking 3D printers' resolutions with geometric element test targets. *J. Imaging Sci. Technol.* 62(1):010504-1–010504-8.

Lu, Y., Badarinath, R., Lehtihet, E. A., De Meter, E. C., Simpson, T. W. 2018. Experimental sampling of the Z-axis error and laser positioning error of an EOSINT M280 DMLS machine. *Addit. Manuf.* 21:201–216.

Matthews., M. J., Guss, G., Khairallah, S. A., Rubenchik, A. M., Depond, P. J., King, W. E. 2016. Denudation of metal powder layers in laser powder bed fusion processes. *Acta Mat.* 114:33–42.

McGlauflin, M., Moylan, S. 2016. Powder bed layer geometry. *Proc. ASPE*, Raleigh, June, vol. 64:108–113.

Metel, A. S., Stebulyanin, M. M., Fedorov, S. V., Okunkova, A. A. 2019. Power density distribution for laser additive manufacturing (SLM): Potential fundamentals and advanced applications. *Technologies* 7(5):1–29.

MIL-STD-150A. 1951. Military standard photographic lenses. Armed Forces Supply Support Center.

Moylan, S., Cooke, A., Jurrens, K., Slotwinski, J., Donmez, M. A. 2012. *A Review of Test Artifacts for Additive Manufacturing*. NIST 7858. National Institute of Standards and Technology.

Moylan, S., Drescher, J., Donmez, M. A. 2014a. Powder bed fusion machine performance testing. *Proc. ASPE*, Berkeley, April, vol. 57:123–126.

Moylan, S., Slotwinski, J., Cooke, A., Jurrens, K., Donmez, M. A. 2014b. An additive manufacturing test artifact. *J. Res. NIST.* 119:429–459.

MSFC-STD-3716. 2017. Standard for additively manufactured spaceflight hardware by laser powder bed fusion in metals. National Aeronautics and Space Administration.

MSFC-SPEC-3717. 2017. Specification for control and qualification of laser powder bed fusion metallurgical processes. National Aeronautics and Space Administration.

Muralikrishnan, B., Phillips, S., Sawyer, D. 2016. Laser trackers for large-scale dimensional metrology: A review. *Precis. Eng.* 44:13–28.

NIST Additive Manufacturing Test Artifact. https://www.nist.gov/el/intelligent-systems-division-73500/production-systems-group/nist-additive-manufacturing-test (accessed 30-8-2019).

Rebaioli, L., Fassi, I. 2017. A review on benchmark artifacts for evaluating the geometrical performance of additive manufacturing processes. *Int. J. Adv. Man. Tech.* 93(5–8):2571–2598.

Richter, J., Jacobs, P. 1992. *Rapid Prototyping & Manufacturing*. Society of Manufacturing Engineers.

Rivas Santos, V. M., Thompson, A., Sims-Waterhouse, D., Maskery, I., Woolliams, P., Leach, R. K. 2019. Design and characterisation of an additive manufacturing benchmarking artefact following a design-for-metrology approach. *Addit. Manuf.* 32:100964.

Scaravetti, D., Dubois, P., Duchamp, R. 2008. Qualification of rapid prototyping tools: Proposition of a procedure and a test part. *Int. J. Adv. Man. Tech.* 38(7):683–690.

Schwenke, H., Knapp, W., Haitjema, H., Weckenmann, A., Schmitt, R., Delbressine, F. 2008. Geometric error measurement and compensation of machines – an update. *Ann. CIRP* 57(2):660–675.

Slocum, A.H. 1992. *Precision Machine Design*. Prentice-Hall Inc.

Tang, Y., Loh, H. T., Fuh, J. Y. H., Wong, Y. S., Lu, L., Ning, Y., Wang, X. 2004. Accuracy analysis and improvement for direct laser sintering. *Proc. SMA Symp*, Singapore, January.

Toguem, S.-C., Rupal, B., Mehdi-Souzani, C., Qureshi, A. J., Anwar, N. 2018. A review of AM artefact design methods. *Proc. ASPE and EUSPEN*. 69:132–137.

Weikert, S. 2004. R-test, a new device for accuracy measurement on five axis machine tools. *Ann. CIRP* 53(1):429–432.

9

Non-Destructive Evaluation for Additive Manufacturing

Ben Dutton and Wilson Vesga

CONTENTS

9.1 Introduction

The complexity of additive manufactured (AM) parts creates challenges for non-destructive evaluation (NDE). AM parts, as in other production processes, can contain defects, and understanding the cause of these defects is important. Different manufacturing processes can produce very different defects, but it is also worth noting that some AM defects are similar to those observed in casting and welding (Sharratt 2015, Frazier 2014, Dutton et al. 2020, Lewandowski and Seifi 2016).

AM differs fundamentally from conventional formative or subtractive manufacturing processes in that it produces parts that are close to the final geometry during manufacture, where the part is built to a near-net-shape design through the 'layer-by-layer' approach (Kraussa et al. 2014, Lu and Wong 2017). However, to be successful, the relationships between build process, microstructure, mechanical properties and integrity of the AM part need to be understood (AMSC 2018, Waller et al. 2019, Russell et al. 2019, Seifi et al. 2016, Seifi et al. 2017). From the business point of view, qualification and certification of components have been identified as a challenge 'break-point'. The acceptance of AM for structurally critical components is not currently possible due to the high costs and durations of conventional qualification routes (Frazier 2014, Tofail et al. 2018, Zerbst and Hilgenberg 2017). Therefore, there is a requirement for the development of reliable monitoring and NDE methods to validate and certify the integrity and quality of components built with AM technology. This is particularly pertinent in industrial sectors such as aerospace, defence and medical, where quality demands are stringent (Lewandowski and Seifi 2016, ENIQ 2007, MSFC 2017, Gorelik 2017, Fousová et al. 2018, Bhavar et al. 2017).

This chapter focuses on the applicability of NDE techniques to AM, highlighting some less conventional NDE methods, such as the resonance acoustic method (RAM), non-linear resonance testing (NLR) and process concentrated resonance testing (PCRT), which are showing potential for the inspection of AM parts.

This section describes, in brief, the main types of defects and their effect on the quality and integrity of AM parts. A more extensive review is presented in Chapter 13. This section concentrates on defect detectability, which does not depend on the cause of the defect but rather on the size and geometry (and potentially the morphology) of the defect. The NDE challenges, limitations and advantages for AM are also discussed.

Standardisation of novel methods is discussed and the highlights of the outcome from work carried out under ISO/ASTM technical committees (TCs) J59 and J60 are included. NDE methods that suit detectability of typical AM defects are analysed and compared, including for post-process and in-process inspection. Finally, NDE reliability is discussed; the foremost aspects are covered, including methodologies established for the probability of detection (PoD) and model-assisted PoD (MAPoD).

9.2 Typical Defects in AM

There are seven distinct AM process categories (see Section 1.2.2): vat photopolymerisation, material jetting, material extrusion, binder jetting, sheet lamination, powder bed fusion (PBF) and direct energy deposition (DED). PBF and DED based technologies represent the two major metal AM processes to build parts from powder feedstock (Tofail et al. 2018, Bhavar et al. 2017, ISO/ASTM 52900 2015, Gebhardt and Hötter 2016, Pang et al. 2014, Kempen et al. 2014). Both PBF and DED can also be classified based on the type of energy source used. Laser beam and electron beam are representative energy sources of PBF-based technologies. Laser engineered net shaping, direct metal deposition, electron beam freeform fabrication and arc-based AM are some of the commonly employed DED-based technologies (Sharratt 2015, Lewandowski and Seifi 2016, Everton et al. 2016, NIST 2013, Antony and Arivazhagan 2015).

Depending on the different input materials, power sources, energies and process parameters, manufacture can yield a variety of flaw types. Explanations of the mechanisms by which these flaws/defects are generated can be linked to the process parameters selected and the resulting processing conditions (Gong et al. 2015, Collins et al. 2017). Note that in this chapter, the terms *flaw* and *defect* can be used interchangeably. Understanding the conditions under which defects are generated and simplifying the terminology used to describe them will hopefully aid the drive for quality improvements as are required for widespread implementation of the technology (Dutton et al. 2020).

The causes and effects of a number of AM defects have been reported in the European project AMAZE (2017) and in several publications, and are covered by draft specification standards ASTM E3166 (2020) and ISO/ASTM DTR 52905 (2020). A more comprehensive description of defects relevant to NDE for AM will be covered in Dutton et al. (2020) and ISO/ASTM DTR 52905 (2020), to be published in 2020. A complementary description of defects along with their morphology and microstructure implications is covered in Chapter 12 and Chapter 13 and elsewhere (Fousová et al. 2018, Collins et al. 2017, Awd et al. 2017, Liu and Shin 2019, Gong et al. 2014, Kasperovich et al. 2016, Singh et al. 2017, Brown et al. 2017, Yasa et al. 2011, Li et al. 2011).

The flowchart shown in Figure 9.1 presents an overall picture of the complexity of defect generation within the PBF processes, including both laser and electron beam, and some of these are also applicable to DED (AMAZE 2017). Figure 9.1 shows that the generation of one defect type can result in an anomalous processing condition, which in turn generates a second defect. For example, the presence of a thick layer, or low laser (or electron beam) power, can lead to under-melting, which in turn can lead to un-consolidated powder. Coupled with the tendency of the power source to decrease the surface energy of un-consolidated powder under the action of surface tension, ball formation may arise due to shrinkage and worsened wetting, leading to pitting, an uneven build surface or an increase in the levels of surface texture (Li et al. 2011). Therefore, even when there are

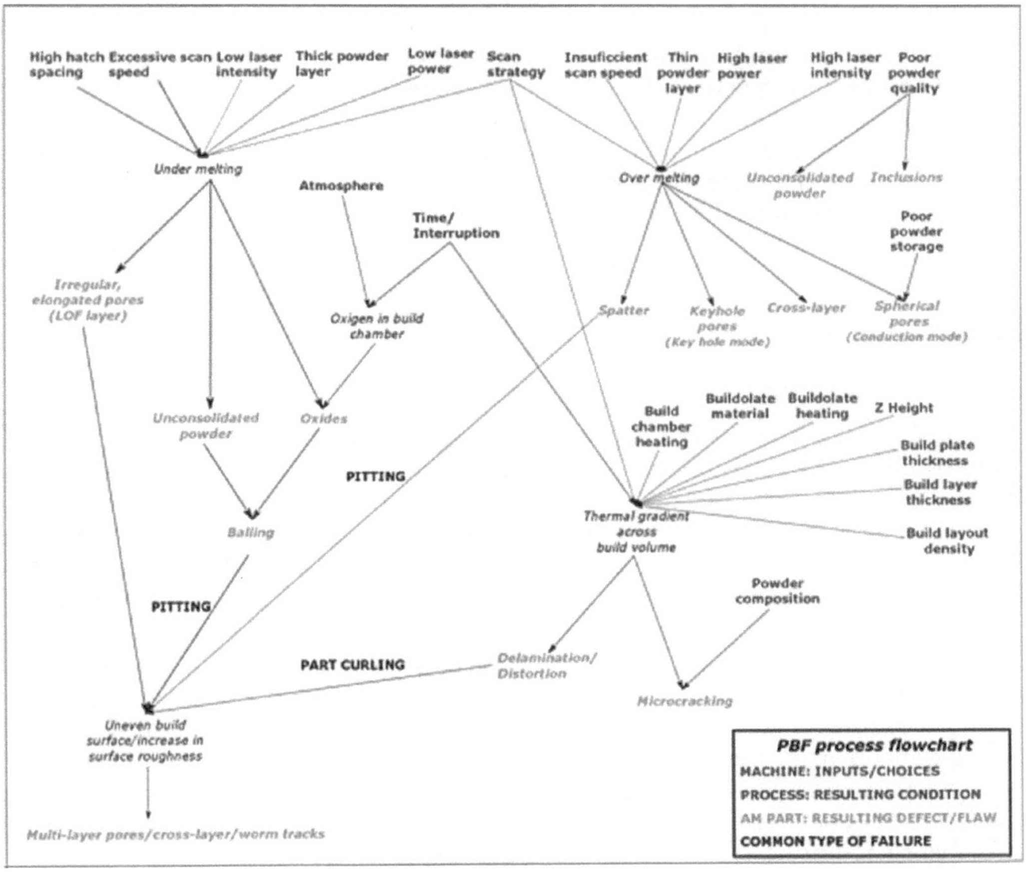

FIGURE 9.1
Flowchart showing the complexity of defect generation within the L-PBF process. In the legend, PBF machine inputs (green), processing conditions (red), AM resulting defect/flaws (blue) and common type of failure (black). (Courtesy of AMAZE EU Project.)

multiple causes, a single defect type or condition (for example, excessive surface texture) can be generated, causing failure by a single mode (surface cracking leading to reduced fatigue properties) (Nicoletto 2017, Wycisk et al. 2014, Mohr 2016). Alternatively, it is also conceivable that a single defect type or condition can cause failure by several different modes (Lewandowski and Seifi 2016, Singh et al. 2017, Yadollahi and Shamsaei 2017).

In general, parts manufactured via AM may have any of the following defects: porosity, layer or horizontal lack of fusion, cross-layer or vertical lack of fusion, trapped powder, unconsolidated power, inclusions, stop/start-type flaws, cracks, residual stress and poor dimensional accuracy (ISO/ASTM DTR 52905 2020, ASTM E3166 2020). Defects can also be introduced by post-processing or damage caused by qualification testing before being placed into service (Fousová et al. 2018, Gong et al. 2015, Awd et al. 2017, Greitemeier et al. 2017). Once in service, additional damage can be incurred due to the impact of mechanical damage, cyclic loads, thermal cycling, physical ageing and environmental effects. Specific defects, such as cracking, have many causes, but these are generally related to the grain boundary (apart from solidification cracking). Note that the issue of spattering, which is believed to be caused by prominent indirect deposition (or indeed welding), is still a significant issue in PBF (Collins et al. 2017, Gong et al. 2014, Hrabe et al. 2017). For laser PBF (L-PBF), there are issues of ablation at the surface of the melt pool caused by the large thermal gradients. For electron beam PBF (EB-PBF), challenges occur from two mechanisms: ablation and charging of the powder (Dutton et al. 2020, Qiu et al. 2015).

Some defects are unique to metal AM processes such as PBF or DED, while others are common across all metal manufacturing techniques, including closely related conventional manufacturing processes such as welding and casting. A detailed summary of the key defects produced during AM is given elsewhere (Dutton et al. 2020) and can be found in the upcoming standards ISO/ASTM DTR 52905 (2020) and ASTM E3166 (2020).

9.3 NDE Challenges in AM

Some of the defects that typically occur in AM can be difficult to detect with conventional NDE techniques, but the typical variations that can occur (high geometrical complexity, dense material or relatively large parts) add extra challenges. A number of NDE methods have already been trialled for AM part inspection, but for highly complex geometries, X-ray computed tomography (CT) is the preferred option (Thompson et al. 2016, Chapter 12). Other well-established NDE methods have partial capabilities and may need to be combined to achieve full coverage (Everton et al. 2016). These methods include visual testing (VT), radiographic testing (RT), ultrasonic testing (UT), eddy current testing (ET), penetrant testing (PT), thermographic testing (TT), acoustic testing and magnetic testing (MT) (ASTM E3166 2020). However, as stated in NDE procedures for AM by ISO (Dutton et al. 2020, ISO/ASTM DTR 52905 2020), the NDE methods do not cover all the inspection requirements that AM parts often need.

Both practical and material considerations must be taken into account when discussing the detectability of an NDE method. Practical considerations include a) special equipment and/or facility requirements, b) cost of inspection, c) personnel and facilities qualification, d) geometric complexity of the part, e) part size and accessibility of the inspection surface or volume relative to the NDE method used and f) process history and post-processing. Materials considerations embrace i) defect type and sensitivity of the NDE method for that defect type, ii) the presence of dissimilar metals, iii) the presence of ferrous or nonferrous

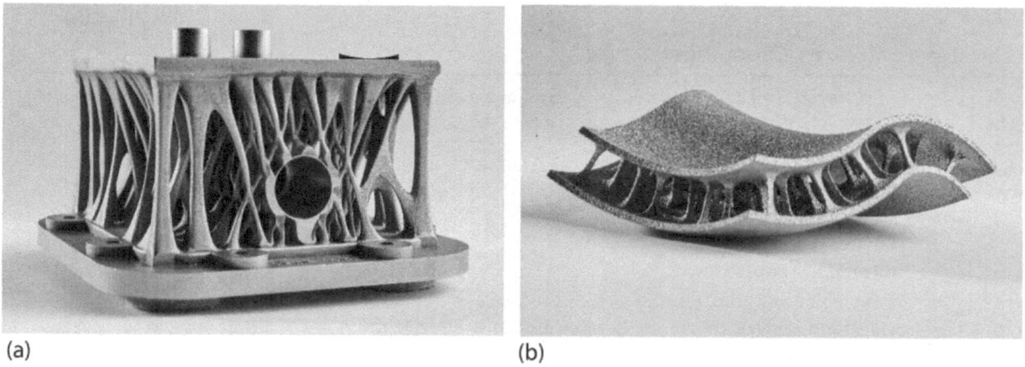

FIGURE 9.2
Example of complex shape and rough surfaces texture parts manufactured by AM. (a) Part complexity. (Courtesy of 3T RPD Ltd [www.3trpd.co.uk[.) (b) Part surface finish. (Courtesy of MTC's National Centre for Additive Manufacturing [NCAM].)

metals, iv) surface texture, v) residual stress, vi) unique, non-equilibrium microstructures caused by rapid cooling, vii) density variations due to thickness variations in parts and viii) anisotropy due to layer structure in parts (Dutton et al. 2020, ASTM E3166 2020).

The geometric complexity is a primary factor governing the ability to apply post-process NDE to AM parts (Hunter 2009, Livings et al. 2018, Todorov et al. 2014). While the application of conventional NDE techniques is possible for AM parts with simple geometries, topology-optimised AM parts with more complex geometries require specialised NDE techniques. Complex designs can have regions that cannot be inspected by NDE methods that require line-of-sight access to the inspection surface or volume. An example of complex parts and the quality of their surfaces is shown in Figure 9.2, and more examples are given in Chapter 2.

The effect of design complexity on NDE method selection is summarised in Table 9.1, while Table 9.2 lists industrial NDE procedures that are used to inspect AM parts (ISO/ASTM 3166 2020, ISO/ASTM DTR 52905 2020, Chauveau 2018, Todorov et al. 2014).

TABLE 9.1
AM Design Complexity Groupings

Group	Design	Description
1	Simple parts	These parts are simple with well-established designs that do not capitalise on the advantages of AM. Such parts may already have consensus NDE inspection procedures.
2	Optimized standard parts	These parts are based on conventional designs, but some the advantages of AM, such as lighter weight or fewer parts, are incorporated into the design.
3	Parts with embedded features	The added features add complexity to a part, thereby decreasing NDE inspectability. Access is limited to internal inspection surfaces.
4	Design-to-constraint parts	These parts appear to be freeformed without straight lines or parallel surfaces and have no analogue made by conventional subtractive techniques. The presence of detailed external and internal features greatly reduces NDE inspectability because the amount of inspection surface has increased and the vast majority of the structure is detailed and embedded.
5	Lattice structures	These parts consist of freeform metallic lattices that have a high strength-to-weight ratio, increased surface areas, and tailored stiffness and damage tolerance. The structures pose the greatest challenge for existing NDE technologies, requiring the use of new or creative NDE techniques.

TABLE 9.2

AM Design Complexity Groupings

Method	Material and defect types detected	Surface or interior defect sensitivity	Global screening or detect location
XCT, macro	In any solid material, any condition and/or defect affecting X-ray absorption, with the exception of reduced mechanical properties	Surface and subsurface, >200 μm resolution	Detects and images defect location; field of view dictated by detector size and distance between test article and imaging plane
XCT, microfocus	In any solid material, any condition and/or defect affecting X-ray absorption, with the exception of reduced mechanical properties	Surface and subsurface, typically 10–200 μm resolution for parts 10 to 200 mm thick	Detects and images defect location; field of view minimized; the focal spot may be optimized for resolution at the expense of scan speed
CT, SX	In any solid material, any condition and/or defect affecting hard and soft X-ray absorption, with the exception of reduced mechanical properties	Surface, sub-surface and bulk, usually 5 μm resolution for a field-view of 10 mm in 42 mm thick	Detects and images defect location; the field of view minimized; the focal spot may be optimized for resolution at the expense of scan speed
CT, NI	In any solid material, any condition and/or defect affecting neutron transmission, with the exception of reduced mechanical properties	Surface, sub-surface and bulk, usually 10 μm resolution in 42 mm thick. In general, it is the size of the field of view divided by 2000, all other elements being optimal	Detects and images defect location; resolution affects the scanning time. Neutrons have higher penetration to most metals (for example, aluminium, titanium, nickel) and are isotope sensitive (light elements)
ET	In electrically conducting and/or magnetic materials for local defects (for example, cracks) and distributed flaws (for example, porosity)	Surface and near sub-surface	Detects and images location
VT	In any solid material, any condition and/or defect affecting visible, structured and laser light reflection	Surface	Detects and images location
PCRT	Any solid material. Any defect or condition	Surface and sub-surface	Global screening
PT	Any solid material. Discontinuities – cracks, pores, nicks, others	Surface breaking	Detects and identifies the location
RT	In any solid material, any condition and/or defect affecting X-ray absorption, with the exception of reduced mechanical properties	Surface and sub-surface	Detects and images location but no depth information
TT	In any solid material, any condition and/or defect affecting heat conduction	Surface breaking and potential sub-surface	Detects and images location but limited penetration in metals
UT	In any solid material, any condition and/or defect affecting sound attenuation, propagation, acoustic velocity and/or sensor-part juxtaposition	Surface and sub-surface	Detects and location

[a] Abbreviations used: XCT = X-ray Computed Tomography, ET = Eddy Current Testing, VT = Visual testing, PCRT = Process Compensated Resonance Testing, PT = Penetrant Testing, RT = Radiographic Testing, TT = Thermographic Testing, UT = Ultrasonic Testing. NI=Neutron Imaging, SX = Synchrotron X-Ray.

The ability of each technique to detect different types of defects as well as locate them on the interior or exterior of a part is also presented in Table 9.2.

For the cases where parts are too large and/or complex, an alternative, or perhaps the only solution, would be to perform an in-process inspection. Advanced in-process inspection methods, such as laser ultrasonic testing (LUT), acoustic spectroscopy and infrared (IR) thermographic melt pool monitoring hold promise for qualifying parts during the build (Li et al. 2011; see also Chapter 13).

9.4 NDE Methods – Advantages and Limitations

NDE techniques can generally be categorised into two main groups: contact or non-contact methods (Hunter 2009, Chauveau 2018). Since the evaluation of the AM part integrity can be carried out after the build and potentially during the build, it is important to define which NDE methods would be applicable, either contact or contactless, or if there are specific techniques for post-process or in-process inspection (AMAZE 2017).

Figure 9.3 shows the relationship between AM defects and the capability of NDE methods to detect them. When part geometries are not internally intricate and the external surface is not highly complex, NDE methods such as UT will have good capability for defect detection (Na et al. 2018, Lu and Wong 2017). Additionally, if the part has been machined, it will increase the capability of UT and allow for other NDE methods, such as dye PT, to be capable of detection of surface-breaking defects (Yasa et al. 2011). A useful tool to initially understand the potential of NDE methods (both post-build and in-process) to detect AM defects has been developed in the European AMAZE project (AMAZE 2017). The study has produced a software tool to estimate the potential of NDE methods to detect AM defects, depending on part complexity and surface texture.

The AMAZE tool considers the micro- and macro-scale complexity of the part and the potential defects of interest. In addition, the tool analyses a range of component structures and surface textures, ranging from simple machined blocks/cylinders to inspection of as-built EB-PBF (very rough) lattice components. For each type of defect, the inspection method is given a score based on its capability for detecting that type of defect in each of the different build categories. Each build category is colour coded to facilitate simple identification. The matrix of capabilities is used to populate the selector tool matrices and is used for reference only. There are very few NDE techniques that are capable of detecting microstructure variations and residual stresses. These defect types are not often encountered in the majority of NDE inspections, and this is reflected in the reduced range of applicable inspection methods for these defects (AMAZE 2017).

The tool is demonstrated in two extreme cases. When the part is a simple block with machined surface finish, Table 9.3 shows that commonly used NDE methods, such as contact/immersion UT and dye PT for bulk and surface-breaking defects, respectively, could be applied. Conversely, when the part is a complex lattice structure with as-built surface finish, Table 9.4 shows a reduction in the number of NDE methods available to mainly X-ray CT and RT. It is important to highlight that these predictions have been generally confirmed from the analysis performed using some of these NDE methods, which is summarised in Table 9.5 (ISO/ASTM DTR 52905 2020).

For post-built inspection, X-ray CT along with acoustic and ultrasonic resonance methods (RAM and PCRT) are the most suitable NDE methods for inspecting highly complex geometries, such as topology optimised parts and lattices in the as-built condition, which

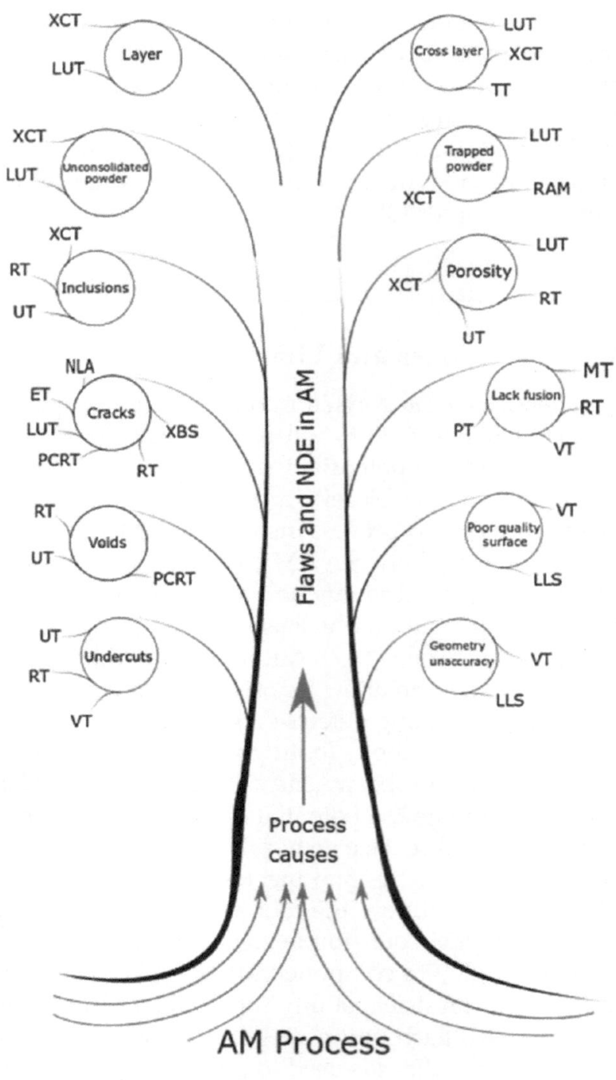

FIGURE 9.3
Diagram displaying the relationship between NDE methods and defects detectability in AM processes. (Courtesy of the MTC.)

are characteristically produced via AM process (Albakri et al. 2015, Kiefel et al. 2018, Aloisi and Carmignato 2016). For an in-depth analysis of X-ray CT capability for defect detection, metrology and surface characterisation, see Chapter 12.

9.5 NDE Standardisation for AM

A comprehensive review of current and upcoming specification standards, including for NDE, is covered in Chapter 6. This section mainly covers the origins of and new developments in upcoming NDE standards.

TABLE 9.3

Review of NDE Technique Potential For AM Parts. Post Built NDE Method Potential for Lowest Surface Roughness and Least Complexity Geometry, for Example, Machined Block/Cylinder

Class	Type	Sub-type	Including...	Surface breaking cracks / lack-of-fusion	Surface breaking voids	Internal cracks / lack-of-fusion / layer defects	Isolated / clustered porosity	Internal voids, incl. cross-layer defects	Inclusions	Trapped powder (Powder Bed Fusion only)	Near surface microstructure variation	Sub-surface microstructure variation	Near surface residual stress	Sub-surface residual stress
Mechanical	Ultrasonic	Contact or near-contact (air-coupled)	Single / twin / array probe, Time of Flight Diffraction											
		Immersion												
	Vibration analysis	Resonance testing	Acoustic pattern recognition											
Optical / visible light	Simple		Aids such as lighting / boroscope etc.											
	Dye-penetrant		Fluorescent / visible											
Radiographic	X-ray	Conventional, 2D	Film / Computed / Real-time / Digital											
		Computed Tomography	2D (fan beam) / 3D (cone beam) CT / Laminography											
		Diffraction												
Thermal	Optically excited	Flash												
		Laser												
	Electrically excited		Induction-heated											
	Vibrationally excited		Thermosonics											
Electromagnetic	Eddy current		Single / array probe											
	Magnetic field	Magnetic particle												
		Barkhausen												
		Alternating Current Field Measurement												
Mixed	Eletromagnetic-Mechanical	Electromagnetic Acoustic Transducer Ultrasound												
	Optical-Mechanical	Laser Ultrasound												
		Spatially Resolved Acoustic Spectroscopy												
		Shearography	Electronic speckle pattern interferometry											
		Laser Speckle Photometry												
		Grazing Incidence Ultrasound Microscopy												

Source: Courtesy of AMAZE EU Project.

TABLE 9.4

Review of NDE Technique Potential for AM Parts. Post Built NDE Method Potential for a Complex Lattice Structure with As-Built Surface Finish

Class	Type	Sub-type	Including...	Surface breaking cracks / lack-of-fusion	Surface breaking voids	Internal cracks / lack-of-fusion / layer defects	Isolated / clustered porosity	Internal voids, incl. cross-layer defects	Inclusions	Trapped powder (Powder Bed Fusion only)	Near surface microstructure variation	Sub-surface microstructure variation	Near surface residual stress	Sub-surface residual stress
Mechanical	Ultrasonic	Contact or near-contact (air-coupled)	Single / twin / array probe, Time of Flight Diffraction											
		Immersion												
	Vibration analysis	Resonance testing	Acoustic pattern recognition											
Optical / visible light	Simple		Aids such as lighting / boroscope etc.											
	Dye-penetrant		Fluorescent / visible											
Radiographic	X-ray	Conventional, 2D	Film / Computed / Real-time / Digital											
		Computed Tomography	2D (fan beam) / 3D (cone beam) CT / Laminography											
		Diffraction												
Thermal	Optically excited	Flash												
		Laser												
	Electrically excited		Induction-heated											
	Vibrationally excited		Thermosonics											
Electromagnetic	Eddy current		Single / array probe											
	Magnetic field	Magnetic particle												
		Barkhausen												
		Alternating Current Field Measurement												
Mixed	Eletromagnetic-Mechanical	Electromagnetic Acoustic Transducer Ultrasound												
	Optical-Mechanical	Laser Ultrasound												
		Spatially Resolved Acoustic Spectroscopy												
		Shearography	Electronic speckle pattern interferometry											
		Laser Speckle Photometry												
		Grazing Incidence Ultrasound Microscopy												

Source: Courtesy AMAZE EU Project.

TABLE 9.5

NDE Method with Most Potential for Post Processing and In-Process Inspection for AM

Inspection type[a]	NDE methods				
Post-process	X-ray	X-ray CT Neutron CT	Thermography	UTPA & TFM	PCRT, RAM and NLR
In-process	Laser scan line	Optical imaging	Thermography	X-ray backscattered Digital X-ray	LUT

In circumstances where it is necessary to examine critically the integrity of a part, in new or existing AM builds, by the use of NDE methods, it is also necessary to establish the acceptance levels for the defects revealed. The derivation of acceptance levels for defects is based on the concept of 'fitness for purpose'; however, a distinction has to be made between acceptance based on quality control and acceptance based on fitness for purpose.

Quality control levels are, of necessity, both arbitrary and usually conservative. Defects that are at the level of, or less severe than, the given quality control levels are acceptable without further consideration. If defects more severe than the quality control levels are revealed, rejection is not necessarily automatic; decisions on whether rejection and/or repairs are justified may be based on fitness for purpose, either in the light of previously documented experience with similar material, stress and environmental combinations, or on the basis of 'engineering critical assessment'.

Although current NDE terminology distinguishes between a *flaw* (an imperfection or discontinuity that is not necessarily rejectable) and a *defect* (a flaw that does not meet specified acceptance criteria and is rejectable) (ASTM E1316 2019b), harmonised terminology for specific AM flaw types is still emerging (and, again, in this chapter, the two terms have essentially the same meaning). As discussed elsewhere (AMSC 2018, Seifi et al. 2017), there is a high priority need for industry-driven standardisation with input from experts in metallurgy, NDE and AM fabrication to identify defects or defect concentrations that can jeopardise an AM part's intended use. Accepted definitions for individual defect types or classifications that are needed to accept or reject AM parts by NDE have been compiled in ISO/ASTM 52900 (2015). These definitions will ultimately become accepted terminology under the joint jurisdiction of ASTM TC F42 and ISO TC 261 and will be included in future revisions of the ISO/ASTM 52900.

Existing standards for welding and castings have been reviewed as the manufacturing process (or the end products) are similar to DED and PBF (Chauveau 2018). A summary of some recent research work is presented in Table 9.6. It is important to highlight that defects 8–11 are not typical defects occurring in the current processes and are AM specific, therefore requiring special attention in NDE standards for AM.

The summary of the review on the current standards (Table 9.6) is shown in Table 9.7. Defects that would be covered by other types of inspection, for example, dimensional measurement or material characterisation, are categorised as 'non-NDE'. All defects listed in the table for DED are generally covered by current NDE standards, except for the non-NDE instance. For PBF, seven defects are not covered by current NDE standards. Three of these are non-NDE, and four are defects unique to AM (unconsolidated powder, layer, cross-layer and trapped powder). The unique defects require new NDE recommendations, which will be addressed and will also refer to newly developed standards in other sectors, such as aerospace (ISO/ASTM DTR 52905 2020). A new structure for NDE standards for AM has been proposed, as shown in Figure 9.4. This structure is built on the current welding standards structure shown by the white boxes, while the required development of new standards tailored specifically

TABLE 9.6

Review of NDE Methods for Detection of Post-Build AM Defects[a]

No.	Defect in as-built AM part	Defect in welded or cast part	Condition	NDE	Applicability to AM defects
1	Void	Void	As-built	RT	Yes
			Post-machined	RT	Yes
				UT	Simple geometry is possible, but complex geometry is limited
2	Porosity	Porosity	As-built	RT	Yes
			Post-machined	RT	Yes
				UT	Simple geometry is possible, but complex geometry is limited
3	Surface defect (cracks, crater, voids, porosity). This excludes poor surface finish, which should be covered by dimensional measurement (surface roughness)	Surface defect (cracks, voids, inclusions, undercut, overlap, incomplete fusion, spatter, sagging, excess penetration, crater crack, incomplete root penetration, groove, excess weld metal, steps in a part)	As-built	VT	Yes
			Post-machined	VT	Yes
				PT	Yes
				MPT	Yes, ferromagnetic
4	Incomplete fusion (DED only)	Non-uniform weld bead and fusion characteristic	As-built	VT	Applicable for external defects only
				RT	Yes
			Post-machined	VT	External defects only
				RT	In DED, the direction of radiography beam is important
				UT	For simple geometry; limited for complex geometry
5	Undercuts at the toe of the welds between adjoining weld beads	Undercuts	As-built	VT	Applicable for external defects only
				RT	In DED, direction of radiographic beam is important
			Post-machined	VT	Applicable for external defects only
				RT	Direction of X-ray beam is important to get coverage of the defects
				UT	For simple geometry; limited for complex geometry
6	Inclusion in part (from contamination in powder or from wire feedstock)	Inclusion (slag, flux, oxide and metallic inclusion other than copper)	As-built	RT	Inclusions with the same density as the part will be difficult to detect

(*Continued*)

TABLE 9.6 (CONTINUED)

Review of NDE Methods for Detection of Post-Build AM Defects[a]

No.	Defect in as-built AM part	Defect in welded or cast part	Condition	NDE	Applicability to AM defects
				UT	Not possible due to surface roughness
			Post-machined	RT	Inclusions with the same density as the part will be difficult to detect
				UT	For simple geometry; limited for complex geometry
7	Non-uniform weld bead and fusion characteristic	Non-uniform weld bead	As-built	VT	Applicable for external defects only
				RT	Direction of the X-ray beam is important to get coverage on the defects
			Post machined	VT	Applicable for external defects only
				RT	Direction of the X-ray beam is important to get coverage of the defects
				UT	Only for simple external and internal geometry; limited for complex geometry.
8	Unconsolidated powder	None	As-built	None	TBD
			Post-machined	None	TBD
9	Cross layer	None	As-built	None	TBD
			Post-machined	None	TBD
10	Layer	None	As-built	None	TBD
			Post-machined	None	TBD
11	Trapped powder	None	As-built	None	TBD
			Post-machined	None	TBD

[a] Abbreviations used: AM = additive manufactured, DED = Directed Energy Deposition, MPT = magnetic particle testing, NDE = non-destructive evaluation, PT = penetrant testing, NDE = non-destructive evaluation, RT = radiographic testing, UT = ultrasonic testing, VT = visual testing, TBD = to be developed.

for AM is shown by the grey boxes (ISO/ASTM DTR 52905 2020). ISO TC 261 and ASTM TC F42 have initiated work to support the standard development covering the design, seeding of defects, fabrication and characterisation of artefacts to demonstrate NDE capability. Currently, the publication of one standard is in progress (ISO/ASTM DTR 52905 2020), while a second is under the balloting process (ISO/ASTM DTR 52906, ASTM WK56649).

In ISO/ASTM DTR 52905, AM artefact specimens were developed to address the types of AM defects not covered by current NDE standards and that are unique for additive processes, such as layer, cross-layer and unconsolidated powder/trapped powder. Some of these defects (layer and cross-layer) are also being covered in ASTM E3166.

A two-level approach has been proposed. One uses a generic star geometry that will provide defect detectability using a specific NDE method, and may be used for relatively

TABLE 9.7
Classification of Directed Energy Deposition (DED) and Powder Bed Fusion (PBF) Flaws

	Flaw type	Non- NDT	Common in DED & PBF	Covered by current standards	Unique to AM
DED	Poor surface finish	X	X		
	Porosity		X	X	
	Incomplete fusion			X	
	Lack of geometrical accuracy/steps in part	X	X		
	Undercuts			X	
	Non-uniform weld bead and fusion characteristic			X	
	Hole or void		X	X	
	Non-metallic inclusions		X	X	
	Cracking		X	X	
PBF	**Unconsolidated powder**				X
	Lack of geometrical accuracy/steps in part	X	X		
	Reduced mechanical properties	X			
	Inclusions			X	
	Void		X	X	
	Layer				X
	Cross layer				X
	Porosity		X	X	
	Poor surface finish	X	X		
	Trapped powder				X

simple geometries (Figure 9.5). The second is an 'à la carte' design with a more representative geometry of the production part (Figure 9.6). The generic star geometry has the cross-section of a five-point star. The three final versions of the star artefact were created with defects varying between 0.1 mm and 0.7 mm in diameter and between 2 mm and 5 mm in length. The three designs are identified as Reference S0 – No nominal defects; design S1

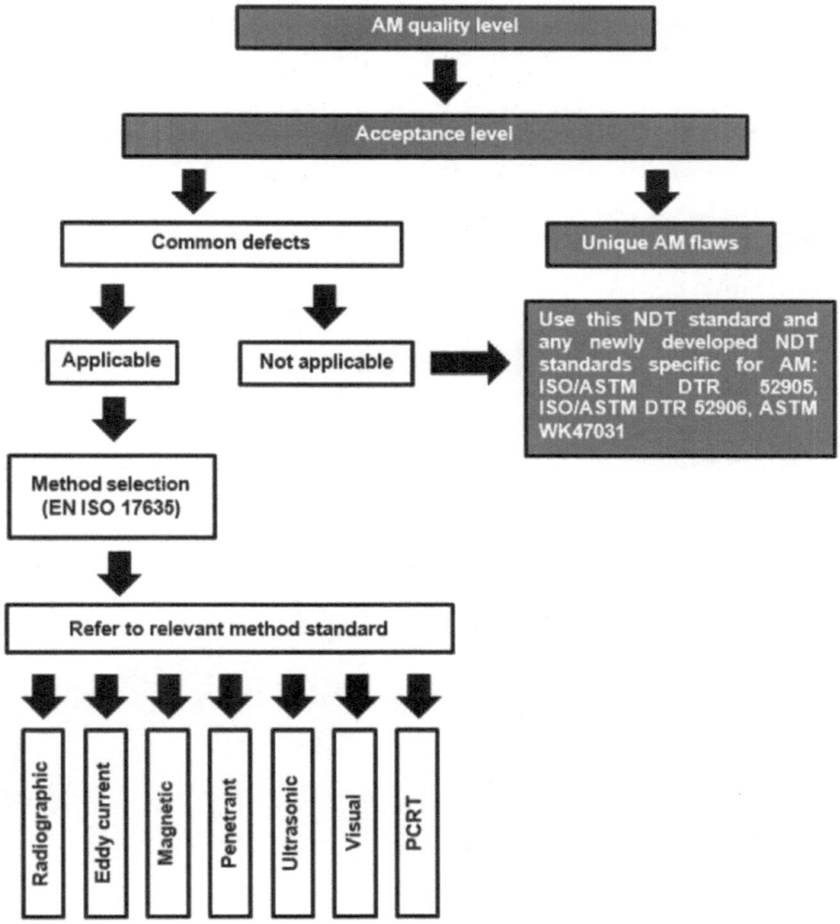

FIGURE 9.4
Proposed NDE standards structure for AM. The boxes indicate development of new NDE standards for AM.

– Nominal defects positioned close to the tip and on the surface of the internal pentagon; and design S2 – nominal defects as S1 but positioned at the corner, away from the tip.

The dimensions of the artefact are dependent on a number of factors, including the material, accelerating voltage and the type of filter, based on 10% transmission of X-ray energy achieved by the X-ray CT method. The à la carte design has defects seeded at geometrically or structurally critical locations, based on the required defect sizes and defect types. Details of the designs are given elsewhere (Dutton et al. 2020, ISO/ASTM DTR 52905 2020). Using the part geometry, the following four steps are recommended to create the à la carte framework: i) identify structurally critical defect locations and sizes to be required for detection, ii) seed those defects to the part in the identified critical areas, iii) build the part which is now the NDE test/calibration artefact and iv) test the built artefact with NDE methods to qualify if it is capable to assess the part quality requirements.

These artefacts will be available after the ISO/ASTM DTR 52905 standard is published so that capability comparisons can be performed. Moreover, under ISO TC 261 JG60, a new standard for seeded defects is currently being developed: the ISO/ASTM DTR 52906. The scope of this document is intended to serve as a good practice guide for the identification and 'seeding'

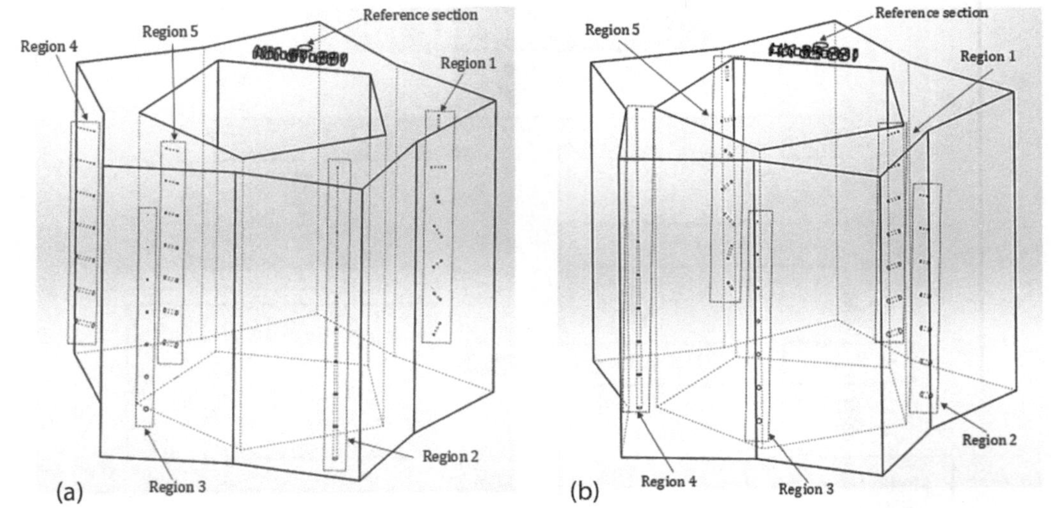

FIGURE 9.5
Star artefact design developed for ISO/ASTM DTR 52905 standard. (a) Star artefact design version S, (b) Star artefact design version S2. (Courtesy of NOSFAM project.)

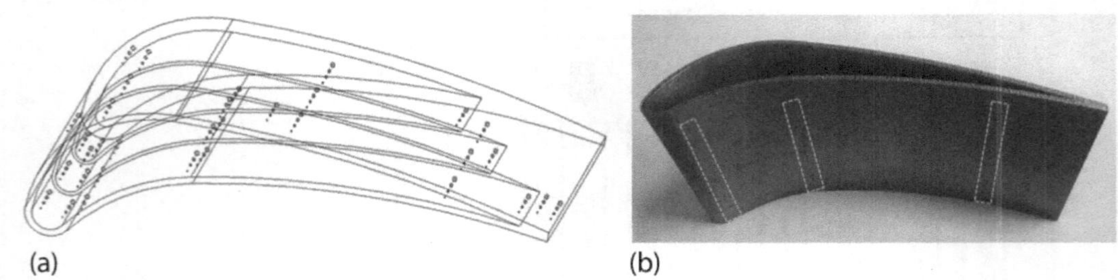

FIGURE 9.6
Example of 'à la carte' design based on generic airfoils: (a) airfoil model showing CAD seeded defects, (b) As-built 'à la carte' generic airfoil. The rectangles show simulated defects manufactured by PBF.

of non-destructively detectable defect replicas of metal alloy PBF and DED processes. Three seeding categories are covered: Category 1 – process defects through a CAD design, Category 2 – build parameter manipulation and Category 3 – subtractive manufacturing.

Although the development of NDE standards for AM components, through ISO TC 261 and ASTM F42 joint groups 59 and 60 and other ASTM participations, has been successfully funded through projects such as AMAZE (2017), more joint efforts are required to deliver what the industry urgently requires. Without NDE standards, the AM industry will be hindered or at least delayed in its current development. Novel NDE methods covering in-process, post-build and/or further development of existing methods are also needed to generate specific NDE standards that can cope with the level of complexity that AM demands.

9.6 NDE for Qualification in AM

In order for AM processes to be successful, the product quality must first be ensured. Before the NDE methods for post-process and in-process inspection are described, it is necessary to harmonise some concepts that are currently not clearly defined. A number of terms are used to describe how quality assurance and quality control are assessed during the manufacturing process. The definition of these terms is included to set the context of any measurement that is made in relation to the manufacture of a product.

The terms are divided into five sub-themes: in-process, on-machine, in-line, in-situ and off-line, which are described as follows (see also Chapter 13 for similar terms in the context of process monitoring).

- *In-process*: Real-time measurements that could be used to predict or detect issues at any stage of the manufacturing process. These measurements must be synchronised with the manufacturing process to allow it to be monitored.
- *In-line*: This is considered as in-process; however, in-line measurements are regular and must be taken before the next stage of the process can continue, either for the part moving to the next station or as a sample of the batch.
- *In-situ*: Measurements are collected directly from the location where the manufacturing process is occurring on the machine. This measurement is used to monitor how the machine is interacting with the part. This can be done either at the station/point of contact or through measurement of the workpiece after the process has passed before the task at the station is finalised.
- *On-machine*: Measurement and recording of data is performed on a station during manufacturing and can be determined from any variable that provides information about the process. On-machine measurements are generally in-process; however, these can be off-line if the data is not being used for process monitoring and control. Conversely, *off-machine* would describe any measurement that is not taken on the manufacturing system.
- *Off-line*: Describes any measurements that are not in-process and not synchronised with the manufacturing process route and, therefore, cannot be used for process monitoring.

Typically, quality inspections are performed after the build of the full part, which becomes difficult for complex geometries when conventional NDE methods are used. Despite combinations of commonly used methods, these NDE methods would not cover all the requirements that an AM part requires (IAEA 2017, Waller et al. 2014, Waller et al. 2015, Waller 2018). A holistic approach to AM quality is presented in Figure 9.7, which displays the main areas for quality improvements, where in-process monitoring alongside in-process inspection are highlighted as having the potential to reduce or eliminate the need for the processes that follow.

Potential NDE methods for in-process and post-process inspection in PBF and DED have been evaluated. Both inspection paths were assessed using a balanced number of criteria that included capability and difficulty. The capability was weighted based on defect detection, part coverage, speed and geometry independence, while cost, development required, integration and automation were assessed for difficulty against capability. The main conclusions obtained from this work are summarised in Table 9.8. The total score of the in-process NDE methods is plotted in Figure 9.8a (web spider) and Figure 9.8b (visualisation graph), while the same approach is used for post-process NDE methods displayed in Figure 9.9a and Figure 9.9b.

The NDE methods that have been analysed previously and shown to have significant advantages over conventional NDE methods to inspect parts manufactured via AM, both

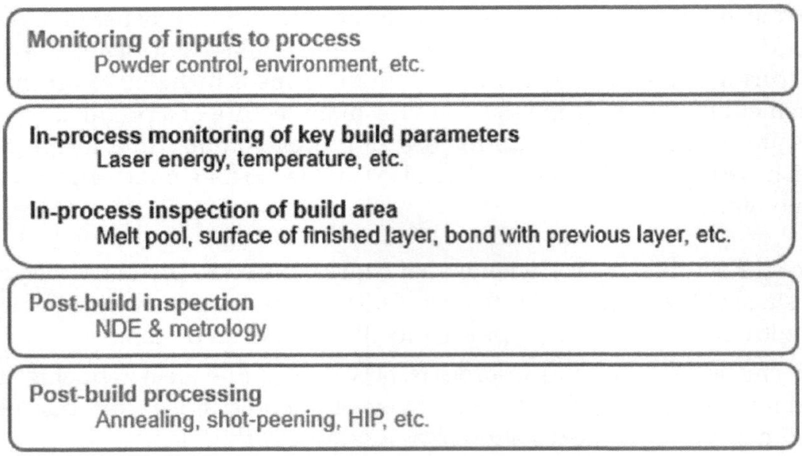

FIGURE 9.7
Holistic approach to assessing AM quality.

TABLE 9.8
Summary of NDE Methods Capability for Post-Process and In-Process Inspection

	NDE methods		
Inspection type[a]	**Capable and low difficulty**	**Intermediate (but require machining)**	**High capability and difficulty**
Post-process	MPI/DPI (but require machining) PCRT	Contact ultrasound (UT) Immersion ultrasound	X-ray CT Digital X-ray
In-process	Optical Laser scan line IR/Thermography	Contact ultrasound (UT)	Laser ultrasound (LUT) X-ray CT (XCT) Digital X-ray (D X-ray) X-ray backscattered (BX-ray)

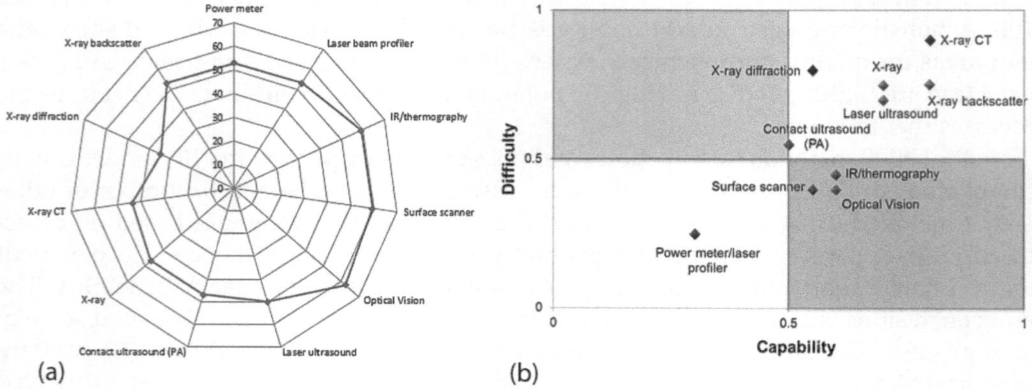

FIGURE 9.8
NDE technologies for in-process inspection of AM parts. (a) Spider web in-process NDE methods total score, (b) vectorial representation of the NDE potential for in-process inspection.

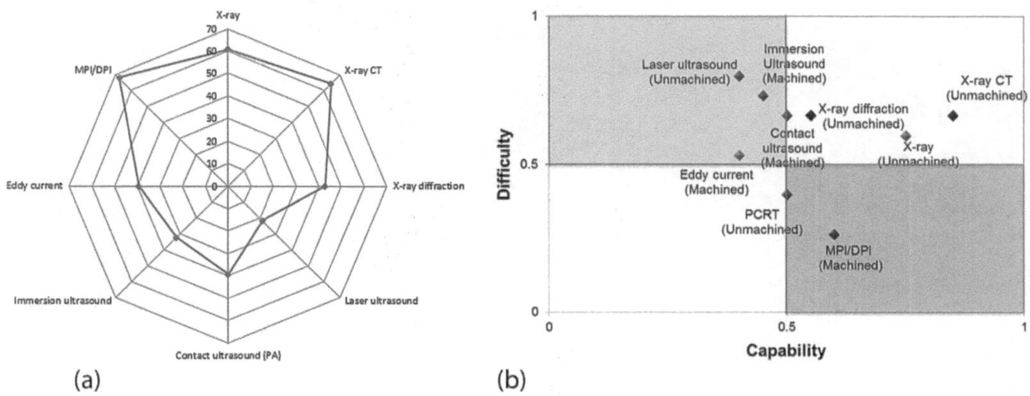

FIGURE 9.9
NDE technologies for post-build inspection of AM parts. (a) Spider web post-build NDE methods total score, (b) vectorial representation of the NDE potential for post-build inspection.

post-process and in-process, are covered separately in the following two sections. These methods include X-ray CT, acoustic methods (PCRT, NLR and RAM), advanced thermography, X-ray synchrotron and neutron imaging/diffraction characterisation (microstructure, residual stress) and LUT (Tofail et al. 2018, ASTM E3166 2020, Hirsch et al. 2017, Livings et al. 2018, Todorov et al. 2014). Since evaluation of AM part integrity can be carried out either after or during the build, it is important to define which NDE methods would be applicable in both conditions or if there are specific techniques for each application.

9.6.1 Post-Process Inspection

Although AM is considered to have the potential for significant market growth in many industrial sectors, its progress is being constrained due to three main factors: material characterisation, process control and quality assurance during the manufacturing process. The inspection capability to assess the quality and integrity of the AM build must be an integral part of the AM process. One of the difficulties in quality assurance, mainly post-process NDE, is the geometrical complexity of AM parts. Conventional NDE approaches can be applied for simpler geometries; however, optimised AM parts with more complex designs or geometries require advanced NDE techniques. The effect of design complexity on the selection of the NDE method is further addressed elsewhere (Todorov et al. 2014, Waller et al. 2014, Murr et al. 2009), where industrial NDE methods were linked with the ability of each technique to detect different types of defects located either on the inside or outside of the part.

Methods such as X-ray CT, RT, UT and neutron radiography/tomography (NR) have varying degrees of additional utility as metrological tools. X-ray CT in particular is well established to measure internal dimensions of complex, topology-optimised metal AM parts; synchrotron X-ray CT (SX) is considered the most capable method (Zhou et al. 2015) but is limited, particularly, due to the high cost involved for industrial applications.

The conventional post-build NDE techniques for AM parts include X-ray CT, ET, VT, PCRT, RAM, NLR, PT, RT, thermographic testing (TT), UT, LU, SX and neutron imaging (NI). Typical methods for external surface inspection are PT, VT, TT, UT and ET. However, the surface quality of the part can impact the capability of detection, particularly with PT and VT. Traditionally, X-ray CT and RT are the main inspection systems

FIGURE 9.10
Computed tomogram (left) and corresponding micrograph of a cross-section (right) of a titanium alloy artefact showing location of defects present in both (circled and pointed at by black arrow) and present only on the cross-section (squared region). Voxel resolution of 41 µm. (Courtesy of AMAZE project.)

utilised for many AM manufacturers and are used to detect gross defects and confirm internal geometry. However, RT and X-ray CT can have reduced resolution for large geometries and artefacts due to the high density of the typical materials, both limiting defect detection capabilities.

An example of an X-ray CT resolution limit is shown in Figure 9.10 left, showing a comparison of X-ray CT slice data with a corresponding optical micrograph of a polished cross-section taken from the same location. Figure 9.10 right shows that X-ray CT was found not to be capable of identifying defects smaller than approximately 132 µm. Also, the CT results on an Inconel® star artefact presented in Figure 9.11 indicate

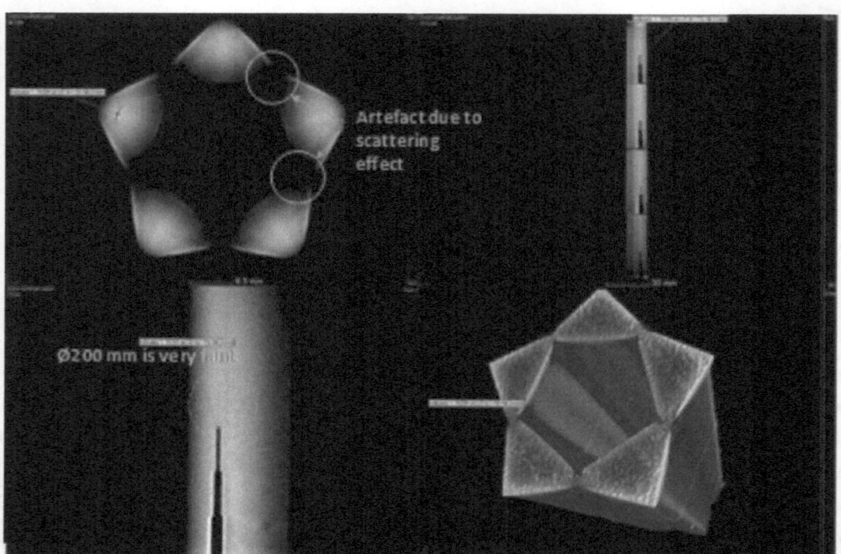

FIGURE 9.11
X-ray CT example from Inconel® star artefact for ISO/ASTME DTR 52905 standard. (Courtesy of AMAZE project.)

that the 3D CT scan data show inspection limitations for feature sizes smaller than about 200 μm. This highlights the importance of knowing what the minimum detectable defect size is that can be reliably detected relative to the critical initial defect size, thus ensuring defects larger than this are detected and appropriate part screening by NDE is performed.

The second CT capability limitation is image artefacts, typically beam hardening for denser materials (see Chapter 12), especially at hard corners, as indicated by shadowing and haloing effects. This is exemplified in an à la carte design (airfoil) which was built and tested to demonstrate NDE capability for complex AM parts. Figure 9.12 shows the geometry where defects were incorporated at geometrically critical locations and at required defect sizes. The 3D scan images show that the majority of the defects (holes) on the airfoils can be detected (Figure 9.12a), except for the smallest holes (0.1 mm diameter) seeded on the part, which cannot be observed at a number of positions (Figure 9.12c). An additional reason for an à la carte design is that all AM designs are different, and it is not feasible to generate a specific image quality indicator or physical reference artefact for all possible geometries.

Due to the layer building process of AM, it can introduce defects that do not form in the conventional manufacturing processes, such as casting and forging. Each individual layer is prone to having defects, and each layer is so thin that the ability of NDE systems to detect defects is affected by sensitivity and resolution.

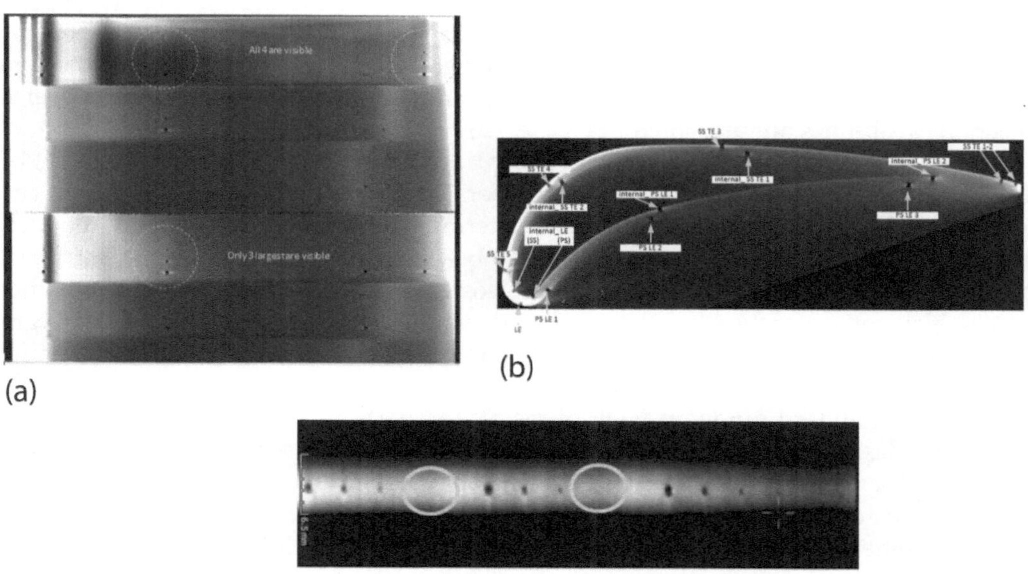

FIGURE 9.12
X-ray CT of the generic airfoil. Cylindrical holes representing layer defects with diameter 0.1 mm, 0.3 mm, 0.5 mm and 0.7 mm are shown, and capability of detection on features with 0.1 mm diameter. (a) X-ray CT images of the airfoil showing the three thickness section 2 mm, 3 mm and 4 mm. Top view, (b) X-ray CT scan of the airfoil showing detailed evaluation of the holes 0.1 mm diameter. Labelling is as follows: SS is 'suction side' = convex side of the airfoil; PS is 'pressure side' = concave side of the airfoil; LE is 'leading edge' and TE is 'trailing edge'. (c) X-ray CT scan showing the thinnest wall section at SS TE 5 region where 0.1 mm hole diameter was not visible. The holes on the middle (3 mm) and on the thick section (4 mm) were not visible. (Courtesy of GE Power.)

Under a comparison run by ISO/ASTM JG59, several materials, such as Hastelloy®, Maraging alloys, titanium, stainless steel, aluminium and cobalt-chrome, were examined using NDE techniques selected for their expected capability to inspect AM parts. These techniques included X-ray CT, PCRT, TT, NLR, NI and SX (Dutton et al. 2020, ISO/ASTM DTR 52905 2020). This experimental study aimed to establish the baseline and confirm the capabilities of the selected NDE techniques to detect defects from the star artefacts and à la carte airfoil described above.

Acoustic methods, such as PCRT or RAM, have demonstrated the capability to differentiate between star designs with and without defects by differentiating the natural frequencies of the part, which are sensitive to the material, integrity and dimensional characteristics. The results and comparisons of the three groups of samples tested are clearly identifiable, as shown in Figure 9.13, and an example of the set-up and acoustic spectrum output of RAM is presented in Figure 9.14. NLR has shown potential to detect very close-touching cracks, where the frequency shifts for the defective part compared to the non-defective one. An example is presented in Figure 9.15. However, NLR is limited by the seeded features, the result of which is either that the samples do not exhibit significant non-linear resonance features, or that higher excitation levels are required in order for them to do so. This could limit NLR's applicability for inspection of AM parts unless some improvements in the means to excite the samples are carried out.

TT is limited by the fact that its capability to detect sub-surface and bulk defects is constrained to the detection of open surfaces defects, as discussed elsewhere (Waller et al. 2015, Rummel 2010). Defects with a diameter greater than 0.4 mm are detected using pulse flash thermography, while the step-heating method has the capability to detect open surface defects not less than 0.2 mm in diameter, where some limitations of the method are determined by the high diffusivity and thermal conductivity of metallic materials. Defect detection capabilities are presented in Figure 9.16 and Figure 9.17.

Residual stress is an important factor in AM produced parts, in particular in L-PBF. Measuring the distribution of the stresses generated during the manufacturing process is necessary. There are different methods to determine residual stress, grouped into three categories: destructive, semi-destructive and non-destructive. For AM, non-destructive approaches are desirable; however, some are costly and not suitable for the production process. Neutron diffraction has the capability to enhance the visualisation of strain gradients from the near-surface into the bulk. A parabolic and symmetric tendency of stress distribution is observed in Figure 9.18.

The ASTM standard guide on NDE of metal AM parts covers dimensional measurements using laser triangulation, fringe projection and photogrammetric methods (ASTM E3166 2020, Townsend et al. 2016), while dimensional measurements are not currently considered in the analogous ISO/ASTM TDR 52905 standard (but see Chapter 10).

Finally, some of the emerging methods mentioned above (RAM, PCRT, NLA) have potential for automation as they can be fast, are relatively inexpensive and have no health and safety issues, unlike X-ray methods, for example. In addition, a combination of methods has the potential to cover both macro- and micro-scale defects. Lastly, X-ray CT has limited capability on micro-cracks, where resonant methods are expected to be capable. To establish the capabilities and limitations of these NDE methods, there has been some work in the development of an automated hybrid system using thermography and non-linear resonance, along with data fusion, under the Rate Scalable NDT for AM (RASCAL) project. The software package and system developed in RASCAL is presented in Figure 9.19.

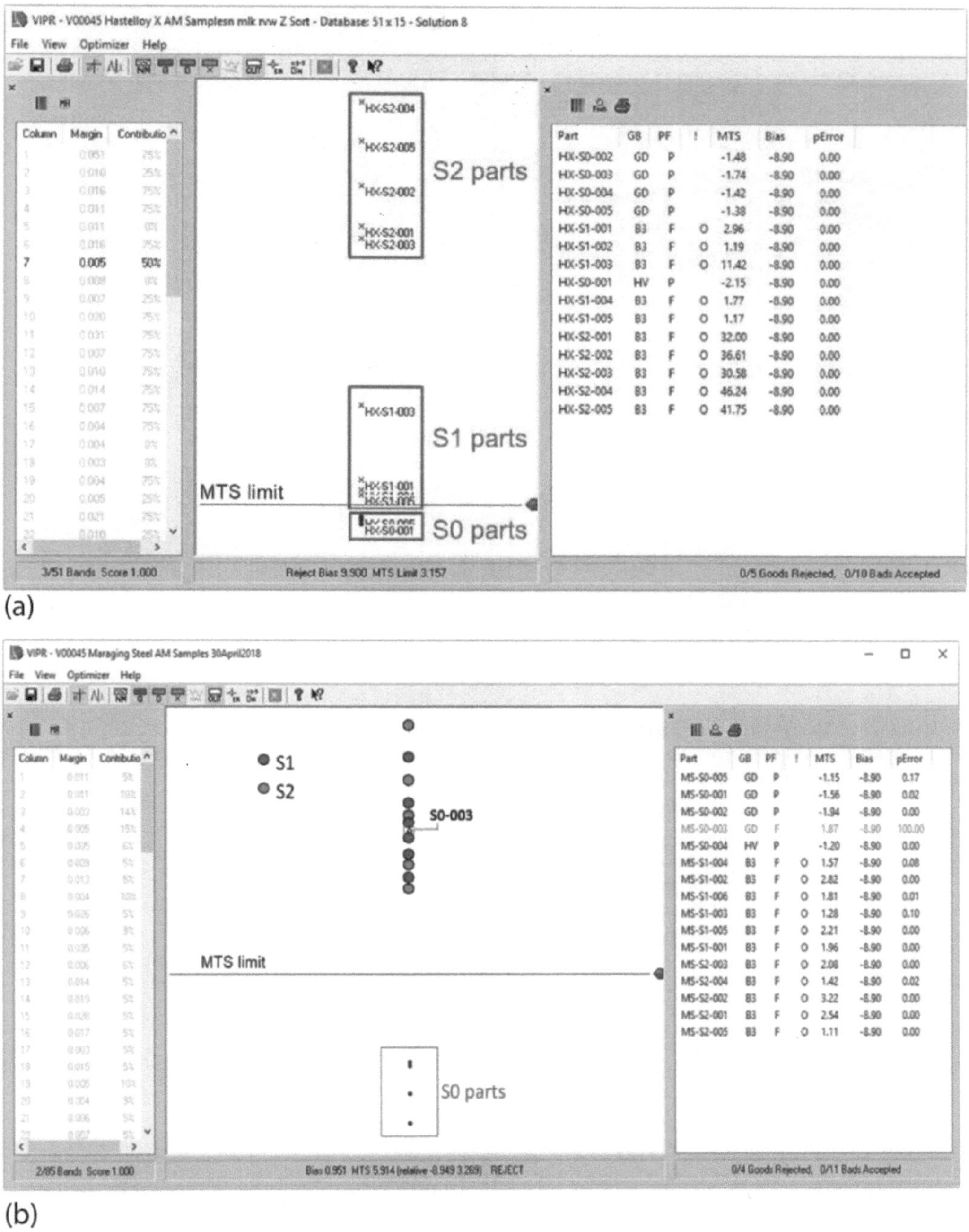

FIGURE 9.13
Process compensated resonance test results obtained on Hastelloy® alloy and Maraging steel star artefact for ISO/ASTM DTR 52905 standard using VIPR analysis. (a) Hastelloy alloy, (b) Maraging steel. (Courtesy of NoSFAM project.)

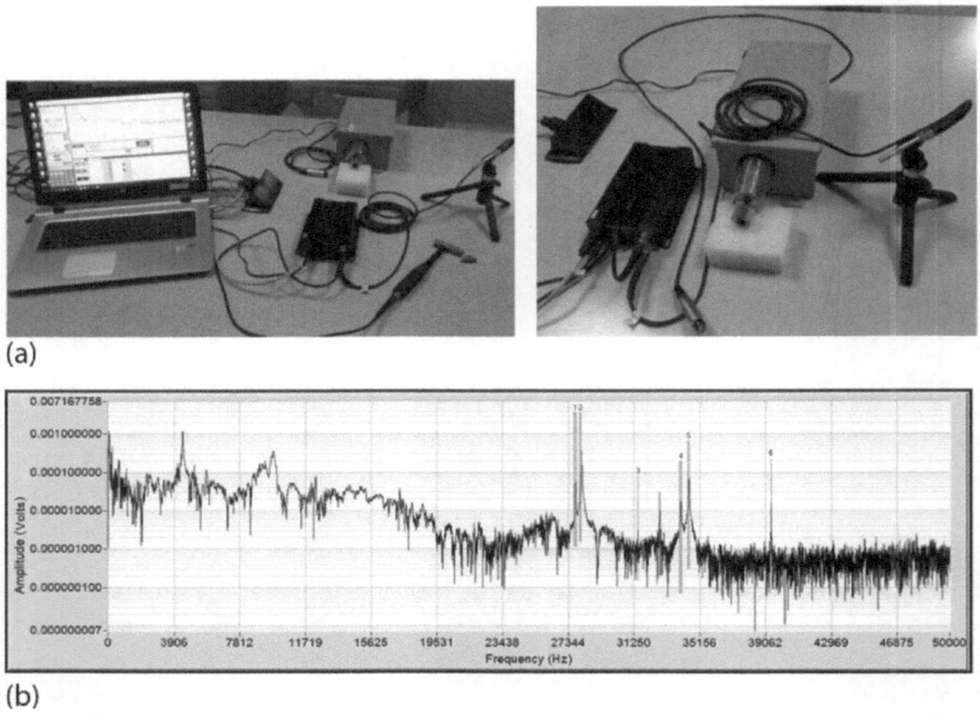

FIGURE 9.14
Experimental set-up used for RAM trials: (a) Experimental set up for RAM tests, (b) typical half-size stainless steel star artefact RAM spectrum showing the sub-set of the six resonant peaks used for comparison criteria. (Courtesy of LNE (France) and NIST (USA).)

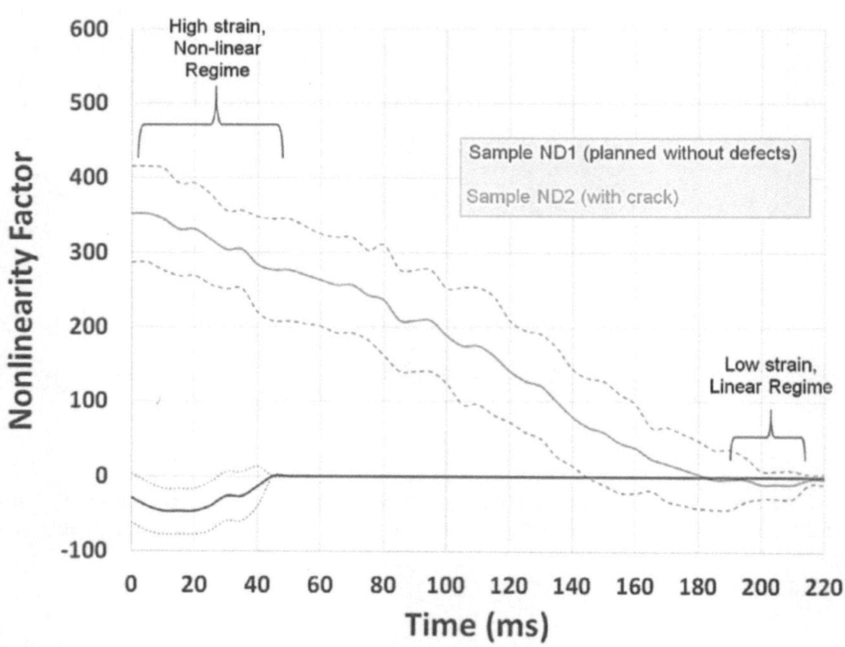

FIGURE 9.15
Non-linear resonance curves of two nacelles tested and labelled as ND1 and ND2. (Courtesy of RASCAL project.)

FIGURE 9.16
Star artefacts tested by PT and SHT thermography methods: (a) Thermal data of configuration MS-S1 for three different views (bottom, side and top); (b) thermal data of configuration HS-S2 for three different views (bottom, side and top). (Courtesy of NoSFAM project.)

9.6.2 In-Process Inspection

In-process inspection may be the only viable method to inspect complex and/or large geometries and can offer cost savings over post-build inspection and, furthermore, can enable in-process control. Optical, infrared and pyrometer technologies are some of the most common methods for in-process monitoring and control (see Chapter 13). However,

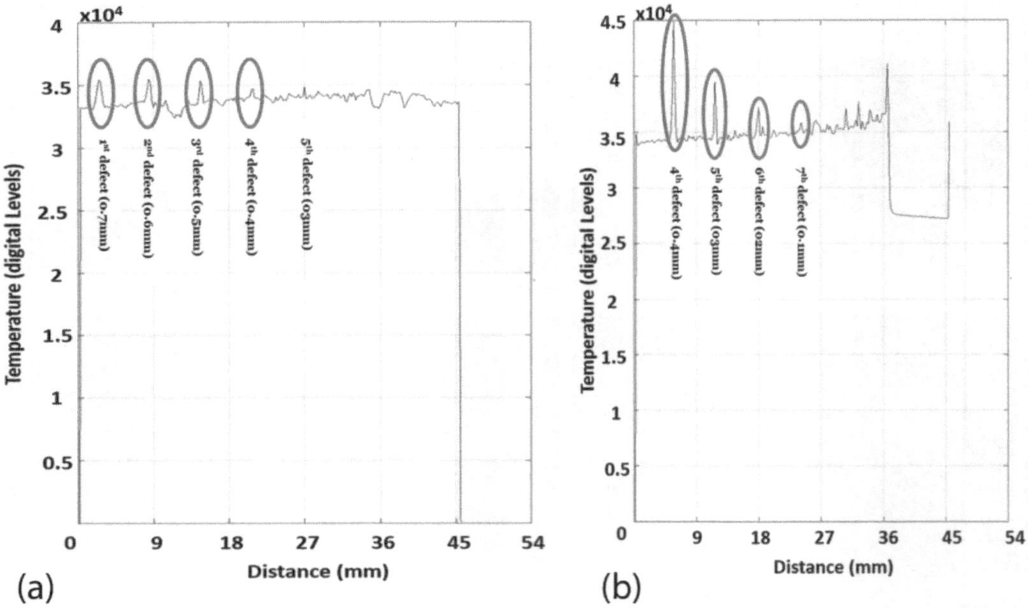

FIGURE 9.17
Curve showing distance–temperature variation during cooling phase: (a) Maraging steel star artefact MS-S1. Temperature versus Distance plot during cooling phase at the region 4 (edge line of the artefact) by SHT. The peak highlighted in grey oval indicated open surface defect detected. (b) Hastelloy star artefact HX-S2. Temperature versus Distance plot during cooling phase at the region 2 obtained by SHT method. The open surface defects detected are shown by grey ellipse. (Courtesy of NoSFAM project.)

there are also less widely used technologies which have shown potential for off-line (post-build inspection) but which show potential for in-process inspection, such as LUT, back-scatter X-ray, LT, acoustic emission and laser triangulation (external geometry only).

In-process inspection has important implications for the manufacturing sector for validation of parts as they are being built, but it has until now been difficult to achieve. Conventional NDE methods cannot cope with the complicated geometries and small sizes typically produced by AM. Most of the NDE approaches are used for inspection of the process after parts or components have been built; therefore, defective parts would only be rejected after the manufacturing process is completed.

For critical requirements of part quality, the desired solution is an NDE technique that allows the in-process inspection and the detection of defects as the layers are deposited. The capability may also allow the process to be controlled and corrected (Kraussa et al. 2014, Hirsch et al. 2017).

L-PBF and DED AM processes operate at elevated temperatures; therefore, contactless NDE technologies are considered the most suitable. Taking advantage of the layer-by-layer build method (Kraussa et al. 2014, Kelly and Kampe 2004), an ideal place to verify the part quality is after a layer or number of layers have been built. This offers the advantage of reducing or eliminating the need to inspect after the full build or after machining. Current AM in-process monitoring relies mainly on surface measurements, potentially missing sub-surface defects (see Chapter 13).

LUT is a method that has the capability to detect surface, close-to-surface and bulk defects since it generates surface and bulk waves at the same time. LUT has shown the capability

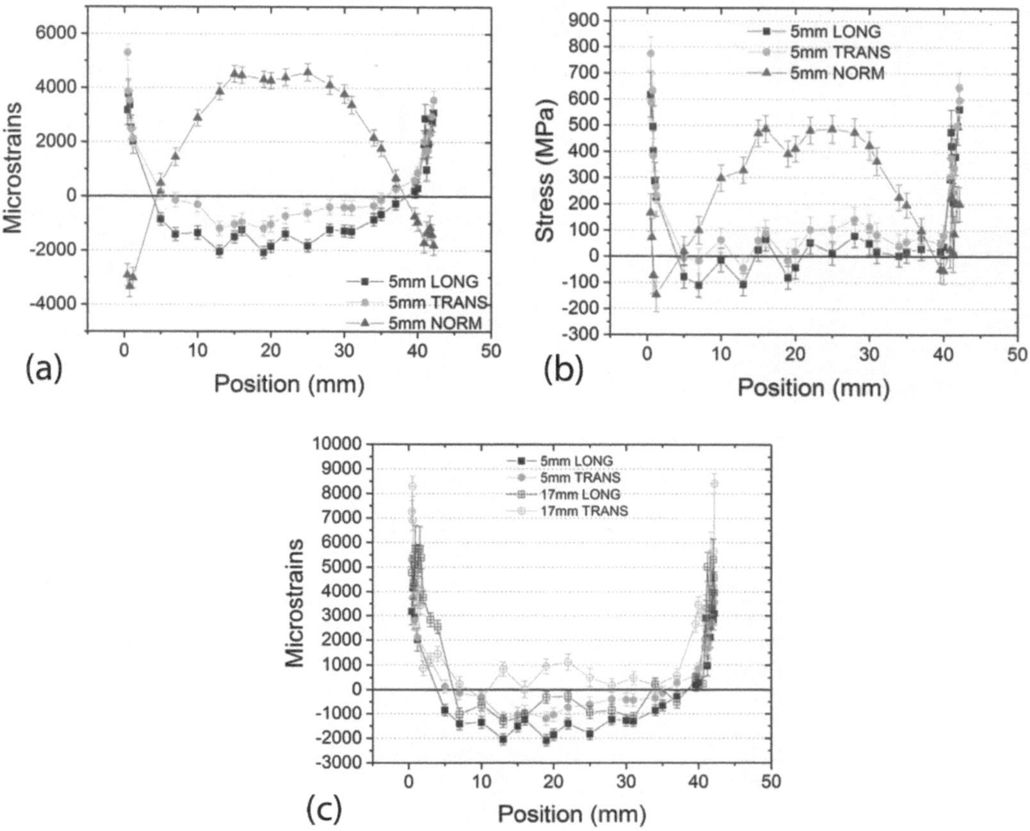

FIGURE 9.18
Residual stress measurement of the three geometrical principal directions on star artefacts manufactured using Ti6Al4V (Figures a to c). (a) Strain for normal, longitudinal and transverse directions from the bottom surface (0 ref Position) towards top surface (42.6 mm). Errors $\pm$300 $\mu\varepsilon$. (b) Related stress results. Errors $\pm$50 MPa. (c) Strain distribution for longitudinal and transverse directions from bottom towards top surface for lines scans at 5 mm and 17 mm from the tip of the star sample. Errors $\pm$300 $\mu\varepsilon$. (Courtesy of ILL, France.)

to detect sub-surface defects as small as 400 μm in diameter on as-built L-PBF surfaces, and of naturally occurring void defects as small as 200 μm, see Figure 9.20a and Figure 9.20b respectively. Figure 9.21 shows an example of the capability of LUT for post-process inspection, where natural defects on a cladded sample are detected at 0.33 mm depth. Other examples of the LUT capability are presented in Figure 9.22, showing its capability to detect fusion irregularities between side by side and top and bottom beads in a DED sample. Other industrial applications of in-process LUT inspection are shown in Figure 9.23 (a–c).

9.7 NDE Reliability in AM

NDE has become an integrated component of production systems to assess the quality and integrity of the parts manufactured or in-service. The capabilities of various NDE

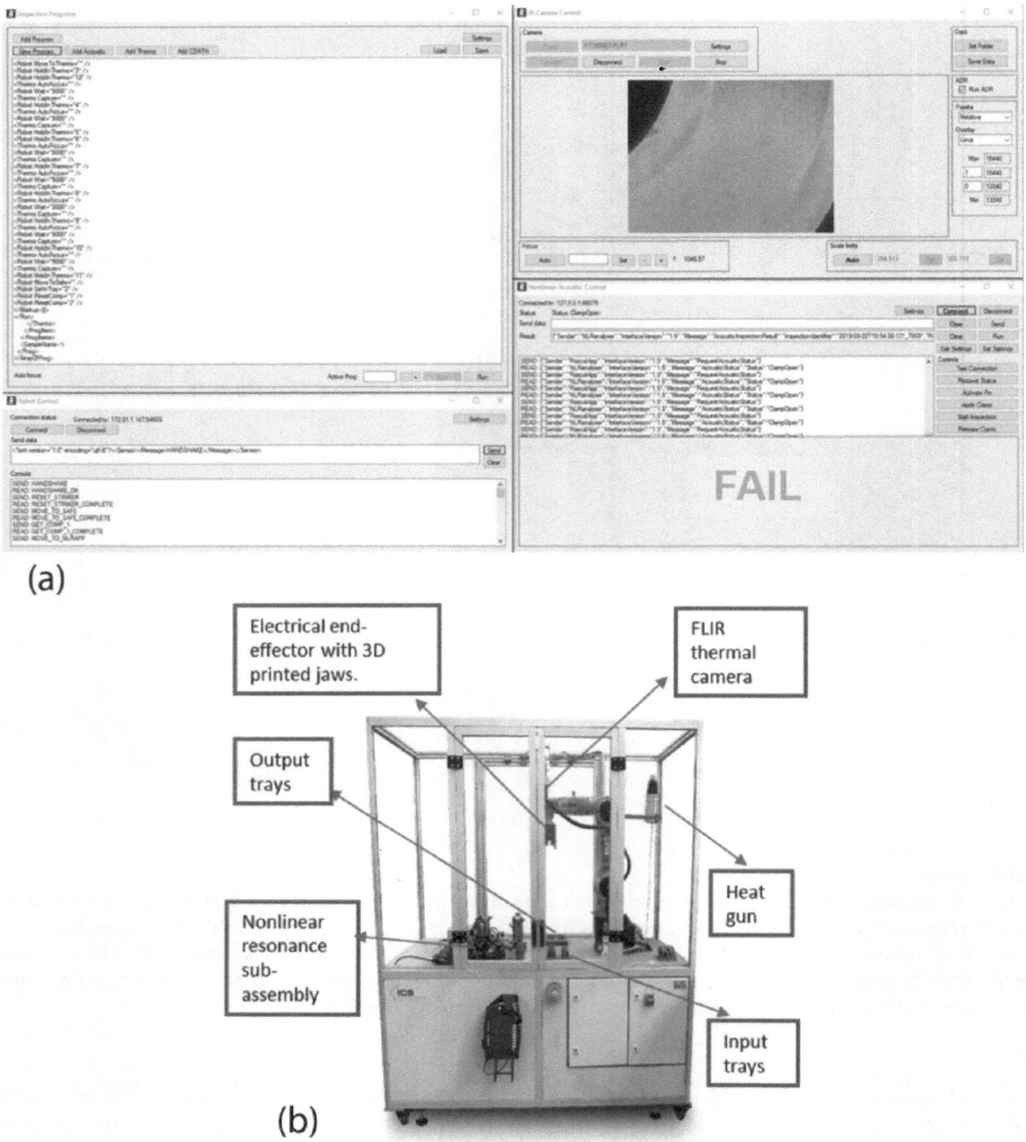

FIGURE 9.19
Software interface developed within RASCAL project and automated NDE cell demonstrator. (a) Screenshot of NI Vision Builder AI software used to develop defect detection algorithms and automated sentencing. (b) RASCAL – Automated NDE cell demonstrator. (Courtesy of RASCAL project.)

methods are quantified by their reliability, which is defined as the capability of detecting a defect/crack in a given size group under the inspection conditions and procedure specified (Rummel 1982). The reliability of NDE is based on the following elements: applicability, reproducibility, repeatability and capability. Capability is accounted for by the PoD of a defect for the method used (Dobmann et al. 2007, Berens 1989).

The use of PoD functions as a metric for quantifying NDE capability underwent considerable development around the late 1960s and early 1970s. The concept of PoD was

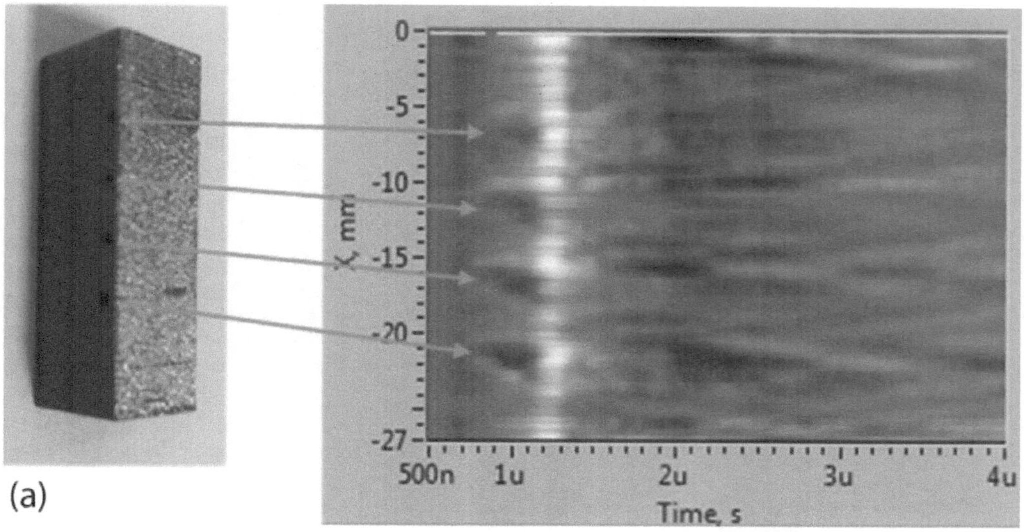

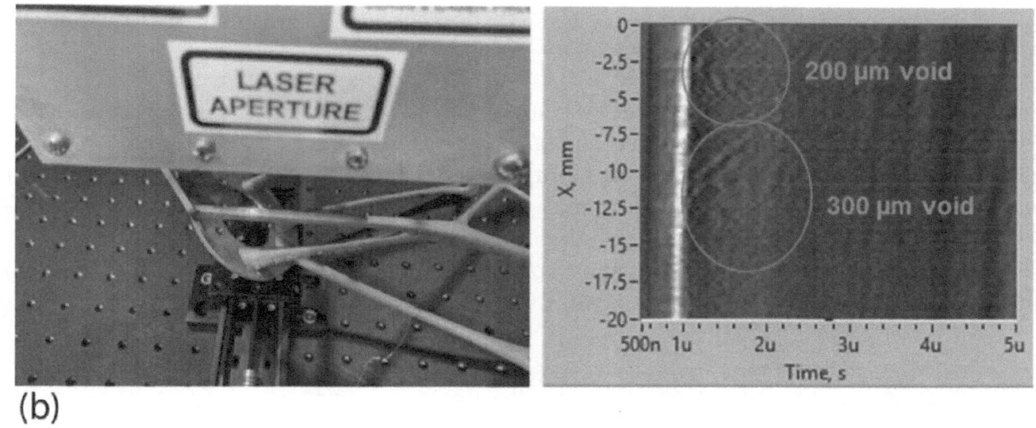

FIGURE 9.20
LUT capability for in-process inspection. (a) Calibration samples consisted of Ti64 cuboids with dimensions (10 mm × 10 mm × 30 mm), left, and B-Scan showing them be detected by LUT. (b) A complex and realistic part from an aluminium geometry with internal defects. (Courtesy of AMAZE project.)

established mainly at NASA and became a fundamental element of engineering design, risk and lifetime assessment and prediction (Dobmann et al. 2007, Berens and Hovey 1989, Keprate 2016).

A PoD curve estimates the capability of detection of an inspection technique in regard to discontinuity size. In an ideal technique, the PoD for discontinuities smaller than an established critical size would be zero. On the other hand, discontinuities greater than the critical size would have PoD equal to unity, or 100% probability of detection. Such an ideal technique could never exist, and what is known as False Positive (rejection of acceptable components) or False Negative (approval of defective components) can occur in a real example (Annis et al. 2013, Kanzler 2017, Keprate 2016, Georgiou 2006).

The reliability of NDE is expressed in terms of the defect or crack size (***a***) having a detection probability of 90%, known as the ***a90*** crack size. However, there is certain statistical

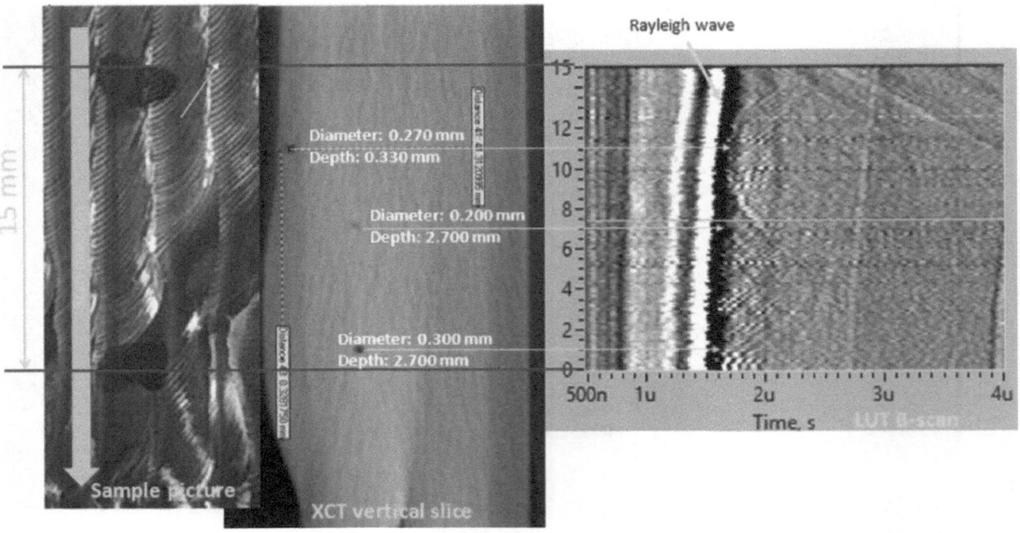

FIGURE 9.21
Natural defects generated during directed energy deposition process, where left image is the part photo, middle is a vertical CT slice and right is the LUT B-Scan. (Courtesy of Open Hybrid project.)

uncertainty associated with the value of ***a90***, which is represented by a 95% confidence interval. This crack size is represented as ***a90/95*** and serves as an important parameter to quantify NDE reliability (Keprate 2016, Georgiou 2006, Berens 1989) for cast and wrought metals parts. However, this criterion may not apply to AM, as discussed elsewhere (Dobmann et al. 2007, Kanzler 2017, Kanzler and Müller 2016a).

PoD curves are determined either via experimental approaches or through numerical simulation. Experimental approaches can involve a high number of experiments, making

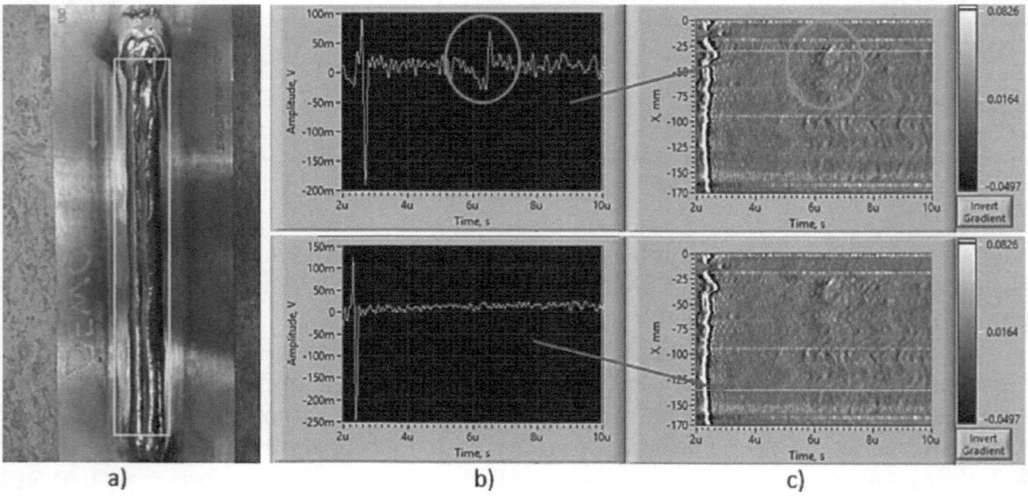

FIGURE 9.22
LUT scan results on a DED AM titanium sample with LOF defects, side to side and top and bottom beads. The figure shows the interpretation of the scan curves with and without defects detected by LU. (Courtesy of AMAZE project.)

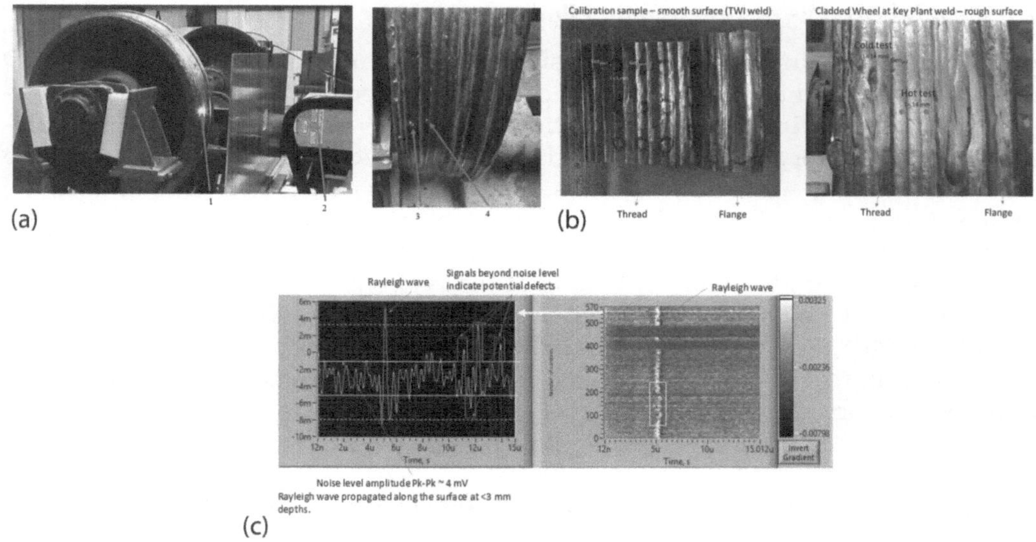

FIGURE 9.23
LUT in-process inspection of cladded wheel of the trials carried out in AURORA project, which is displayed in Figures (a–c). (a) LUT in-process inspection: (1) Plasma generated, (2) laser detection spot, (3) laser generation spot. (b) Comparison between TWI and key plant claddings surface finish of the cladded wheel. (c) Results of cold test (3.6 °C–6.5 °C) performed at 20 Hz repetition laser 8 averages, speed rate 600 mm/min, scan at left edge.

them often time-consuming and costly. Numerical simulation, known as MAPoD, involves less time and cost but is not widely recognised as a reliable methodology (Thompson et al. 2009, Calmon et al. 2018).

Since its introduction by NASA, PoD has been improved progressively by the aerospace industry in the USA to generate guide MIL-HDBK-1823A, which was published in 2009 (Berens and Hovey 1989, MIL-HDBK-1823A 2009). Currently, MIL-HDBK-1823A is considered a state-of-the-art guide for conducting PoD studies. For example, the nuclear industry under the European Network for Inspection and Qualification (ENIQ) has released several reports to introduce PoD in Europe (ENIQ 2007), while in the oil and gas industry, the qualification of NDE is performed by Det Norske Veritas (DNV 2010).

There are two related probabilistic methods for analysing reliability data and producing PoD curves as a function of the defect size $\boldsymbol{a}$. The way of recording the data defines the method to be used. If the data is recorded in terms of whether the defect was detected or not, it is discrete data, called *hit/miss* data. If there is more information, such as voltage, signal amplitude or light intensity, this signal response can be interpreted as the perceived defect size. This latter type of data is called $\hat{\boldsymbol{a}}$ (*a hat data*) or *signal response* data, called *continuous data* (Georgiou 2006, Dobmann et al. 2007, Child et al. 1999). The hit/miss analysis is useful due to its simplicity; however, some NDE techniques relate a signal with response $\hat{\boldsymbol{a}}$ to the real dimension $\boldsymbol{a}$ of the discontinuity. Both methods may be used to implement PoD curves, but different results are obtained when applied to the same data set. More detail covering the theoretical PoD methodology can be found elsewhere (Keprate 2016, Georgiou 2006, Berens 1989, Schneider et al. 2012, Virkkunen et al. 2019, Generazio 2011, Generazio 2014, Generazio 2015). Hit/miss data have recently been standardised by ASTM E2862 (2018), while $\hat{\boldsymbol{a}}$ vs $\boldsymbol{a}$ is under ASTM E3023 (2015), which are congruent with the current MIL-HDBK-1823A methodology (2009).

9.7.1 General Aspects of Experimental Pod Curves

Experimental PoD curves are plotted using a high volume of inspection data. PoD curves can be applied using test artefacts containing defects with controlled characteristics, such as type, dimensions and location. PoD curves are also applied with equipment where the history of the inspection is completely recorded on the same reference artefacts containing well-known defects. Real defects are difficult to manufacture; therefore, to simulate defects or flaws, it is common practice to manufacture block artefacts containing a substantial number of seeded defects which provide a range of size features large enough to allow estimation of the PoD curves. However, these defects may not represent realistic defects, and it can be costly to get a sufficient amount of data (ASTM E2862 2018, ASTM E3023 2015, Kanzler and Müller 2016b). Therefore, simulation or modelling approaches are considered to be the most feasible and economical solution for PoD estimation.

9.7.1.1 General Aspects of PoD Curves Modelled through Experimental Data

If only a small sample of experimental data is available, it is possible to plot a PoD curve using a mathematical model. The main benefits of this methodology are low cost, convenience and relatively fast generation of PoD curves. The disadvantages can be that, in the case of extrapolation of larger defects from inspection data to smaller defects, the PoD obtained may be too low and would not represent the real situation (Maztkanin and Yolken 2001, Gratiet et al. 2016, Cobb et al. 2009).

9.7.1.2 Mathematical Simulation of PoD Curves

Recently, research investigating the modelling of PoD curves has increased considerably. The low computational cost of simulation, compared to the fabrication of test artefacts and acquisition of resources for inspection and use of equipment, is driving the use of this methodology (Keprate 2016). Currently, various approaches are implemented to predict the PoD curves of a range of NDE methods by computational means. Some of these approaches include Monte Carlo simulation, MAPoD, receiver operating characteristic (ROC) curves, Bayesian and non-Bayesian approaches and bootstrap methods (Keprate 2016, Annis et al. 2013, Georgiou 2006, Karl Härdle et al. 2015, Boos and Brownie 1989, Visser 2000). A general view of other approaches for modelling of PoD curves is summarised in Table 9.9 (Wall and Wedgwood 1998).

MAPoD is receiving attention in the AM community, and despite its limitations, there are numerous demonstrations of MAPoD that indicate that the method offers potential; however, applications often have unique considerations or constraints (Georgiou 2006, Thompson et al. 2009). The model-assisted POD working group (MAPOD, www.cnde.iastate.edu/mapod/) has established two distinct approaches: the transfer function approach (TFA) and the full model-assisted approach (FMA). TFA is a fully empirical approach which uses physics-based models of the inspection process to transfer results to another situation. In the FMA approach, the factors that control the variability of inspection are systematically identified and, using physics-based models, FMA predicts the signal and noise distributions that are influenced by those factors that represent well-understood physical phenomena (Thompson et al. 2009).

TABLE 9.9

Approaches to Modelling of PoD

Modelling approach	Basic principle	General comments
Physical PoD models	These models are based on established and well verified physical models. The models also incorporate variable factors such as noise, geometry, defect visibility and detection criteria necessary to make predictions in reliability terms.	• The prediction of PoD combines estimates of signals expected from specific defects and transducers, estimates of background noise and a threshold criterion. • The simulation of defects is being constantly improved based on the analysis of real defect data from an operating plant. • The models allow an analysis of image-based data and allow correction for human and environmental effects.
Signal/noise models	• The signal and noise values derived from models or experiment are converted to PoD and PoFA using statistical methods. • A modular approach may be adopted with input data to the PoD model derived from a physical model or experiment.	• This method avoids the statistical difficulties associated with conventional PoD trials and allows predictions to be made for new inspection techniques that may be too complex to physically model. • This approach has been used to make PoD predictions for magnetic methods and forms the basis for a basic spreadsheet model derived for member organizations of the Harwell Offshore Inspection Service (HOIS).
Image classification models	• These represent methods for the analysis of image-based inspection data, such as radiographs, to give information in terms of the PoD and PoFA.	Inspection simulators are used to simulate the inspection process by presenting simulated inspection results to the operator. The programme present a series of simulated images to the inspector like a 'spot the ball' contest, essentially simulating a PoD trial.
Expert judgement	It is used where input on PoD is required for the fracture mechanics (FM) or risk-based assessments and is not available from experiment	Provided the judgement comes from trained inspectors and sensitivity analysis is used, this can be an effective method.
Statistical models	These use methods for statistical analysis or curve fitting to experimental data, with the aim of making this data accessible for use in other applications (such as FM).	These do not model the inspection process as such.
Human reliability Index	These take account of the effects of human error in the inspection process and correctly predict PoD values for these effects.	An example is a methodology applied by AEA Technology to utilise human error data from the PISC 3 work.

9.7.2 Estimation of Experimental PoD

The methodology of PoD reliability studies has been widely applied in the aerospace industry and adopted by other industries, such as oil and gas, rail, steelmaking and nuclear. Despite the considerable studies carried out and data collected, the industry faces certain challenges regarding the PoD metric. The discrepancy in the PoD definition of the confidence level and the number of defects used is driving different PoD curves in relation to various NDE methods (Visser 2000). The PoD metric is derived empirically from the data produced experimentally using the round-robin test method, making it time-consuming, expensive and prone to human error, whilst the use of poor statistics leads to considerable scatter that can jeopardise the method. A lack of understanding relating to the derivation of the PoD curves from experimental data and misconceptions about PoD amongst NDE personnel, such as assuming that 'repeated inspections improve the PoD', are other aspects that negatively impact the use of the PoD methodology.

In general, the large majority of the literature on PoD methodology covers applications where the size of the defects, particularly cracks, are of the order of millimetres; however, for AM, the size of the defects, especially for parts produced by powder-based processes, are of the order of micrometres rather than millimetres. The lack of PoD data available for AM is one of the topics that NASA has identified as a challenge to establishing NDE reliability in the AM sector (Waller et al. 2014, Hirsch et al. 2017).

Whilst several advanced NDE methods are available for use, the majority of the NDE techniques used for quality control and in-service inspection are conventional. For AM, it is critical to compare the capability of these methods on various attributes, as it is clear that reliable 'special' NDE methods are needed to assess manufactured parts or components made by AM which contain smaller defect sizes. It is foreseen that modelling and simulations would be the link between experimental work and model-driven PoD. As the obtained PoD curves are based on a probabilistic modelling of system input parameters, these parameters have to be validated. Having experimental data enables the development of a model that represents the reality, and further enables development to predict or generate PoD assessments of parts built by AM.

9.8 Current PoD Performed in AM

There has been a substantial investigation on PoD in conventional manufacturing; however, PoD has not yet been fully developed for AM due to two main factors: lack of adequate or routine NDE processes integrated into AM, and the unique features of parts made by AM. The corresponding lack of reliable NDE methods currently available is a key barrier to AM expansion, due to the need to validate and certify the quality of the components fabricated through AM technologies (Rummel 2010, Keprate 2016). Some efforts to reduce this gap have been addressed by NASA and their partners through the Office of Safety and Mission Assurance (OSMA) NDE programme, which aims to reduce the gaps associated with NDE applications for AM in the aerospace sector (Waller et al. 2015). X-ray CT is the NDE technology most widely used to assess the quality of parts made by AM, and despite its limitations, it is still the method that has the most potential for PoD application.

PoD determinations based on signal response data ($\hat{a}$ vs. a) have already been carried out for a variety of NDE methods, and the general applicability of ultrasound data for such a PoD analysis is already documented. However, ultrasonic data usually do not provide a direct correlation between the UT signal and the defect size. In some cases, the reflected amplitude can be used as an estimate for the defect size, but a reasonably accurate determination of crack depths requires analysis techniques, such as evaluation of crack tip signals (Kanzler and Müller 2016).

A comparison of metallographic cross grindings and RT indications has been carried out to show that the handling of real defects by PoD evaluation creates further challenges in order to get adequate information about the defect size (Kanzler and Müller 2016). This comparison was combined with the use of a multi-scale filtering approach for the PoD calculation, which enabled the consideration of the dependence of the indication size in the evaluation, without increasing the amount of data needed. Therefore, linking experimental data with modelling is considered promising for the evaluation of NDE systems with real defects for PoD assessment.

Kim et al. (2019, 2019) carried out PoD studies by X-ray CT. In this work, the thresholded defect volume as a function of the signal response was chosen as the NDE instrument detection limit. For an $\hat{a}$ *vs* a PoD study, an arbitrary decision threshold needs to be chosen based on the objective of the PoD study. If the threshold changes, the curve does too. Therefore, for choosing a threshold, it is necessary to balance the true positive rate and the false positive rate. Once a satisfactory decision threshold and a corresponding PoD curve are found, they can be used to decide whether the threshold should be accepted or rejected for an actual part during a future X-ray CT inspection. If there is a defect with a signal response higher than the decision threshold, the part can be rejected with the determined PoD and confidence level.

A process for determining a PoD curve for X-ray CT measurements was also demonstrated based on the measured data by Kim et al. (2019). The results indicate that, due to the complex external and internal geometry of the AM component, X-ray CT is considered a promising NDE technique. X-ray CT, however, is a relatively new NDE technique with many possible instrument settings, which greatly affect the inspection results. A process for determining a PoD curve for X-ray CT measurements was demonstrated based on the measured data and a signal response ($\hat{a}$ *vs* a) model was used. For the $\hat{a}$ value, the X-ray CT images were thresholded to measure the volume of the defects.

X-ray CT is widely recognised as the approach most capable to assess the quality and integrity of the AM parts; however, some limitations exist which are linked with its limitations in detecting defects in some critical locations. Simulating the capability of X-ray CT to determine these critical areas or locations is one of the areas of current interest. Brierley et al. (2019) developed an algorithm capable of mapping out the variations in X-ray CT inspection performance. This new model-assisted inspection qualification capability should enable the more widespread usage of X-ray CT and hence advanced manufacturing techniques that necessitate the use of this inspection technique for safety-critical components. However, it was concluded that it is not safe to assume that because a certain defect was detected in one location of a scanned volume, this feature would be equally detectable in a different place.

Reduction of an experimental test to estimate the PoD curves is one of the areas that AM is eager to develop where MAPoD can play an important role in this development. Meyer et al (2014) carried out a survey to assess specific approaches that may be appropriate for application to improve estimates of PoD for NDE of nuclear power plant components. Despite the area of application, this survey provides an introduction to parametric PoD models and gives an overview of the literature, presenting demonstrations of MAPoD using non-Bayesian and Bayesian approaches. The document also covers the effect of the human factors considered relevant to the NDE of nuclear power plant components along with a discussion and analysis of how this concept can potentially be integrated into MAPoD.

9.9 Conclusions and Future Research

This chapter has covered the main aspects of quality and integrity in AM from the NDE perspective. Particular attention is given to defects and NDE methods, including standardisation and reliability aspects, based on the available literature and current research. Representative research conducted to date, including characteristic data acquired using consensus NDE methods, are summarised, focusing primarily on NDE applied to post-process and in-process of built metal parts fabricated using PBF and DED processes.

X-ray CT is still the preferred option for AM, but there are some NDE techniques that are showing potential, for example, PCRT, RAM and NLR. Hybridisation and the level of automation in both inspection and sentencing are necessary to perform NDE faster and more economically and reliably.

Performing NDE inspection of parts or components after they are built is not ideal for AM since this may result in higher part cost and scrap rate, in addition to generating difficulties in identifying some defects. Nevertheless, AM's layer-by-layer manufacturing process can be exploited for inspection by inspecting each layer as it is built. The part will then be fully inspected by the end of the build, therefore reducing cost and scrap. It has been recognised that, for some very complex and large AM parts, layer-by-layer is probably the only NDE-based solution to expand AM applicability. Additionally, in-process monitoring has the potential to lead to feedback control of the process to reduce quality issues. However, in-process inspection methods have not yet fully matured, and some commercial systems have only recently become available, are being evaluated and are limited to surface 'views'. LUT is an emerging in-process inspection method with potential, but it is still immature.

Standardisation is considered one of the most important and critical topics for quality and certification of AM parts. The review of the current NDE standards for post-processing and in-process inspection in DED and PBF processes has found that there are some similarities between defects from these two processes. Most defects in DED can potentially be covered by current standards; however, for PBF, some defects are unique, including layer, cross-layer, trapped powder and unconsolidated powder defects. Generic (star) and à la carte artefacts have been developed to address the defects not included by current NDE standards, which are unique for AM processes, such as layer, cross-layer and unconsolidated powder/trapped powder. The star artefact geometries can be used as an initial NDE technique verification and are only applicable to relatively simple AM geometries; however, the à la carte artefacts and/or approach can be useful to assess more complex parts and applications.

There is significant concern about ensuring the reliability of metal parts produced by AM, which is impeding the pace of implementation of this technology. More work is needed on these methods to assess their capability, such as PoD studies. The future for NDE is to improve current methods, such as X-ray CT, and develop more resonance methods capability with automation, but the best future scenario will be for NDE to be integrated within the AM processes, enabling immediate fault detection and feedback for corrective actions.

Acknowledgements

The authors would like to acknowledge the support from the Manufacturing Technology Centre (MTC) through AMAZE, Core Research Program NoSFAM & RAMPID, AURORA, OpenHybrid and RASCAL projects. Acknowledgement is also extended to David Wimpenny (chief engineer at AM Group), Joseph Darlington (chief engineer at DE Group), Andrew Clough (manager of M&NDT at DE) and colleagues at the M&NDT team for their support. Special thanks to Dominic Nussbaum at GE Power for his continued interest and support, and for providing the generic airfoil for the a la carte design. To Anne-Francoise Obaton at LNE, Alkan Donmez and Felix Kim at NIST for their support and contribution through the NosFAM and RAMPID projects.

Special thanks to Jess Waller at NASA-JSC White Sands Test Facility for his continued contribution to the standardisation topic and at the Institut Laue-Langevin (ILL) through Duncan Atkins with his Synchrotron work and to Sandra Cabezas for her very valuable support in residual stress work. The additional support received through the standard committees ISO TC261/ASTM F42 JG59, ASTM E07 and ASTM Centre of Excellence (CoE) at the MTC are also appreciated. Finally, thanks to Eric Biermann at Vibrant NDT for their work on PCRT and Bryan Butsch at The Modal Shop for his contribution with RAM.

References

Albakri, M., Sturm, L., Williams, C. B., and Tarazaga, P. 2015. Non-destructive evaluation of additively manufactured parts via impedance-based monitoring. *Solid Freeform Fabrication Symp.* 8:1475–1490.

Aloisi, V., and Carmignato, S. 2016. Influence of surface roughness on X-ray computed tomography dimensional measurements of additive manufactured parts – Case Studies. *Nondestr. Test. Eval.* 6:104–110.

AMAZE. 2017. Additive manufacturing aiming towards zero waste and efficient production of high-tech metal products. The Manufacturing Technology Centre, [Online]. Available: http://www.the-mtc.org/our-projects/amaze. [Accessed 25 January 2019].

AMSC – Additive Manufacturing Standardization Collaborative. 2018. *Standardization Roadmap for Additive Manufacturing*, 2nd ed., ANSI and NDCMM/America Makes.

Annis, C., Gandossi, L., and Martin, O. 2013. Optimal sample size for probability of detection curves. *Nucl. Eng. Des.* 262:98–105.

Antony, K., and N. Arivazhagan, N. 2015. Studies on energy penetration and Marangoni effect during laser melting process. *J. Eng. Sci. Tech.* 10, 4:509–525.

ASTM E1316. 2019b. Standard Terminology for Nondestructive Examinations. ASTM International.

ASTM E2862. 2018. Standard Practice for Probability of Detection Analysis for Hit/Miss Data.

ASTM E3023. 2015. Standard Practice for Probability of Detection Analysis for â Versus a Data.

ASTM E3166. 2020. Standard Guide for Nondestructive Examination of Metal Additively Manufactured Aerospace Parts After Build,. ASTM International.

ASTM WK56649. 2020. New Guide for Standard Practice/Guide for Intentionally Seeding Flaws in Additively Manufactured (AM) Parts. ASTM International. In development.

Attar, H., Bönisch, M., Calin, M., Zhang, L. C., Scudino, S., and Eckert, J. 2014. Selective laser melting of in situ titanium–titanium boride composites: Processing, microstructure and mechanical properties. *Acta Mate.* 76:13–22.

Awd, M., Tenkamp, J., Hirtler, M., Siddique, S., Bambach, M., and Walther, F. 2014. Comparison of microstructure and mechanical properties of scalmalloy produced by selective laser melting and laser metal deposition. *Materials* 11:17.

Berens, A. 1989. NDE reliability data analysis. In *ASM Handbook*, Vol. 17. Nondestructive Evaluation of Materials.

Berens, A. P., and Hovey, W. 1989. Statistical methods for estimating crack detection probabilities. In *STP798 Probabilistic Fracture Mechanics and Fatigue Methods: Applications for Structural Design and Maintenance*. ASTM International.

Berens, P., and Hovey, P. W. 1981. AFWAL-TR-81-4160-Evaluation of NDE reliability characterisation. *DTIC Selected:1–98.*

Bhavar, V., Kattire P., Patil, V., Khot, S., Gujar, K., and Singh, R. 2017. A review on powder bed fusion technology of metal additive manufacturing. In *Additive Manufacturing Handbook: Product Development for the Defense Industry*. Boca Raton, FL: CRC Press.

Boos, D. D., and Brownie, C. 1989. Bootstrap methods for testing homogeneity of variances. *Technometrics* 31:69–82.

Brierley, N., Nye, B., and McGuiness, J. 2019. Mapping the spatial performance variability of an X-ray computed tomography inspection. *ND & E Int.*107:1–11.

Brown, A., Jones, Z., and Tilson, W. 2017. Classification, effects, and prevention of build defects in powder-bed fusion printed Inconel® 718. NASA Marshall Space Flight Center.

Calmon, P., Chapuis, B., Jenson, F., and Sjerve, E. 2018. *Best Practices for the Use of Simulation in POD Curves Estimation*. Springer.

Chauveau, D. 2018. Review of NDT and process monitoring techniques usable to produce high-quality parts by welding or additive manufacturing. *Weld. World*. 62:1097–1118.

Child, F. R.., Phillips, D. H., Liese, L. W., and Rummel, W. D. 1999. Quantitative assessment of the detect ability of ceramic inclusions in structural titanium castings by X-radiography. *Rev. Prog. in QNDE*. 18B:2311-2317.

Cobb, A. C., Fisher, J., and Michaels, J. E. 2009. Model-assisted probability of detection for ultrasonic structural health monitoring. *4th European-American Workshop on Reliability of NDT*, 1–9.

Collins, P. C., Bond, L. J., Taheri, H., Bigelow, T. A, Shoaib, M. R. B. M., and Koester, L. W. 2017. Powder-based additive manufacturing – A review of types of defects, generation mechanisms, detection, property evaluation and metrology. *Int. J. Addit Subtractive Mat. Manuf.* 1(2):172–209.

DNV. 20100. RP-G101– Risk based inspection of offshore topsides static mechanical equipment.

Dobmann G., Cioclov, D., and Kurz, J. H. 2007. The role of probabilistic approaches in NDT defect-detection, classification, and sizing. *Weld. World* 51:9–15.

Dutton, B., Vesga, W., Waller, J., James, S., and Seifi, M. 2020. Metal additive manufacturing defect formation and nondestructive evaluation detectability, in *Structural Integrity of Additive Manufactured Parts*, ed. N. Shamsaei, S. Daniewicz, N. Hrabe, S. Beretta, J. Waller, and M. Seifi, West Conshohocken, PA: ASTM International, 1–50.

ENIQ. 2007. European methodology for qualification of non-destructive testing. EUR 22906 EN.

Everton, S. K., Hirsch, M., Stavroulakis, P. I., Leach R. K., and Clare, A. T. 2016. Review of in-situ process monitoring and in-situ metrology for metal additive manufacturing. *Mater. Des.* 95:431–445.

Fousová, M., Vojtěch, D., Doubrava, K., Daniel M., and Lin, C. F. 2018. Influence of inherent surface and internal defects on mechanical properties of additively manufactured Ti6Al4V alloy: Comparison between selective laser melting and electron beam melting, *Materials* 11:537.

Frazier, W. E. 2014. Metal additive manufacturing: A review. *J. Mat. Eng. Perform.* 23:1917–1928.

Gebhardt, A., and Hötter, J. S. 2016. Direct manufacturing: Rapid manufacturing. In *Additive Manufacturing*: 3D *Printing for Prototyping and Manufacturing* (Munich, Germany: Verlag), Ch 6: 395–450.

Generazio, E. R., 2011. NASA/TP–2011-217176 – Binomial test method for determining probability of detection capability for fracture critical applications.

Generazio, E. R. 2014. NASA/TM–2014-218183 – Interrelationships between receiver/relative operating characteristics display, binomial, logit, and Bayes' rule probability of detection methodologies.

Generazio, E. R. 2015. Directed design of experiments for validating probability of detection capability of NDE systems (DOEPOD).

Georgiou, G. 2006. Report 454 – Probability of Detection (PoD) curves – derivation, applications and limitations. HSE.

Gong, H., Rafi, K., Gu, H., Ram, G. D. J., Starr, T., and Stucker, B. 2015. Influence of defects on mechanical properties of Ti6Al4V components produced by selective laser melting and electron beam melting. *Mater. Des.* 86:545–554.

Gong, H., Rafi, K., Gu, H., Starr, T., and Stucker, B. 2014. Analysis of defect generation in Ti-6Al-4V parts made using powder bed fusion additive manufacturing processes. *Addit. Manuf.* 1:87–98.

Gorelik, M. 2017. Additive manufacturing in the context of structural integrity. *Int. J. Fatigue.* 94:168–177.

Gratiet, L. L., Iooss, B., Blatman, G., Browne, T., Cordeiro, S., and Goursaud, B. 2016. Model assisted probability of detection curves: New statistical tools and progressive methodology. *J. NDE* 36:1–12.

Greitemeier, D., Palm, F., Syassen, F., and Melz, T. 2017. Fatigue performance of additive manufactured TiAl6V4 using electron and laser beam melting. *Int. J. Fatigue.* 94:211–217.

Hirsch, M., Patel, R., Li, W., Guan, G., Leach, R. K., Sharples, S. D., and Clare, A. T. 2017. Assessing the capability of in-situ nondestructive analysis during layer based additive manufacture. *Addit. Manuf.* 13:135–142.

Hrabe, N., Gnäupel-Herold, T., and T. Quinn, T. 2017. Fatigue properties of a titanium alloy (Ti6Al4V) fabricated via electron beam melting (EBM): Effects of internal defects and residual stress. *Int. J. Fatigue.* 94:202–210.

Hunter, L. 2009. Reliable modern NDT. *4th European-American Workshop on Reliability of NDE*, Berlin, Germany.

IAEA. 2017. Non-destructive testing: A guidebook for industrial management and quality control personnel.

ISO/ASTM 52900. 2015. Additive manufacturing – General principles – Terminology. ASTM International.

ISO/ASTM DTR 52905. 2020. Additive manufacturing – General principles – Non-destructive testing of additive manufactured products. ASTM International. In balloting.

ISO/ASTM DTR 52906. 2020. Additive manufacturing – Non-destructive testing and evaluation – Standard guideline for intentionally seeding flaws in metallic parts. In development.

Kanzler, D. 2017. How reliable are the results of my NDT process? A scientific answer to a practical everyday question. *ESIS TC24 Workshop: Integrity of Railway Structures.*

Kanzler, D., and Müller, C. 2016a. Evaluating RT systems with a new POD approach. *19th World Conference on Non-Destructive Testing*, Berlin, Germany.

Kanzler, D., and Müller, C. 2016b. How much information do we need? A reflection of the correct use of real defects in POD-evaluations. *AIP Conf. Proc.* 1706:1–8.

Karl Härdle, W., Ritov, Y., and Wang, W. 2015. Tie the straps: Uniform bootstrap confidence bands for semiparametric additive models. *J. Multivariate Anal.* 134:129–145.

Kasperovich, G., Haubrich, J., Gussone, J., and Requena, G. 2016. Correlation between porosity and processing parameters in TiAl6V4 produced by selective laser melting. *Mater. Des.* 105:160–170.

Kelly, S. M., and Kampe, S. L. 2004. Microstructural evolution in laser-deposited multilayer Ti-6Al-4V builds: Part I. Microstructural characterization. *Metall. Mater. Trans. A.* 35(6):1861–1867.

Kempen, K., Vrancken, B., Buls, S., Thijs, L., Van Humbeeck, J., and Kruth, J. P. 2014. Selective laser melting of crack-free high density M2 high speed steel parts by baseplate preheating. J. Manuf. Sci. Eng. 136:061026–061032.

Keprate, A. 2016. Probability of detection: History, development and future. *Pipeline Techn. J.* 8:41–45.

Kiefel, D., Scius-Bertrand, M., and Stößel, R. 2018. Computed tomography inspection of additive manufactured components. *Aeronautic Industry, 8th Conference on Industrial Computed Tomography*, Wels, Austria

Kim, F. H., Adam, J., Pintar, L., Fox, J., Tarr, J., Donmez, A. M., and Anne-Françoise Obaton, A. F. 2019. Probability of detection of X-ray computed tomography of additive manufacturing defects. *Review of Progress in Quantitative Nondestructive Evaluation.*

Kim, F. H., Pintar, L., Moylan, S. P., and Garboczi, E. J. 2019. The influence of X-ray computed tomography acquisition parameters on image quality and probability of detection of additive manufacturing defects. J. Manuf. Sci. Eng. 141:11.

Kraussa, H., Zeugnerb, T., and Zaeha, M. F. 2014. Layer-by-layer NDT for additive manufacturing: A new paradigm in quality assurance. *Physics Procedia*, 56: 64–71.

Lewandowski, J. J., and Seifi, M. 2016. Metal additive manufacturing: A review of mechanical properties. *Ann. Rev. Mat. Res.*, 46:151–186.

Li, P. H., Guo, W. G., Huang, W. D., Su, Y., Lin, X., and Yuan, K. B., 2015. Thermomechanical response of 3D laser-deposited Ti-6Al-4V alloy over a wide range of strain rates and temperatures. *Mat. Sci. Eng. A*. 647:34–42.

Li, R.., Liu, J., Shi, Y., Wang, L., and Jiang, W. 2011. Balling behavior of stainless steel and nickel powder during selective laser melting process. *Int. J. Adv. Manuf. Tech.* 59:1025–1035.

Liu, S., and Shin, Y. C. 2019. Additive manufacturing of Ti6Al4V alloy: A review. *Mater. Des.* 164:107552.

Livings, R. A., Biedermann, E. J., Wang, C., Chung, T., James, S., Waller, J. M., S. Volk, S., Krishnan, A., and Collins, S. 2018. Nondestructive evaluation of additive manufactured parts using process compensated resonance testing. *STP1620 Symposium on Structural Integrity of Additive Manufactured Parts*, Washington, DC, USA.

Lu, Q. Y., and Wong, C. H. 2017. Applications of non-destructive testing techniques for post-process control of additively manufactured parts. *Virtual Phys. Prototyping* 12:301–321.

Maztkanin, G., and Yolken, H. T. 2001. NTIAC-TA-00-01 – Probability of detection (PoD) for nondestructive evaluation (NDE).

Meyer, R. M., Crawford, S. L., Lareau, J. P., and Anderson, M. T. 2014. PNNL-23714 – Review of literature for model assisted probability of detection.

MIL-HDBK-1823A. 2009. Nondestructive evaluation system reliability assessment.

Mohr, W. 2016. Assessment of literature on fatigue performance of additively manufactured metals. *EWI*. Available at: https://ewi.org/wp-content/uploads/2016/07/Mohr-Literature-on-Fatigue-Performance-of-AM-Metals.pdf.

MSFC. 2017. MSFC-STD-37-Standard for additively manufactured spaceflight hardware by laser powder bed fusion in metals.

Murr, L. E., Quinones, S. A., Gaytan, S. M., Lopez, M. I., Rodela, A., Martinez, E. Y., Hernandez, D. H., Martinez, E., Medina, F., and Wicker, R. B. 2009. Microstructure and mechanical behavior of Ti-6Al-4V produced by rapid-layer manufacturing, for biomedical applications. *J. Mech. Behav. Biom. Mat.* 2(1):20–32.

Na, J. M., Middedorf, J. K., Lander, J., and Waller, M. 2018. Nondestructive evaluation of programmed defects in Ti-6Al-4V L-PBF ASTM E8 compliant dog-bone sample. *STP1620 Symposium on Structural Integrity of Additive Manufactured Parts*, Washington, DC, USA.

Nicoletto, G. 2017. Anisotropic high cycle fatigue behavior of Ti-6Al-4V obtained by powder bed laser fusion. *Int. J. Fatigue*. 94:255–262.

NIST. 2013. Measurement science roadmap for metal-based additive manufacturing.

Pang, S., Chen, W., and Wang, W. 2014. A quantitative model of keyhole instability induced porosity in laser welding of titanium alloy. *Metall. Mat. Trans. A*. 45:2808–2818.

Qiu, C., Ravi, G., and Attallah, M. M. 2015. Microstructural control during direct laser deposition of a β-titanium alloy. *Mater. Des.* 81:21–30.

Rummel, W. 1982. Recommended practice for a demonstration of NDE reliability on aircraft production parts. *Mater. Eval.* 40:1–11.

Rummel, W. D. 2010. Nondestructive inspection reliability – History, status and future path. *18th World Conference on Nondestructive Testing*, Durban, South Africa.

Russell, R., Wells, D., Waller, J., Poorganji, B., Ott, E., Nakagawa, T., Sandoval, J., Shamsaei, N., and Seifi, M. 2019. Qualification and certification of metal additive manufactured hardware for aerospace applications. In *Additive Manufacturing for the Aerospace Industry*. ed. F. Froes and R. Boyer (Amsterdam: Elsevier), Ch3: 33–66.

Schneider, C. R. A., Sanderson, R. M., Carpentier, C., Zhao, L., and Nageswaran, C. 2012. Estimation of probability of detection curves based on theoretical simulation of the inspection process. *Annual BINDT Conference on NDT*, Cambridge, UK.

Seifi, M., Gorelik, M., Waller, J., Hrabe, N., Shamsaei, N., Daniewicz, S., and Lewandowski, J. J. 2017. Progress towards metal additive manufacturing standardization to support qualification and certification. *JOM* 69:439–455.

Seifi, M., Salem, A., Beuth, J., Harrysson, O., and Lewandowski, J. J. 2016. Overview of materials qualification needs for metal additive manufacturing, *JOM* 68:747–764.

Sharratt, B. M. 2015. Non-destructive techniques and technologies for qualification of additive manufactured parts and processes: A literature review. Contract Report DRDC-RDDC-2015-C035.

Singh, S., Ramakrishna, S., and Singh, R. 2017. Material issues in additive manufacturing: A review. *J. Manuf. Proc.* 25:185–200.

Thompson, A., Maskery, I., and Leach, R. K. 2016. X-ray computed tomography for additive manufacturing: A review. *Meas. Sci. Technol.* 27:1–18.

Thompson, W., Bruce, R., Brasche, L. J., Forsyth, D. S., Lindgren, E., Swindell, P., and Winfree, W. 2009. Recent advances in model-assisted probability of detection. *4th European-American Workshop on Reliability of NDE*, Berlin, Germany.

Tisseur, D. 2013. POD calculation on a radiographic weld inspection with CIVA 11 RT module. *10th International Conference on NDE in Relation to Structural Integrity for Nuclear and Pressurized Components*, Cannes, France.

Todorov, E., Spencer, R., Gleeson, S., Jamshidinia, S., and Ewi, S. M. K. 2014. AFRL-RX-WP-TR-2014-0162. Nondestructive Evaluation (NDE) of complex metallic additive manufactured.

Tofail, S. A., Koumoulos, E. P., Bandyopadhyay, A., Bose, S., O'Donoghue, L., and Charitidis, C. 2018. Additive manufacturing: Scientific and technological challenges, market uptake and opportunities. *Materials Today* 21:22–37.

Townsend, A., Senin, N., Blunt, L., Leach, R. K., and Taylor, J. S. 2016. Surface texture metrology for metal additive manufacturing: A review. *Precis. Eng.* 46:34–47.

Virkkunen, I., Koskinen, T., Papula, S., Sarikka, T., and Hänninen, H. 2019. Comparison of â versus a and Hit/Miss POD-Estimation methods: A European viewpoint. *J. NDE.* 38:89.

Visser, W. 2000. *POD/POS Curves for Non-Destructive Examination.* Surrey, UK: HSE Books.

Wall, F. A. M., and Wedgwood, S. B. 1998. Modelling of NDT reliability (POD) and applying corrections for human factors. *NDT.net.* 3:1–7.

Waller, J. M. 2018. Nondestructive testing of additive manufactured metal parts used in aerospace applications. *ASTM International Webinar.*

Waller, J. M., Parker, B. H.., Hodges, K. L., Burke, E. R., Walker, J. L., and Generazio, E. R. 2014. Nondestructive evaluation of additive manufacturing state-of-the-discipline report. NASA-TM -218560:4–7.

Waller, J. M., Saulsberry, R. L., Parker, B. H.., Hodges, K. L., Burke, E. R., and Taminger, K. M. 2015. Summary of NDE of additive manufacturing efforts in NASA. *Rev. Prog. QNDE* 1650:51–62.

Waller, J., Burke, E., Nichols, C., Wells, D., Born, M., Brandão, A., Gumpinger, J., Nakagawa, T., Itoh, T., and Mitsui, M. 2019. NDT-Based quality assurance of metal additive manufactured aerospace parts at NASA, JAXA, and ESA. *STP1620 Symposium on Structural Integrity of Additive Manufactured Parts*, Washington, D.C, USA.

Wycisk, E., Solbach, A., Siddique, S., Herzog, D., Walther, F., and Emmelmann, C. 2014. Effects of defects in laser additive manufactured Ti-6Al-4V on fatigue properties. *Physics Procedia.* 56:371–378.

Yadollahi, A., and Shamsaei, N. 2017. Additive manufacturing of fatigue resistant materials: Challenges and opportunities. *Int. J. Fatigue.* 98:14–31.

Yasa, E., Deckers, E., and Kruth, J. 2011. The investigation of the influence of laser re-melting on density, surface quality and microstructure of selective laser melting parts. *Rap. Prot. J.* 17:312–327.

Zerbst, U., and Hilgenberg, K. 2017. Damage development and damage tolerance of structures manufactured by selective laser melting – A review. *Proc. Struct. Integrity.* 7:141–148.

Zhou, X., Wang, D., Liu, X., Zhang, D., Qu, S., Ma, J., London, G., Z. Shen, Z., and Liu, W. 2015. 3D-imaging of selective laser melting defects in a Co–Cr–Mo alloy by synchrotron radiation micro-CT. *Acta Mater.* 98:1–16.

10

Post-Process Coordinate Metrology

Richard Leach and Amrozia Shaheen

CONTENTS

10.1 Introduction

This chapter focuses on off-line measurement of the external shape of metal additive manufactured (AM) products. Together with surface texture measurement (see Chapter 11) and internal feature measurement (see Chapter 12), measurement of the external shape is

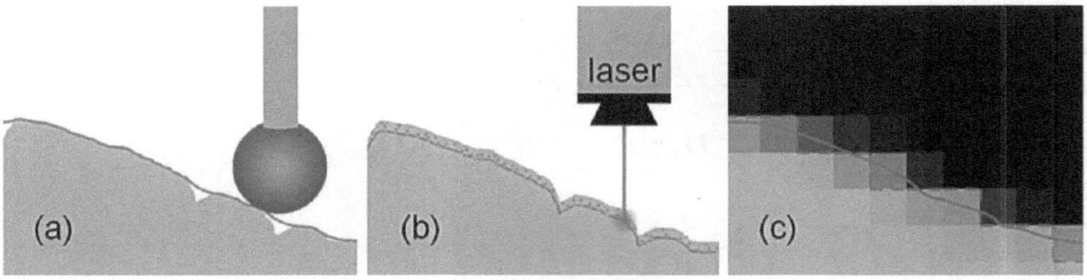

FIGURE 10.1
Different measurement techniques produce different measurement results: schematic representation of (a) tactile, (b) optical and (c) X-ray computed tomography scanning of a surface. The lines represent extracted points on the measured surface. (From Carmignato and Savio, 2011. With permission.)

fundamental for quality control of metal AM products, as well as for providing feedback for AM process optimisation (Leach et al. 2019). The challenges that metal AM brings to coordinate metrology (see Section 10.2 for definitions), compared to traditional manufacturing, are due to the specific characteristics of metal AM products, including: (i) complex freeform shapes, (ii) characteristic surface texture with typically high roughness values, (iii) multiple occlusions and difficult-to-access features and (iv) wide material range with different surface and optical properties. All these challenging characteristics may contribute to measurement errors and deviations between different sensors measuring the same AM surface (see Figure 10.1). For example, several studies have demonstrated that AM surfaces with high surface roughness cause inherent systematic errors between dimensional measurements obtained from contact and non-contact measuring systems (Aloisi and Carmignato 2016, Boeckmans et al. 2015, Carmignato et al. 2017a, Rivas Santos et al. 2019). The above issues also mean that it is still not clear how to obtain measurement traceability for coordinate measurements, and this will be addressed in Section 10.6.

To select the appropriate coordinate measuring sensor for a specific metal AM product, the performance of the sensor has to be compared to the requirements of the specific

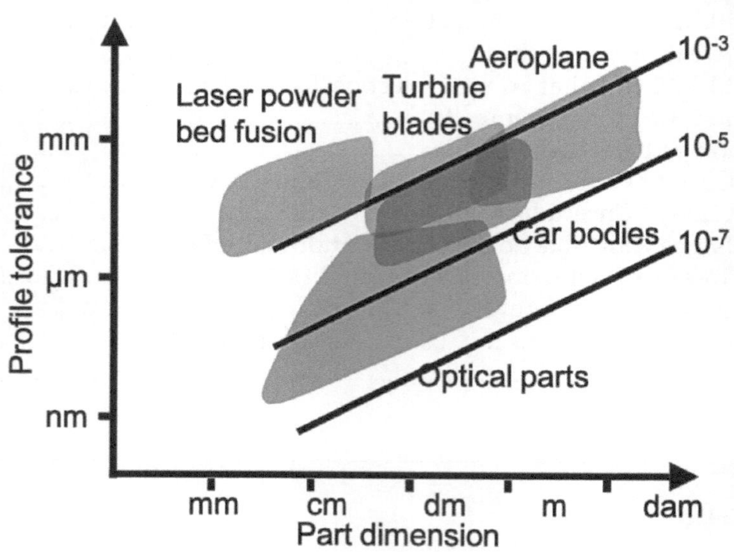

FIGURE 10.2
Typical tolerances of freeform parts against part dimension, for various industrial applications. (Adapted from Savio et al. 2007.)

application. The requirements in terms of profile tolerance are reported in Figure 10.2 for different industrial applications of freeform shaped parts.

Contact probing coordinate measuring systems (CMSs), such as tactile coordinate measuring machines (CMMs), have been used for decades in many different industrial applications and can measure form and dimensions with high accuracy – usually higher than current non-contact CMSs. However, they are generally slower than optical instruments and can collect a lower number of points on the measured surface. Principles and characteristics of tactile probing systems are reviewed elsewhere (Weckenmann et al. 2004, Leach and Smith 2018) and discussed in Section 10.4.

When metal AM parts are produced with rough surface textures, the highest probing accuracy offered by the most precise CMMs is typically not needed (Rivas Santos et al. 2019). However, when post-processing steps and conventional subtractive manufacturing techniques are used to finish AM surfaces (for example, to achieve tight dimensional tolerances not achievable by AM alone), higher accuracy contact probes are more frequently needed.

10.2 Basic Definitions

10.2.1 Surface and Coordinate Metrology Terms and Definitions

A CMS is an instrument for measuring the physical geometry of an object, and coordinate metrology is the science and application of such measurements. Typically, a CMS has the ability to move a probing system (sensor) that detects the location of the surface on an object and the capability to determine spatial coordinate values on the surface relative to a reference coordinate system. Before looking at the sub-systems and use of CMSs, a number of important terms need to be defined.

The following definitions are used in this book:

- *Surface form* – underlying shape of a part (Leach 2014) or fit to a measured surface (ISO 10110-8 2010).
- *Surface topography* – all the surface features treated as a continuum of spatial wavelengths (Leach 2014). Basically, here surface topography is everything that makes up the geometry of the object's surface, that is, it is the surface form plus the surface texture.
- *Surface texture* – the geometrical irregularities present at a surface. Surface texture does not include those geometrical irregularities contributing to the form or shape of the surface (Leach 2015). Simply put, surface texture is what is left of the surface topography once the surface form has been removed.

Note that form and texture make up topography, but it is not always obvious where to draw the distinction between them. Many modern CMSs are equipped with sensors for measuring both form and texture. AM surfaces can also blur the lines between what is considered texture and what is considered form, as the fine-scale structure is often of the same order in size as the features that make up the shape of the object. Despite this blurred distinction, coordinate and surface metrology tend to be very different fields, including different specification standards and ways in which measurement traceability is demonstrated. Hence, in this book, they are treated with separate chapters, and it is suggested that the reader becomes familiar with both chapters before starting to apply the principles contained within them.

The geometrical form of a part can be critical for its effective implementation. Form is mathematically described as the deviation of the measured surface from the fundamental definition of the corresponding geometrical primitive, also known as the *reference geometry* feature. The following form parameters are common to all geometries presented here.

1. Peak-to-reference deviation: maximum deviation of measured points in the positive direction with respect to the reference geometry.
2. Reference-to-valley deviation: maximum deviation of measured points in the negative direction with respect to the reference geometry.
3. Peak-to-valley deviation: a measure of the maximum variation in measured points, sum of 1 and 2.
4. Root-mean-square deviation: a measure of the distribution of measured points with respect to the reference geometry.

Form deviations are often quoted only as positive values since form parameters are defined for a specific direction from the reference geometry. In some cases, reference-to-valley deviations are quoted as negative values, in which case its absolute value is used in determining peak-to-valley deviation.

10.2.2 General Metrology Terms and Definitions

There are a number of terms relating to the field of metrology that need to be discussed briefly – these terms can also be used when reading Chapters 11 and 12. Any of these terms are used almost indistinguishably in practice, which can often lead to confusion when specifying instruments. The terms used in this chapter are taken from the latest version of the BIPM International Vocabulary of Metrology (BIPM 2012).

- *Traceability* – The concept of traceability is one of the most fundamental in metrology and is the basis upon which all measurements can be claimed to be accurate. Traceability is defined as follows:

  Traceability is the property of the result of a measurement whereby it can be related to stated references, usually national or international standards, through a documented unbroken chain of comparisons all having stated uncertainties.

It is important to note the last part of the definition of traceability that states *all having stated uncertainties*. This is an essential part of traceability as it is impossible to usefully compare, and hence calibrate, instruments without a statement of uncertainty (Carmignato 2014). Uncertainty and traceability are inseparable (Haitjema 2013). Traceability applied to coordinate metrology is discussed in Section 10.6.

- *Calibration* is defined as follows:

  Operation that, under specified conditions, in a first step establishes a relation between the quantity values with measurement uncertainties provided by measurement standards and corresponding indications with associated measurement uncertainties and, in a second step, uses this information to establish a relation for obtaining a measurement result from an indication.

In simpler terms, calibration is a comparison between two measurements, one of which is a reference or standard value, and the other which is being tested. Again, note the use of the term *uncertainty* in the formal definition of calibration. Commonly the term *calibration* is misused, which has led to confusion in understanding the aim of the calibration process. The frequent misuse of the calibration term is when it is confused with *adjustment*. Also, in optical form measurement using camera-based approaches (see Section 10.5.3), *calibration* is often used to refer to the process of determining the intrinsic and extrinsic system parameters; whilst this is not strictly calibration (uncertainty is often not considered), almost the entire industry uses the term, so care must be taken. Strictly speaking, the term *adjustment* should be used (software constants are changed).

- *Adjustment* is defined as follows:

  Set of operations carried out on a measuring system so that it provides prescribed indications corresponding to given values of a quantity to be measured.

The adjustment process physically changes some parameters of a metrological tool (it can be a mechanical adjustment or it could be the result of changing the value of a software constant) to provide an indication that is closer to a known value. The adjustment process does not provide information about measurement uncertainty. Similar results could be obtained by correcting the measurement result using the results from a calibration certificate. A meaningful measurement result can be presented without adjustment, but it must have an associated uncertainty.

An example of adjustment of a CMS is when the manufacturer adjusts the coordinate data based on a map of the degrees of freedom of the machine axes (known as *software error correction*, see Section 10.3.4). All modern tactile CMSs (and those with multiple sensors) have software error correction applied. The adjustment cannot account for the uncertainty associated with the measurement result; it only uses a value from the range of possible values that are within the limits given by the measurement uncertainty. After this adjustment, the measurement of an object will provide a different result. The basic difference between calibration and adjustment is also illustrated by the requirement in ISO 17025 (2005) that an instrument should be calibrated before and after adjustment.

- *Verification* is defined as follows:

  Provision of objective evidence that a given item fulfils specified requirements.

A verification test is designed to check whether a particular instrument attribute meets its specification. Verification, therefore, does not necessarily imply that measurement uncertainty is part of the test, but usually some form of quantitative measure is required. Often, an assertion that an instrument is within specification assumes that the test result is inside the specification by at least a 'guard band', for example, the expanded uncertainty. Verification in coordinate metrology is discussed in detail in Section 10.6.

10.3 Basics for Coordinate Metrology

10.3.1 Coordinate Metrology System Configurations

A CMS is an instrument for measuring the physical geometry of an object. Typically, a CMS has the ability to move a probing system that detects the location of the surface

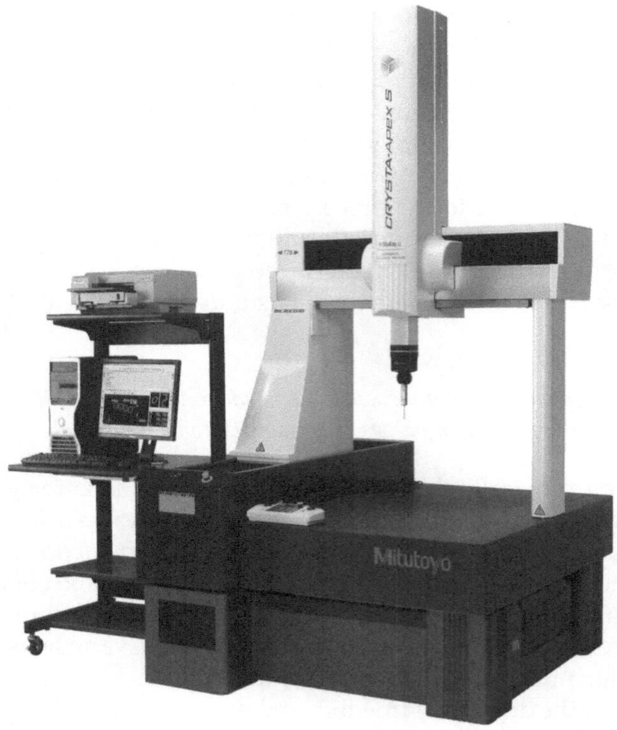

FIGURE 10.3
A bridge-type contact CMS. (Courtesy of Mitutoyo.)

on an object (see Section 10.4) and the capability to determine spatial coordinate values on the surface relative to a reference coordinate system. However, modern camera-based optical systems can operate in completely different configurations (see Section 10.5). The first CMSs were contact types, similar to that shown in Figure 10.3, and were previously referred to as CMMs. However, the relevant ISO technical committee (ISO 213 working group 10) has introduced the term *coordinate measuring system* to cover all types of instruments for measuring geometry, including those using optical and X-ray techniques (these are covered in Chapter 12). Contact type CMSs come in a number of configurations (see Figure 10.4 for a non-exhaustive list) and a range of sizes, from those able to measure something the size of a large aerospace component to micro-scale versions (these are not covered in this book, but see Thalmann et al. 2016 for a recent review). However, the majority of industrial CMSs have working volumes of cubes of sides of approximately 0.5 m to 2 m. By far the most common configuration of contact-type CMS is the moving-bridge type (B in Figure 10.4). Conventional contact-type CMSs generally incorporate three linear axes and use Cartesian coordinates, but CMSs are available with four (and more) axes, where the fourth axis is generally a rotary axis (referred to as the *C axis*). Some optical CMS even have zero axes, as they are essentially 3D imaging systems and do not require either part or machine motion. CMSs are often housed in temperature-controlled rooms held close to 20 °C. The first CMSs became available in the late 1950s and early 1960s (see Hocken and Pereira 2011 for a thorough description of CMSs and some history, and Flack and Hannaford 2005 for an overview of CMS use).

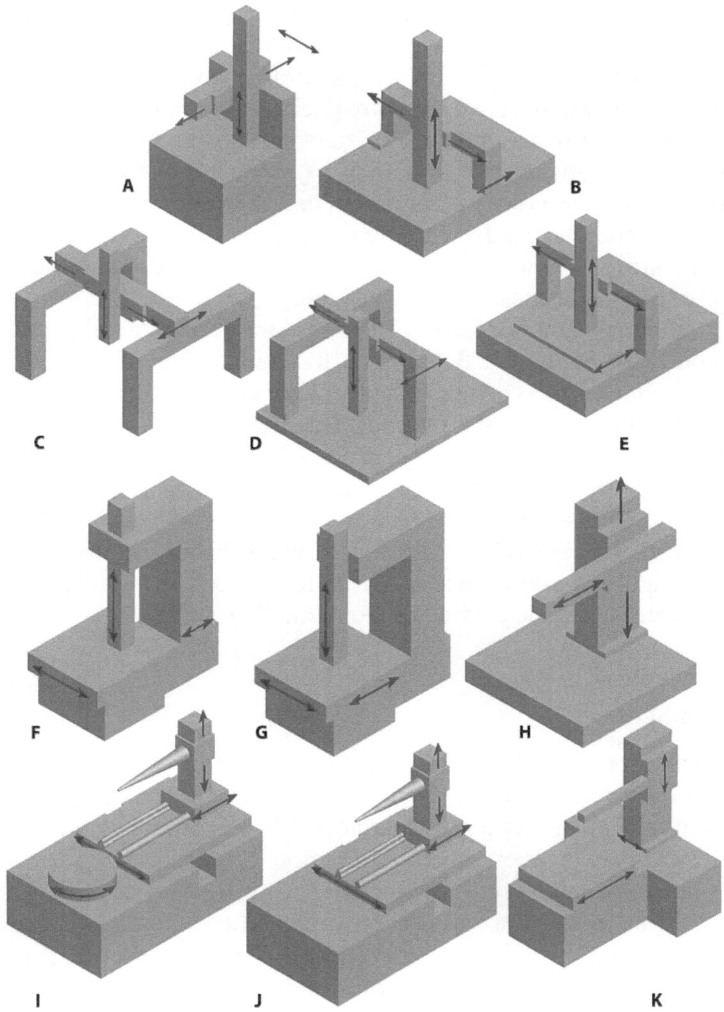

FIGURE 10.4
Typical CMS configurations. A: fixed table cantilever, B: moving bridge, C: gantry, D: L-shaped bridge, E: fixed bridge, F: moving table cantilever, G: column, H: moving ram horizontal-arm, I and J: fixed table horizontal-arm, K: moving table horizontal-arm.

10.3.2 Coordinate Metrology Software

The measured data from a CMS forms what is referred to as a *point cloud*, which is a set of data points with (*x*,*y*,*z*) coordinates (note that some instruments produce polar coordinates and many optical instruments produce a map of *z* height data at [*x*,*y*] positions, often called *range data*). The point cloud needs to be aligned with either the component engineering drawing or, more usually these days, a computer-aided design (CAD) model. This alignment is often carried out with reference to defined datum features on the drawing or model. However, for complex components, such as those with freeform geometry (i.e. geometry not conforming to a regular shape) common to additive components, alignment is usually carried out by a mathematical best-fit operation, most commonly a least-squares fit (Forbes and Minh 2012). Once data are collected from the probe and machine axes (or from cameras in some optical systems), they are analysed by a software package. Along with the geometry data, the software will also collect data from various environmental sensors on or around the CMS, for example, temperature, barometric pressure and humidity sensors. Note that all modern contact or multi-sensor CMS systems have software packages for point cloud analysis, but some camera-based optical systems may not – in this case, there are a host of software packages available for point cloud manipulation and analysis, some freely available to download.

The CMS software mathematically fits associated features (circles, planes, etc.) to the collected data that can then be used to calculate intersection points, distances between features, locations of features in the workpiece coordinate frame, distances between features and form errors such as roundness, cylindricity, etc.

In addition to the above functions, the CMS software will create alignments relating to the part in question (see Section 10.3.3), report the data and compare against CAD data where necessary. Modern CMS software is capable of being programmed directly from a CAD model. Furthermore, once data are collected, the actual points can be compared to the nominal points and pictorial representations of the errors created. Point clouds can also be best-fitted to the CAD model for alignment purposes. Figure 10.5 shows a point cloud that has been compared to its CAD model.

CMS software needs to be tested, and this is covered in ISO 10360 part 6 (2001). These tests use reference data sets and reference software to check the ability of the software to calculate the parameters of basic geometric elements (Forbes 2013).

10.3.3 CMS Alignment

To measure an object on a CMS, its alignment relative to the coordinate system of the machine needs to be determined (see Flack 2014 for more detailed good practice advice on CMS alignment and measurement principles). The alignment process confirms to the software the physical part location on the CMS; more advanced alignments use the CMS to iterate to provide more accuracy in the process. Alignment is usually carried out using datum features on the object – datum features are essentially locations used as guides to tell the machine where it is or as directions on how to get to a particular position (note that these physical datum features may or may not be the same as the datums on the associated CAD drawing). More advanced alignments, for example, best-fit alignments and reference point alignments, are used for freeform geometries.

CMS can be used to measure highly complex objects, but it is useful to split the geometry types into the following:

- Prismatic components, examples of which include cylindrical engine block cavities, brake components and flat pad brake bearings.
- Freeform components, examples of which include car door panels, aircraft wing sections and mobile phone covers. Note that many additive components are highly freeform.

Prismatic components can be broken down into elements that are readily defined mathematically, for example, planes, circles, cylinders, cones and spheres. A measurement will consist of breaking down the component into these geometries and then looking at their inter-relationships, for example, the distance between two holes or the diameter of a circle. Freeform components cannot be broken down as with prismatic components. Generally, the surface is contacted at a large number of points and a surface approximated (fit) to the data.

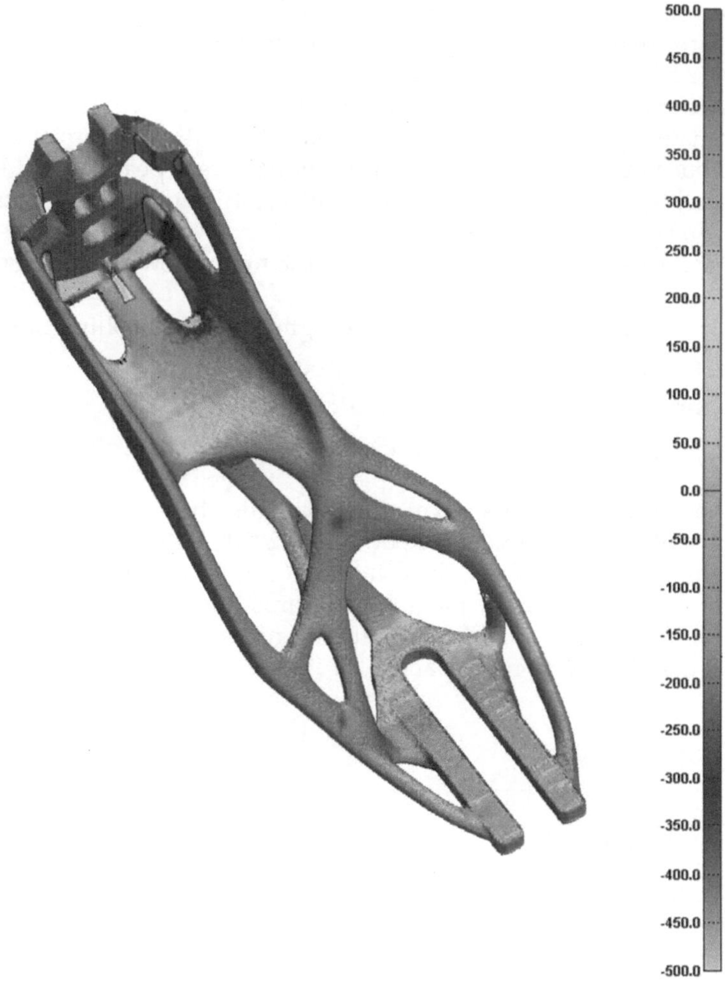

FIGURE 10.5
Comparison of measured point cloud with CAD data. Units: micrometres. (Courtesy of BMW.)

Many real-world additive components are a mixture of freeform surfaces and geometric features; for example, a mobile phone cover may have location pins that need to be measured.

If a CAD model exists, then the point cloud of data can be compared directly against the CAD model (see Figure 10.5). Having a CAD model is an advantage for freeform surfaces, as the nominal local slope at the contact point is known in advance. The local slope is needed to determine the point of contact on the surface based on the centre location of the probe sphere at contact and surface normal vector at that point. This section has mainly covered the alignments necessary for contact CMSs, but dedicated alignment processes will need to be developed for non-contact CMSs (although they are often available in the control software).

10.3.4 CMS Errors

A typical CMS with three linear axes has twenty-one sources of geometric error (see Chapter 5). These can be accounted for by noting that each axis has three linear errors, three rotation errors (six per axis gives eighteen). In addition, there are orthogonality errors between any two pairs of axes (three more makes twenty-one). The twenty-one geometrical errors are minimised during manufacture of the CMS but can also be error-mapped (volumetric error compensation) with corrections to geometric errors made in software (Hocken and Pereira 2011, Schwenke et al. 2008). All commercial CMS manufacturers carry out such error mapping procedures, which are repeated (or at least verified) often on an annual basis during regular service visits. It is not usually expected that the user carries out this procedure.

CMS geometric errors are determined in one of the four following manners:

- Using artefacts and instruments, such as straight edges, autocollimators and levels.
- Using a laser interferometer system and associated optics.
- Using a calibrated-hole plate (Lee and Budekin 2001).
- Using a tracking laser interferometer (Schwenke et al. 2008).

In addition to the geometric error sources, CMSs can also be subject to significant errors due to thermal stability and gradients (Kruth et al. 2000). Often, CMSs are kept in thermally stable environments, but if not, they should not be housed in areas where there are significant thermal gradients or the maximum permissible error specification cannot be assured. The probing systems on CMS are also significant potential error sources, and these will be discussed in Sections 10.4 and 10.5.

10.4 Contact Methods

CMSs measure either by discrete probing, where data from single points on the surface are collected; or by scanning, where data are collected continuously as the mechanical stylus tip is moved across the surface. The stylus tip in contact with the surface is usually a synthetic ruby ball, although other geometries and materials are possible, for example, cylindrical or ceramic stylus tips.

10.4.1 Contact Probe Types

The majority of CMSs use a touch-trigger probe (see Hocken and Pereira 2011 and Flack 2001 for further details about all mechanical probe types). The touch-trigger probe employs a form of kinematic location to retain a stylus in a highly repeatable manner. A typical mechanism (see Figure 10.6 for a photograph) consists of three cylindrical rods, each pressed in the groove formed between a pair of spheres. This action constrains all six degrees of freedom of the stylus so that it returns to the same position after deflection has taken place. An electrical circuit is made through all six contacts such that a trigger signal is generated whenever the stylus is deflected in any direction, in this case by the continuity of the circuit being broken. At the moment of contact, this trigger signal notifies the computer to record the machine position. Because the stylus can relatively freely move following contact, this can accommodate over-travel during the necessary time for the CMS to decelerate. When the probe is retracted, the stylus moves off the surface and the spring force causes the stylus mounting to

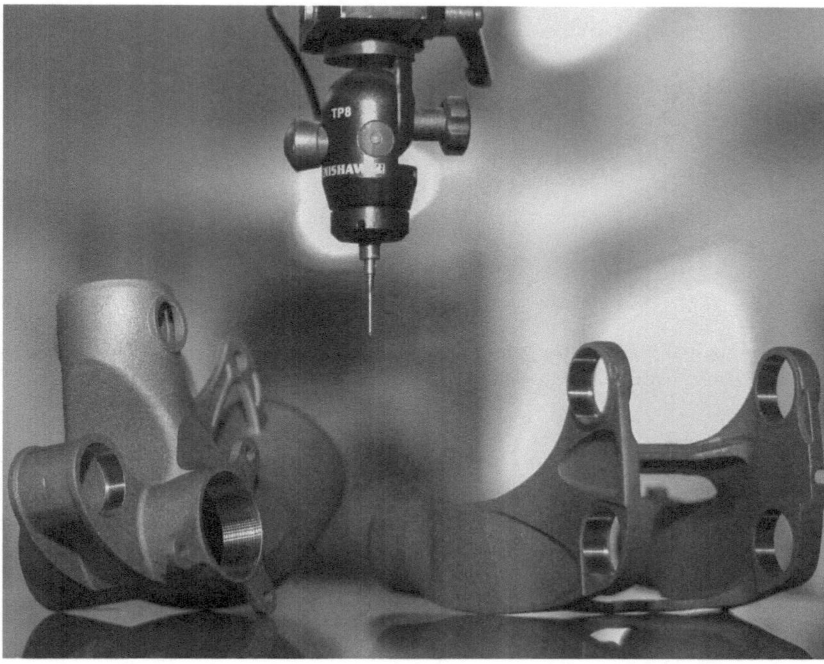

FIGURE 10.6
A touch-trigger probe. (Courtesy of Renishaw.)

reseat. The mechanism allows the probe to detect surface contact with sub-micrometre repeatability.

For higher accuracy applications, analogue probes tend to be the preferred option. On a typical analogue probe, the probing system consists of three orthogonal spring parallelograms and sensors to measure their deflection. Each parallelogram is clamped in its neutral position; the zero points of the sensors are adjusted to this position. Typically, a moving coil system generates the measuring force when contact is made with the object being measured. When the probing system has adjusted into a near-zero position, the machine coordinates and the digitised residual deflections of the probe head are transferred to the computer. Often, the probe head is pre-deflected in the probing direction to ensure that the probe head can be stopped within its deflecting range in case of a contact or collision. Modern analogue probes can scan in a variety of manners that allow access into complex components – some examples of probing actions are shown in Figure 10.7. Scan path planning is typically carried out with the machine software which makes use of the CAD information, although in some cases, feedback mechanisms can be applied to scan objects without a priori knowledge (see Flack 2001 for more on probe scanning path planning).

Measurement with an analogue probe is static, which results in a considerable increase in accuracy when compared to a touch-trigger probe. A typical touch-trigger probe has a repeatability of around 0.5 μm and a form error of around 1 μm; a typical analogue probe can be five times better. In some cases, contact can be continuously measured so that a plot of deflection as a function of contact force can be determined. It is then is possible to extrapolate to identify more precisely the point of first contact – particularly important when measuring materials of low elastic modulus or compliant mechanisms.

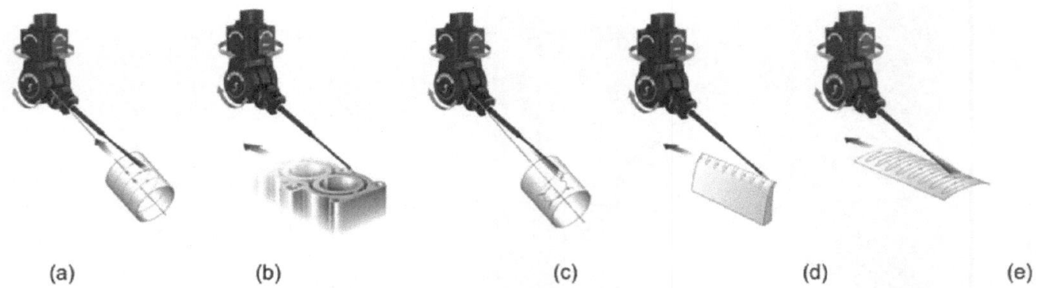

FIGURE 10.7
Example analogue probe scanning modes: (a) circular, (b) gasket, (c) head touch, (d) sweepscan – edges, (e) sweepscan – surfaces. (Courtesy of Renishaw.)

The data collected by the CMS are essentially ball centre data. Therefore, the stylus tip in contact with the surface needs to be qualified to determine the effective stylus radius and the position of the centre of the tip relative to the coordinate reference in the CMS. Stylus qualification is carried out by measuring a known artefact, usually a high-precision ceramic reference sphere (Flack 2001).

To reduce the hardware cost and simplify measurements, Renishaw have developed a contact CMS based on parallel kinematics (Karin and Weber 2018) that acts as a 3D comparator; see Figure 10.8. The Equator system's structure, in the form of a parallel kinematic machine, allows for improved repeatability, reduced inertial effects and lower power consumption when compared to the traditional Cartesian structures used for CMSs. However, the accuracy of the system is difficult to quantify absolutely, which is why the system is operated as a comparator. The Equator has three axes (extending/contracting struts), all of which must be moved in order to position the vertically inclined probe body and attached stylus – which may be vertical or cranked – in any position in x, y and z. The platform to which the axes and probe body holding a scanning probe, for example, are connected is constrained to remain parallel to the base surface by means of two struts associated with each axis – these prevent the platform from twisting or tilting as the axes extend/contract, controlling the rotational degrees of freedom. The Equator first measures a calibrated reference object, which needs to match the object being measured within a defined tolerance, then subsequently acts as a 3D comparator. Maximum permissible error (see Section 10.6.1) values of a few micrometres can be obtained, depending on the size and complexity of the object being measured (Papananias et al. 2017). Renishaw use the Equator to measure test artefacts (see Chapter 5) for benchmarking the performance of their AM systems.

Contact probes are subject to a number of sources of error, which are only summarised here; detailed information on contact probing errors and their determination and adjustment are given in Flack (2001).

10.4.2 Contact Probe Errors

The short distance travelled by the probe between the instant at which actual contact with the object surface takes place and the instant at which the measurement point is recorded is called *pre-travel*. The pre-travel can vary considerably over a range of different stylus configurations and different tensions for the elastic mechanism used to keep the probe tip in contact with the object surface. The longer the stylus, for instance, the more it is allowed to bend before the probe records contact. Touch-trigger probes are also subject

FIGURE 10.8
3D comparator CMS measuring an AM part. (Courtesy of Renishaw.)

to complex lobing errors due to their inherent three-fold symmetry (Yang et al. 1996), i.e. the kinematic configuration used, and analogue probes have similar albeit different issues (Nawara and Kowalski 1984). In order for the position of the probe tip to be known relative to the machine axes for different stylus offsets, a probe calibration procedure must be undertaken before any measurements can be done with that specific offset, and this is usually carried out by probing a reference sphere with calibrated dimensions. This probe calibration procedure is detailed in Flack (2001) and can be quite involved for probes with multiple styli, rotating heads and other complex configurations.

10.4.3 AM Roughness Issues with Contact Probing

As briefly discussed in Section 10.5 and highlighted in Figure 10.1, coordinate measurements of surface form and dimensions are complicated by specific characteristics of metal AM products, including complex shape and surface texture with typically high roughness. For example, the roughness of AM surfaces causes significant deviations between dimensional measurements performed using different contact or non-contact CMSs (Rivas Santos et al. 2019). Lou et al. (2019) have carried out a comprehensive study of the morphological filtering effects of the mechanical probe tip when measuring AM surfaces and established quantitative relationships between surface texture parameters and the difference between contact and non-contact dimensional measurement results. However, there is still work to do, and there is still an open research question regarding measurement traceability. Whilst there have been significant developments to allow comparison of various instruments, the 'ground truth' is still missing. The complexity and rough nature of AM surfaces means that the use of contact CMS is problematic. Whilst an AM surface is 'just another surface', the added complexity brings extra challenges when establishing the traceability route. This traceability issue needs urgent attention as AM processes establish quality systems and industry compliance with established standards (Leach et al. 2019). More on calibration and traceability is given in Section 10.6.

10.5 Optical Methods

The contact-type CMS has been used extensively in advanced manufacturing and engineering due to several advantages (including robustness, accuracy and capability of probing the mechanically relevant surface points), but it has two major limitations. Firstly, there is a need to physically contact the object being measured. The finite contact force always causes some deflection of the surface and may cause damage in the case of plastic deformation, especially with low modulus or low hardness objects, for example polymer AM parts. Secondly, the need to physically contact the surface at each measurement point means that contact CMSs are fundamentally serial measurement devices, which results in potentially long measurement times and/or only a small number of data points. These two limitations can be overcome by the use of optical CMSs, which are covered in this chapter; and X-ray CMSs, which are covered in Chapter 12.

There are many different types of optical instrument that are increasingly becoming available to be used either in conjunction with a CMS platform or as stand-alone instrument. All such optical instruments will be referred to hereafter as *optical probes*, regardless of how they are mounted or used. Optical probes have a number of advantages over contact probes. They do not physically contact the surface being measured and hence do not present a risk of damaging the surface. This non-contact nature can also lead to much faster measurement times for the optical scanning instruments. However, care must be taken when interpreting the data from an optical probe (compared to that from a contact probe). Whereas it is relatively simple to predict the output of a contact probe by modelling it as a ball of finite diameter moving across the surface, it is not such a trivial matter to model the interaction of an electromagnetic field with the surface. Often many assumptions are made about the nature of the incident beam or the surface being measured that can be difficult to justify in practice. Optical probes have a number of limitations, some of which are generic, and some that are specific to instrument types. This section briefly discusses some of the generic limitations, while specific limitations can be found in Leach (2014) and Harding (2014).

Many optical probes use a microscope objective to magnify the features on the surface being measured. It is worth noting that the magnification of the objective is not the value assigned to the objective, but the combination of the objective and the microscope's tube length. The tube length may vary between 160 mm and 210 mm, and thus, if the nominal magnification of the objective assigned by the manufacturer is based on a 160 mm tube length, then the magnification of this objective on a system with 210 mm tube length will be about 30% greater, as magnification equals tube length divided by the focal length of the objective. Magnifications vary from 1.0× to 200× depending on the application and the type of object being measured.

Instruments employing a microscope objective will have two fundamental limitations. Firstly, the numerical (or angular) aperture (NA) determines the largest slope angle on the surface that can be measured and affects the optical resolution. The NA of an objective is given by

$$A_N = n \sin\alpha \tag{10.1}$$

where n is the refractive index of the medium between the objective and the object (usually air, so n can be approximated by unity) and α is the acceptance angle of the aperture (see Figure 10.9, where the objective is approximated by a single lens). The acceptance angle will determine the slopes on the surface that can specularly reflect light back into the objective lens and hence be measured. Note that, if there is some degree of diffuse reflectance (scattering) from a rough surface (for example, with an unfinished metal AM surface), some light can still reflect back into the aperture, allowing larger angles than those dictated by Equation (10.1) to be detected (see Figure 10.9).

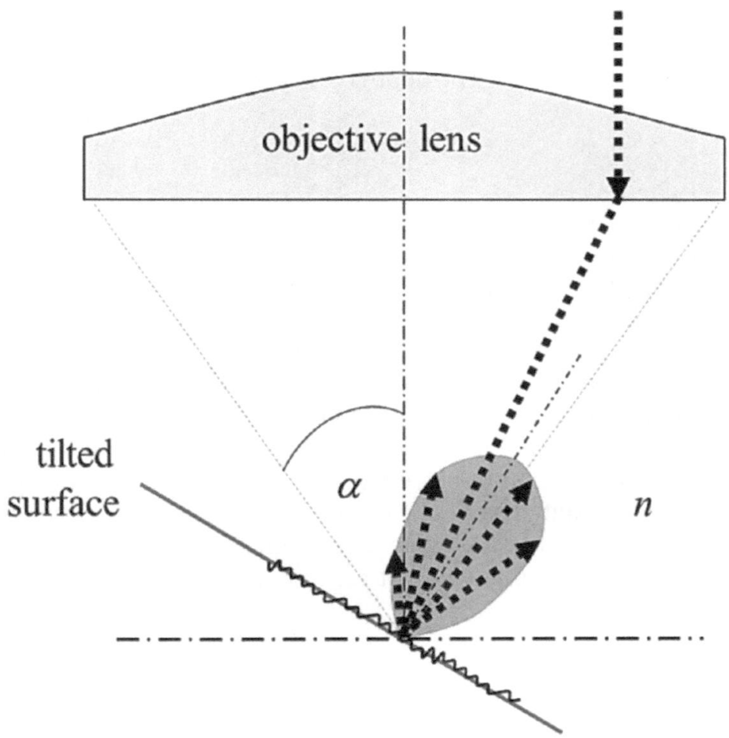

FIGURE 10.9
Microscope aperture imaging a tilted surface.

The second limitation is the optical resolution of the objective. The resolution determines the minimum distance between two lateral features on an object that can be measured. The resolution is given by

$$r = \frac{k\lambda}{A_N} \tag{10.2}$$

where λ is the wavelength of the incident radiation and k is a constant value depending on the definition applied. For a theoretically perfect optical system with a filled objective pupil, the optical resolution is given by the Rayleigh criterion, where k in Equation (10.2) is replaced by 0.61. Yet another measure of the optical resolution is the Sparrow criterion, or the spatial wavelength where the instrument response drops to zero and where the k in Equation (10.2) is replaced by 0.47. Equation (10.2) and the Rayleigh and Sparrow criteria are often used almost indiscriminately, so the instrument user should always check which expression has been used where optical resolution is a limiting factor. Also, Equation (10.2) sets a minimum value (although the Sparrow criteria will give a smaller numerical value – this is down to the manner in which 'resolved' is defined). If the objective is not optically perfect (i.e. aberration free) or if a part of the beam is blocked (for example, when a steep edge is measured), the value becomes higher (worse).

For optical probes that do not employ microscope objectives, for example, fringe projection, it may be the distance between the pixels (determined by the image size and the number of pixels in the camera array) in the microscope camera array that determines the lateral resolution.

As well as lateral resolution, there is a similar effect due to diffraction in the direction of propagation, i.e. the diffraction disk has a finite thickness. The depth of field in the object plane is the thickness of the optical section along the principle axis of the objective lens within which the object is in focus. The depth of field Z is given by

$$Z = \frac{n\lambda}{A_N^2} \tag{10.3}$$

where n is the refractive index of the medium between the lens and the object. Depth of field is effected by the optics used, lens aberrations and the magnification. Note that as depth of field increases, lateral resolution proportionally decreases.

Finally, it is important to note that surface texture plays a significant role in measurement quality when using optical probes, especially the typical surface textures encountered with AM objects (see Chapter 11 and Leach et al. 2019). Although it may be argued that the local gradients of rough surfaces exceed the limit dictated by the NA of the objective and, therefore, would be classified as beyond the capability of optical instrumentation, measured values with high signal-to-noise ratio are often returned in practice.

There are many more limitations of optical instruments that are outside the scope of this book. The user is encouraged to investigate these limitations for their specific instrument type; see Leach (2014) and Harding (2014) for further reading.

10.5.1 Vision Systems

Vision systems (Coveney 2014) usually comprise a two- or three-axis motion system with a long stand-off microscope-type objective equipped with a camera as the probe. They can also incorporate rotary axes and a rotatable probe head. Typical accuracies with vision CMSs are of the order of 0.05 mm over a distance of up to a metre, although they can be

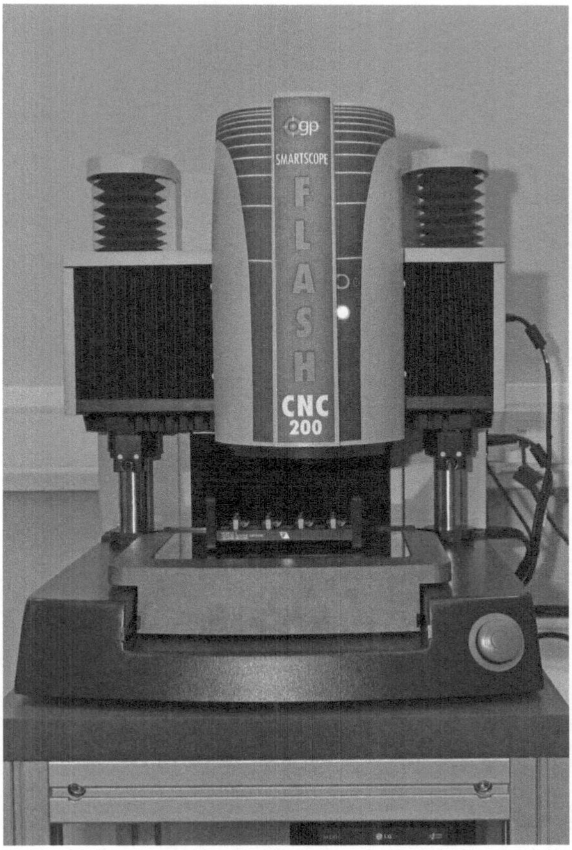

FIGURE 10.10
Typical vision system CMS. (Courtesy of Jon Banner, Verus Metrology Partners.)

much higher than this with specialised instruments. Vision systems are 2D imaging systems and are, therefore, limited in their 3D capabilities. They typically measure dimensional aspects by referencing to edges in a component or other dominant 2D features, and so they have not seen much use in the metal AM field due to the roughness of the components. Figure 10.10 is a photograph of a typical commercial vision system CMS.

10.5.2 Scanning Optical Probes

Laser scanning technology is widely used in 3D optical metrology. It can be categorised as either the time-of-flight method, which requires no triangulation; triangulation-based laser scanning; or interferometric (not discussed here, as there are no commercial systems based on this principle that can be applied to AM parts).

10.5.2.1 Time-of-Flight Method

Time of flight (TOF) is an optical method that calculates the time difference of a round trip of light being emitted and bounced back from the object surface to the sensor (Kolb et al. 2010). Light travels at a very high speed of $c = 3 \times 10^8$ m/s, and the distance travelled

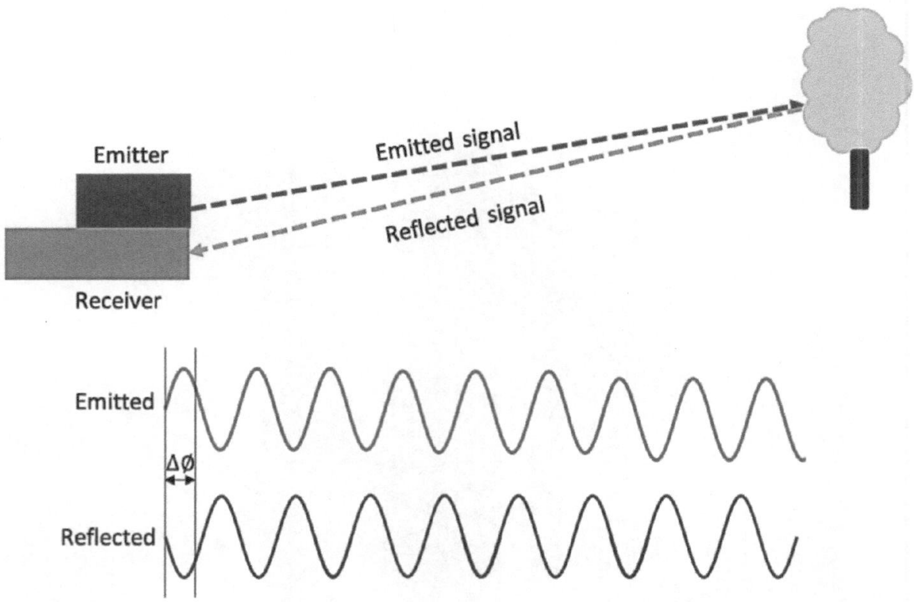

FIGURE 10.11
Working principle of time of flight sensor.

is equal to $ct/2$ (where t is time); therefore, in order to resolve small height changes (for example, 1 mm), the sensor's time resolution should be of the order of a picosecond, which is difficult to achieve in practice. Rather than directly measuring the time difference, the TOF technique usually measures the phase change for the round trip; see Figure 10.11.

A typical TOF assembly consists of a light source and a light sensor. A modulated light beam with a frequency f_m is emitted by the light source, reflected back from the object surface and captured by the sensor. The phase difference $\Delta\varphi$ between the emitted and reflected light beam is calculated and the height information z of the object can be obtained using (Foix et al. 2011)

$$z = \frac{c\Delta\varphi}{4\pi f_m}. \tag{10.4}$$

The TOF sensor usually measures the intensity and the phase information simultaneously. The compact design of the TOF laser scanning technique makes it suitable for applications in the construction, robotics and automotive industries, but rarely with AM. TOF laser scanners are limited by spatial and depth resolution, which are a function of the complexity in manufacturing the sensor.

10.5.2.2 Laser Triangulation

In laser triangulation, the distance between the laser and the object at each specific point is determined, as depicted in Figure 10.12. A typical laser triangulation system comprises a laser source, a detector and a lens that focuses the laser beam on the photodetector. The laser light is incident on the surface of the object, and the intensity of the reflected light is detected by the photodetector. Height information can be retrieved by the triangular

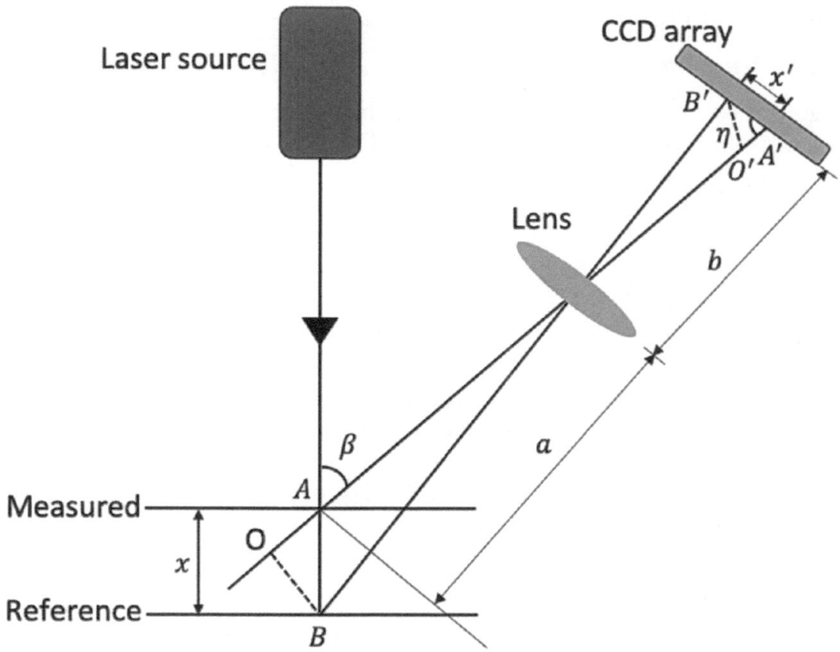

FIGURE 10.12
Schema of laser triangulation. x is the object point displacement on the measured surface, x' is the image point displacement on the camera sensor, η is the angle between the camera sensor and the optical axis of the lens, β is the angle between the laser beam and the lens, a is the distance of the object from the lens, b is the distance of the image from the lens, A and B are the intersection points of the laser beam at the measured and the reference surfaces respectively, A' and B' are the intersection points at the camera sensor, O and O' are the intersection points of the perpendicular lines from points B and B' respectively.

geometry of the laser source point, the reflecting point and the sensing point on the detector.

The distance between the object point and the image point on the detector is given by (Li et al. 2018)

$$x = \frac{ax'\sin\eta}{b\sin\beta - x'\sin(\beta + \eta)}, \tag{10.5}$$

where x' is the image point displacement on the detector, β is the angle between the incident laser beam and the receiving lens and η is the angle between the detector and the optical axis of the lens; the other terms are shown in Figure 10.12.

Laser triangulation scanners are capable of achieving high spatial and depth resolutions (down to single-digit micrometres) and measuring large objects, even buildings or ships, albeit with corresponding losses in resolution as the size increases. However, the method can be time-consuming as it requires a point-by-point scan of the object, although line scanning is now more common on CMS platforms. Another drawback is that the use of coherent sources means that system accuracy is significantly affected by speckle noise. Lastly, laser triangulation cannot be used with highly specular surfaces without the application of a diffuse coating, although this is rarely an issue for as-built AM parts.

10.5.3 Areal Optical Probes

10.5.3.1 Digital Fringe Projection Technique

Digital fringe projection (DFP) is an optical method that is widely used for the dimensional characterisation of complex manufactured objects due to its fast acquisition rates and its non-destructive and non-contact nature (Stavroulakis and Leach 2016, Zhang 2012, Zhang 2010, Nguyen et al. 2015). DFP has been used in a variety of applications, such as AM quality control (Yen et al. 2006), medical science (Genovese and Pappalettere 2006), reverse engineering (Lin et al. 2005) and the aerospace (Heredia-Ortiz and Patterson 2003) and automotive industries (Kus 2009). A DFP system is comprised of a camera, a projector and a computer, see Figure 10.13. Computer-generated 'fringe' patterns are projected onto the surface of the object using the projector, and the camera captures the distorted optical patterns that are reflected off the surface of the 3D object. Height information from the distorted patterns can be retrieved with image processing.

Figure 10.14 shows the typical measurement pipeline of the DFP. The process of reconstructing the 3D shape of the object being measured involves the following steps:

1. Fringe generation and projection onto the object surface.
2. Image acquisition of the distorted patterns.
3. Fringe analysis, which includes phase calculation and phase unwrapping.

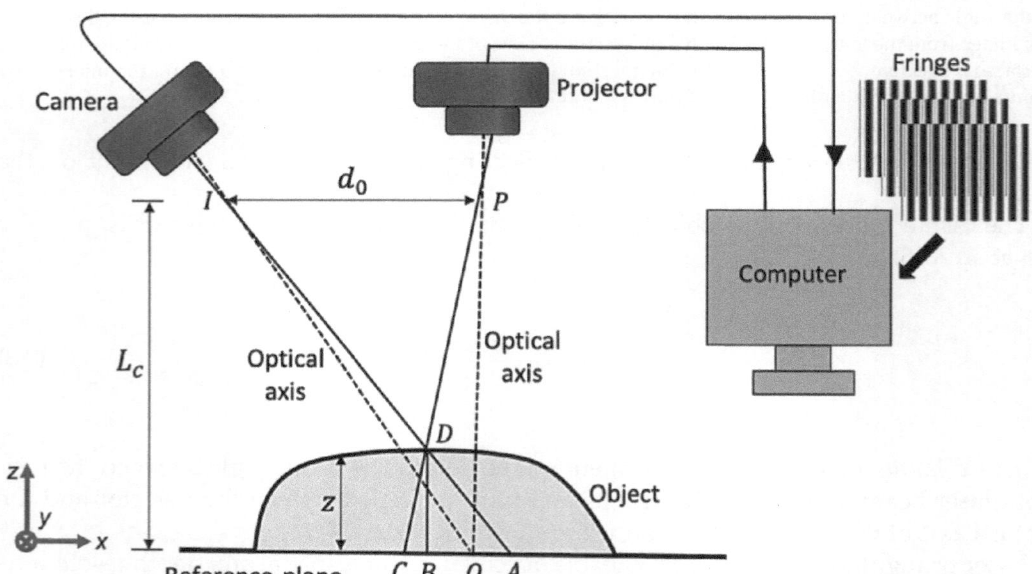

FIGURE 10.13
Schema of a fringe projection system. d_0 is the distance between the optical centres of the camera (I) and the projector (P), L_c is the distance between the optical centre of the camera and point O on the reference plane, z is the height of object, O is intersection point of the optical centres of the projector and the camera at the reference plane when there is no object, D and C are the intersection points at the object surface and the reference plane respectively for a ray generated from the projector, A is the intersection point at the reference plane for the ray reflected off to the camera, B is intersection point for a perpendicular line from point D to the reference plane.

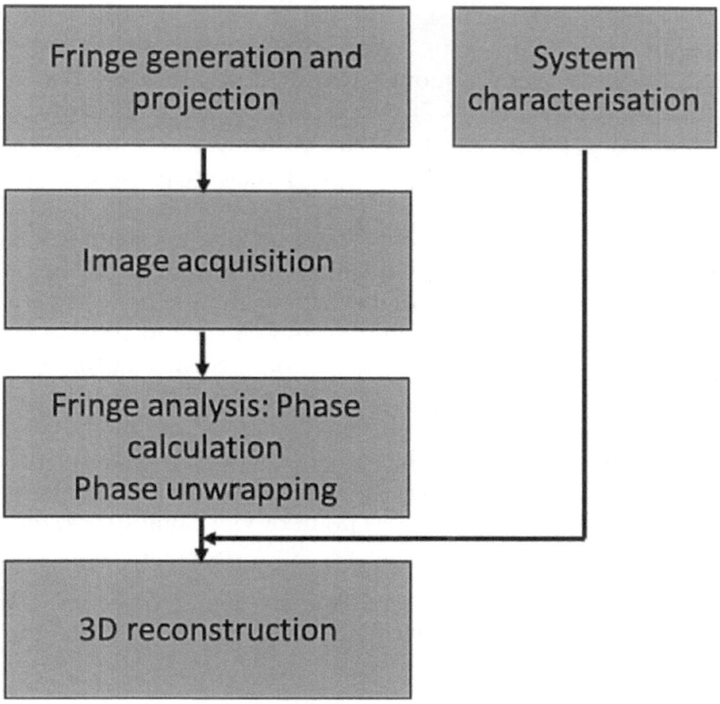

FIGURE 10.14
Pipeline of the fringe projection system.

4. 3D reconstruction, which determines the 3D coordinates for all camera pixels from the absolute phase distributions using the geometrical characterisation results. The process is based on the one-to-one mapping between the absolute phase maps to 3D world coordinates.

In general, the phase contains the height information of the scanned object that needs to be retrieved from the captured fringe images. A minimum of three fringe images are required to uniquely solve for the phase per pixel (Zhang 2016). For higher measurement accuracy, more fringe images can be used, but this clearly reduces the measurement time.

For N phase shifted patterns, the intensity of the ith image with a phase shift of δ_i can be represented as

$$I_i(x,y) = I'(x,y) + I''(x,y)\cos\left(\varphi(x,y) + \delta_i\right), \tag{10.6}$$

where $I'(x,y)$ is the average intensity of the fringe image, $I''(x,y)$ is the modulated intensity, which corresponds to the quality of phase map in the range of [0, 1]; 1 being the best, $\varphi(x,y)$ is the phase and $\delta_i = (2\pi i/N)$ is the phase difference. The phase can be written as

$$\varphi(x,y) = -\tan^{-1}\left(\frac{\sum_{i=1}^{N} I_i \sin\delta_i}{\sum_{i=1}^{N} I_i \cos\delta_i}\right). \tag{10.7}$$

Equation (10.7) indicates that the phase value lies in the range of $[-\pi, \pi]$. A continuous phase map can be obtained by applying a phase unwrapping algorithm. Phase unwrapping is a process of detecting and removing the 2π discontinuities that result from the trigonometric functions in Equation (10.7), which can be achieved by adding or subtracting 2π at each pixel position in the image. There are multiple phase unwrapping algorithms that can be applied depending on the surface type, specification required (resolution, accuracy, range, etc.), noise level and many other factors (see Zhang 2016 for a full treatment).

A schematic diagram of a typical 3D CMS based on DFP is shown in Figure 10.13. The optical axes of the camera and the projector intersect at point O on the reference plane. The distance between the optical centres of the camera and the projector is d_0, the distance between the optical centre of the camera and point O is L_c, and $z(x,y)$ denotes the height of test object. The points P and I represent the virtual aperture of the optical axis of the projector and the camera, respectively.

When setting up the instrument (which is usually carried out by the manufacturer, not the user), a flat plane is measured, the phase map of which is used as a reference phase map (Zhang 2016). The height of the object is measured relative to this reference plane. Referring to Figure 10.13, from the projector's point of view, point D on the object surface has the same phase value as point C on the reference plane $\left(\varnothing_D = \varnothing_C^{ref}\right)$. Likewise, from camera's perspective, points C and A have the same pixels for the phase C map $\left(\varnothing_D = \varnothing_A^{ref}\right)$. Subtracting the reference phase map from the object phase map yields the phase difference at this specific pixel

$$\varnothing_{DA} = \varnothing_D - \varnothing_A^{ref} = \varnothing_C^{ref} - \varnothing_A^{ref} = \Delta\varnothing_{CA}^{ref}. \tag{10.8}$$

Using the property of similar triangles ΔPID and ΔCAD, the height DB of the point D on the object surface depends on the distance between points A and C; therefore,

$$z(x,y) = \overline{DB} = \frac{\overline{AC} \cdot L_c}{d_0 + \overline{AC}}. \tag{10.9}$$

Using the assumption $d_0 \gg \overline{AC}$ and combining equations (10.8) and (10.9), the relative height of the object can be obtained from

$$z(x,y) \approx \frac{L_c}{2\pi f_0 d_0} \Delta\varnothing_{CA}^{ref} \approx K_0 \cdot \Delta\varnothing_{CA}^{ref}, \tag{10.10}$$

where f_0 is the spatial frequency of the fringes, K_0 is a constant that can be found by determining the system intrinsic and extrinsic parameters (a process carried out by the manufacturer of the instrument, see Zhang 2016) and $\Delta\varnothing_{CA}^{ref}$ is the phase containing the height information.

In DFP-based CMS, factors such as surface texture, occlusions, varying surface reflectivity, environmental noise, phase errors, image noise and the depth-of-field of the camera and projector will affect the measurement accuracy of the system. All these factors make DFP a difficult system to set up and use in practice, although commercial systems have user interfaces and software that can help to reduce the complexity for the user. DFP systems are commercially available that are robot mounted and allow full measurement coverage of complex objects.

10.5.3.2 Photogrammetry

Photogrammetry uses methods of image measurement to retrieve the 3D shape of an object from one or more photographs (Luhman et al. 2006). Photogrammetry is well-established in topographic mapping; however, in recent years, it has been extensively used in the fields of architecture, industry, engineering, medicine, geology and now in coordinate metrology for the generation of 3D data (Sims-Waterhouse et al. 2017).

The fundamental principle of photogrammetry is based on triangulation, and typical photogrammetry methods employ the stereo technique (Brunoa et al. 2011). By capturing images from different perspectives (at least two different locations) and measuring the same target in each image, a 'line of sight' is defined from each camera to the target. By determining the camera location and orientation, the intersection of rays emerging from the cameras is used to produce the 3D coordinates of the point on the object (Fraser 1992).

Figure 10.15 shows a schematic illustration of the stereovision principle. The stereovision technique is suitable for high-speed applications since it measures the whole area at once. The basic idea behind the stereovision technique is the epipolar geometry and geometric optics, as depicted in Figure 10.16. A single point R on one camera corresponds to many points on the other camera (P_1,P_2,P_3) depending on how far an object point (Q_1,Q_2,Q_3) is from a 3D space (x,y,z) to the 2D space (u,v), and the inverse is a one-to-many mapping. The projection from a 3D world coordinate to a 2D image plane can be represented as

$$s\begin{pmatrix} u \\ v \\ 1 \end{pmatrix} = A[R,t]\begin{pmatrix} x \\ y \\ z \end{pmatrix} \tag{10.11}$$

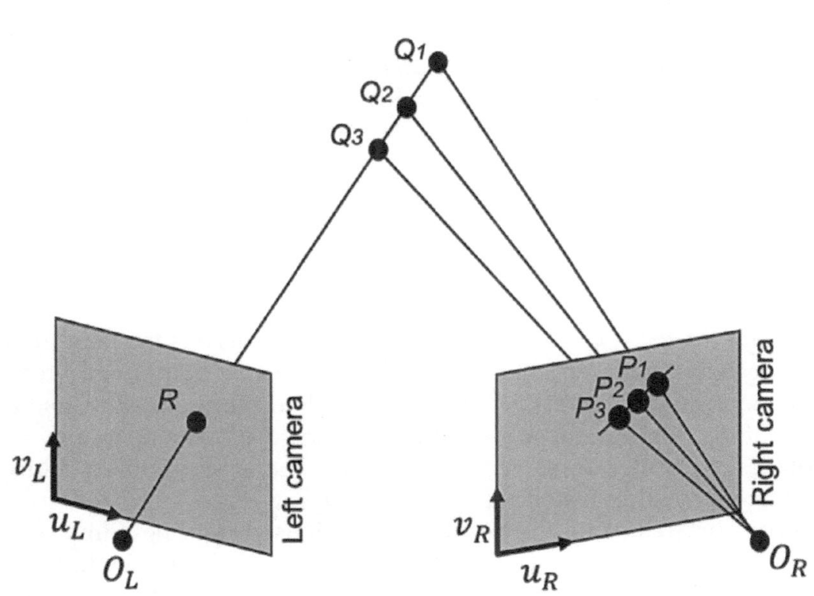

FIGURE 10.15
Schema of binocular stereovision.

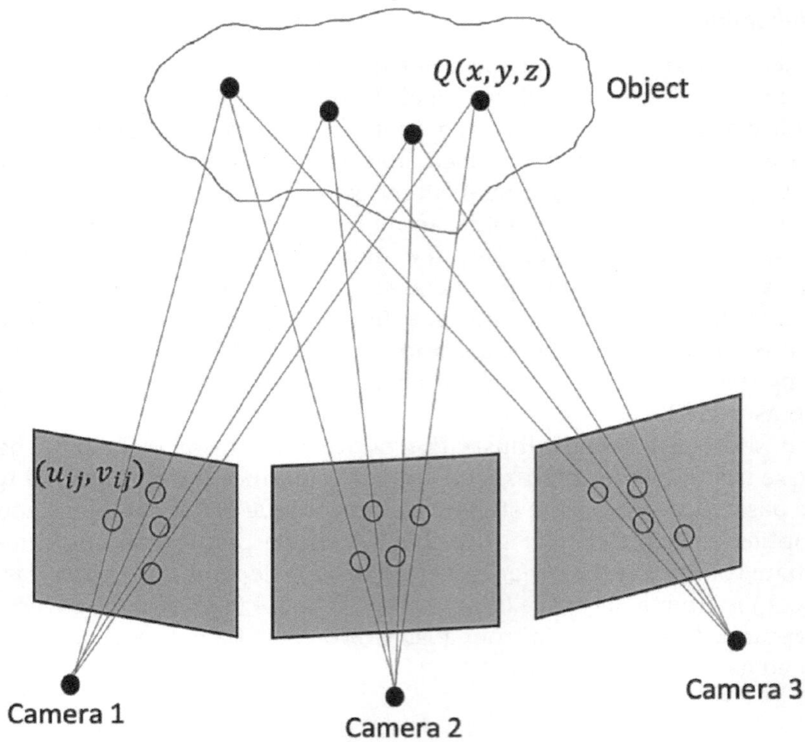

FIGURE 10.16
Close-range photogrammetry.

where s is a scaling factor which represents the depth of projection for a 3D object, R is a 3×3 rotation matrix and t is a (3×1) translation vector. $[R,t]$ define the extrinsic parameters of the system, basically, a conversion from 3D real world coordinates to 3D camera lens coordinates, and A is the intrinsic parameter of the camera, represented as

$$A = \begin{pmatrix} f_u & \alpha & u_0 \\ 0 & f_v & v_0 \\ 0 & 0 & 1 \end{pmatrix} \tag{10.12}$$

where f_u and f_v are the focal lengths in the u and α directions, α is the perpendicularity of the camera's horizontal and vertical pixels, and the principle point (u_0, v_0) represents the point where the optical axis traverses the camera. By determining the intrinsic and extrinsic parameters of the cameras, the 3D coordinates of the object can be found.

The advantages of photogrammetry are cost-effectiveness, achievable resolution comparable to the fringe projection systems, fast acquisition times and suitability for measuring rough to smooth surfaces using the benefits of specialised lighting (Sims-Waterhouse et al. 2017). However, photogrammetry struggles with low depth-of-field and optical errors associated with camera lens distortion, shot noise, divergence and non-linearities. Despite fast acquisition times, the processing speed can be slow since the 3D reconstruction is dependent on processing of a large amount of data.

10.6 Calibration and Traceability

10.6.1 Current Performance Evaluation Framework

Performance verification is a series of tests that allows the manufacturer or user of the CMS to demonstrate that an individual machine meets the manufacturer's specification (see Flack 2001 for a thorough description of the various methods for contact CMS). Note that calibration can be part of the performance verification.

The ISO 10360 series of specification standards defines the procedures for performance verification of CMSs. The series is broken down into seven parts, which are listed below and, with the exception of those parts which are still at draft stage, described in detail in Hocken and Pereira (2011). The development and content of Part 11 and Part 13 can be found in Chapter 12, with a short discussion on some of the issues below.

- Part 1: Vocabulary (2000).
- Part 2: CMMs used for measuring linear dimensions (2009).
- Part 3: CMMs with the axis of a rotary table as the fourth axis (2000).
- Part 4: CMMs used in scanning measuring mode (2000).
- Part 5: CMMs using single and multiple-stylus contacting probing systems (2010).
- Part 6: Estimation of errors in computing Gaussian associated features (2001).
- Part 7: CMMs equipped with imaging probing systems (2011).
- Part 8: CMMs with optical distance sensors (2013).
- Part 9: CMMs with multiple probing systems (2013).
- Part 10: Laser trackers for measuring point-to-point distances (2016).
- Part 11: CMMs using the principle of computed tomography (2020) – still at committee draft stage.
- Part 12: Articulated arm coordinate measurement machines (2016).
- Part 13: Optical 3D CMS (2020) – still at committee draft stage.

The criterion mainly used to quantify the performance of a CMS is the 'maximum permissible error' (MPE), which is defined (in the ISO 10360 standards) as the largest error or deviation of a measurement from a reference quantity value. The MPE of a CMS is often specified as a constant value plus a length-dependent term, for example MPE = 2.4 + $2L/1000$ μm, where L is the measured length in millimetres. A CMS's MPE serves as a benchmark for its performance when compared to other similar instruments (Carmignato et al. 2010). Often, prior to purchasing a measuring instrument, several tests with dedicated reference objects are performed to determine if the particular instrument satisfies its specified MPE. For a specific CMS, typically more than one MPE can be specified; the one above is the MPE for length measurement, but there are others related to probing errors and different measurands (see relevant part of ISO 10360 standard specific to the CMS type being used).

An instrument's MPE is often mistaken for the uncertainty of measurements performed on the instrument. While MPE is a maximum expected deviation of measurements by a particular instrument from a reference value, uncertainty is a statistical dispersion of values that can be attributed to the result of a measurement (see Section 10.6.2). Figure

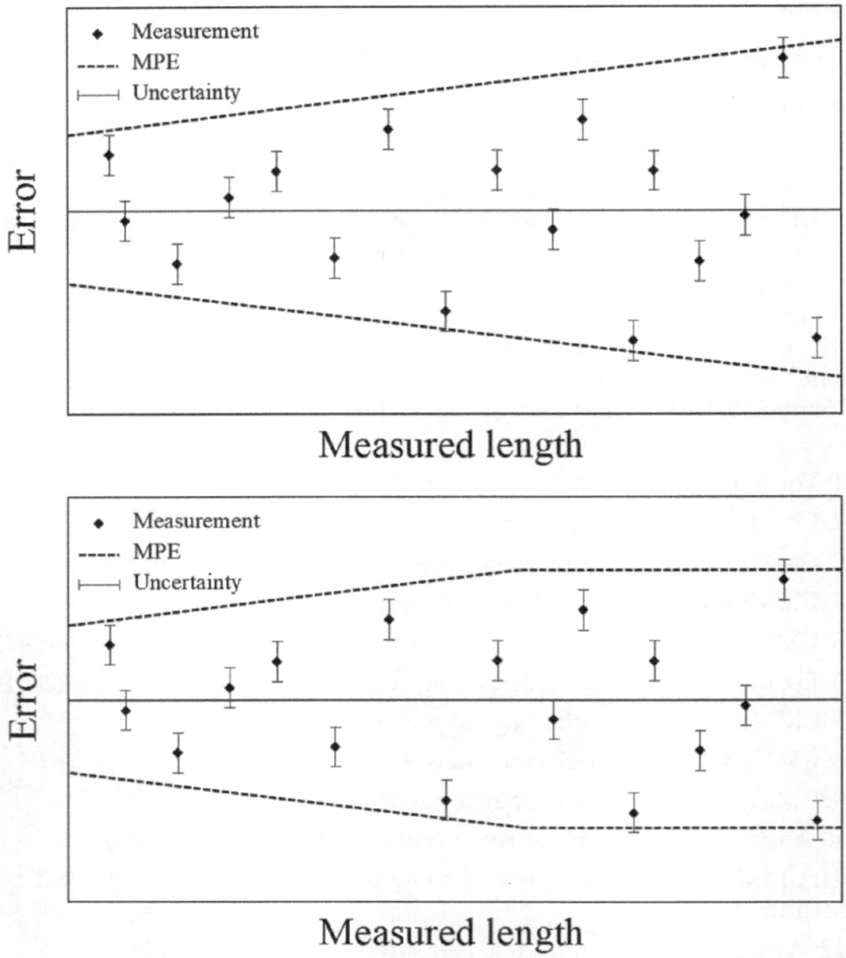

FIGURE 10.17
Two types of maximum permissible error (MPE) plots. (Courtesy of Dr Xiaobing Feng, Shanghai Jiatong University.)

10.17 illustrates the concepts of MPE and uncertainty as they pertain to measurements on features with various lengths. In this plot, the error of each measurement result from the corresponding reference value is plotted as a function of the measured length. Some of the measurement points consist of uncertainty intervals that exceed the MPE (dotted line). The use of these measurement results in deciding whether or not the instrument satisfies the MPE would prove inconclusive. Finally, unlike the MPE of a measuring instrument, measurement uncertainty is a task-specific quantity (see Section 10.6.2).

10.6.2 Current Uncertainty Framework

It is a difficult task to estimate the uncertainty for a CMS measurement from first principles, even for measurement of a simple prismatic component, for example a smooth sphere. The large number of influence factors (see Section 10.3.4) and complex nature of CMS probing strategies mean that, in many cases, an analytical expression for the measurement model cannot easily be found. For this reason, ISO Technical Committee 213 working group 10 have developed a number of strategies for uncertainty estimation, at least for contact CMS, and these are presented in this section. Strategies for non-contact CMS remain an open research area, although there has been some recent work highlighting the issues (Rivas Santos et al. 2019, Lou et al. 2019, Leach et al. 2019, Gayton et al. 2019).

As with measurement uncertainty, traceability of measurements carried out by CMSs is difficult to demonstrate. It used to be the case that the only way to demonstrate traceability was to carry out ISO 10360-type performance verification tests on the CMS. However, if a CMS is performance-verified, this does not automatically mean that measurements carried out with this CMS are calibrated and/or traceable. A performance verification only demonstrates that the machine meets its specification for measuring simple lengths, i.e. it is not task specific.

A better method to achieve at least a degree of traceability is described in ISO 15530 part 3 (2011). This specification standard makes use of calibrated artefacts to essentially use the CMS as a comparator. The uncertainty evaluation is based on a sequence of measurements on a calibrated object or objects, performed in the same way and under the same conditions as the actual measurements (this is known as the 'substitution method'). The differences between the results obtained from the measurement of the objects and the known calibration values of these calibrated objects are used to estimate the uncertainty of the measurements. As an example, if an external dimension needs to be measured, a calibrated length bar of similar length can be mounted adjacent to the object being measured, and the measurement carried out as a comparison. This ensures that many of the systematic influence factors are common to both measurements and, therefore, reduces combined uncertainty. Uncertainty estimations for CMSs using the substitution method are not usually performed on complex and individual AM parts due to the difficulty and expense of developing a calibrated artefact to compare with the workpiece.

Alternative methods can be used to determine the task-specific uncertainty of coordinate measurements. One such method that evaluates the uncertainty by Monte Carlo methods is described in ISO/TS 15530 part 4 (2008). To allow CMS users to easily create uncertainty statements, CMS suppliers and other third-party companies have developed uncertainty-evaluating software, also known as 'virtual CMMs' (Balsamo et al. 1999, Flack 2013), but such software is only available for contact probes.

As briefly discussed in Section 5 and highlighted in Figure 10.1, as-built AM surfaces present a significant issue for contact probing in terms of traceability. Determination of measurement uncertainty using a virtual CMS is carried out by performing repeated simulated measurements with varying inputs (influence factors) on a simulated CMS and determining how those inputs affect the measurand. The software determines the variability that will occur with the physical CMS measurements with a modelling of the uncertainty contributions for each point probed by the CMS. Known systematic uncertainty contributions remain constant, while unknown systematic and random contributions are varied in each simulated measurement throughout their ranges. This simulation is repeated a significant number of times until a statistical evaluation of these virtual measurements is made, and the expanded uncertainty is reported. The peak-to-valley deviation of the measured surface from the nominal is an important input for the uncertainty calculations. For commercial virtual CMS software packages, the ISO 4287 (2000) *Rz* parameter can be used to define peak-to-valley deviation (Rivas Santos et al. 2019). Previous reviews showed that *Rz* values for metal AM surfaces lie between tens and hundreds of micrometres (see Chapter 11), and results show a high dependence of CMM measurement uncertainties on surface texture (Rivas Santos et al. 2019). This surface texture with AM issue is significant;

it is not clear that contact CMSs can be used for traceable reference measurement on AM artefacts without the use of a high degree of filtering – this remains an open research question (Leach et al. 2019).

10.6.3 Issues with Non-Contact CMS Performance Evaluation

At the time of writing, ISO Technical Committee 213 working group 10 is still developing a part that addresses the performance verification of optical CMS (ISO 10360 part 13). Due to the current lack of a published ISO standards in this area, the German VDI/VDE 2634 part 3 (2011) guideline is often used in practice for performance verification of optical CMS. For X-ray computed tomography CMS, the ISO working group is developing ISO 10360 part 11, although the exact contents of the draft are still a subject of some debate (Carmignato et al. 2017, Leach et al. 2019).

10.6.4 Towards Calibration of Non-Contact CMSs

How to estimate measurement uncertainty and apply the ISO 15530 framework to non-contact CMS is still an open research field. The already-discussed AM surface issues make the use of the substitution method problematic, and the complexity of modelling non-contact sensors makes the development of virtual non-contact CMSs a challenge. There has been some recent activity on modelling optical systems (Zhang et al. 2017, Gayton et al. 2019, Sims-Waterhouse et al. 2020) but there is still a lot to do, and the issues are not yet under development in ISO Technical Committee 213 working group 10.

10.7 Current Research and Future Look

Coordinate measurements of surface form and dimensions are complicated by specific characteristics of metal AM products, including complex shape and surface texture, with typically high roughness. For example, the roughness of AM surfaces causes significant deviations between dimensional measurements performed using different contact or non-contact CMSs. Besides tactile CMMs, optical methods – such as laser triangulation, photogrammetry and fringe projection – as well as X-ray computed tomography are increasingly used for coordinate measurements of metal AM parts; however, the use of such non-contact measuring systems brings new challenges for traceability and comparability of measurements. Current research efforts are focused on improving the measurement results in terms of both accuracy and measuring time. Research is under way to maximise object coverage whilst at the same time minimising both measuring time and the amount of manual intervention in the measurement planning. Recently, machine learning methods have been applied to help solve the object coverage issues and allow sensors to be repositioned without the need for recalibration of the extrinsic parameters (Stavroulakis et al. 2017, Stavroulakis et al. 2019), but there still remains much work to do in developing CMS principles and instruments that are optimised for AM components.

Acknowledgements

The author would like to thank the Engineering and Physical Sciences Research Council [EPSRC grant number P/M008983/1] for funding the work.

References

Aloisi, V., and Carmignato, S. 2016. Influence of surface roughness on X-ray computed tomography dimensional measurements of additive manufactured parts. *Case Studies in Nondestructive Testing and Evaluation* 6/B:104–110.

Balsamo, A., Di Ciommo, M., Mugno, R., Rebaglia, B. I., Ricci, E., and Grella, R. 1999. Evaluation of CMM uncertainty through Monte Carlo simulations. *Ann. CIRP* 48:425–428.

BIPM, IEC, IFCC, ILAC, ISO, IUPAC, IUPAP and OIML, 2012. *International Vocabulary of Metrology – Basic and General Concepts and Associated Terms*. Bureau International des Poids et Mesures: Saint-Cloud, France, JCGM 200.

Boeckmans, B., Tan, Y., Welkenhuyzen, Y., Guo, F., Dewulf, W., and Kruth, J. P. 2015. Roughness offset differences between contact and non-contact measurements. In *Proc. euspen*, Leuven, Belgium, Jun. 189–190.

Brunoa, F., Biancoa, G., Muzzupappa, M., Barone, S., and A. V. Razionale, A. V. 2011. Experimentation of structured light and stereo vision for underwater 3-D reconstruction. *ISPRS J. Photogramm. Remote Sens.* 66:508–518.

Carmignato, S. 2014. Calibration. In: Laperrière, L., and Reinhart, G. (Ed.), *CIRP Encyclopedia of Production Engineering*. Springer-Verlag: Berlin Heidelberg.

Carmignato, S., Voltan, A., and Savio, E. 2010. Metrological performance of optical coordinate measuring machines under industrial conditions. *Ann. CIRP* 59:497–500.

Carmignato, S., and Savio, E. 2011. Traceable volume measurements using coordinate measuring systems. *Ann. CIRP* 60:519–522.

Carmignato, S., Aloisi, V., Medeossi, F., Zanini, F., and Savio, E. 2017a. Influence of surface roughness on computed tomography dimensional measurements. *Ann. CIRP* 66:499–502.

Carmignato, S., Dewulf, W., and Leach, R. K. 2017b. *Industrial X-ray Computed Tomography*. Springer.

Coveney, T. 2014. Dimensional measurement using vision systems. *NPL Good Practice Guide No. 39*. National Physical Laboratory.

Dorsch, R. G., Häusler, G., and Herrmann, J. M. 1994. Laser triangulation: Fundamental uncertainty in distance measurement. *Appl. Opt.* 33:1306–1314.

Flack, D. R. 2001. CMM probing. *NPL Good Practice Guide No. 43*. National Physical Laboratory.

Flack, D. R. 2013. Co-ordinate measuring machines task specific measurement uncertainties. *NPL Good Practice Guide No. 130*. National Physical Laboratory.

Flack, D. R. 2014. CMM measurement strategies. *NPL Good Practice Guide No. 41*. Teddington, UK: National Physical Laboratory.

Flack, D. R., and Hannaford, J. 2005. Fundamental good practice in dimensional metrology. *NPL Good Practice Guide No. 80*, National Physical Laboratory.

Foix, S., Alenya, G., and Torras, C. 2011. Lock-in time-of-flight (ToF) cameras: A survey. *IEEE Sens. J.* 11:1917–1926.

Forbes, A. B. 2013. Areal form removal. In: Leach, R. K. (Ed.), *Characterisation of Areal Surface Texture*. Berlin: Springer, Ch 5.

Forbes, A. B., and Minh, H. D. 2012. Generation of numerical artefacts for geometric form and tolerance assessment. *Int. J. Metrol. Qual. Eng.* 3:145–150.

Fraser, C. S. 1992. Photogrammetric measurement to one part in a million. *Photogramm. Eng. Remote Sens.* 58:305–310.

Gayton, G., Su, R., and Leach, R. K. 2019. Model-based uncertainty estimation of uncertainty for fringe projection. *Proc. ISMTII*, Niigata, Japan, Sep.

Genovese, K., and Pappalettere, C. 2006. Whole 3D shape reconstruction of vascular segments under pressure via fringe projection techniques. *Opt. Laser Eng.* 44:1311–1323.

Haitjema, H. 2013. Measurement uncertainty. In: Laperrière, L., and Reinhart, G. (Ed.), *CIRP Encyclopedia of Production Engineering*. Berlin Heidelberg: Springer-Verlag.

Harding, K. 2014. *Handbook of Optical Dimensional Metrology*. Boca Raton, FL: CRC Press.

Heredia-Ortiz, M., and Patterson, E. A. 2003. On the industrial applications of Moiré and fringe projection techniques. *Strain* 39:95–100.

Hocken, R. J., and Pereira, P. 2011. *Coordinate Measuring Machines and Systems*, 2nd ed. Boca Raton, FL: CRC Press.

ISO 4287. 2000. Geometrical product specification (GPS) – Surface texture: Profile method – Terms, definitions and surface texture parameters, International Organization of Standardization.

ISO 10110 part 8. 2010. Optics and photonics – Preparation of drawings for optical elements and systems - Part 8: Surface texture; roughness and waviness, International Organization for Standardization.

ISO 10360 part 1. 2000. Geometrical Product Specifications (GPS) – Acceptance and Reverification Tests for Coordinate Measuring Machines (CMM) – Part 1: Vocabulary, International Organization for Standardization.

ISO 10360 part 2. 2009. Geometrical Product Specifications (GPS) – Acceptance and Reverification Tests for Coordinate Measuring Machines (CMM) – Part 2: CMMs Used for Measuring Size, International Organization for Standardization.

ISO 10360 part 3. 2000. Geometrical Product Specifications (GPS) – Acceptance and Reverification Tests for Coordinate Measuring Machines (CMM) – Part 3: CMMs with the Axis of a Rotary Table as the Fourth Axis, International Organization for Standardization.

ISO 10360 part 4. 2000. Geometrical Product Specifications (GPS) – Acceptance and Reverification Tests for Coordinate Measuring Machines (CMM) – Part 4: CMMs Used in Scanning Measuring Mode, International Organization for Standardization.

ISO 10360 part 5. 2010. Geometrical Product Specifications (GPS) – Acceptance and Reverification Tests for Coordinate Measuring Machines (CMM) – Part 5: CMMs Using Single and Multiple-Stylus Contacting Probing Systems, International Organization for Standardization.

ISO 10360 part 6. 2001. Geometrical Product Specifications (GPS) – Acceptance and Reverification Tests for Coordinate Measuring Machines (CMM) – Part 6: Estimation of Errors in Computing Gaussian Associated Features, International Organization for Standardization.

ISO 10360 part 7. 2011. Geometrical Product Specifications (GPS) – Acceptance and Reverification Tests for Coordinate Measuring Machines (CMM) – Part 7: CMMs Equipped with Imaging Probing Systems, International Organization for Standardization.

ISO 10360 part 8. 2013. Geometrical Product Specifications (GPS) – Acceptance and Reverification Tests for Coordinate Measuring Machines (CMM) – Part 8: CMMs with Optical Distance Sensors, International Organization for Standardization.

ISO 10360 part 9. 2013. Geometrical Product Specifications (GPS) – Acceptance and Reverification Tests for Coordinate Measuring Machines (CMM) – Part 9: CMMs with Multiple Probing Systems, International Organization for Standardization.

ISO 10360 part 10. 2016. Geometrical Product Specifications (GPS) – Acceptance and Reverification Tests for Coordinate Measuring Machines (CMM) – Part 10: Laser Trackers for Measuring Point-To-Point Distances, International Organization for Standardization.

ISO/CD 10360 part 11. 2019. Geometrical Product Specifications (GPS) – Acceptance and Reverification Tests for Coordinate Measuring Machines (CMM) – Part 11: CMMs Using the Principle of Computed Tomography (CT), International Organization for Standardization.

ISO 10360 part 12. 2016. Geometrical Product Specifications (GPS) – Acceptance and Reverification Tests for Coordinate Measuring Machines (CMM) – Part 12: Articulated Arm Coordinate Measurement Machines (CMM), International Organization for Standardization.

ISO/CD 10360 part 13. 2019. Geometrical Product Specifications (GPS) – Acceptance and Reverification Tests for Coordinate Measuring Machines (CMM) – Part 13: Optical 3D CMS, International Organization for Standardization.

ISO 15530 part 3. 2011. Geometrical product specifications (GPS) – Coordinate measuring machines (CMM): Technique for determining the uncertainty of measurement – Part 3: Use of calibrated workpieces or measurement standards, International Organization for Standardization.

ISO/TS 15530 part 4. 2008. Geometrical product specifications (GPS) – Coordinate measuring machines (CMM): Technique for determining the uncertainty of measurement – Part 4: Evaluating CMM uncertainty using task specific simulation, International Organization for Standardization.

ISO 17025. 2005. General requirements for the competence of testing and calibration laboratories, International Organization of Standardization.

Karin, S., and Weber, U. 2018. Kinematic design. In: Leach, R. K., and Smith, S. T. (Ed.), *Basics of Precision Engineering*. Boco Ranton: CRC Press.

Kolb, A., Barth, E., Koch, R., and Larsen, R. 2010. Time-of-flight cameras in computer graphics. *Comput. Graph. Forum* 29:141–159.

Kruth, J. P., Vanherck, P., and Van den Bergh, C. 2000. Thermal compensation for a CMM based on interferometer measurements and testing its applicability with a thermal stable artefact. *Proc. 33rd Int. MATADOR Conf.* 223–228.

Kuş, A. 2009. Implementation of 3D optical scanning technology for automotive applications. *Sensors* 9:1967–1979.

Leach, R. K. 2014. *Fundamental Principles of Engineering Nanometrology*, 2rd edn. Elsevier.

Leach, R. K. 2015. Surface texture. In: Laperrière, L., and Reinhart, G. (Ed.), *CIRP Encyclopaedia of Production Engineering*. Berlin: Springer-Verlag.

Leach, R. K., Bourell, D., Carmignato, S., Donmez, A., Senin, N., and Dewulf, W. 2019. Geometrical metrology for metal additive manufacturing. *Ann. CIRP* 68:677–700.

Lee, E. S., and Burdekin, M. 2001. A hole plate artifact design for volumetric error calibration of a CMM. *Int. J. Adv. Manuf. Technol.* 17:508–515.

Li, S., Jia, X., Chen, M., and Yang, Y. 2018. Error analysis and correction for color in laser triangulation measurement. *Optik* 168:165–173.

Lin, C-H., He, H-T., Guo, H-W., Chen, M-Y., Shi, X., and Yu, T. 2005. Fringe projection measurement system in reverse engineering. *J. Shanghai Univ.* 9:153–158.

Lou, S., Brown, S. B., Sun, W., Zeng, W., Jiang, X., and Scott, P. J. 2019. An investigation of the mechanical filtering effect of tactile CMM in the measurement of additively manufactured parts. *Measurement*. 144:173–182.

Luhman, T., Robson, S., Kyle, S., and Harley, I. 2006. *Close Range Photogrammetry: Principles, Methods and Applications*. Dunbeath, UK: Whittles Publishing.

Nawara, L., and Kowalski, M. 1984. The investigations on selected dynamical phenomena in the heads of multi-coordinate devices. *Ann. CIRP*. 33:373–375.

Nguyen, H., Nguyen, D., Wang, Z., Kieu, H., and Le, M. 2015. Real-time, high-accuracy 3D imaging and shape measurement. *Appl. Opt.* 54:A9–A17.

Papananias, M., Fletcher, S., Longstaff, A. P., Mengot, A., Jonas, K., and Forbes, A. B. 2017. Modelling uncertainty associated with comparative coordinate measurement through analysis of variance techniques. *Proc. euspen*, Hannover, Germany, Jun.

Rivas Santos, V. M., Thompson, A., Sims-Waterhouse, D., Maskery, I., Woolliams, P., and Leach, R. K. 2019. Design and characterisation of an additive manufacturing benchmarking artefact following a design-for-metrology approach. *Addit. Manuf.* 32:100964.

Savio, E., De Chiffre, L., and Schmitt, R. 2007. Metrology of freeform shaped parts. *Ann. CIRP* 56:810–835.

Schwenke, H., Knapp, W., Haitjema, H., Weckenmann, A., Schmitt, R., and Delbressine, F. 2008. Geometric error measurement and compensation for machines – An update. *Ann. CIRP* 57:660–675.

Sims-Waterhouse, D., Piano, S., and Leach, R. K. 2017. Verification of micro-scale photogrammetry for smooth three-dimensional object measurement. *Meas. Sci. Technol.* 28:055010.

Sims-Waterhouse, D., Isa, M. A., Piano, S., and Leach, R. K. 2020. Uncertainty model for a traceable stereo-photogrammetry system. *Prec. Eng.* 63:1–9.

Stavroulakis, P. I., and Leach, R. K. 2016. Review of post-process optical form metrology for industrial-grade metal additive manufactured parts. *Rev. Sci. Instrum.* 87:041101-1-15.

Stavroulakis, P., Sims-Waterhouse, D., Piano, S., and Leach, R. K. 2017. A flexible decoupled camera and projector fringe projection system using inertial sensors. *Opt. Eng.* 56:104106.

Stavroulakis, P., Chen, S., Derlome, C., Bointon, P., Tzimiropoulos, G., and Leach, R. K. 2019. Rapid calibration tracking of extrinsic projector parameters in fringe projection using machine learning. *Opt. Lasers Eng.* 114:7–14.

Thalmann, R., Meli, F., and Küng, A. 2016. State of the art of tactile micro coordinate metrology. *Appl. Sci.* 6(5):150.

VDI/VDE 2634 part 3. 2011. *Optical 3D-Measuring Systems Multiple View Systems Based on Area Scanning*. Düsseldorf, Germany: VDI/VDE.

Weckenmann, A., Estler, T., Peggs, G., and McMurtry, D. 2004. Probing systems in dimensional metrology. *Ann. CIRP* 53:657–684.

Yang, Q., Butler, C., and Baird, P. 1996. Error compensation of touch trigger probes. *Measurement* 18:47–57.

Yen, H-N., Tsai, D-M., and Yang, J-Y. 2006. Full-field 3-D measurement of solder pastes using LCD-based phase shifting techniques. *IEEE Trans. Electron. Packag. Manuf.* 29:50–57.

Zhang, S. 2010. Recent progresses on real-time 3D shape measurement using digital fringe projection techniques. *Opt. Lasers Eng.* 48:149–158.

Zhang, Z. H. 2012. Review of single-shot 3D shape measurement by phase calculation-based fringe projection techniques. *Opt. Lasers Eng.* 50:1097–1106.

Zhang, S. 2016. *High-Speed 3D Imaging with Digital Fringe Projection Techniques*. Boca Raton, FL: CRC Press.

Zhang, B., Davies, A., Ziegert, J., and Evans, C. 2017. Application of instrument transfer function to a fringe projection system for measuring rough surfaces. *Proc. SPIE* 10373:103730S.

11

Post-Process Surface Metrology

Nicola Senin and François Blateyron

CONTENTS

11.1 Introduction

This chapter focuses on off-line measurement of the surface topography of metal additive manufactured (AM) products. Together with measurement of external shape (see Chapter 10) and internal feature measurement (see Chapter 12), measurement of surface topography is fundamental for quality control of metal AM products, as well as for providing feedback for AM process optimisation (Leach et al. 2019a).

Metal AM surfaces may be inspected in their 'as-built' state, that is, as they appear when they come out of the additive fabrication process. Alternatively, metal AM surfaces may be inspected after being subjected to further finishing operations, for example by material removal or local plastic deformation. The inspection of an AM surface after finishing is a more common occurrence if the surfaces are required to conform to strict design specifications (Gibson et al. 2009).

In this chapter, the term *post process* means that the measurements are performed 'after' the additive fabrication process, regardless of whether the surfaces are considered in their as-built state or they have been subjected to further finishing operations. Surface measurement and characterisation are intended in this chapter as off-line processes, as they are performed on the surface when the AM process is terminated, not whilst the part is being fabricated. In-process measurement is covered in Chapter 13.

The off-line measurement and characterisation of as-built AM surfaces plays a couple of important roles. Firstly, it may be helpful to better understand the AM process itself. Secondly, it may be useful to assess the amount of modification that the as-built surfaces may require for finishing. The investigation of AM surfaces after the application of finishing operations is useful to assess whether the surfaces conform to the specifications that were originally requested.

11.1.1 The Nature of Metal AM Surfaces

Metal AM surfaces in their as-built states are often very irregular, feature topographical content at multiple scales and may exhibit local high aspect-ratio features as well as heterogeneity of material properties. Example images taken from the same metal AM surface at different magnifications are shown in Figure 11.1. These images give a qualitative idea of the heterogeneity and complexity of surface features that are encountered when measuring AM surfaces.

The topographic complexity of AM surfaces is often accompanied by an equivalently complex underlying form. Intricate geometries can be created by AM, such as freeform

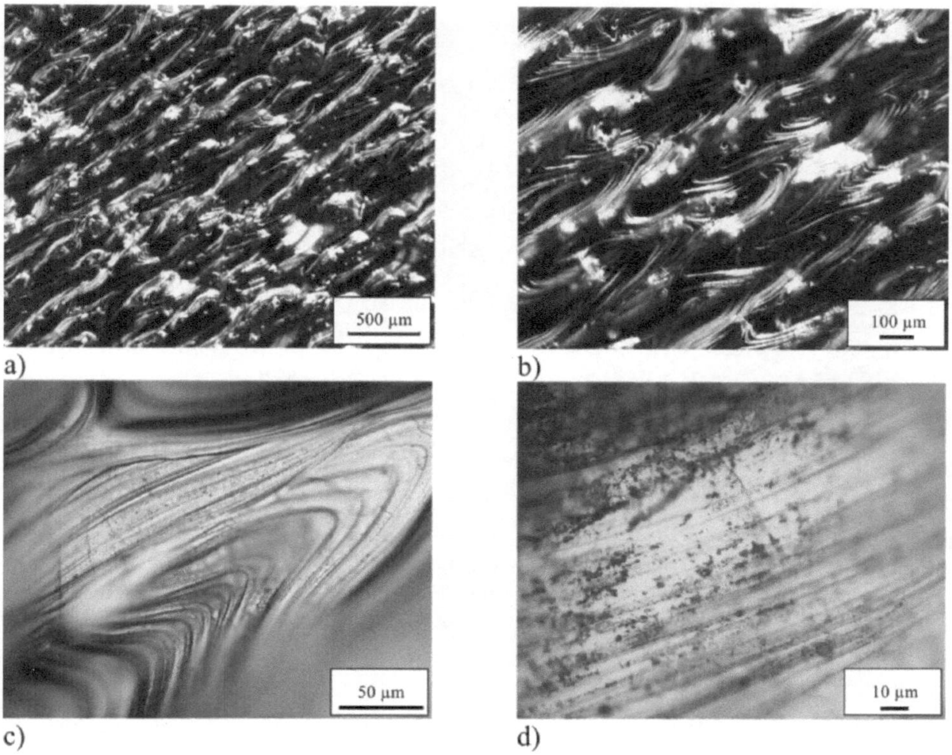

FIGURE 11.1
Topography of a metal AM surface observed at multiple scales: top surface of the last layer of a Ti4Al4V part fabricated by laser powder bed fusion (L-PBF) and imaged via a digital optical microscope at different magnifications: (a) 5×, (b) 10×, (c) 50×, (d) 100×.

surfaces, hollow features, and lattice or irregular, trabecular-like structures. The sudden step-up in terms of the geometrical complexity of the parts that can be produced is also creating significant challenges in the measurement of surfaces, as the conventional paradigm of surface texture seen as small-scale, uniformly distributed irregularities over an otherwise regular and smooth substrate, is often lost. New analysis and processing methods are required to turn surface measurement into a meaningful and useful assessment of surface states in metal AM parts (for example, see the surfaces in Figure 11.2).

11.2 Definitions and Standards

11.2.1 Surface Metrology Terms and Definitions

The following terms are here only given a summary description. Please refer to the following sections of this chapter for further explanations.

- *Surface topography* – the set of all the features present on a surface, treated as a continuum of spatial wavelengths. Surface topography is comprised of *surface form* and *surface texture.*

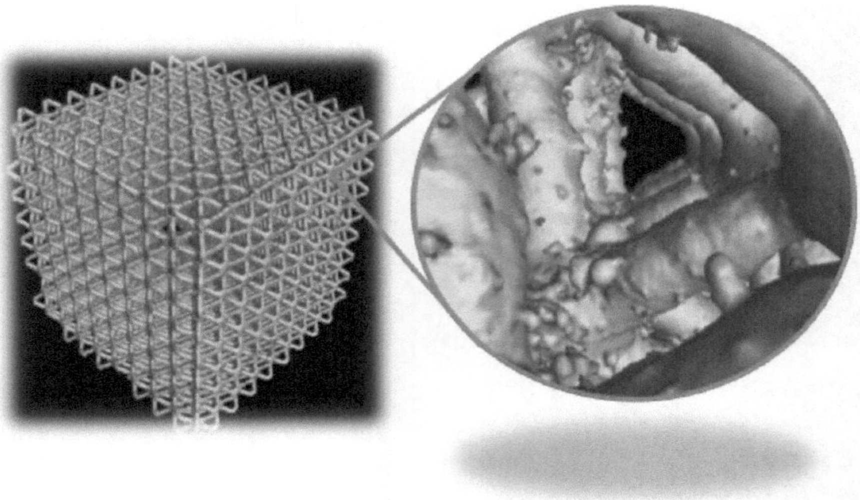

FIGURE 11.2
Three-dimensional reconstruction of the surface topography of a lattice-like structure of Ti6Al4V (10 × 10 × 10) mm part fabricated by L-PBF, obtained by X-ray computed tomography. The complex topography of the surface is overlaid onto an equivalently complex geometry, creating significant measurement challenges. (Courtesy of Dr Adam Thompson, University of Nottingham.)

- *Surface form* – the underlying shape of a part or fit to a measured surface within the region of surface that has been measured.
- *Surface texture* – the geometrical irregularities present on a surface. Surface texture does not include those geometrical irregularities contributing to the form or shape of the surface (Leach 2015). Surface texture is what remains after form and form error have been removed from surface topography. Simply put, surface texture is comprised of all those smaller-scale features that can be adequately characterised within the extents of the measured region, once any underlying, larger-scale component related to the shape of the part (i.e. form) has been removed.
- *Surface metrology* – the science and application of measurement and characterisation of surface texture.
- *Surface topography measurement system* – a measurement system dedicated to capturing surface topography information. Notice that in surface metrology, topography (form + texture) is usually measured, albeit the goal is the characterisation of texture.
- *Areal topography measurement system* – a surface topography measurement system that captures topography by areal sampling, resulting in a topographical image of a surface (ISO 25178-6 2010), which may be represented mathematically as a height function $z(x,y)$ of two independent variables (x,y). When sampled in a discrete space, the height function is often referred to as an *elevation map* or *height map*.
- *Profile measurement system* – a surface topography measurement system that captures topography by scanning the surface along an individual, straight line (a process referred to as *line-profiling* in ISO 25178-6, 2010), resulting in a one-dimensional height function $z(x)$.
- *Measured surface* – the result of a surface measurement process (from now onwards assumed as areal measurement). Often this term is more specifically referred to as

the mechanical surface, electromagnetic surface, etc. depending on the measurement technology.

- *Scale* – a recent redesign of the concept of spatial wavelengths, currently incorporated by the specification standard ISO 25178-2 (2012) for areal topography measurement and soon also by the new standard in multiple parts ISO 21920 for profiles (see Section 11.2.2 for further details). The concept of scale extends the concept of cut-off to include new tools, such as morphological filters and multi-scale or fractal analyses (Brown et al. 2018), and makes it possible to specify a range of scales or a scale limitation in the same way that bandwidth limitation with a cut-off can be specified (see Section 11.4.1).
- *Primary surface* – measured surface after application of an S-filter (see Section 11.4.1 and the specification standard in multiple parts ISO 16610, Section 11.2.2).
- *Filtration* – operation that isolates and extracts topographic components of a surface, based on their wavelength or scale. Often used to separate small-scale/short wavelength components from large-scale/long wavelength components on a primary surface. Filtration can be accomplished in multiple ways (see the multiple parts of the specification standard ISO 16610 in Section 11.2.2 and Section 11.4.1).
- *F-operator* – a mathematical operation that removes form from a topography, ultimately producing texture (ISO 25178-2 2012).
- *S,L filters* – terms used to refer to specific filtration operations. An S-filter removes small-scale/short-wavelength components (i.e. it is a low-pass filter in terms of spatial frequencies). An L-filter removes the large-scale/long-wavelength components (i.e. it is a high-pass filter in terms of spatial frequencies). S-filters with very small cut-offs (defined next) are often used to remove small-scale/high spatial frequency noise). L-filters with large cut-offs are often used to remove large-scale/long-wavelength components.
- *Cut-off* – the wavelength value that controls what components are kept and what are attenuated or removed by a filtration operation (superseded and generalised by the concept of nesting index).
- *Roughness* – component of surface texture that contains the shortest wavelengths, below the cut-off value. The term is mainly used for profiles, as a surface may exhibit roughness-like texture in one direction and waviness-like texture in the orthogonal direction. However, the term is sometimes used in colloquial form also to refer to areal topographic content comprised solely of the short wavelengths.
- *Waviness* – component of surface texture that contains the longest wavelengths, above the cut-off value. Again the term is mainly used for profiles but is colloquially used for areal topography data as well.
- *Nesting index* – a generalisation of the concept of cut-off wavelength suitable to be applied also to filters that do not operate directly with wavelengths (for example, for morphological filters, the nesting index is the size of the structuring element, see ISO 14406 [2010]). For linear filters, the nesting index is equivalent to the cut-off wavelength.
- *Scale-limited surface* – any surface whose topographic content has been reduced by application of one or multiple filtering operations, therefore limiting the range of scales/wavelengths represented within the surface (ISO 25178-2 2012).
- *SF-surface* – a scale-limited surface resulting from the application of an S-filter to a measured surface followed by application of an F-operator (form removal operator).

- *SL-surface* – a scale-limited surface resulting from the application of an L-filter (a filter which removes the largest-scale components, for example, waviness removal) to a SF-surface.

11.2.2 Relevant Specification Standards

The relevant specification standards for the measurement and characterisation of surfaces are categorised as profile standards or areal standards. A third set of standards encompasses specific types of filtration operators (further detail in Section 11.4.1) and subjects specific to optics and photonics applications (see Section 11.2.2.3).

11.2.2.1 Profile Standards

- *ISO 3274 (1996) – Nominal characteristics of contact (stylus) instruments*: Defines the main components of a contact stylus instrument (more details in Section 11.3.3.1).
- *ISO 4287 (1997) – Terms, definitions and surface texture parameters*: Main profile standard that defines vocabulary and gives parameter equations.
- *ISO 4288 (1996) – Rules and procedures for the assessment of surface texture*: This standard defines how parameters are configured, estimated and compared with tolerances. It defines the 16% and Max rules.
- *ISO 5436 – Material measures*: Standard in two parts with the first part dealing with physical material measures for calibration/verification, and the second part dealing with software measurement standards. Part 1 is being replaced by ISO 25178-70 (2014).
- *ISO 12085 (1996) – Motifs parameters*: Specific standard, mainly used in the automotive industry, which defines a segmentation method on profiles to characterise width and height of motifs.
- *ISO 12179 (2000) – Calibration of contact (stylus) instruments*: This standard explains the calibration procedure of a contact stylus instrument.
- *ISO 13565 – Surfaces having stratified functional properties*: Standard in three parts that defines a robust (double-Gaussian) filter to be used with stratified surfaces, a set of parameters (*Rk*) based on the Abbott-Firestone curve and a set of parameters (*Rpq*) based on the material probability curve.
- *ISO/DIS 21920 (2020) – Profile surface texture*: New set of three standards aimed at replacing all previous profile and areal characterisation standards. Currently under development by ISO Technical Committee 213 Working Group 16.

11.2.2.2 Areal Standards

- *ISO 25178 – Areal surface texture*: standard in multiple parts that defines terminology, measurement methods and parameters for use in areal topography measurement and characterisation (Table 11.1).

11.2.2.3 Other Standards

- *ISO 16610 – Filtration*: Standard series on filters for surfaces and roundness profiles, organised in multiple parts (Table 11.2).

TABLE 11.1

Parts of the Standard ISO 25178

1	2016	Indication of surface texture
2	2012	Terms, definitions and surface texture parameters (*)
3	2012	Specification operators
6	2010	Classification of methods for measuring surface texture
70	2014	Material measures
71	2017	Software measurement standards
72	2017	XML file format x3p
73	2019	Terms and definitions for surface defects on material measures
600	2019	Metrological characteristics for areal topography measuring methods
601	2010	Nominal characteristics of contact (stylus) instruments
602	2010	Nominal characteristics of non-contact (confocal chromatic probe) instruments
603	2013	Nominal characteristics of non-contact (phase-shifting interferometric microscopy) instruments
604	2013	Nominal characteristics of non-contact (coherence scanning interferometry) instruments
605	2014	Nominal characteristics of non-contact (point autofocus probe) instruments
606	2015	Nominal characteristics of non-contact (focus variation) instruments
607	2019	Nominal characteristics of non-contact (confocal microscopy) instruments
700	CD	Calibration, adjustment and verification of areal topography measuring instruments
701	2010	Calibration and measurement standards for contact (stylus) instruments

(*) this standard is under revision – CD: committee draft (not yet published)

TABLE 11.2

Parts of the Standard ISO 16610

1	2015	Filtration – Overview and basic concepts
20	2015	Profile linear filters – Basic concepts
21	2011	Profile Gaussian filter (replaces ISO 11562)
22	2015	Profile spline filter
28	2016	Profile end effects
29	2015	Profile spline filter
30	2015	Robust profile filters – Basic concepts
31	2016	Profile Gaussian regression filter
32	2009	Profile robust spline filter (TS)
40	2015	Profile morphological filters – Basic concepts
41	2015	Profile morphological disk and line segment filters
45	CD	Profile segmentation
49	2015	Profile morphological scale-space techniques
60	2015	Areal linear filters – Basic concepts
61	2015	Areal Gaussian filter
71	2014	Areal Gaussian regression filter
85	2013	Areal segmentation

- *ISO 10110-8 (2010) – Optics and photonics – Surface texture, roughness and waviness:* Specific standard for optics. Defines specification operators based on the power spectral density.

11.3 Surface Topography Measurement

11.3.1 System Architectures

11.3.1.1 Profile and Areal Topography Measurement Systems

Surface topography measurement instruments all share the common convention where the surface under inspection is imagined as aligned to an *x,y* plane (*x* and *y* are referred to as the lateral axes) and surface height values are acquired by probing the surface from above, along the *z* axis (referred to as the vertical axis). Profile measuring systems are only capable of sampling height points at locations that lay along a straight line (Figure 11.3a). Whilst in general, multiple coordinate systems may be considered (for example, the instrument coordinate system or the specimen coordinate system) in this chapter to simplify the illustration, the *x* axis is always defined as the axis along which the profile scan has taken place. The result of the measurement process is, therefore, a profile, that is, a series of height (*z*) values at specific *x* locations. On the contrary, areal measurement systems (Figure 11.3b) can sample height at multiple locations on the *x,y* plane. Also in this case, to simplify the illustration, a specific set of *x,y* axes will always be considered, as the one aligned to the areal sampling pattern.

In both the profile and the areal configurations, only one height value is captured for any given *x,y* (or *x*) position; thus, vertical walls, re-entrant surfaces or any other feature occluded by other features along the direction of probing cannot be captured. For this reason, profile datasets are often colloquially referred to as '1.5D' geometry and areal datasets as '2.5D' geometry, both missing the capability of representing undercuts required to become full two-dimensional (2D) and full three-dimensional (3D) geometries, respectively.

A partial workaround to the limitation of acquiring re-entrant surfaces is represented by five-axis measurement systems, as discussed in Section 11.3.1.2. Another solution is provided by X-ray computed tomography, discussed in Section 11.3.5.3 (and in more detail in Chapter 12).

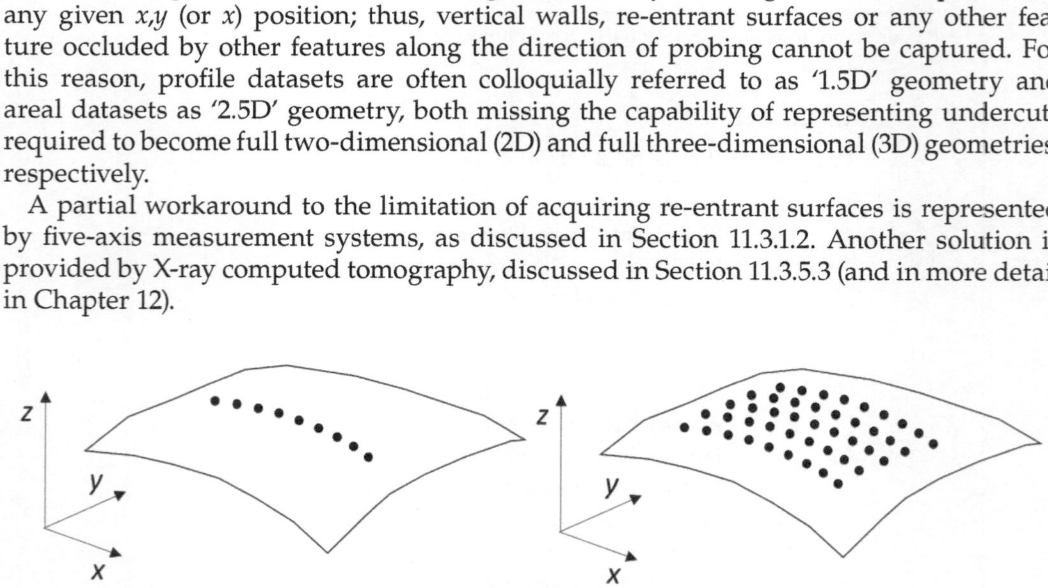

FIGURE 11.3
Profile and areal sampling of a surface topography: (a) profile sampling along a sampling line; (b) areal sampling (in the figure, height values are acquired at *x,y* positions of a regular grid).

11.3.1.2 Three-Axis And Five-Axis Topography Measurement Systems

Acquisitions as shown in Figure 11.3 can be accomplished by a three-axis measurement system: translations along the x and y axes allow any point on the surface to be reached, whilst translation along z allows the probe to approach the surface in order to reach the appropriate operating position. Instruments operating in 'imaging' modes (see Section 11.3.4 on non-contact, optical methods) are often capable of performing areal sampling without any physical translation in the x,y directions, in which case the x,y axes may still be needed to correctly place the probe on top of the investigated region or to perform multiple measurements in different regions of the surface (see Section 11.3.4). Similarly, the z axis may be needed to simply place the probe at the correct working distance from the surface, or it may be required to perform a vertical scanning as a necessary part of the height determination mechanism implemented by the probe (see Section 11.3.4). Recent advances in the design of areal topography measurement solutions have seen the introduction of motorised four- and five-axis architectures where one or two additional rotation axes are introduced, so that the relative orientation between the probe and the surface can be changed within the same setup. Four-axis and five-axis movements are typically achieved by tilting or rotating either the height-measurement probe or the specimen holder/stage, so that point-based height sampling can take place from multiple approach directions. Five-axis architectures are typically found in combination with non-contact, optical sensors (optical measurement technologies are illustrated in Section 11.3.4). An example five-axis system is reported in Helmli et al. (2011). The optical probes mounted on five-axis instruments are usually not single-point measurement devices, but rather areal topography measurement devices operating from a single probing direction (see Section 11.3.4). Five-axis instruments typically operate by acquiring datasets for each approaching direction, and then by aggregating them into a fully 3D model representing the surface (Figure 11.4). The aggregation is known as *stitching* and is generally performed via software, using the

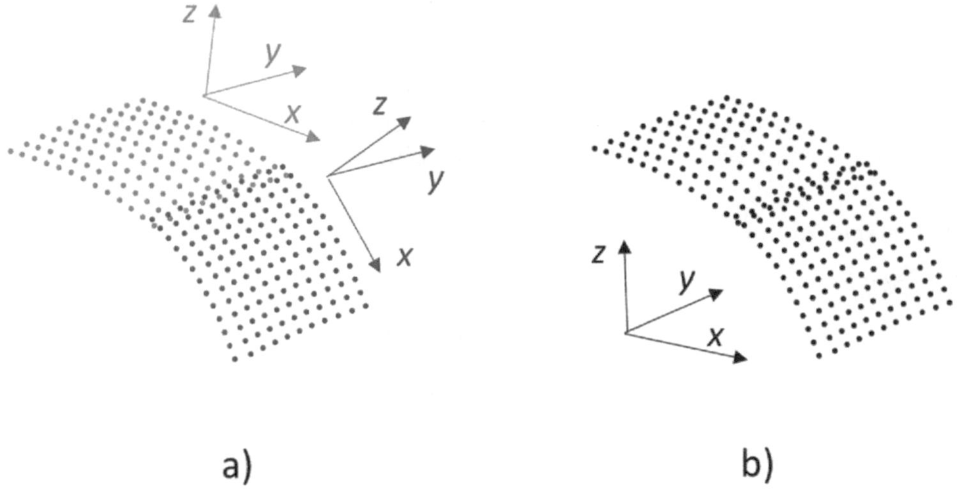

FIGURE 11.4
Five-axis surface measurement based on stitching multiple datasets acquired at different probing directions: (a) two datasets acquired independently and referred to their respective local probe coordinate systems; (b) the same datasets after co-localisation within the same global coordinate system. A different sampling density is typical in the overlapping region and is generally removed after the two datasets are merged into the final one by stitching.

co-ordinates of the controlled axes to coarsely identify the relative positions and orientations of the individual datasets (Wang et al. 2015). Alignment algorithms are used to refine the relative position before the datasets are merged into a single triangle mesh. Alignment is typically based on finding correspondences between overlapping regions of contiguous individual datasets (Figure 11.4).

Five-axis areal measurement systems can be used either to capture a larger range of slopes within a single surface or, more commonly, to measure multiple adjacent surfaces of different orientations within the same measurement setup. Capturing a larger range of slopes from the same surface is sometimes attempted, for example to measure the side walls of microelectronic devices on wafers (Ukraintsev and Banke 2012). However, it is generally difficult to accomplish, because the possibility of tilting the probe is severely limited by the presence of the surface itself, which limits manoeuvrability, as in most cases, the probe is already placed a very small distance from the surface in order to operate correctly.

11.3.1.3 Three- and Five-Axis Measurement on AM Surfaces

Most AM surfaces have very complex topographies. An example detail from a surface fabricated by laser powder bed fusion (L-PBF) is shown in Figure 11.5. The detail shows a particle that has remained attached to a slope after processing. The surface is a digital model obtained by an X-ray computed tomography measurement (more details in Section 11.3.5.3) and clearly shows an example of a re-entrant surface that would not be accessible from a single probing direction. Figure 11.5 is meant to give an idea of the challenges involved with sampling a complex topography characterised by high aspect-ratio features, deep recesses and re-entrant surfaces. The dots shown on the triangle mesh of are obtained by simulation, and indicate the regions that can be reached by three-axis areal topography measurement assuming a high-density probing takes place by approaching the surface from above. Whilst in theory a five-axis solution may increase the extent of coverage, in reality, there is generally little room for tilting the sample or the probe, given the proximity of the probe itself to the surface (as required by contact and by most optical measurement technologies, see Section 11.3.4). Moreover, for optical technologies, topographical complexity usually creates illumination challenges (for example, it is difficult to reach deep recesses), which further limit the possibilities of tilting.

11.3.2 Surface Topography Datasets

11.3.2.1 Height Maps

Surface topography information acquired by profile or areal topography measurement systems from a single probing direction is usually stored in digital form as height maps (also known as *elevation maps*). A height map (Figure 11.6a) consists of a matrix of heights where each height value (for example, z_{ij} in Figure 11.6a) is stored in a cell of a rectangular matrix (for example, the ith, j^{th} cell) mapped to a location on the x,y plane defined as $x_i = (i-1)dx$, $y_i = (j-1)dy$, where dx and dy represent constant lateral spacing values between x,y locations, and the coordinates x_1, y_1 are assumed as {0, 0}. More complex data structures may allow for storing additional information in correspondence of each position of the map, for example, colour (for optical measurement technologies) or quality of the measurement result (for example, local repeatability error, for instruments that are capable of estimating it).

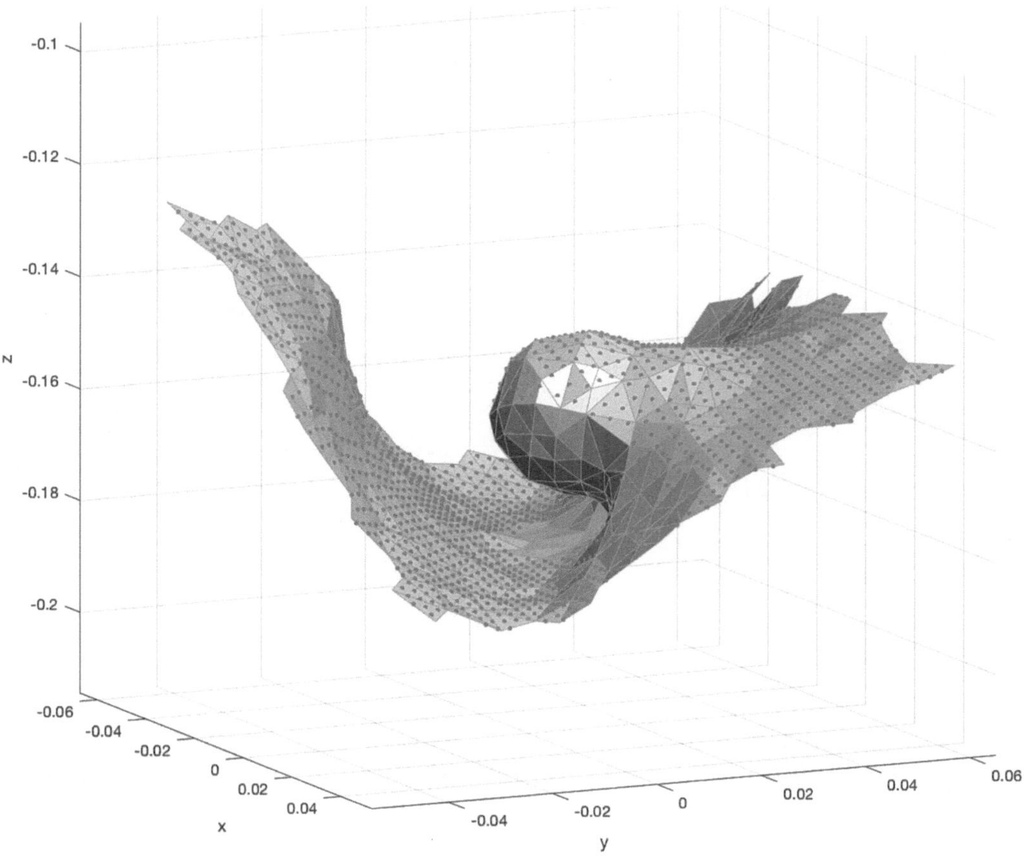

FIGURE 11.5
Simulation of topography coverage when the surface is observed through a three-axis areal topography measurement instrument capable of high-density sampling. The triangle mesh has been obtained by X-ray computed tomography measurement (size of the plotted volume approximately (0.1 × 0.1 × 0.1) mm). The dots represent points that are reachable when the surface is approached from above.

When height information is considered, a height map can also be expressed by a function of the type $z = z(x,y)$, which becomes $z = z(x)$ when measurement takes place along a profile (i.e. profile measuring systems – as illustrated in Section 11.3.1). The mathematical formulation used for height maps reiterates the concept that only one z value is possible at each x,y position (i.e. no vertical surfaces, undercuts or occluded features can be represented). Compare with the example topographies shown in Figure 11.1 to appreciate the issues of using height maps to encode the complex topographic formations typically encountered on AM surfaces.

The main advantage of using height maps is that mathematically they are equivalent to digital images (because of the matrix format) with added, known lateral scaling information (because rows and columns are mapped to real x,y coordinates) and known vertical scaling information (because the 'intensities' encoded in the pixels are actual height values measured by a distance measurement probe). Being equivalent to images, height maps can be analysed and manipulated by using of a wide variety of solutions (for example, filtering) which have already been developed for digital images. In the same way, profile data

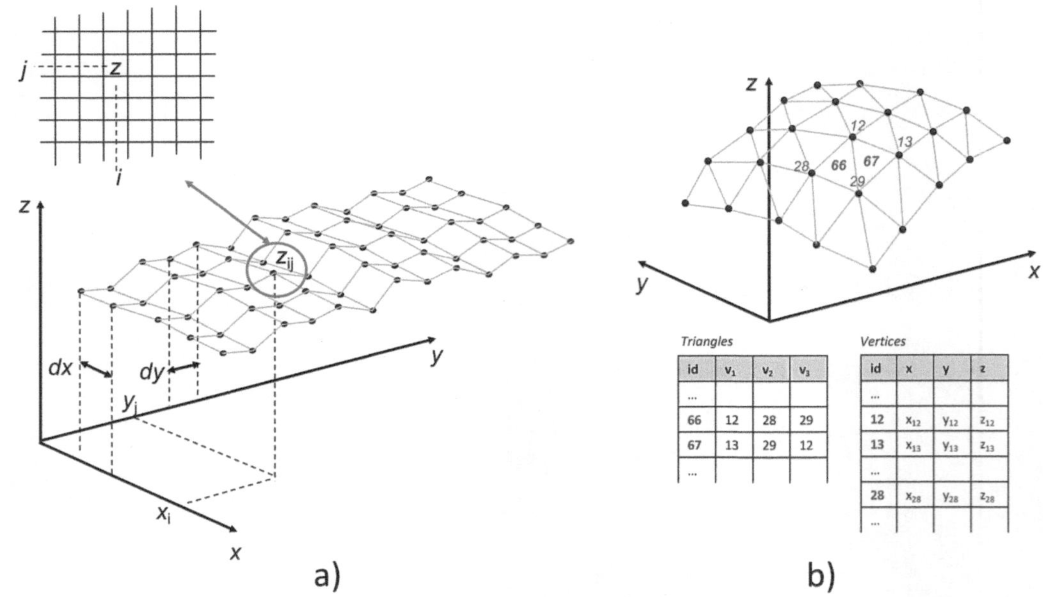

FIGURE 11.6
Typical data structures used to encode surface topography information: (a) height map; (b) triangle mesh.

is mathematically equivalent to a time series with uniform temporal spacing, and thus can be manipulated with numerous solutions developed in the domain of signal processing.

11.3.2.2 Triangle Meshes

Topography information can also be stored as a triangle mesh, i.e. a series of topologically connected triangles, where the vertices of the mesh are the measured data points themselves (typically, but not always, as sometimes the mesh is processed, for example simplified, via algorithmic generation of additional vertices). One of the most basic representations for a triangle mesh is shown in Figure 11.6b: vertex information is saved as a table, each row containing the identifier (*id*), and the *x,y,z* coordinates of an individual vertex. Triangles are stored as rows of a separate table, each row listing the *id* values of the vertices forming the triangle, as well as the triangle's own *id* (Figure 11.6b). More complex representations are possible, where additional information is stored to allow for more efficient navigation of the data structure (for example, triangles sharing the same vertex, edges connected to the current edge). Triangle meshes do not have the limitations of height maps (i.e. only one height [z] value per *x,y* location) and can be used to encode fully 3D geometric information. However, the use of triangle meshes to encode topography information has become popular only recently (despite having been used in dimensional metrology for a long time) with the introduction of fixe-axis measurement architectures (see Section 11.3.1.2) and X-ray computed tomography solutions (see Section 11.3.5.3). Thus, no consolidated methods exist yet on how to analyse and manipulate topography information defined as a triangle mesh in surface metrology applications, aside from recent research work (Pagani and Scott 2018, Pagani et al. 2019, Zanini et al. 2019a). Approaches to converting triangle meshes into height maps have been investigated, for example, based on sampling height information from a triangle mesh using a

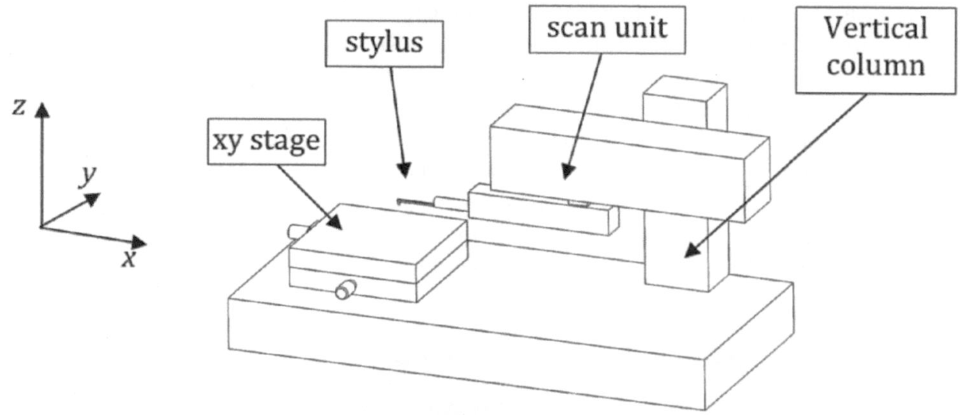

FIGURE 11.7
Typical architecture of a stylus instrument. The vertical column is used to place the stylus at different heights depending on the specimen. Profiles are scanned in the x direction using the scan unit, whist an optional x,y stage can be used to control sample placement. The y-axis of the x,y stage can be motorized to perform multiple parallel scans and thus turn the profilometer into a three-axis areal topography measurement instrument.

regular x,y grid and a constant z-probing direction, using approaches conceptually similar to ray-tracing (Thompson et al. 2017).

11.3.3 Contact Methods

11.3.3.1 Stylus Instruments – Basics

Stylus instruments (ISO 25178-601 2010) have been used for more than one hundred years to measure surface topography. The typical architecture of a stylus instrument is shown in Figure 11.7. Heights are recorded as the vertical displacements of a stylus sliding on top of a surface and following its reliefs. The stylus travels along a scan line defined by the drive unit reference axis (usually called the x-axis). The tip of the stylus is typically a diamond hemisphere attached to a cone, in turn mounted on a shaft that is connected to a pivot point. The other end of the shaft is interfaced to a detection system (usually an inductive, capacitive or optical sensor) that converts the height variations of the shaft into an electrical signal that is later digitised. The history, background and more technical details about stylus instruments can be found elsewhere (Leach 2014, Whitehouse 2010).

11.3.3.2 Stylus Instruments – Uses in AM

Due to the contact between the stylus tip and the surface, scans have to be slow to avoid damage to the surface or to the tip. Stylus-based measurement is, therefore, suitable for relatively flat surfaces, whilst the presence of obstacles, such as steps, deep holes or high aspect-ratio shapes, is likely to block the stylus and could break it. Unfortunately, topographic obstacles are quite common in AM surfaces; therefore, stylus instruments should be used with care (Leach et al. 2019, Townsend et al. 2016). Another issue with stylus measurement of AM surfaces stems from the realisation that not all topographies are suitable to be described by individual profile scans. For example, a machined surface with parallel machining marks can be adequately described by a scan laying orthogonally to the marks. Analogously, a highly random texture with no apparent dominant orientation of its reliefs (for example, a sandblasted surface) may be adequately described by a profile sampled in any direction. Conversely, the topography of an AM surface is often characterised by relevant information content in multiple directions, and thus is typically unsuitable to be adequately characterised by a single profile scan, regardless of scan direction (for example, see Figure 11.8). Additional care should, therefore, be exercised when measuring as-built AM surfaces with a stylus instrument, as the risk of failing to capture all the relevant information is high. A way to perform a more informative scan of an AM surface with a stylus instrument is to use a stylus-based areal sampling architecture. In this case, the instrument is equipped with an additional y-axis that allows execution of multiple parallel

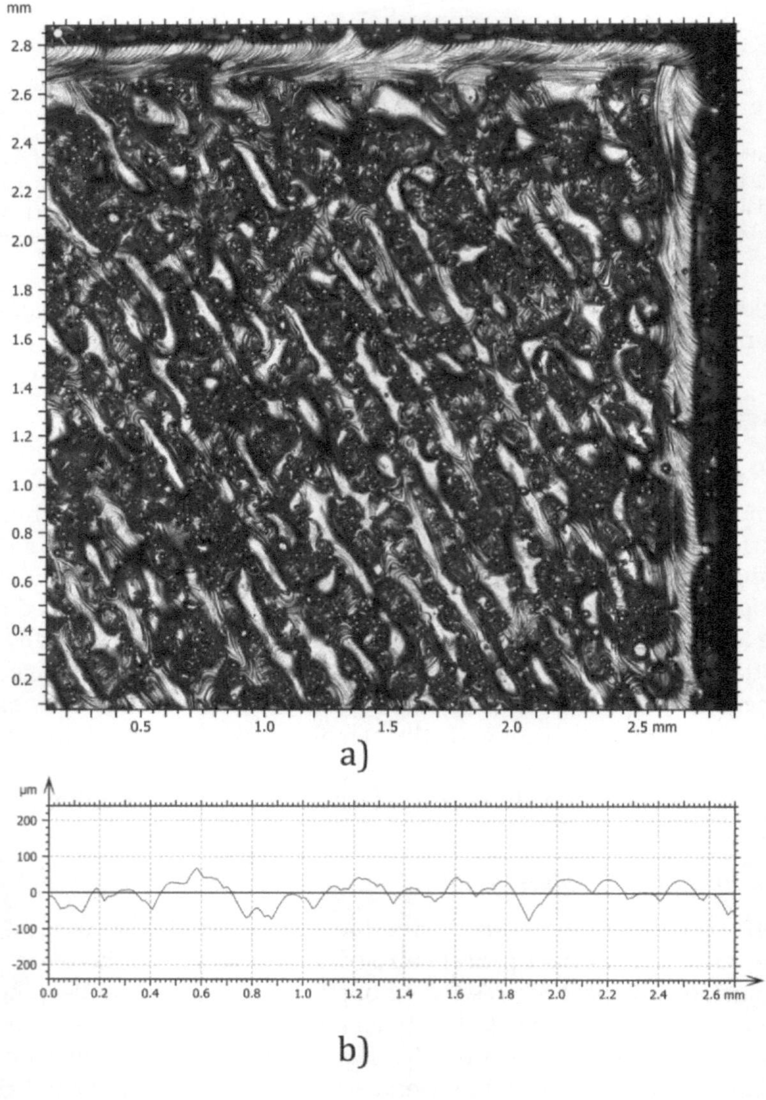

FIGURE 11.8
Example profile vs. areal measurement of a L-PBF surface: (a) optical image acquired by confocal microscopy (see Section 11.3.4.3); (b) profile scan form the same surface acquired with a stylus instrument. The profile conveys little information about the nature of the topography and does not allow full appreciation of the surface complexity. However, in some cases, it may still be good enough to obtain an estimate of surface roughness.

x-scans (a process referred to as *raster scanning*), ultimately resulting in an areal sampling compatible to be encoded as a height map. The downside of stylus-based raster scanning, in addition to the aforementioned issues related to stylus–surface contact, is that measurement in raster scanning mode requires a long time (frequently hours). Moreover, point samples are acquired sequentially, thus are potentially influenced by drift effects; so, for example, it should be ensured that the surface is thermally stable, which may often not be the case for AM surfaces, in particular if freshly fabricated.

11.3.4 Non-Contact, Optical Methods

In non-contact measurement, there is no mechanical contact between the probing sensor and the measured region of the surface. The most common type of non-contact measurement is optical measurement. All optical methods are based on the idea of exploiting visible light reflected from a surface as a means to reconstruct topography. Most methods use a light source to illuminate the surface in order to control which type of light is reflected from the surface. Different ideas have been explored on how to generate and analyse reflected light, and different optical principles can be exploited to extract information from reflected light and use it to reconstruct topography. In general, though, all optical technologies require a significant amount of software processing of raw optical data before the measurement result can be generated, and thus an instrument cannot be fully understood or characterised by the sole knowledge of what optical measurement method it implements; raw data processing internal to the instrument (and often proprietary) becomes a very important factor, more so than with stylus-based contact measurement. Because of the significant presence of the data processing component, all optical instruments have the possibility of returning a 'non-evaluated' flag corresponding to some of the sampled surface points, indicating that there was not enough optical information for the internal algorithms to obtain a reliable result. Such points are commonly referred to as *void* or *non-measured* points and are typically handled by Boolean masks associated to the final height maps produced by measurement (Senin and Blunt 2013, Medeossi et al. 2016).

The default configuration for an optical surface topography measurement system is the three-axis configuration, as previously illustrated in Section 11.3.1.2. A typical architecture for a three-axis optical system is illustrated in Figure 11.9. Fixe-axis configurations usually have two additional rotation axes connected to the x,y stage. The most established and widespread optical methods currently adopted to measure surface topography, and their uses on AM surfaces are described in the following.

11.3.4.1 Optical Methods for Topography Measurement – Basics

There are a number of optical methods for surface topography measurement, and they will not be covered in detail here. Leach (2011) presents detailed background on all the techniques which have been recently standardised: *phase-shifting interferometric microscopy* (ISO 25178-603 2013, de Groot 2011a), *coherence scanning interferometry* (ISO 25178-604 2013, de Groot 2011b), *confocal microscopy* (ISO 25178-607 2019, Artigas 2011), *confocal chromatic probe instruments* (ISO 25178-602 2010, Blateyron 2011), *focus variation instruments* (ISO 25178-606 2015, Helmli 2011) and *point autofocus probe instruments* (ISO 25178-605 2014, Miura and Nose 2011). Though most instruments typically belong to one of the above classes, a few commercial offerings have begun to appear embedding multiple optical modes within the same instrument body so that they can switch across methods and select the most suitable one for each specific test case. In the following sections, the discussion will focus on

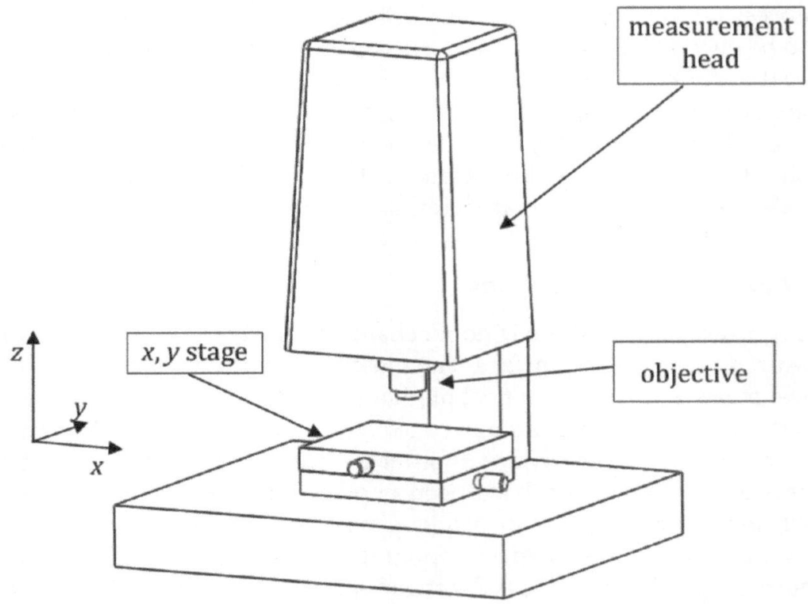

FIGURE 11.9
Typical architecture of a three-axis, optical surface topography measurement system. If the optical sensor is a single-point sensor (i.e. it is capable of capturing only one height point at a time) then the *x,y* stage is used to perform areal scanning and thus it is typically motorised. If the optical sensor is capable of simultaneously capturing a grid of height points (the most frequent case with current technologies) then the *x,y* stage can be used to acquire multiple images to be stitched at a later time and the *x,y* stage is either manual or motorised.

the use of such methods on AM surfaces and on the advantages and limitations of each method in the specific context of AM related applications.

11.3.4.2 Optical Measurement of Surface Topography – Uses in AM

Phase-shifting interferometric (PSI) microscopy is better suited for smooth, high precision surfaces, thus it is usually not suitable to the irregular topographies produced by additive technologies, in particular when analysed in their as-built states. Coherence scanning interferometry (CSI) (Gomez et al. 2017) can handle more irregular topographies and allows measurement of highly detailed topographic reconstructions, despite the presence of several non-measured points (for example, see Figure 11.10).

Non-measured points are typically found in correspondence to scarcely illuminated areas, or high aspect-ratio features, steep walls, and more in general, to any surface region not reflecting enough light back into the instrument detector. Naturally, such topographic formations are relatively common with AM surfaces, although this is an issue shared by all optical technologies when applied to AM surfaces. Instruments implementing CSI are significantly faster than a stylus instrument in acquiring areal topography data.

Confocal microscopy (CM) offers similar speeds to CSI and results in similar topography reconstructions. An areal topography measurement realised with a confocal microscope is shown in Figure 11.11. The confocal measurement was performed in approximately the same region as that with the CSI previously shown in Figure 11.10, with many non-measured points falling in similar places.

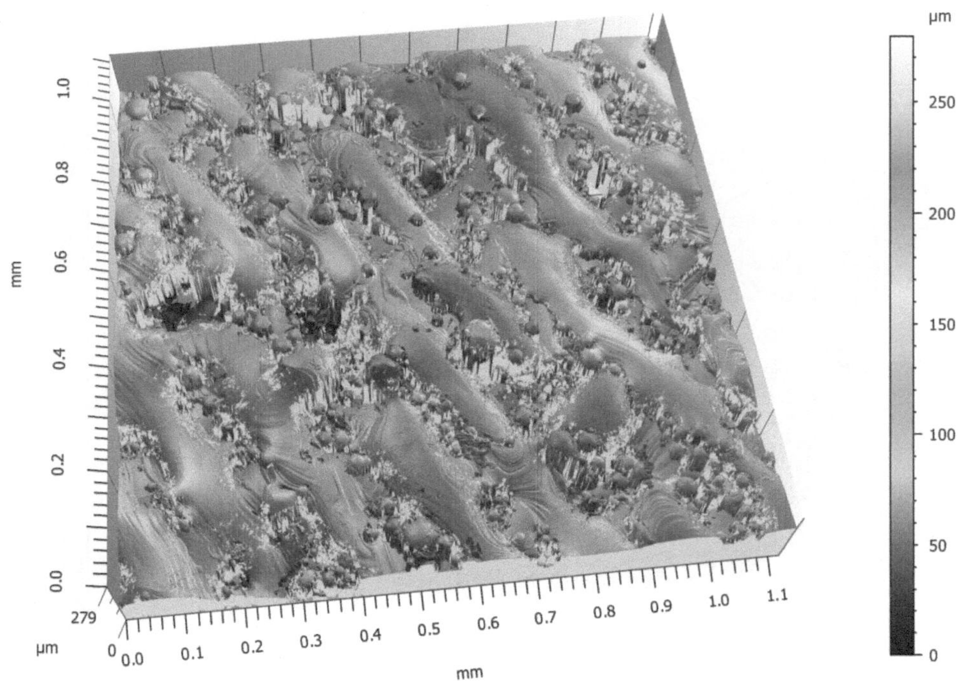

FIGURE 11.10
Example topography data obtained by coherence scanning interferometry at 20× (L-PBF top surface of TI6Al4V part).

Focus variation (FV) instruments offer similar measurement speeds to CSI and CM. As with the other optical methods, FV suffers when measuring scarcely illuminated areas typical of AM surfaces (for example, deep recesses, high slopes). Increasing illumination intensity may improve the measurement of dark regions but often also results in over-saturation of the already bright regions, thus creating additional non-measured points in different places (Newton et al. 2019a). An example FV measurement of an electron beam powder bed fusion (EB-PBF) surface is shown in Figure 11.12.

Confocal chromatic probe instruments are typically single-point measurement devices; therefore, in order to acquire areal topography data, they need to perform a raster scanning process, which makes them intrinsically slow compared to native areal devices, such as those based on CSI, CM or FV. Confocal chromatic probe instruments are relatively sensitive to local slopes, in particular on highly reflective, smooth surface patches often found in metal AM topographies, for example on powder bed fusion (PBF) melted tracks. Point autofocus probe instruments are also single-point measurement devices that need to perform raster scanning in order to acquire an areal topography. Point autofocus is more robust than chromatic confocal against high slopes and can approach most AM surfaces even when they are very irregular, with good lateral and vertical accuracy. Similar to the other scanning techniques, the main issue with point autofocus is related to the long measurement times, due to the intrinsically sequential nature of the measurement process. For AM surfaces, this implies that often, a compromise on sampling range and resolution must be made in order to optimise measurement time. Raster scanning also has the additional risk of misalignment between parallel scans, due to the mechanical nature of the lateral

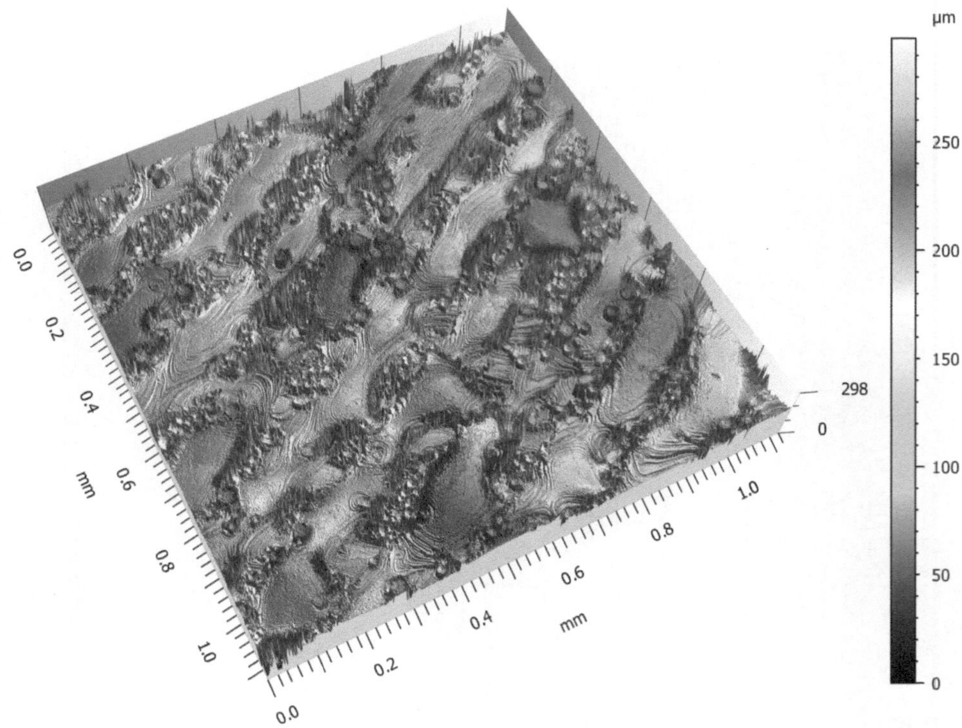

FIGURE 11.11
Example topography data obtained by confocal microscopy at 20× (L-PBF top surface of TI6Al4V part).

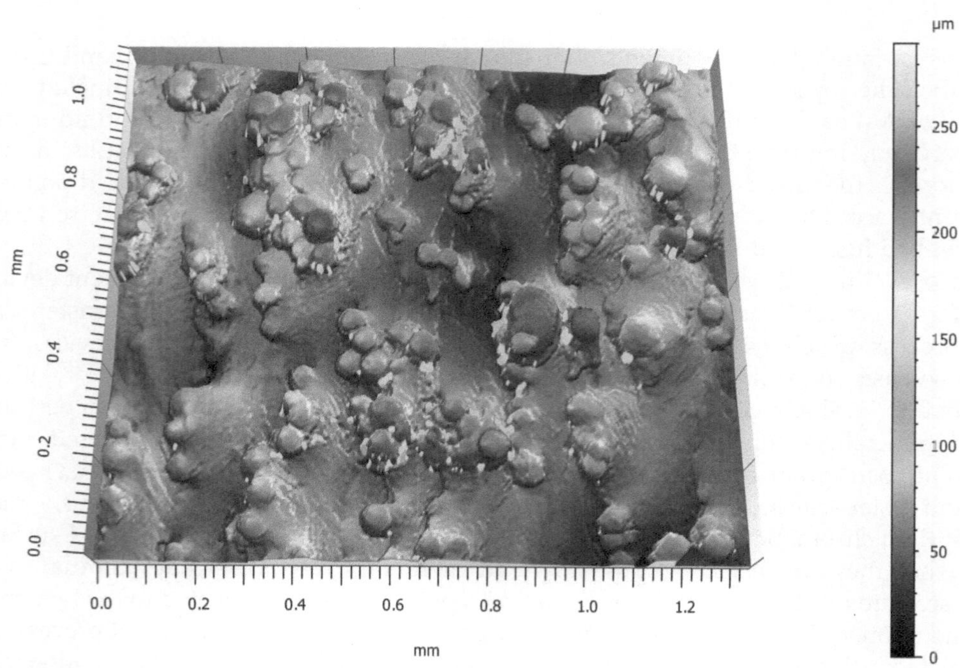

FIGURE 11.12
Example topography data obtained by focus variation microscopy at 20× (EB-PBF side surface of TI6Al4V part).

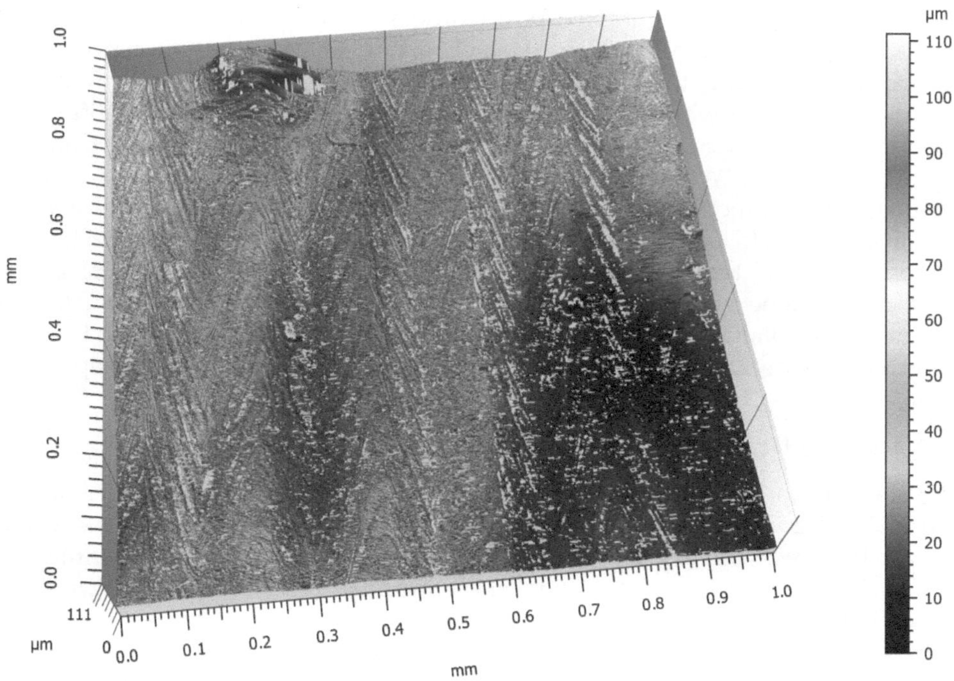

FIGURE 11.13
Example topography data obtained by point autofocus profilometry (EB-PBF top surface of Inconel part).

movement (Figure 11.13), a risk which is also shared with all the other technologies that perform single-point raster scanning, including stylus instruments.

11.3.5 Non-Contact, Non-Optical Methods

Amongst the non-optical methods that could potentially be used for non-contact measurement of surface topography, the most notable are scanning electron microscopy and X-ray computed tomography. These two methods are discussed in this section.

11.3.5.1 Scanning Electron Microscopy – Basics

Scanning electron microscopes (SEMs) usually produce greyscale, 2D digital images of a surface topography by focusing an electron beam on the surface and capturing the resulting electron emissions with a detector. Two main types of emissions are usually captured: *back-scattered electrons* (BSEs) and *secondary electrons* (SEs). In both cases, the local intensity of the emissions is correlated to the local slope of the topography relative to the incident angle of the colliding electron beam; thus, both BSE and SE images can be inspected to acquire information about topography. In addition, BSE provides an indication of how atomic properties of the surface may very across the image, thus allowing mapping of the presence of multiple materials/phases on the surface. Because the result of SEM is a 2D digital image, it is not possible to directly generate height maps or triangle meshes. However, at least two methods are have been developed to obtain 3D topography data from the analysis of SEM images (Mignot 2018): stereophotogrammetry and shape-from-shading. In stereophotogrammetry, multiple images are taken from slightly different viewpoints, usually by having a fixed position detector, but mounting the specimen on a tilting stage (De Chiffre et al. 2011, Marinello et al. 2008). Correspondences between images are algorithmically found, then their distance on the image plane is used to solve a triangulation problem which leads to reconstruction of point height and, ultimately, to the generation of the entire height map (as with optical photogrammetry, see Chapter 10). In shape-from-shading (Worthington and Hancock 2001), the colour of a surface point in a SEM image (more precisely, the grey level, or 'shade', in the greyscale intensity image) must be proportional to local slope (see discussion above). The specimen is held by a fixed-orientation stage, and two or four symmetrically mounted detectors are typically used (four to estimate slope in two directions), so that four shade values for each surface point are obtained. These values are used to estimate local slope for the point. When the slope of all points has been estimated, they can be combined to compute the height map.

11.3.5.2 Scanning Electron Microscopy – Uses in AM

SEM images have a much greater depth of field compared to optical images at similar magnifications, leading to wider in-focus regions, which is useful to visually inspect high-aspect ratio features, such as those on metal AM surfaces. Compare the SEM images of a L-PBF surface shown in Figure 11.14 with the optical images of the same surface, previously shown in Figure 11.1.

However, several issues must still be addressed to achieve reliable 3D topography reconstruction of AM surfaces from SEM images. The main challenge of using SEM stereophotogrammetry is the identification of correspondences between image points, necessary to solve the triangulation problem. For very irregular surfaces, such as many AM surfaces, identification is challenging, and no established solution exists. The other main issue is the sampling density resulting from stereophotogrammetry, which is typically low, as the topography between the points that have been solved by triangulation must be obtained by interpolation of the triangulated points. The main drawback of shape-from-shading is that the method is more reliable when applied to smooth topographies, which is rarely the case for AM surfaces. Shape-from-shading may be a viable option only at very high magnifications, where only very small, but smooth, details are imaged.

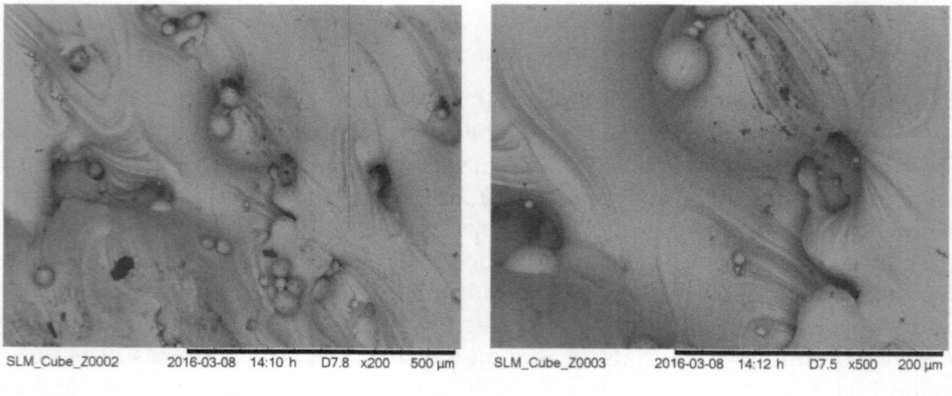

FIGURE 11.14
SEM imaging of PBF surface at different magnifications obtained in secondary electron mode; (a) 200×; (b) 500×.

11.3.5.3 X-Ray Computed Tomography – Basics

X-ray computed tomography (CT) will be covered in greater depth in Chapter 12. However, as X-ray CT can also be used to acquire surface topography information, some details are provided in this chapter. In X-ray CT, a series of radiographic projections of the imaged part are obtained at multiple angular orientations (see Section 12.2 and Figure 12.2). The projections are algorithmically recombined into a volumetric dataset (a discretised voxel volume) representing material density at different positions inside the part. The volumetric dataset can be analysed to search for changes in density, representing either the transition from one material to another or the transition from outside to inside the part. The latter is used to reconstruct the part surfaces and, if the volumetric dataset possesses sufficient resolution, the surface topography (see Section 12.4.3).

11.3.5.4 X-Ray Computed Tomography – Uses in AM

Because of the way a part surface is obtained, X-ray CT is the only measurement method that nowadays allows a full 3D reconstruction of the interfacial surface of an object to be obtained regardless of the irregularity of the surface. CT is capable of capturing re-entrant surfaces and even internal pores/cavities which would be otherwise inaccessible to any measurement technology approaching the surface from the outside. CT is the only method which gives access to deep recesses and otherwise difficult-to-reach surfaces of an AM component; this not only applies to generally planar AM surfaces that are challenging to measure because of high aspect ratio features (particles, deep recesses, etc.), but also to the surfaces of AM parts of complex geometry (lattice structures, hollow parts, etc.) which would not be accessible with any other measurement method (De Chiffre et al. 2014, du Plessis et al. 2018, Thompson et al. 2018a, Thompson et al. 2018b, Townsend et al. 2017, Townsend et al. 2018, Zanini et al. 2019b). An example topography of a L-PBF surface measured by CT is given in Figure 11.15. Because of the fully 3D nature of the topography captured by the method, triangle mesh representations are required to store the

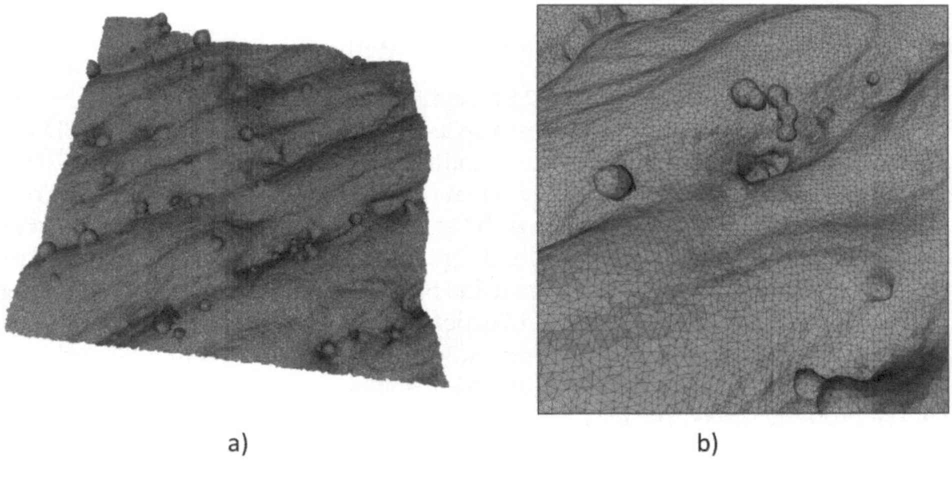

FIGURE 11.15
Three-dimensional surface topography of PBF surface, as obtained by X-ray CT measurement: (a) reconstructed region, approximately (0.75 × 0.75) mm; (b) detail showing particle clusters and cavities.

FIGURE 11.16
Variation of surface topography when the surface is generated by using increasing threshold values (from left to right) from the same X-ray CT volumetric dataset, covering approximately a (1.40 × 1.40) mm area (axes shown in voxel units). The assessment of the boundaries of internal cavities is also affected (as visible in cross-section).

measurement result (see Section 11.3.2.2), and conversion to height map is required before texture parameters can be computed.

The main challenge with CT measurement of surface topography is related to understanding what amount of uncertainty is associated firstly to the generation of the voxel volume, and secondly to the determination of a part surface starting from the volume. Surface determination errors are relevant in dimensional metrology, and even more so when the interest is in the analysis of geometric formations existing at smaller scales (for example, surface topography) that are comparable with the minimum observational scales currently possible with CT measurement of dense metals (Zanini and Carmignato 2017, Lifton et al. 2015). An example of the effects of surface determination in the topographic reconstruction of the same surface is given in Figure 11.16, where a simple change of threshold value for determining whether a voxel can be considered inside or outside of the part produces appreciable topographic differences. Surface determination in CT measurement is a significant open challenge, as discussed in Chapter 12.

11.3.6 Performance Comparison of Non-Contact Methods

In their current state of development, most non-contact technologies are capable of performing high-density surface sampling, and most also lead to detailed reconstruction of 3D surface topography. Full three-dimensionality (i.e. capability of capturing re-entrant surfaces) is only possible with CT, whilst optical technologies are mostly limited by the probing direction, and five-axis measurement cannot really cope with the unfavourable combination of short probe working distances and irregular surfaces, limiting the options for tilting (although, see recent developments in Zangl et al. 2018). The downside of CT (apart from the costs and numerous usability challenges – see Section 12.2 and Section 12.3) is that the lateral sampling resolutions compared to those of optical measurement can only be achieved in a limited number of cases, typically related to achievable volumetric resolutions, itself limited by specimen size and material properties (see Chapter 12). Despite being much easier and affordable to operate, optical technologies also have downsides, mainly because each optical technology reacts differently to reflected light from the same topographic features, leading to diversified metrological performance. A recent series of studies has shown that measurement artefacts are often the same order of magnitude as the surface features being investigated, leading to potentially

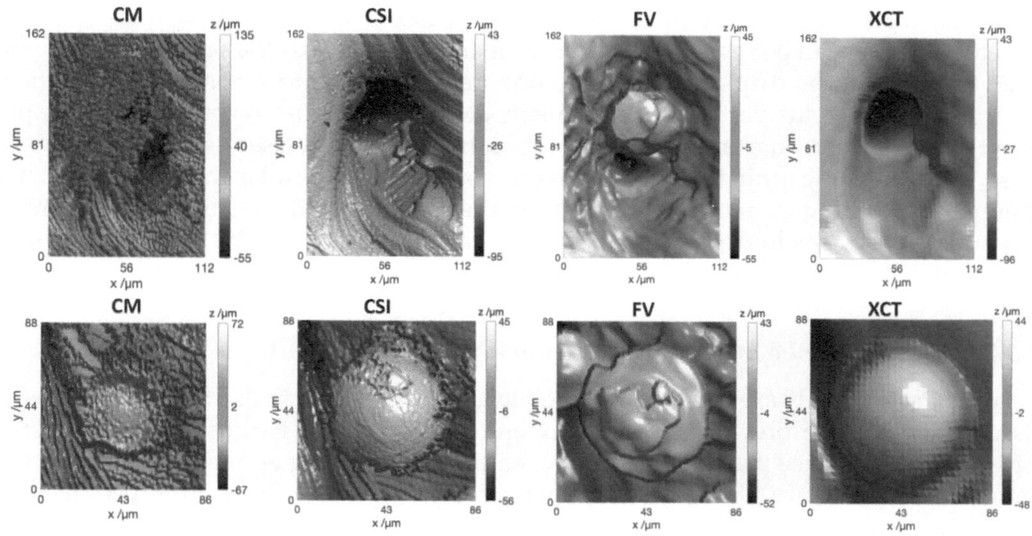

FIGURE 11.17
Topographic features of PBF surfaces and how they appear when measured by means of the main optical technologies. Above: recess. Below: particle. CM: confocal microscopy; CSI: coherence scanning interferometry; FV: focus variation; XCT: X-ray computed tomography.

significant misinterpretation of the topography being investigated (Thompson et al. 2017, Senin et al. 2017, Cabanettes et al. 2018). Examples involving typical topographic features of PBF surfaces and their rendition through different measurement technologies are shown in Figure 11.17. In the figure, both recesses and particles from the same PBF surface may result in diverse reconstructions; thus, geometric assessment of individual features should be handled with care. In terms of what technologies perform better with AM surfaces, there is no clear winner; the properties of the materials and the nature of the topography requires that each application is independently investigated. The significant influence of measurement error in the topographic reconstruction of AM surfaces measured with non-contact technologies suggests also that a thorough assessment of measurement uncertainty is necessary to better understand and compare measurement results. Uncertainty in measurement and characterisation of AM surfaces is discussed in Section 11.5.

11.3.7 Pseudo-Contact Methods

Pseudo-contact measurement methods are essentially embodied by scanning probe microscopy (SPM), a generalisation of the stylus instrument concept, but translated to very small scales. In SPM, a cantilever with a sharp tip (probe) travels across the surface following its reliefs; its vertical displacements are recorded in order to reconstruct topography. However, at very small scales, the probe–surface interaction cannot be simply described as mechanical contact because it is governed by attractive or repulsive forces operating at the atomic level (Leach 2014).

11.3.7.1 Scanning Probe Microscopy – Basics

SPMs are classified based on the nature of the interactions between the probe and the surface. In atomic force microscopy (AFM), the tip of the probe (typical height less than 20 µm, radius less than 10 nm) is placed so close to the surface that the attractive and repulsive interatomic forces can be sensed. As the probe is translated across the surface, interatomic forces and cantilever displacement are monitored and used to reconstruct the reliefs underneath. The stylus vertical displacements are measured and recorded, for example by optical means, through measuring the deflection angle of a laser beam incident onto the stylus shaft. In scanning tunnelling microscopy (STM), a similar mechanism is used where the tunnelling current created between the stylus tip and the surface (which must be conductive) is measured and kept constant by raising or lowering the tip so that it keeps a constant distance from the surface.

11.3.7.2 Scanning Probe Microscopy – Uses in AM

Because SPM methods are designed to operate at much smaller scales than all the methods described in the previous sections (except for SEM), SPM is thus primarily useful to inspect local surface properties (as opposed to characterising entire AM surfaces), for example microstructure. SPM measurement is also rarely possible on as-built AM surfaces, as the latter are too irregular to allow for the probe to safely reach the working distances, let alone scanning even over short lengths. The lateral field of view of a typical SPM instrument is usually limited to a few hundred micrometres, or even only tens of micrometres, as in the case of STMs. SPM is, therefore, complementary to, but cannot replace, optical, X-ray CT or stylus-based surface measurement.

11.4 Surface Topography Analysis

The conventional way to analyse topography data acquired by measurement is through the computation of a series of scalar indicators, each designed to capture a specific property of the topography in summarised form. These scalar indicators are known as *texture parameters* (see Section 11.2.1 for a list of related international standards). The most famous and widely adopted texture parameter is *Ra*, the arithmetical mean deviation of the roughness profile, a parameter that was developed with stylus-based measurement and fits well with the conceptual approach of characterising topography in terms of its randomness, sampled through a profile scan. Many other parameters have been introduced for profiles and later for areal topography data, as discussed in Section 11.4.3. Before a surface topography can be characterised through the computation of texture parameters, pre-processing is needed to remove topographic components whose characterisation is not required, as illustrated in Section 11.4.1.

11.4.1 Topography Data Pre-Processing

The most important pre-processing operations are a) form removal and levelling, b) cleaning and c) filtering. These operations exist for both profile and areal data. Areal pre-processing is discussed in the following; profile versions of the same operations should be seen as simplified implementations of the same concepts.

Form removal is the removal of the underlying overall shape of the surface, referred to as *form* (Forbes 2013). For example, if the topography has been measured from a cylindrical surface, a cylindrical shape must be removed so that the analysis can focus on the smaller scale irregularities. Form removal is typically implemented by fitting the measured topography to the ideal geometry believed to better represent the shape of the region under observation (in the previous example, a cylindrical surface) and then by removing said fitted geometry from the original topography datasets (by subtraction of local height values, assuming the height map formal representation – see Section 11.3.2.1), leading to a topography that only contains the residuals. Levelling is a special case of a form removal operation used to compensate for surface tilts due to specimen placement within the measurement instrument. Levelling is performed by fitting the topography to a plane, which is then subtracted from the dataset.

Cleaning (Senin and Blunt 2013) encompasses a) processing of non-measured points (also known as *voids*, for measurement solutions that produce them) (Pawlus et al. 2917) and b) processing of isolated measurement artefacts (spikes, pits, plateaus and other topographic formations generated by measurement error). Non-measured points do not need identification, as they are flagged as such by the instrument itself during measurement. Voids are typically handled by interpolation of non-void neighbours, the contribution of each neighbour usually being weighed by distance to the interpolated point. Measurement artefacts can be handled in a variety of ways: firstly, they must be recognised as such (differently from voids, artefacts appear as valid measured points, as with any other point in the topography). The identification of artefacts appearing as spikes (protruded or recessed) is often implemented by testing for local outliers (using moving windows and comparing each point with the average of its neighbours). Outlier correction is then implemented by replacement with predicted points obtained by using the same weighted interpolation methods adopted to remove voids (see for example, Le Goïc et al. 2012, Podulka et al. 2014).

Finally, filtering operations are typically applied to remove additional, unwanted topographic features (Seewig 2013). In the most common case, topography components existing at specific scales deemed not relevant to the investigation are assumed as modellable by periodic waves and are removed by mathematical operators working by convolution, or in the domain of spatial frequency. Convolution-/frequency-based filters are governed by the specification of a cut-off value, determining which wavelengths are transmitted and which are blocked/attenuated by the filter. As new filtering methods have been developed for the characterisation of surface topography, for example morphological filters (Jiang et al. 2011), as well as new surface decomposition methods in scale-space, for example, fractal analysis (Brown 2013), the concept of cut-off has been superseded by the more general concept of nesting index, which still plays the same role of defining scale-related thresholds to control the behaviour of the filter (Seewig 2013).

11.4.2 Topography Data Pre-Processing for AM Surfaces

The application of form removal operators to AM topographies does not present particularly notable elements of specificity with respect to other applications when the surfaces are either planar or characterised by elementary underlying geometry. However, when freeform surfaces are considered, an increasingly frequent scenario as AM technologies favour the realisation of complex geometries, the problem of form removal may become a challenging one. The characterisation of surface topography over freeform shapes is still an open challenge (Scott and Jiang 2014, Jiang and Scott 2020). In some cases, frequently with AM of metals, a complex underlying freeform detected by surface measurement may not be due to the surface nominal design but to part warping instead. In many such circumstances, form removal may be achieved by application of high-pass filters, circumventing

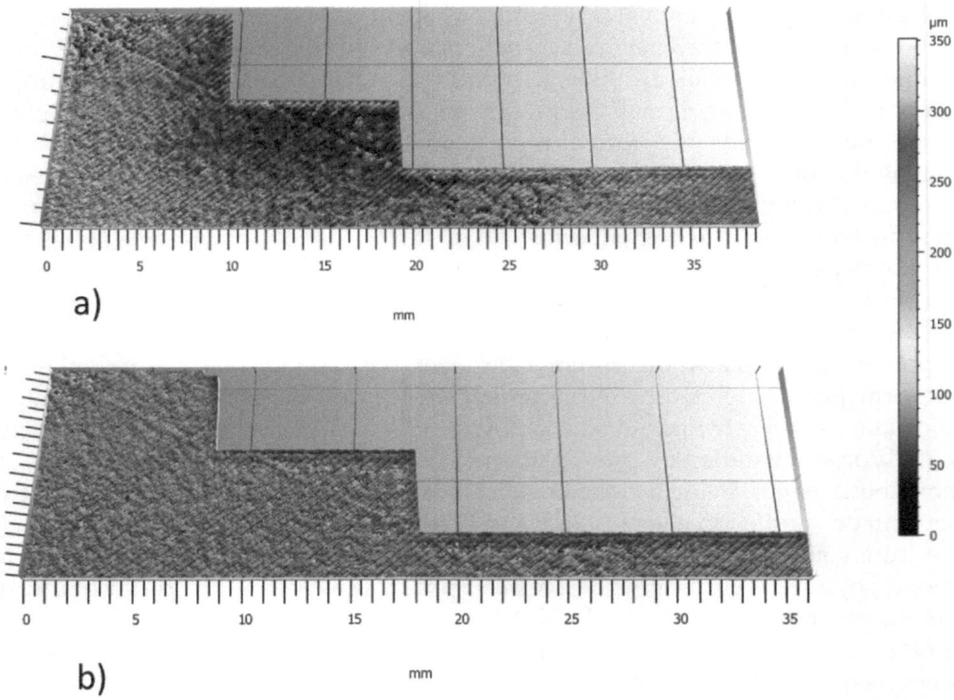

FIGURE 11.18
Example AM (L-PBF) surface where the underlying form error is due to warping; (a) original topography; (b) result of form removal by high-pass filtering.

entirely the need to find a suitable mathematical model to represent form, as needed by the form removal operator (Figure 11.18).

Removal of voids from optical measurements and removal of measurement artefacts from any measurement technology also create significant issues, which manifest themselves particularly with AM surfaces, given their complex nature. Again, no consolidated solution exists for the application of cleaning operations to such surfaces, and effects of void and artefact removal should be understood for each specific test case before relying on the interpretation of texture parameter values (Figure 11.19). As a general rule of thumb, the larger the region affected by the problem (void or artefact), the less reliable is any replacement by interpolation of valid neighbours.

For AM surfaces, filtering may be driven by two primary needs: a) the characterisation of a surface functional performance, in which case the metrologist should operate according to an underlying understanding of how the surface is supposed to operate (i.e. at what scales topographic features contribute to function); or b) the investigation of signature features left by the manufacturing process, in which case, knowledge of the expected sizes of signature features should be acquired. An example filtering operation to isolate signature features on a L-PBF surface is shown in Figure 11.20: a robust Gaussian filter (Seewig 2013) with cut-off 280 µm is applied to separate weld tracks (appearing as elongated protrusions), particles and spatter formations on one side (Figure 11.20b) from the larger-scale underlying waviness (Figure 11.20c). It should be noted that filtering is almost never expected to produce ideal separation, both because of the non-perfectly periodic nature of topography and because of the typical filter responses, which do not produce

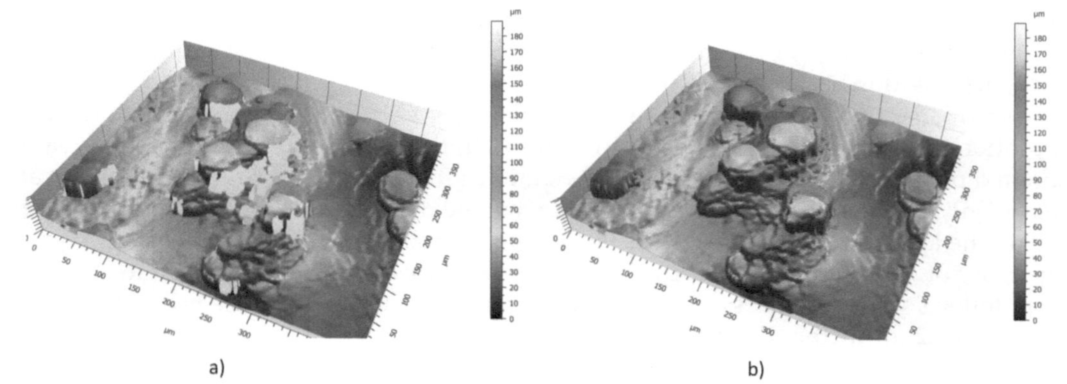

FIGURE 11.19
Effect of void filling on L-PBF topography measured by confocal microscopy (detail): (a) original region; (b) void-filled, by weighted interpolation of non-void neighbours.

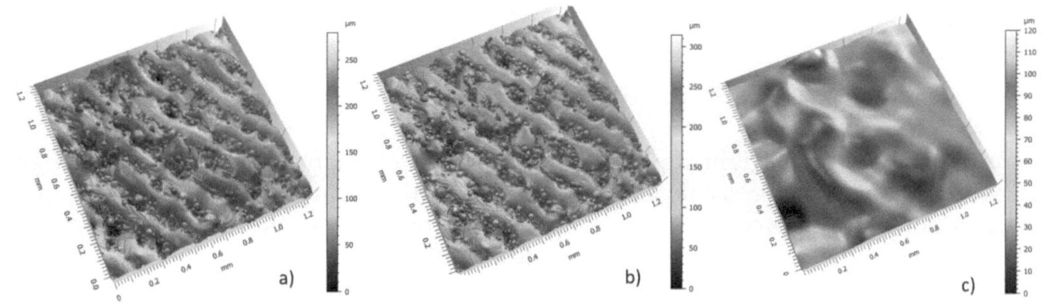

FIGURE 11.20
Robust Gaussian filtering of L-PBF surface with 280 μm cut-off: (a) original topography (CSI at 20×); (b) high-pass result (weld tracks, particles and spatter formations); (c) low-pass result (underlying large-scale waviness).

a clear cut in wavelength removal but rather introduce an attenuation effect that alters several spatial wavelengths before and after the cut-off by differing amounts. For such reasons, filtering typically alters the morphology of the features it tries to separate, as visible, for example, when comparing the weld tracks in the original topography (Figure 11.20a) to those appearing in the filtered one (Figure 11.20b).

Other filtering techniques that have been investigated more recently for component separation in surface topography include the use of wavelets (Jiang et al. 2000, Senin et al. 2015), Gabor filters and pyramid decompositions (Senin et al. 2015) and morphological filters (Lou et al. 2013).

11.4.3 Surface Texture Parameters

After a surface topography has been pre-processed, the conventional approach to topography characterisation sees the computation of quantitative indicators known as *texture parameters*, which summarise widespread topographic properties through a series of scalar values. There are many parameters to choose from, dedicated to the assessment of profiles and to the assessment of areal topography. A recent, comprehensive overview for areal parameters is provided in (Blateyron 2013a, Blateyron 2013b), whilst further information for profile and areal parameters can be found in the international standards, previously discussed in Section 11.2.2 and in Leach (2014).

From this section onwards, this chapter will target areal parameters, with the notable exception of the profile parameter *Ra*, which is still the most commonly and universally adopted parameter to evaluate surface topography (Townsend et al. 2016, Todhunter et al. 2017). Only the most widely used parameters will be discussed, with considerations specific to the application to AM surfaces. As pointed out in Section 11.2.2, there is an incoming series of new standards for profile parameters; thus, the reader is advised to consider what follows with awareness that some terminology will change in the near future (see Section 11.2.2 for further details).

11.4.3.1 The Arithmetical Mean Deviation of the Roughness Profile – Ra

As stated in Section 11.4 and Section 11.4.3, by far the most widely known and adopted parameter is the *Ra*, from the standard ISO 4287 (1997). The *Ra* parameter is meant to be computed on a profile (originally measured by a stylus instrument, nowadays also extracted from an areal topography dataset). The definition of the parameter is the following:

$$Ra = \frac{1}{l}\int_0^l |z(x)|\,dx, \tag{11.1}$$

where $z(x)$ is the profile height at position x along the profile, and l is the sampling length over which the parameter is calculated. Note that whilst Equation (11.1) gives a continuous definition of *Ra*, in practice, it (and every other parameter) would always be calculated discretely due to the finite sampling of any measuring instrument.

The *Ra* parameter describes the mean distance of any profile height point from a mean line which has been placed at zero-height and thus provides a general indication of vertical scatter across the profile mean. It is important to point out that no value of the *Ra* parameter (or of any other texture parameter for that matter) bears any meaning if it is not reported with an indication of what filtering operations were performed on the surface. As illustrated previously (for example, see Figure 11.20), filtering significantly modifies the topography; thus, any texture parameter will capture completely different aspects of the same surface depending on what filtering (and other pre-processing) operations were performed on it. The *Ra* parameter is by far the most widely adopted texture parameter even when specifically considering AM applications (Townsend et al. 2017), and there is sometimes confusion around its interpretation, in particular when filtering operations are not documented. Regardless, it is important to highlight the limited descriptive power of *Ra*. Firstly, it has already been shown in this chapter that an individual profile bears little representativeness of the complexity of the topography from which it was (see Figure 11.8). Secondly, the *Ra* parameter only quantifies the dispersion of the height values with respect to a mean line irrespective of where such height values are located along the profile. These issues result in many visually dissimilar AM surfaces returning very similar *Ra* values, thus highlighting the limited descriptive power of the parameter.

11.4.3.2 Areal Height Parameters

Areal height parameters are a family of ISO 25178-2 (2012) parameters dedicated to describing the probability distribution formed by all the height values which have been collected from an areal surface measurement. A scale-limited surface is chosen as the starting point (i.e. a surface which has been subjected to levelling/form removal and filtering), and topography height values are assumed as measured from a zero-valued mean plane acting as reference. The texture parameters discussed in this section are known as *Sa* (arithmetical mean height), *Sq* (root mean square height), *Ssk* (skewness) and *Sku* (kurtosis). *Sa* is the areal equivalent of *Ra* (see Section 11.4.3.1) and analogously describes the mean dispersion of height values about the zero-height plane:

$$Sa = \frac{1}{A}\iint_A \left|z(x,y)\right| dxdy, \tag{11.2}$$

where $z(x,y)$ is the surface height at position x,y and A is the definition area over which the parameter is calculated.

The parameters *Sq*, *Ssk* and *Sku* are the second-, third- and fourth-order moments of the probability distribution of heights, respectively. *Ssk* and *Sku* are also normalised with respect to *Sq*:

$$Sq = \sqrt{\frac{1}{A}\iint_A z^2(x,y)dxdy}, \tag{11.3}$$

$$Ssk = \frac{1}{Sq^3}\frac{1}{A}\iint_A z^3(x,y)dxdy, \tag{11.4}$$

$$Sku = \frac{1}{Sq^4}\frac{1}{A}\iint_A z^4(x,y)dxdy. \tag{11.5}$$

The *Sq* parameter is similar to *Sa* in that it captures the standard deviation of the probability distribution of heights, thus measuring dispersion with respect to the zero-height plane. *Ssk* indicates whether the distribution of heights is symmetrical or skewed above or below the mean plane. *Sku* looks at the peakedness of the probability distribution of heights, compared to a Gaussian (i.e. whether the distribution is sharper/peakier or flatter than the typical bell shape of a Gaussian distribution). Other notable height parameters are the maximum peak height – *Sp*, the highest point on the topography; the maximum pit height – *Sv*, the lowest height point of the topography; and the maximum height – *Sz*, the vertical distance between the highest and the lowest point in the topography. The three parameters *Sp*, *Sv* and *Sz* describe the extremal values of the probability distribution of heights and can be significantly influenced by isolated peaks/valleys on a surface (for example, scratches or spatter particles).

11.4.3.3 Areal Height Parameters – Uses in AM

Height parameters capture only those aspects of an AM surface that are adequately captured through the shape of the probability distribution of heights. Consider the two EB-PBF surfaces shown in Figure 11.21, fabricated at different angles with respect to the build direction. The surfaces are shown together with their probability distributions of heights and related height parameters. Clearly, some aspects of the topographies shown in Figure 11.21 cannot be adequately represented by the height parameters, for example, the distribution, shape and number of particles. It may be argued that *Ssk* may provide

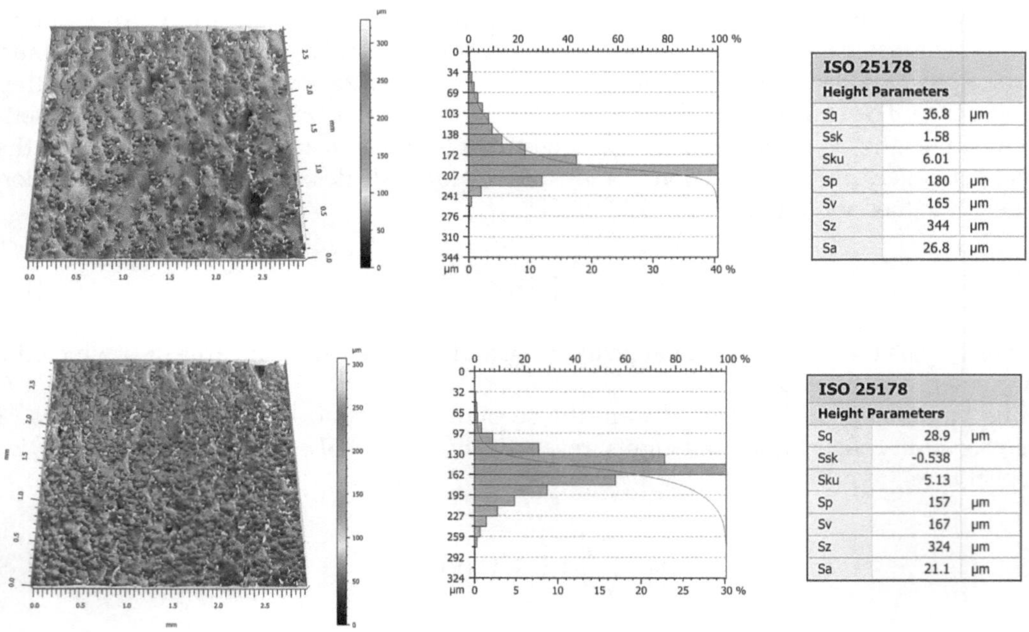

FIGURE 11.21
Two EB-PBF surfaces fabricated in Ti6Al4V at different angles with respect to the build orientation. The probability distributions of heights are shown as bar diagrams and the related height parameters are reported in tabular form: (top) 50° surface; (bottom) 100° surface.

indirect indication of whether particles are present on the surface or not, because the presence of a large amount of particles alters the height at which most of the surface will be found (expected value of the probability distribution shifted upwards). However, such an indication is disputable and needs to be confirmed by visual inspection, as there are many other reasons, related to different topographic configurations, for the expected height value to shift upwards.

11.4.3.4 The Areal Material Ratio Curve and Related Parameters

The areal material ratio curve is obtained by computing the cumulative probability distribution of heights. Examples have already been shown in Figure 11.21. The curve is named as such because it can be obtained by virtually slicing the topography from above with increasingly lower horizontal planes, whilst quantifying the amount of material encountered at each height into a value on the curve. Represented in this way, the areal material ratio curve is effective in describing all those functional behaviours of a surface that imply mechanical removal of material from the top (for example, abrasive wear phenomena) and fluid retention behaviour considering the valleys as basins, which also get increasingly worn. Following the same approach adopted to characterise the shape of the probability distribution of heights, a series of parameters has been defined in ISO 25178-2 (2012) to quantify the shape of the material ratio curve (equivalent parameters were defined in ISO 13565-2 [1996] for profiles). The parameters related to the material ratio curve for areal topography are all derived from the geometric construction highlighted in Figure 11.22. In the construction, the topography is imagined to be characterised by three regions that span its vertical range: a peak region occupied by the tallest

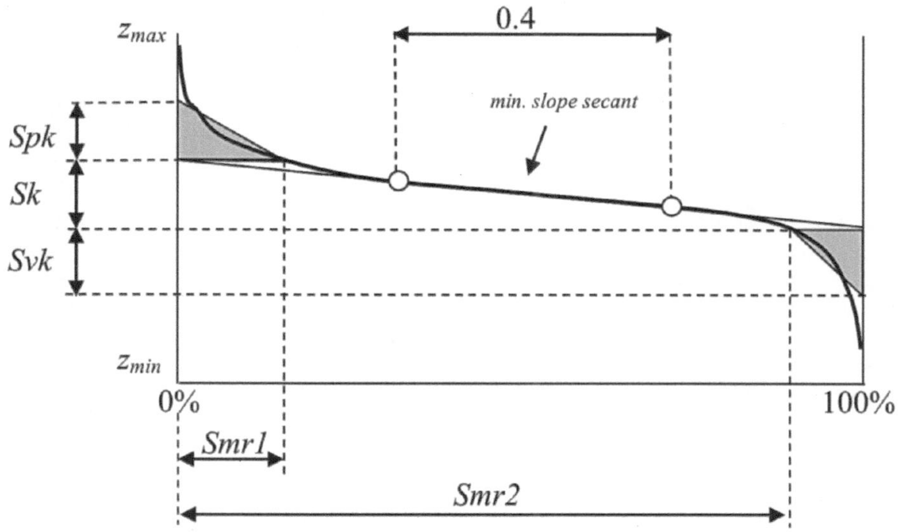

FIGURE 11.22
The areal material ratio curve and the geometric construction that defines its principal texture parameters. The curve is represented with height in the vertical axis and material ratio on the horizontal axis (as a percentage of the total projected area covered by the surface). The diagonal line crossing the curve is the minimum-slope secant passing through two curve points at 0.4 distance from each other.

peaks that will be erased during the early mechanical interactions; a core region, i.e. the bulk of the surface that will characterise most of its functional life; and a bottom region occupied by the deepest pits. The vertical extents of these three regions are captured by parameters: reduced peak height – *Spk*, core height – *Sk* and reduced valley depth – *Svk*. Further details, as well as additional parameters computed from the material ratio curve, can be found in Blateyron (2013a).

11.4.3.5 The Areal Material Ratio Curve and Related Parameters – Uses in AM

Because the shape of the material ratio curve can give information about the amount of material that will be increasingly encountered during a material removal or modification process proceeding downwards into the surface, texture parameters, such as *Spk*, *Sk* and *Svk*, can provide information useful when planning a surface finishing process for as-built AM surfaces (see Chapter 5). The parameters are also typically a better choice than *Sp*, *Sv* and *Sz* (see Section 11.4.3.2) to describe the height range of the surface because, due to their geometrical construction, they are less sensitive to extremes (individual spikes or pits) which are frequent in the measurement of AM surfaces, due to actual features and measurement error. The computation of the material ratio curve and related parameters for the two EB-PBF surfaces in Figure 11.21 is shown in Figure 11.23.

The same considerations as for height parameters can be made for material ratio parameters; the parameters are representative only of those topographic properties that are suitably represented by the shape of the curve on which they are computed, in this case, the material ratio curve. Again, typical topographic properties of AM surfaces, such as the number or density of particles or the number or spacing of the weld tracks on the surface, may not be well captured by the material ratio curve; thus they cannot be captured by material ratio parameters. Some indirect considerations can still be made: the presence

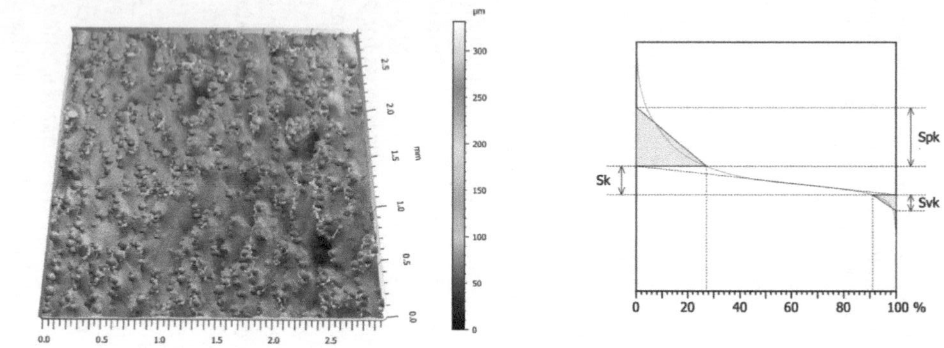

Parameters	Value	Unit
Sk	39.4	µm
Spk	81.4	µm
Svk	22.0	µm

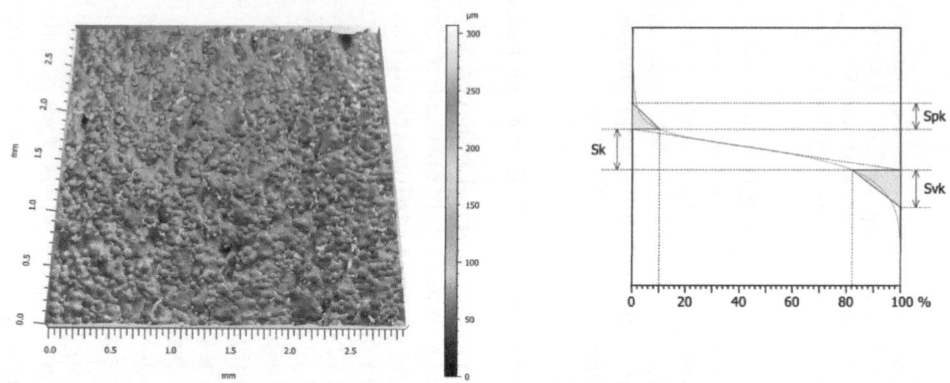

Parameters	Value	Unit
Sk	51.7	µm
Spk	32.6	µm
Svk	47.0	µm

FIGURE 11.23
The same two EB-PBF surfaces shown in Figure 11.21, this time with their material ratio curves and related parameters *Spk*, *Sk* and *Svk*: (top) 50° surface; (bottom) 100° surface.

of particles in the second surface in Figure 11.23 increases the expected value of the probability distribution of heights, influencing *Ssk* (as shown in Figure 11.21). The change is obviously reflected in the material ratio curve (which is the integration of the probability distribution), and consequently in the related parameter, for example *Spk* (Figure 11.23).

11.4.3.6 Spatial Parameters

The parameters belonging to the spatial family, also defined in the ISO 25178-2 (2012), capture spatial correlation between surface points and can be used to infer anisotropy, the presence of textural patterns with visible orientation. The computation of the spatial parameters is based on the areal correlation function; see Blateyron (2013a) for details. For example, the parameter surface autocorrelation length – *Sal* measures the minimum distance at which two surface points can be considered weakly correlated to each other. The parameter surface texture ratio – *Str* describes the ratio between distances measured in the direction of highest and lowest correlation, indicating the presence of anisotropic properties in the surface. Finally, surface texture direction – *Std* indicates the dominant texture direction for topographies that have been found to be consistently anisotropic. Periodicity can be inferred by visual inspection of the autocorrelation function.

11.4.3.7 Spatial Parameters – Uses in AM

Three main aspects influence anisotropy, periodicity and the presence of a dominant texture direction in AM surfaces: a) staircase effects due to layer stacking or traces of layer bonding (on vertical surfaces), visible for all AM processes, everywhere except on top layer surfaces where only one layer is observed; b) weld tracks and similar processing tracks, for AM processes where the layer is built by point-scanning (for example, electron beam and laser PBF); and c) particles/spatter formations (for PBF processes) that may remain attached to the surface, typically with random spatial dispersion, thus obstructing the view on any directionality or periodicity on the surface underneath (see Townsend et al. 2018, Leach et al. 2019a). Spatial parameters and the related analysis of the autocorrelation function cannot discriminate amongst these three aspects but can help detect whether their combination generates a net result of anisotropy, periodicity and/or dominant directionality in textural content. For example, in Figure 11.24 two EB-PBF surfaces are shown. The first one is a top surface with virtually no particles, no staircase effect (it is the surface of a single layer) and well defined parallel weld tracks, which confer a clearly anisotropic behaviour to the surface, notable periodicity and a dominant texture direction. These qualities are reflected in the autocorrelation function (reported in Figure 11.24) and confirmed by the values of the spatial texture parameters (autocorrelation and parameters were computed on a scale-limited surface at 280 μm cut-off). The second surface shown in Figure 11.24 belongs to the same PBF sample, but is vertical (i.e. parallel to the build direction). Though the bonding between layers should generate anisotropy, periodicity and textural orientation, the large number of particles present on top of the surface covers up most of the effects, albeit some slight traces of preferential orientation are still captured by *Str* and *Std* (also barely appreciable by zooming in on the central peak of the truncated autocorrelation function).

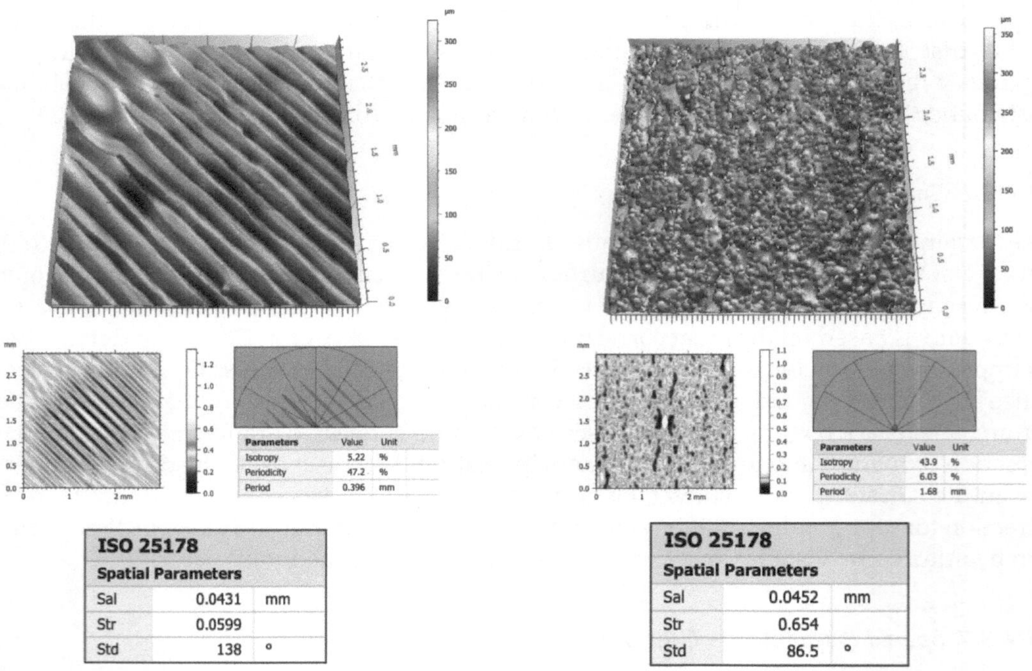

FIGURE 11.24
Two EB-PBF surfaces fabricated in Ti6Al4V at different angles with respect to the build orientation: (left) horizontal surface (i.e. top surface of a layer); (right) vertical surface. Below the surfaces are reported the autocorrelation function; the truncated areal correlation function (with estimates of isotropy, periodicity and period); and the spatial parameters computed on the truncated areal correlation function.

11.4.4 Topography Segmentation and Characterisation of Surface Features

The texture parameters illustrated in the previous sections are all referred to as *field parameters*, in that they capture and summarise properties that belong to the entire region that has been measured (referred to as *field*). As the examples have illustrated, field parameters cannot directly capture properties belonging to individual topographic formations of interest in the topography, so for example, no parameter amongst those that have been illustrated would be capable of capturing the number of geometric properties of the particles present on an AM surface.

The term *surface feature* is typically used to refer to a topographic formation of interest in the context of a specific application. In scenarios concerned with the investigation of AM surfaces, example features may be the aforementioned particles (in PBF processes), or weld tracks (again in PBF processes), or the crests/edges due to the staircase effect (in all AM processes). The characterisation of a surface feature implies a series of non-trivial steps, which ideally should be performed in automated form through algorithmic data processing: a) feature identification, i.e. the recognition of the presence of a feature in the topography under inspection; b) feature isolation, i.e. the delineation of the feature boundaries and the partitioning of the topography along such boundaries (a process typically referred to as *segmentation*) so that the feature can exist as an independent geometric entity; and c) feature characterisation, i.e. the description of the feature in terms of a set of shape, size and localisation attributes. Most methods of feature-based characterisation

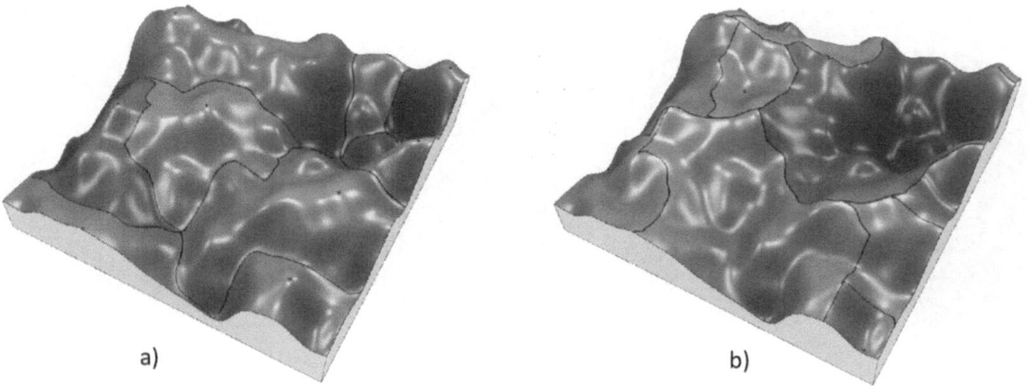

FIGURE 11.25
Example segmentations of the same test topography for the computation of ISO 25178-2 feature parameters: (a) segmentation into hills; (b) segmentation into dales. In the figure, smaller hills/dales have been merged with the largest ones using a method known as *Wolf pruning* (see Blateyron (2013b)).

of surface topography are in their infancy and are still being investigated at the research level. An exception is represented by the ISO 25178-2 (2012) feature parameters (Blateyron 2013b, Scott 2004), a set of parameters specialised to solely capture and characterise regions geometrically recognisable as hills (regions of points sharing the same local maximum) or dales (regions of points sharing the same local minimum). With ISO 25178-2 feature parameters, topographies can be segmented into a series of contiguous hills or dales (Figure 11.25), and the identified features can be described in terms of their main geometric properties (areal footprint, height, volume, etc.). The analysis method is the evolution of an older method for profiles known as *motif analysis* (ISO 12085 1996).

Segmentation, i.e. the act of partitioning a surface topography into regions (segments), is one of the core concepts of feature-based characterisation, and it is being actively studied, although only the watershed method of segmentation has been currently standardised (ISO 16610-85 2013). Several research efforts are in progress in the attempt to generalise the concept of feature-based characterisation so that features of different shapes and sizes (i.e. not necessarily only resembling hills or dales, but also more complex shapes) may be recognised and described through custom sets of properties. Various methods, original and derived from computer vision, are being explored for the identification and isolation of such features, including thresholding, edge detection, template matching, etc.; see Senin and Blunt (2013) for an overview. At the time of writing, though, a comprehensive theoretical foundation to support the general concept of feature-based characterisation is missing, and only application-specific approaches have been presented.

11.4.5 Topography Segmentation and Characterisation of Surface Features – Uses in AM

Typical examples of interesting topographic formations (i.e. surface features) in metal AM surfaces in their as-built state include particles, typical of PBF processes and frequently observed in the previous examples; weld tracks, typically visible on top surfaces of layers of PBF processes; and bonding seams and edge-like steps between consecutive layers, observable on the side surfaces of any AM part (Townsend et al. 2018, Leach et al. 2019).

FIGURE 11.26
Examples of EB-PBF surfaces at different orientations and particles identified via active contours. Regions approximately (3 × 3) mm

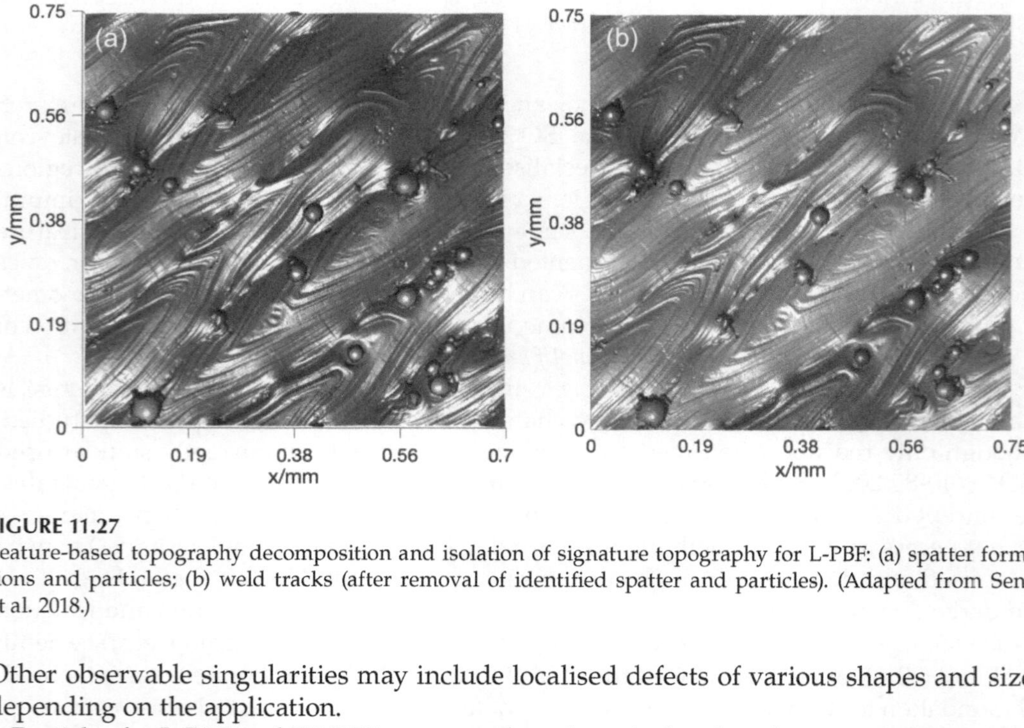

FIGURE 11.27
Feature-based topography decomposition and isolation of signature topography for L-PBF: (a) spatter formations and particles; (b) weld tracks (after removal of identified spatter and particles). (Adapted from Senin et al. 2018.)

Other observable singularities may include localised defects of various shapes and sizes depending on the application.

Particles for L-PBF and EB-PBF processes have been isolated with success using a number of methods, based on ISO 25178 part 2 segmentation (Lou et al. 2019), on the use of active contours (Zhu et al. 2009, Newton et al. 2019) or on the use of custom edge-detection methods (Senin et al. 2018). An example comparison between particles on EB-PBF surfaces fabricated at different orientations is shown in Figure 11.26, where particles have been automatically identified using active contours, so that the surfaces can be differentiated in terms of their number and distribution.

A more general approach to feature-based characterisation of PBF surfaces was recently presented in Senin et al. (2018), where not only particles but also weld tracks and ripple features were identified and characterised (Figure 11.27).

11.4.6 Characterisation of Full-3D Topographies

As discussed earlier in Section 11.3.1.2, although the majority of measurement technologies for surface topography are only capable of performing uni-directional probing of the surface and thus return height maps, a few options are starting to become available that allow fully 3D topography to be obtained, specifically, five-axis optical measurement, despite the limitations discussed in Section 11.3.1.3; and in particular X-ray CT, which in addition to capturing re-entrant surfaces, can also capture entirely occluded sub-surface cavities and porosity, as illustrated in Section 11.3.5.3. The most suitable data format to represent full-3D topographies is the triangle mesh, as discussed in Section 11.3.2. However, since topographies encoded as triangle meshes are no longer representable by $z = z(x,y)$ functions, the whole series of characterisation solutions discussed in Section 11.4.3, including texture parameters and feature-based methods, is no longer applicable. There is no consolidated solution available in terms of how to characterise full-3D topography as of yet; however, some initial research work can be consulted (Pagani et al. 2019, Townsend et al. 2019, Zanini et al. 2019a).

11.5 Uncertainty in Surface Topography Measurement and Characterisation

A significant research effort is being dedicated to the assessment of uncertainty in the measurement of surface topography. For example, in a series of papers, Giusca et al. have introduced solutions specifically dedicated to the calibration of the scales of areal surface topography measuring instruments (Giusca et al. 2012a, Giusca et al. 2012b, Giusca and Leach 2013a). A comprehensive overview of recent literature on the assessment of uncertainty in surface topography measurement can be found in the review paper by Leach et al. (2015). Significant research work is also being carried out to better understand individual topography measurement technologies, for example, focus variation (Giusca et al. 2014, Newton et al. 2019), confocal chromatic (Lu et al. 2019), point autofocus (Maculotti et al. 2019), coherence scanning interferometry (Su et al. 2018) and X-ray CT (Körner et al. 2019, Thompson et al. 2018b, Zanini et al. 2019a, Zanini et al. 2019b). Uncertainty related to X-ray CT measurement of surface topography is also discussed in Section 12.4.3.

Investigations specifically focusing on the assessment of measurement error in repeatability and reproducibility conditions for areal topography measurement of AM surfaces with optical technologies and X-ray CT can be found in (Thompson et al. 2017) and (Senin et al. 2017). This type of approach can be useful to make comparisons of measurements performed with different technologies on the same region of an AM surface.

Important results have also been recently obtained in terms of standardisation efforts: the standard ISO 25178 part 600 (2019) contains general information on a method using metrological characteristics to calibrate topography measurement instruments, and a future part 700 will have more detailed information. Good practice guides from the National Physical Laboratory provide information on how to use the ISO metrological characteristics framework to calibrate topography measurement systems based on optical technologies (coherence scanning interferometry, phase-shifting interferometry and imaging confocal – good practice guides 127 [Giusca and Leach 2013b] and 128 [Giusca and Leach 2013c] and stylus-based instruments – good practice guide 129 [Giusca and Leach 2013d].

Measurement uncertainty propagates through the entire surface topography characterisation procedure and affects its results, for example, the values of texture parameters. The data analysis phase itself may introduce additional error sources that should be considered in the final evaluation of uncertainty associated with the characterisation result. A recent review of the main issues can be found in Haitjema (2015), whilst general concepts related to the application of the metrological characteristics framework to the measurement and characterisation of surface topography can be found in recent work by Leach et al. (2019b).

11.6 Current Research and Future Look

The topographic complexity of the typical AM surface discourages the use of profile-based measurement and characterisation, unless the characterisation goals are very superficial, or there is a specific interest in labelling the surface in terms of traditional profile-based roughness parameters, such as *Ra*. Amongst viable technologies for areal measurement, optical solutions are currently the preferred choice for AM surfaces, because of measurement ease, speed and capability to cope with the challenging topographic formations typically encountered on an AM topography (deep recesses, high aspect-ratio features). Despite the advantages of optical technologies, the actual performance of optical measurement on AM surfaces has not been fully understood yet; measurement results may significantly differ across technologies, even when observing the same surface, and quantification of uncertainty associated with a measurement is a subject of ongoing research and standardisation efforts (for example, see recent work in ISO 25178 part 600 [2019]). It is advisable thus to inspect the same topography with multiple technologies whenever possible, so that a more reliable idea of the nature of an AM topography can be formed through consensus.

In terms of characterisation, the more consolidated amongst the available choices is to go with areal texture parameters, preferably from the ISO 25178 part 2 set. Despite often being unable to quantify or even properly describe aspects related to specific topographic formations that are typical on AM surfaces, such as particles, staircase effect and so on, texture parameters offer a reliable and reproducible description of surface properties, as long as the pre-processing steps, and in particular filtering, are clearly indicated; otherwise the results cannot be correctly interpreted. Feature-based characterisation is gradually being established as a viable complement to texture parameters, in particular for those applications where information related to specific features is requested. Such types of analysis may be particularly important for manufacturing processes that are not fully understood yet, as a means of investigating signature features of the physics involved in the process, as it applies to many AM processes.

To support feature-based characterisation, the current set of segmentation methods (ISO 25178-2 2012) and related feature-based parameters (ISO 16610-85 2013) will be extended to include other segmentation methods and new parameters potentially suitable for the characterisation of AM surfaces.

Future research will also address the measurement of complex AM topographies featuring undercuts or cavities by means of five-axis optical measurement systems and X-ray CT.

As these methods result in fully three-dimensional triangle meshes, one of the future challenges will be about defining new standards adapting surface texture methods to triangle mesh data, i.e. redefining the mathematical background of parameters, filters and form association. ISO has begun working on a set of future standards covering these issues through working group 16 within the technical committee 213.

References

Artigas, R. 2011. Imaging confocal microscopy. In: Leach, R. K. (ed.), *Optical Measurement of Surface Topography*. Springer, Berlin, Heidelberg.

Blateyron, F. 2011. Chromatic confocal microscopy. In: Leach, R. K. (ed.), *Optical Measurement of Surface Topography*. Springer, Berlin, Heidelberg.

Blateyron, F. 2013a. The areal field parameters. In: Leach, R. K. (ed.), *Characterisation of Areal Surface Texture*. Springer, Berlin, Heidelberg.

Blateyron, F. 2013b. The areal feature parameters. In: Leach, R. K. (ed.), *Characterisation of Areal Surface Texture*. Springer, Berlin, Heidelberg.

Brown, C. 2013. Areal fractal methods. In: Leach, R. K. (ed.), *Characterisation of Areal Surface Texture*. Springer, Berlin, Heidelberg.

Brown, C., Hansen, H. N., Jiang, X. J., Blateyron, F., Berglund, J., Senin, N., Bartkowiak, T., Dixon, B., Goic, G. Le, Quinsat, Y., Stemp, W. J., Thompson, M. K., Ungar, P. S., Zahouani, E. H. 2018. Multiscale analyses and characterizations of surface topographies. *Ann. CIRP* 67:839–862.

Cabanettes, F., Joubert, A., Chardon, G., Dumas, V., Rech, J., Grosjean, C., Dimkovski, Z. 2018. Topography of as built surfaces generated in metal additive manufacturing: A multi scale analysis from form to roughness. *Prec. Eng.* 52:249–265.

De Chiffre, L., Carli, L., Eriksen, R. S. 2011. Multiple height calibration artefact for 3D microscopy. *Ann. CIRP* 60:535–538.

De Chiffre, L., Carmignato, S., Kruth. J. P., Schmitt, R., Weckenmann, A. 2014. Industrial applications of computed tomography. *Ann. CIRP* 63:655–677.

de Groot, P., 2011a. Coherence scanning interferometry. In: Leach, R. K. (ed.), *Optical Measurement of Surface Topography*. Springer, Berlin, Heidelberg.

de Groot, P., 2011b. Phase shifting interferometry. In: Leach, R. K. (ed.), *Optical Measurement of Surface Topography*. Springer, Berlin, Heidelberg.

du Plessis, A., Yadroitsev, I., Yadroitsava, I., Le Roux, S. G., 2018. X-ray microcomputed tomography in additive manufacturing: A review of the current technology and applications. *3D Print. Addit. Manuf.* 5:227–247.

Forbes, A. B. 2013. Areal form removal. In: Leach, R. K. (ed.), *Characterisation of Areal Surface Texture*. Springer, Berlin, Heidelberg.

Gibson, I., Rosen, D. W., Stucker, B. 2009. *Additive Manufacturing Technologies: Rapid Prototyping to Direct Digital Manufacturing*. Springer, Boston, MA.

Giusca, C., Leach, R. K., Helery, F., Gutauskas, T., Nimishakavi, L. 2012a. Calibration of the scales of areal surface topography-measuring instruments: Part 1. Measurement noise and residual flatness. *Meas. Sci. Technol.* 23:035008.

Giusca, C., Leach, R. K., Helery, F. 2012b. Calibration of the scales of areal surface topography measuring instruments: Part 2. Amplification, linearity and squareness. *Meas. Sci. Technol.* 23:065005.

Giusca, C., Leach, R. K. 2013a. Calibration of the scales of areal surface topography measuring instruments: Part 3. Resolution. *Meas. Sci. Technol.* 24:105010.

Giusca, C., Leach, R. K. 2013b. Calibration of the metrological characteristics of coherence scanning interferometers (CSI) and phase shifting interferometers (PSI). *NPL Good Practice Guide No. 127*, National Physical Laboratory.

Giusca, C., Leach, R. K. 2013c. Calibration of the metrological characteristics of imaging confocal microscopes. *NPL Good Practice Guide No. 128*, National Physical Laboratory.

Giusca, C., Leach, R. K. 2013d. Calibration of the metrological characteristics of areal contact stylus instruments. *NPL Good Practice Guide No. 129*, National Physical Laboratory.

Giusca, C., Claverley, J., Sun, W., Leach, R. K., Helmli, F., Chavigner, M. 2014. Practical estimation of measurement noise and flatness deviation on focus variation microscopes. *Ann. CIRP* 63:545–548.

Gomez, C., Su, R., Thompson, A., Di Sciacca, J., Lawes, S., Leach, R. K. 2017. Optimization of surface measurement for metal additive manufacturing using coherence scanning interferometry. *Opt. Eng.* 56:111714.

Haitjema, H. 2015. Uncertainty in measurement of surface topography. *Surf. Topogr.* 3:035004.

Helmli, F. 2011. Focus variation instruments. In: Leach, R. K. (ed.), *Optical Measurement of Surface Topography*. Springer, Berlin, Heidelberg.

Helmli, F., Danzl, R., Scherer, S. 2011. Optical measurement of micro cutting tools. *J. Phys. Conf.* 311:012003.

ISO 3274. 1996. Geometrical product specifications (GPS) – Surface texture: Profile method – Nominal characteristics of contact (stylus) instruments, International Organization for Standardization.

ISO 4287. 1997. Geometrical Product Specifications (GPS) – Surface Texture: Profile Method – Terms, Definitions and Surface Texture Parameters, International Organization for Standardization.

ISO 4288. 1996. Geometrical product specifications (GPS) – Surface texture: Profile method – Rules and procedures for the assessment of surface texture, International Organization for Standardization.

ISO 5436 part 1. 2000. Geometrical product specifications (GPS) – Surface texture: Profile method; Measurement standards – Part 1: Material measures, International Organization for Standardization.

ISO 5436 part 2. 2012. Geometrical product specifications (GPS) – Surface texture: Profile method; Measurement standards – Part 2: Software measurement standards, International Organization for Standardization.

ISO 10110 part 8. 2010. Optics and photonics – Surface texture, roughness and waviness: Specific standard for optics, International Organization for Standardization.

ISO 12085. 1996. Geometrical product specifications (GPS) – Surface texture: Profile method – Motif parameters, International Organization for Standardization.

ISO 12179. 2000. Geometrical product specifications (GPS) – Surface texture: Profile method – Calibration of contact (stylus) instruments, International Organization for Standardization.

ISO 13565 part 1. 1996. Geometrical product specifications (GPS) – Surface texture: Profile method; Surfaces having stratified functional properties – Part 1: Filtering and general measurement conditions, International Organization for Standardization.

ISO 13565 part 2. 1996. Geometrical product specifications (GPS) – Surface texture: Profile method; Surfaces having stratified functional properties – Part 2: Height characterization using the linear material ratio curve, International Organization for Standardization.

ISO 13565 part 3. 1998. Geometrical product specifications (GPS) – Surface texture: Profile method; Surfaces having stratified functional properties – Part 3: Height characterization using the material probability curve, International Organization for Standardization.

ISO 14406. 2010. Geometrical product specifications (GPS) – Extraction.

ISO 16610 part 1. 2015. Geometrical product specifications (GPS) – Filtration – Part 1: Overview and basic concepts, International Organization for Standardization.

ISO 16610 part 20. 2015. Geometrical product specifications (GPS) – Filtration – Part 20: Linear profile filters: Basic concepts, International Organization for Standardization.

ISO 16610 part 21. 2011. Geometrical product specifications (GPS) – Filtration – Part 21: Linear profile filters: Gaussian filters, International Organization for Standardization.

ISO 16610 part 22. 2015. Geometrical product specifications (GPS) – Filtration – Part 22: Linear profile filters: Spline filters, International Organization for Standardization.

ISO 16610 part 28. 2016. Geometrical product specifications (GPS) – Filtration – Part 28: Profile filters: End effects, International Organization for Standardization.
ISO 16610 part 29. 2015. Geometrical product specifications (GPS) – Filtration – Part 29: Linear profile filters: Spline wavelets, International Organization for Standardization.
ISO 16610 part 30. 2015. Geometrical product specifications (GPS) – Filtration – Part 30: Robust profile filters: Basic concepts, International Organization for Standardization.
ISO 16610 part 31. 2016. Geometrical product specifications (GPS) – Filtration – Part 31: Robust profile filters: Gaussian regression filters, International Organization for Standardization.
ISO/TS 16610 part 32. 2009. Geometrical product specifications (GPS) – Filtration – Part 32: Robust profile filters: Spline filters, International Organization for Standardization.
ISO 16610 part 40. 2015. Geometrical product specifications (GPS) – Filtration – Part 40: Morphological profile filters: Basic concepts, International Organization for Standardization.
ISO 16610 part 41. 2015. Geometrical product specifications (GPS) – Filtration – Part 41: Morphological profile filters: Disk and horizontal line-segment filters, International Organization for Standardization.
ISO/CD 16610 part 45. Geometrical product specifications (GPS) – Filtration – Part 45: Part 45: Profile Morphological: Segmentation, International Organization for Standardization.
ISO 16610 part 49. 2015. Geometrical product specifications (GPS) – Filtration – Part 49: Morphological profile filters: Scale space techniques, International Organization for Standardization.
ISO 16610 part 60. 2015. Geometrical product specification (GPS) – Filtration – Part 60: Linear areal filters – Basic concepts, International Organization for Standardization.
ISO 16610 part 61. 2015. Geometrical product specification (GPS) – Filtration – Part 61: Linear areal filters – Gaussian filters, International Organization for Standardization.
ISO 16610 part 71. 2014. Geometrical product specifications (GPS) – Filtration – Part 71: Robust areal filters: Gaussian regression filters, International Organization for Standardization.
ISO 16610 part 85. 2013. Geometrical product specifications (GPS) – Filtration – Part 85: Morphological areal filters: Segmentation, International Organization for Standardization.
ISO/CD 21920 part 1. Geometrical product specifications (GPS) – Surface texture: Profile – Part 1: Indication of surface texture, International Organization for Standardization.
ISO/CD 21920 part 2. Geometrical product specifications (GPS) – Surface texture: Profile – Part 2: Terms, definitions and surface texture parameters, International Organization for Standardization.
ISO/CD 21920 part 3. Geometrical product specifications (GPS) – Surface texture: Profile – Part 3: Specification operators, International Organization for Standardization.
ISO 25178 part 1. 2016. Geometrical product specifications (GPS) – Surface texture: Areal – Part 1: Indication of surface texture, International Organization for Standardization.
ISO 25178 part 2. 2012. Geometrical product specifications (GPS) – Surface texture: Areal – Part 2: Terms, definitions and surface texture parameters, International Organization for Standardization.
ISO 25178 part 3. 2012. Geometrical product specifications (GPS) – Surface texture: Areal – Part 3: Specification operators, International Organization for Standardization.
ISO 25178 part 6. 2010. Geometrical product specifications (GPS) – Surface texture: Areal – Part 6: Classification of methods for measuring surface texture, International Organization for Standardization.
ISO 25178 part 70. 2014. Geometrical product specification (GPS) – Surface texture: Areal – Part 70: Material measures, International Organization for Standardization.
ISO 25178 part 71. 2017. Geometrical product specifications (GPS) – Surface texture: Areal – Part 71: Software measurement standards, International Organization for Standardization.
ISO 25178 part 72. 2017. Geometrical product specifications (GPS) – Surface texture: Areal – Part 72: XML file format x3p, International Organization for Standardization.
ISO 25178 part 73. 2019. Geometrical product specifications (GPS) – Surface texture: Areal – Part 73: Terms and definitions for surface defects on material measures, International Organization for Standardization.

ISO 25178 part 600. 2019. Geometrical product specifications (GPS) – Surface texture: Areal – Part 600: Metrological characteristics for areal topography measuring methods, International Organization for Standardization.
ISO 25178 part 601. 2010. Geometrical product specifications (GPS) – Surface texture: Areal – Part 601: Nominal characteristics of contact (stylus) instruments, International Organization for Standardization.
ISO 25178 part 602. 2010. Geometrical product specifications (GPS) – Surface texture: Areal – Part 602: Nominal characteristics of non-contact (confocal chromatic probe) instruments, International Organization for Standardization.
ISO 25178 part 603, 2013. Geometrical product specifications (GPS) – Surface texture: Areal – Part 603: Nominal characteristics of non-contact (phase-shifting interferometric microscopy) instruments, International Organization for Standardization.
ISO 25178 part 604. 2013. Geometrical product specifications (GPS) – Surface texture: Areal – Part 604: Nominal characteristics of non-contact (coherence scanning interferometry) instruments, International Organization for Standardization.
ISO 25178 part 605. 2014. Geometrical product specifications (GPS) – Surface texture: Areal – Part 605: Nominal characteristics of non-contact (point autofocus probe) instruments, International Organization for Standardization.
ISO 25178 part 606. 2015. Geometrical product specification (GPS) – Surface texture: Areal – Part 606: Nominal characteristics of non-contact (focus variation) instruments, International Organization for Standardization.
ISO 25178 part 607. 2019. Geometrical product specifications (GPS) – Surface texture: Areal – Part 607: Nominal characteristics of non-contact (confocal microscopy) instruments, International Organization for Standardization.
ISO 25178 part 701. 2010. Geometrical product specifications (GPS) – Surface texture: Areal – Part 701: Calibration and measurement standards for contact (stylus) instruments, International Organization for Standardization.
Jiang, X., Blunt, L., Stout, K. 2000. Development of a lifting wavelet representation for surface characterization. *Proc. R. Soc. Lond. A*. 456:2283–2313.
Jiang, X., Lou, S., Scott, P., 2011. Morphological method for surface metrology and dimensional metrology based on the alpha shape. *Meas. Sci. Technol.* 23:015003.
Jiang, X. J., Scott, P. J. 2020. *Advanced Metrology: Freeform Surfaces*. Academic Press, Cambridge, MA.
Körner, L., Lawes, S., Bate, D., Newton, L., Senin, N., Leach, R. K. 2019. Increasing throughput in X-ray computed tomography measurement of surface topography using sinogram interpolation. *Meas. Sci. Technol.* 30:125002.
Leach, R. K. 2011. *Optical Measurement of Surface Topography*. Springer, Berlin.
Leach, R. K. 2014. *Fundamental Principles of Engineering Nanometrology*. William Andrew, Norwich, NY.
Leach, R. K. 2015. Surface texture. In: Laperrière, L., Reinhart, G. (eds.), *CIRP Encyclopaedia of Production Engineering*. Springer-Verlag, Berlin.
Leach, R. K., Giusca, C., Haitjema, H., Evans, C., Jiang, X. 2015. Calibration and verification of areal surface texture measuring instruments. *Ann. CIRP* 64:797–813.
Leach, R. K., Bourell, D., Carmignato, S., Donmez, A., Senin, N., Dewulf, W. 2019a. Geometrical metrology for metal additive manufacturing. *Ann. CIRP* 68:677–700.
Leach, R. K., Haitjema, H., Giusca, C. 2019b. A Metrological Characteristics Approach to Uncertainty in Surface Metrology. In: Osten, W. (ed.), *Optical Inspection of Microsystems*. CRC Press, Boca Raton, FL.
Le Goïc, G., Brown, C. A., Favreliere, H., Samper, S., Formosa, F. 2012. Outlier filtering: A new method for improving the quality of surface measurements. *Meas. Sci. Technol.* 24:015001.
Lifton, J. J., Malcolm, A. A., McBride, J. W. 2015. On the uncertainty of surface determination in X-ray computed tomography for dimensional metrology. *Meas. Sci. Technol.* 26:035003.
Lou, S., Jiang, X., Scott, P. 2013. Application of the morphological alpha shape method to the extraction of topographical features from engineering surfaces. *Measurement*. 46:1002–1008.

Lou, S., Jiang, X., Sun, W., Zeng, W., Pagani, L., Scott, P. J. 2019. Characterisation methods for powder bed fusion processed surface topography. *Precis. Eng.* 57:1–15.

Lu, W., Chen, C., Wang, J., Leach, R. K., Liu, X., Jiang, X. 2019. Characterisation of the displacement response in chromatic confocal microscopy with a hybrid radial basis function network. *Opt. Express* 27:22737–22752.

Maculotti, G., Feng, X., Su, R., Galetto, M., Leach, R. K. 2019. Residual flatness and scale calibration for a point autofocus surface topography measuring instrument. *Meas. Sci. Technol.* 30:075005.

Marinello, F., Carmignato, S., Savio, E., Bariani, P., Carli, L., Horsewell, A., De Chiffre, L. 2008. Metrological performance of SEM 3D techniques. *Proc. 17th IMEKO TC 2 Symposium on Photonics in Measurements*, Prague, Czech Republic, August, 120–125.

Medeossi, F., Carmignato, S., Lucchetta, G., Savio, E. 2016. Effect of void pixels on the quantification of surface topography parameters. *Proc. 16th Int. Conf. euspen*, Nottingham, United Kingdom, May.

Mignot, C. 2018. Color (and 3D) for scanning electron microscopy. *Micros. Today* 26: 12–17.

Miura, K., Nose, A. 2011. Point autofocus instruments. In: Leach, R. K. (ed.), *Optical Measurement of Surface Topography*. Springer, Berlin, Heidelberg.

Newton, L., Senin, N., Gomez, C., Danzl, R., Helmli, F., Blunt, L., Leach, R. K. 2019a. Areal topography measurement of metal additive surfaces using focus variation microscopy. *Addit. Manuf.* 25:365–389.

Newton, L., Senin, N., Smith, B., Chatzivagiannis, E., Leach, R. K. (2019b). Comparison and validation of surface topography segmentation methods for feature-based characterisation of metal powder bed fusion surfaces. *Surf. Topogr. Met. Prop.* 7:045020.

Pagani, L., Scott, P. J. 2018. Triangular Bézier surface: From reconstruction to roughness parameter computation. In: *Advanced Mathematical and Computational Tools in Metrology and Testing XI, Ser. Adv. Math. Appl. Sci.* 89:48–57.

Pagani, L., Townsend, A., Zeng, W., Lou, S., Blunt, L., Jiang, X. Q., Scott, P. J. 2019. Towards a new definition of areal surface texture parameters on freeform surface: Reentrant features and functional parameters. *Measurement* 141:442–459.

Pawlus, P., Reizer, R., Wieczorowski, M. 2017. Problem of non-measured points in surface texture measurements. *Metrology and Measurement Systems* 24:525–536.

Podulka, P., Pawlus, P., Dobrzański, P., Lenart, A. 2014. Spikes removal in surface measurement. *J. Phys. Conf. Ser.* 483:012025.

Scott, P. 2004. Pattern analysis and metrology: The extraction of stable features from observable measurements. *Proc. R. Soc. Lond. A*. 460:2050.

Scott, P. J., Jiang, X. 2014. Freeform surface characterisation: Theory and practice. *J. Phys. Conf.* 483:12005.

Seewig, J. 2013. Areal filtering methods. In: Leach, R. K. (ed.), *Characterisation of Areal Surface Texture*. Springer, Berlin, Heidelberg.

Senin, N., Blunt, L. A. 2013. Characterisation of individual areal features. In: Leach, R. K. (ed.), *Characterisation of Areal Surface Texture*. Springer, Berlin, Heidelberg.

Senin, N., Leach, R. K., Pini. S., Blunt, L. 2015. Texture-based segmentation with Gabor filters, wavelet and pyramid decompositions for extracting individual surface features from areal surface topography maps. *Meas. Sci. Technol.* 26:095405.

Senin, N., Thompson, A., Leach, R. K. 2017. Characterisation of the topography of metal additive surface features with different measurement technologies. *Meas. Sci. Technol.* 28:095003.

Senin, N., Thompson, A., Leach, R. K. 2018. Feature-based characterisation of signature topography in laser powder bed fusion of metals. *Meas. Sci. Technol.* 29:045009.

Su, R., Thomas, M., Leach, R. K., Coupland, J. 2018. Effects of defocus on transfer function of coherence scanning interferometry. *Opt. Lett.* 43:82–85.

Thompson, A., Senin, N., Giusca, C., Leach, R. K. 2017. Topography of selectively laser melted surfaces: A comparison of different measurement methods. *Ann. CIRP* 66:543–546.

Thompson, A., Senin, N., Maskery, I., Körner, L., Lawes, S., Leach, R. K. 2018a. Internal surface measurement of metal powder bed fusion parts. *Addit. Manuf.* 20:126–133.

Thompson, A., Senin, N., Maskery, I., Leach, R. K. 2018b. Effects of magnification and sampling resolution in X-ray computed tomography for the measurement of additively manufactured metal surfaces. *Precis. Eng.* 53:54–64.

Todhunter, L. D., Leach, R. K., Lawes, S. D. A., Blateyron, F. 2017. Industrial survey of ISO surface texture parameters. *CIRP J. Manufac. Sci. Technol.* 19:84–92.

Townsend, A., Senin, N., Blunt, L., Leach, R. K., Taylor, J. S. 2016. Surface texture metrology for metal additive manufacturing: A review. *Precis. Eng.* 46:34–47.

Townsend, Andrew, Pagani, L., Blunt, L., Scott, P., Jiang, X. 2017. Factors affecting the accuracy of areal surface texture data extraction from X-ray CT. *Ann. CIRP* 66:547–550.

Townsend, A., Racasan, R., Leach, R. K., Senin, N., Thompson, A., Ramsey, A., Bate, D., Wolliams, P., Brown, S., Blunt, L. 2018. An interlaboratory comparison of X-ray computed tomography measurement for texture and dimensional characterisation of additively manufactured parts. *Addit. Manuf.* 23:422–432.

Townsend, A., Pagani, L., Scott, P., Blunt, L. 2019. Introduction of a surface characterization parameter Sdr_{prime} for analysis of re-entrant features. *J. Nondestruct. Eval.* 38:1–10.

Ukraintsev, V. A., Banke, G. W. 2012. Review of reference metrology for nanotechnology: Significance, challenges, and solutions. *J. Micro/Nanolithography, MEMS, MOEMS* 11:011010.

Wang, J., Leach, R. K., Jiang, X. 2015. Review of the mathematical foundations of data fusion techniques in surface metrology. *Surf. Topog. Metrol. Prop.* 3:023001.

Whitehouse, D. J. 2010. *Handbook of Surface and Nanometrology*. CRC Press, Boca Raton, FL.

Worthington, P. L., Hancock, E. R. 2001. Surface topography using shape-from-shading. *Pattern Recognit*. 34:823–840.

Zangl, K., Danzl, R., Helmli, F., Prantl, M. 2018. Highly accurate optical μcMM for measurement of micro holes. *Procedia CIRP* 75:397–402.

Zanini, F., Carmignato, S. 2017. Two-spheres method for evaluating the metrological structural resolution in dimensional computed tomography. *Meas. Sci. Technol.* 28:114002.

Zanini, F., Pagani, L., Savio, E., Carmignato, S. 2019a. Characterisation of additively manufactured metal surfaces by means of X-ray computed tomography and generalised surface texture parameters. *Ann. CIRP* 68:515–518.

Zanini, F., Sbettega, E., Sorgato, M., Carmignato, S. 2019b. New approach for verifying the accuracy of X-ray computed tomography measurements of surface topographies in additively manufactured metal parts. *J. Nondestruc. Eval.* 38:12.

Zhu, H., Blunt, L., Jiang, X. 2009. An active contour based algorithm for micro-scale surface feature extraction. *The Swedish Production Symposium*, Göteborg, Sweden, December.

12

X-Ray Computed Tomography

Simone Carmignato, Filippo Zanini, Markus Baier and Elia Sbettega

CONTENTS

12.1 Introduction

This chapter focuses on X-ray computed tomography (CT) and its uses in the field of metal additive manufacturing (AM). CT is currently used for many industrial applications (De Chiffre et al. 2014). For example, since the 1980s, CT has been used for non-destructive evaluation and materials analysis, and since the 2000s it has been increasingly applied for dimensional metrology (Kruth et al. 2011). Metrological CT systems are currently available to perform accurate dimensional and geometrical measurements and are becoming fundamental tools not only for product verification but also for product development and process optimisation. The main advantage of CT is its capability of measuring both accessible and non-accessible geometries and features in a non-destructive and non-contact manner.

This capability allows analyses that are difficult and, in many cases, impossible using conventional measuring devices, such as tactile and optical coordinate measuring systems (CMSs). Moreover, when using micro- or nano-focus X-ray sources, CT systems allow high spatial resolutions, enabling the measurement of micro-scale details.

In the field of metal AM, CT can be used to analyse simultaneously material characteristics (such as internal defects, porosity and inclusions – see Chapter 9), form and geometrical properties (see Chapter 10) and surface characteristics (see Chapter 11). Moreover, CT can be fundamental for several analyses throughout the AM process chain, as it can cope with the high degree of freedom of AM design, as well as with the geometrical and structural complexity of AM products. For example, AM enables the production of intricate lattice structures for different applications, including reduction of mass in aeronautics and automotive components (Pyka et al. 2013), medical implants to ease bone in-growth (Taniguchi et al. 2016, Kourra et al. 2018) and internal cavities or channels with complex geometries for conformal cooling purposes (Snyder et al. 2015). All these structures can be difficult to examine using conventional measuring techniques because of the presence of non-accessible features, but they can be successfully measured with CT, as illustrated in this chapter. In addition, CT has the capability of providing holistic three-dimensional (3D) models of the scanned parts by high-density digitisation (Flisch et al. 1999, Orme et al. 2017). Such a capability enables additional uses of CT in conjunction with AM, including reverse engineering and accuracy enhancement of digital twins (see Chapter 4). For example, the CT reconstructed digital model of a part can be used to feed an AM system to print a prototype or copy of the original part, even when its nominal model or drawings are not available. Both the digital model and the physical prototype can be used as starting points for future design iterations. Re-designed parts can again be printed via AM and characterised with CT by comparison with the initial model to investigate experimentally the variations and improvements. CT-reconstructed models can also be used to improve AM process simulations or mechanical/structural simulations (for example, using the finite element method, see Chapter 4) with the aid of the real part geometry. Finally, CT measurements can be compared with in-process analyses (for example, performed by high-speed vision systems or thermal cameras, see Chapter 13), obtaining useful information for effectively improving the quality of processes and products.

This chapter is organised as follows. Section 12.2 presents the fundamentals of industrial X-ray CT, illustrating the evolution of CT technology; the working principles, configurations and components of industrial CT systems; and the typical CT scanning procedure. Section 12.3 discusses error sources and traceability issues of CT dimensional measurements. Section 12.4 provides an overview of the different applications of CT for AM, with a focus on dimensional and geometrical verification of products, analysis of internal defects, characterisation of surface topographies, analysis of powder feedstock and use of CT data for product and process development. Finally, Section 12.5 gives a summary of current research and future developments.

12.2 Fundamentals of Industrial X-Ray Computed Tomography

12.2.1 Evolution of X-Ray Computed Tomography

The discovery of X-rays is attributed to Wilhelm Conrad Röntgen, awarded with the first Nobel Prize in Physics in 1901. X-rays are a specific type of electromagnetic radiation, typically characterised by a wavelength ranging from 0.01 nm to 10 nm. Their ability to penetrate matter leads to their classification as 'soft' (when the wavelength is above 0.1 nm) and 'hard' rays (for smaller wavelengths). The emitted photon is characterised by an energy E given by

$$E = h \cdot f = \frac{h \cdot c}{\lambda} \tag{12.1}$$

where h is Planck's constant, equal to 6.63×10^{-34} J·s, c is the speed of light in a vacuum, equal to 3×10^{8} m·s^{-1} and λ is the radiation wavelength (Hsieh 2015).

A first milestone in the development of CT dates back to 1917, when the mathematician Johann Radon explained the principles – called in his honour the 'Radon transform' and 'inverse Radon transform' – according to which the geometry of an object can be obtained using an ideally infinite number of its projections (Radon 1986). A second fundamental step was accomplished by Allan McLeod Cormack, who came to understand the importance of calculating the attenuation coefficient distribution inside the object, and who performed investigations using the first prototype CT instrument to be physically built (Cormack 1963). Cormack was awarded with the Nobel Prize for Physiology or Medicine in 1979 'for the development of computer assisted tomography' jointly with Godfrey Newbold Hounsfield, the engineer who demonstrated how the internal structure of a body can be reconstructed starting from X-rays acquired in different angular positions (Hounsfield 1976). The first clinical X-ray CT instrument, built in 1971, is attributed to Hounsfield and allowed acquisition of an image of a human patient's head section in 4.5 minutes (Hsieh 2015). From that date, clinical X-ray CT technology has recorded a rapid and constant evolution that can be described through five generations of clinical CT instruments (Hsieh 2015). The development of the technology has focused mainly on the reduction of the acquisition time and the optimisation of the system stability. Approximately in parallel to the evolution of clinical instruments, CT systems for industrial use have been developed. The first applications of CT in industry date back to the 1980s, when CT started to be used for materials characterisation, defect detection and non-destructive evaluation (De Chiffre et al. 2014).

12.2.2 Industrial CT Systems – Configurations and Components

Industrial X-ray CT systems are based on the same principles as clinical instruments but with significant technical differences related to the fields of application. To increase the stability of the system, the X-ray source and detector are commonly fixed in determined positions, and the object under investigation is moved between these positions. In addition, CT systems used for metrological applications are typically characterised by thermally controlled cabinets, micro-focus X-ray sources and precise motion systems to maximise the scan resolution and enhance the measurement accuracy (Carmignato et al. 2017a). The most frequently used industrial X-ray CT systems are characterised by fan-beam or cone-beam configurations, as shown in Figure 12.1.

Systems that employ the fan-beam configuration are equipped with an X-ray source generating a two-dimensional (2D) 'fan' beam of X-rays and a linear detector. In this case, a complete revolution around the rotation axis allows the reconstruction of a single slice of the scanned object. The translation along the vertical direction is hence required to reconstruct the entire volume. In the cone-beam configuration, the source generates a cone beam of X-rays and the detector is typically a 2D flat panel, enabling a complete acquisition of the scanned object in a single full rotation, provided that the entire object fits inside the

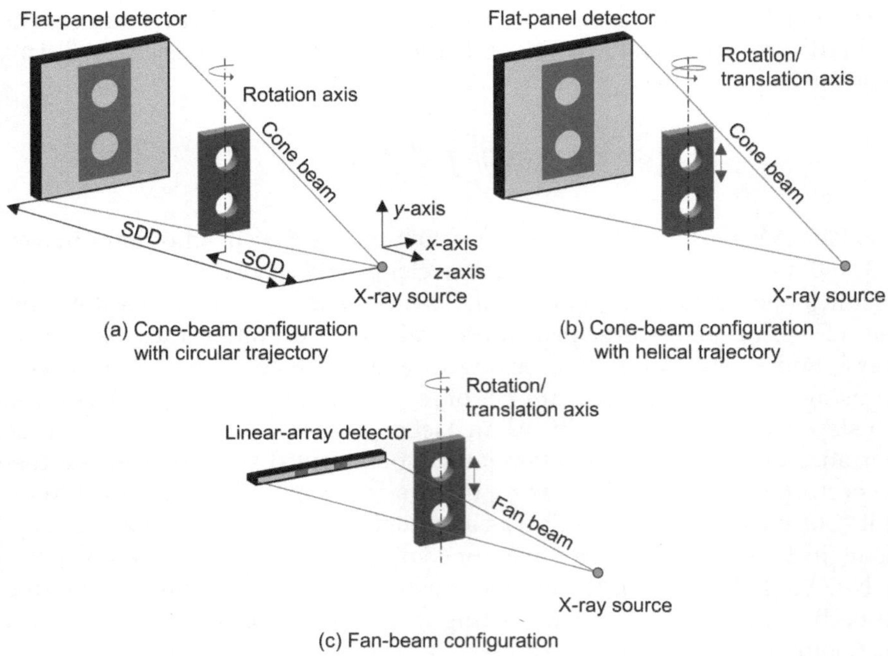

FIGURE 12.1
Typical configurations of industrial X-ray CT systems: (a) X-ray cone-beam configuration, with flat-panel detector and circular trajectory of the scanned object; (b) X-ray cone-beam configuration, with flat-panel detector and helical trajectory of the object; (c) X-ray fan-beam configuration, with linear-array detector and circular trajectory plus vertical translation of the object. The z axis (also called the *magnification axis*) is defined along the direction between the source and the detector; the y axis is parallel to the rotation axis, the x axis is orthogonal to y and z axes, SDD is the source-to-detector distance and SOD is the source-to-object distance.

detector field of view. Both the configurations described above commonly work with circular trajectories, but for special requirements (for example, if the object height exceeds the detector field of view in the vertical direction), a combination of rotation and translation along the vertical direction, namely a helical trajectory, can be implemented. Compared to the fan-beam configuration, the cone-beam configuration leads to a significant reduction of the acquisition time (a single full rotation of the object instead of a number of full rotations), but the acquired projections may be more prone to the presence of image artefacts, such as Feldkamp artefacts and scattering (see Section 12.3.1), which can affect the scan quality (Kruth et al. 2011).

Industrial X-ray CT systems consist of three main hardware components: a radiation source that generates the X-rays, a detector to measure the photon energy remaining after the attenuation as the beam traverses the object and a motion system with a rotary stage to position and rotate the object. In addition, CT systems are commonly supplied with powerful computer units, which are needed for processing the large amount of data generated during the CT scanning procedure (see Section 12.2.3). In the following, the three main hardware components of a typical industrial CT system are described in more detail: the X-ray source, the X-ray detector and the kinematic system. For the sake of simplicity, only the specific case of cone-beam CT systems is addressed, but other configurations are discussed elsewhere (Carmignato et al. 2017a).

12.2.2.1 X-Ray Source

X-rays can be generated in different ways and with different types of sources. The most common type of source for conventional industrial CT systems is the X-ray tube, which is a vacuum tube including a filament (cathode) and a target (anode) that are connected to a high-voltage generator. The filament, connected to the negative pole of the generator, is heated using the Joule effect, and the consequent increase in temperature raises the amount of kinetic energy of the electrons, which are forced to exit the cathode. The target is a metal part, often made of tungsten or molybdenum, and is connected to the positive pole of the generator. The application of an electric potential difference accelerates the electrons generated by the filament toward the target. The interaction between the accelerated electron beam and the metal target emits a large amount of energy. In conventional X-ray tubes, only about 1% of this energy is converted in X-rays, while the remaining part (around 99%) is typically dissipated as heat (Buzug 2008). The electron beam is commonly focused using magnetic deflectors and lenses on a small finite area of the target, namely the focal spot. The dimension of the focal spot strongly affects the quality of the radiographic images: the smaller the focal spot, the sharper the obtained image. However, in the case that a high X-ray power is used (which is common practice in industrial CT scanning of highly absorbing materials), the focal spot size is normally enlarged to dissipate the generated heat over a larger target area, hence avoiding the pitting of the target itself. The X-rays generated after the collision between the accelerated electrons and target's atoms are then typically shaped through a circular aperture or diaphragm, shielded with a window commonly made of beryllium or aluminium. Further details on the X-ray generation mechanisms and the technical solutions for X-ray sources can be found elsewhere (Carmignato et al. 2017a).

12.2.2.2 X-Ray Detector

The X-rays emitted from the source and attenuated by the interaction with the scanned material are captured by the X-ray detector, which is used to convert the radiation energy into electrical signals. The technology behind X-ray detectors has continuously been improved, in parallel to the evolution of the CT instrument generations discussed in Section 12.2.1 (Panetta 2016). Devices based on two different principles have mainly been adopted in industrial CT systems: gas ionisation detectors and scintillator detectors. In gas ionisation detectors, the conversion of X-rays into electrical signals is directly obtained from the ability of X-rays to ionise gas (Hermanek et al. 2017a). In scintillator detectors, instead, the X-ray intensity is first converted into visible light and then, through a photomultiplier, into an electrical signal that is proportional to the radiation intensity (Kruth et al. 2011). A third emerging solution is represented by photon-counting detectors (Taguchi and Iwanczyk 2013), where X-ray photons are directly converted into electric charges that are proportional to the photon energy. The main advantage of photon-counting detectors is their ability to discriminate between different photon energies, with several possible benefits, including the possibility of eliminating beam-hardening artefacts (see Section 12.3.1), of increasing the image contrast for low attenuating materials and of reducing noise due to low energies. However, the application of photon-counting detectors in industrial X-ray CT systems is still limited due to costs and to technology constraints related to the count-rate capability (Ametova et al. 2017). Industrial X-ray CT systems commonly employ flat-panel detectors or linear-array detectors. While linear-array detectors are composed of a single line of diodes, flat-panel detectors are characterised by a 2D matrix of pixels (Figure 12.1). Due to their simpler structure, linear-array detectors guarantee less pixel interaction and scattering effects on the obtained X-ray projection, yielding better image quality; however, they require longer scanning times.

12.2.2.3 Kinematic System

Industrial X-ray CT systems are often equipped with stationary sources and detectors, while the scanned object is positioned within the scanning volume and rotated during the scan using a motion system equipped with a rotary stage. The typical ideal configuration of the mechanical axes is represented in Figure 12.1 and can be summarised as follows (Ferrucci et al. 2015):

1. The magnification axis (z axis) passes through the X-ray source and intersects the detector at its centre.
2. The magnification axis is orthogonal to the detector.
3. The magnification axis intersects the axis of rotation at 90°.
4. The projection of the rotation axis is parallel to the detector column.

The stability of the kinematic system and the correct geometrical alignment of the components and the mechanical axes are fundamental to obtain high-quality CT measurements (Dewulf et al. 2018).

12.2.3 CT Scanning Process

Figure 12.2 is a schema showing the main steps of the typical CT scanning process that can be used for AM applications. A preliminary step is CT system qualification. Industrial CT systems are complex devices that need to be qualified on a regular basis (Bartscher et al. 2017). System qualification typically includes a series of tests targeting each main component of a CT system (see Section 12.2.2), carried out to investigate specific error conditions (see Section 12.3) and to perform fine adjustments/corrections. A typical system qualification procedure may include the following tasks:

- Tests on the system architecture, to ensure a proper geometrical alignment between manipulator, X-ray source and detector (see, for example, Dewulf et al. 2018).
- Determination of the scale factor correction (see, for example, Jiménez et al. 2013).
- Evaluation of the focal spot characteristics, including size, shape and possible drift (Baier et al. 2018, Sun et al. 2016).

After system qualification, the general workflow starts with the acquisition of a number of X-ray projections (radiographs). Such projections are collected during the rotation of the sample, at different angular positions. The number of projections is a typical parameter chosen by the user. A higher number of projections correspond to a smaller angular step and a higher quality of the reconstruction, but to a longer scanning time. An X-ray projection can be described as an *attenuation image*. When the X-ray beam propagates through the object material (before being captured by the X-ray detector to generate the image), its intensity is reduced exponentially according to the Beer-Lambert law

$$I(x) = I_0 e^{-\mu x} \tag{12.2}$$

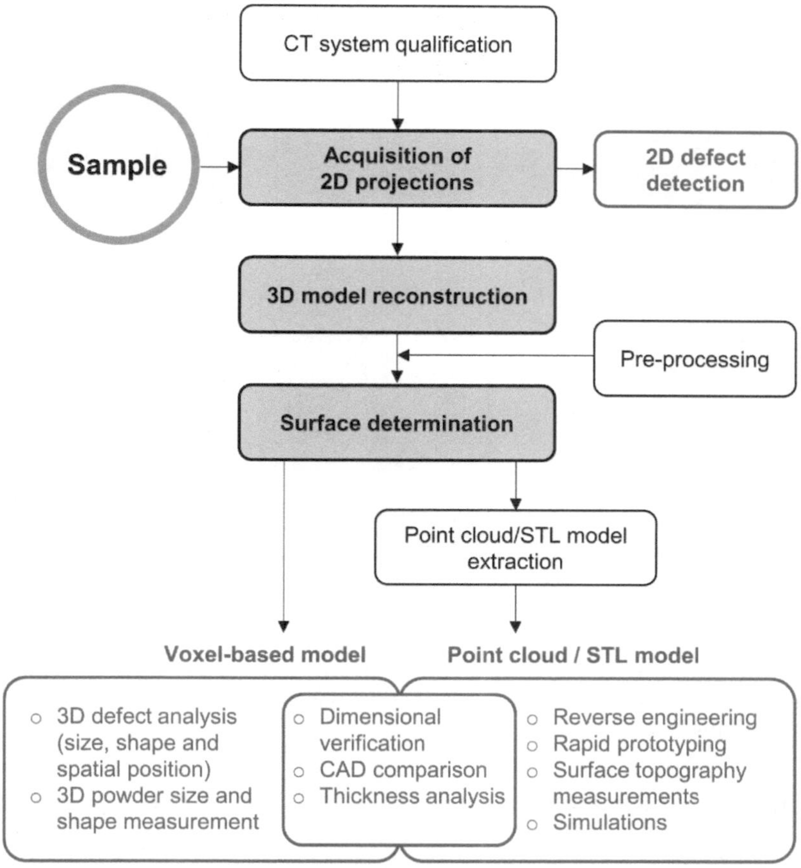

FIGURE 12.2
Schematic representation of the main steps of CT scanning for AM applications (examples of applications are also reported and subdivided depending on the type of used data: 2D projections, voxel-based model and point cloud/STL model).

where I_0 is the incident X-ray intensity, I is the residual intensity after traversing the material, x is the distance travelled through the material and μ is the attenuation coefficient of the specific material. Equation (12.2) is valid for an X-ray beam with a monochromatic spectrum traversing a homogeneous object. However, as the spectrum of industrial CT systems is usually polychromatic, and the material of industrial components is not homogeneous, Equation (12.2) must be modified as follows

$$I = \int I_0(E)\exp\left[-\int \mu(E,x)\,dx\right]dE \tag{12.3}$$

where E is the energy of the X-rays.

As shown in Figure 12.2, the user can perform a 2D defect analysis before reconstructing a 3D CT model by observing the X-ray projections and analysing the local grey value variations; darker regions can usually be explained with a decrease of material density that can be associated with the presence of a void (Lopez et al. 2018). This kind of analysis can require a short acquisition time because the number of projections can be lower than

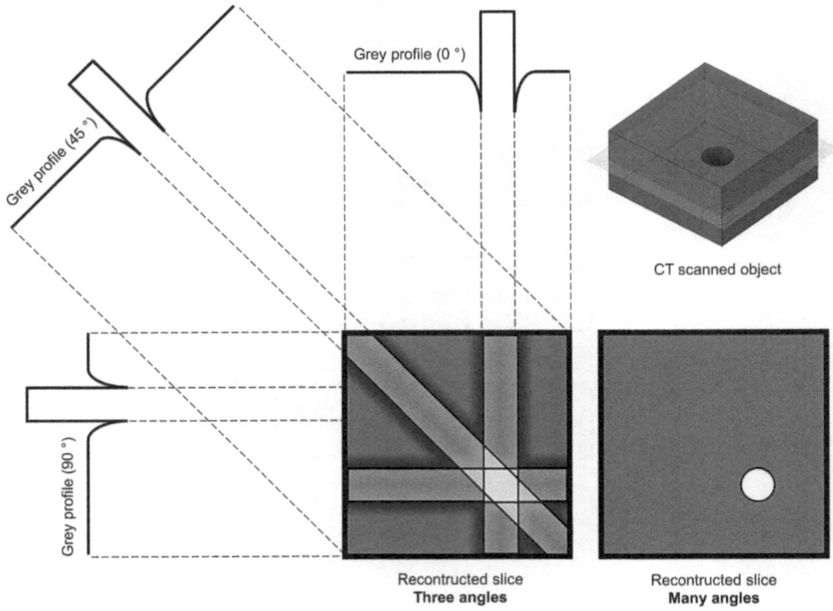

FIGURE 12.3
Schematic representation of the filtered back-projection. For each projection angle, a grey value profile is acquired and filtered to enhance the final reconstructed slice image. Many angles are needed to improve the accuracy of the algorithm.

the number needed for a complete 3D reconstruction, but the data interpretation can be complex.

In order to perform a more complete 3D analysis, a 3D model of the scanned object is reconstructed starting from the acquired projections. In cone-beam industrial CT, this process is often conducted with software implementing the Feldkamp-Davis-Kress (FDK) algorithm (Feldkamp et al. 1984), which is based on the so-called 'filtered back-projection' method (see Figure 12.3). The input for the reconstruction is constituted by the 'grey value profiles', representing the grey value of each pixel located on one line of the detector. In filtered back-projection algorithms, the focal spot is represented as a point, the radiation is considered as monochromatic and noise-free projections are assumed. However, in reality, the actual focal spot size has finite dimensions, the radiation is in most cases polychromatic and the projections are inherently affected by image noise. The assumptions and approximations of the FDK-based algorithms affect the quality of the reconstruction process. Alternative algorithms are available to achieve better quality for the reconstruction (Buzug 2008, Fessler 2000); however, their computational time is normally significantly higher than with the FDK-based algorithms. The reconstruction output is a 3D matrix where the single cell is a volumetric pixel, called a *voxel*. The voxel size v_s of the final reconstructed model is governed by

$$v_s = \frac{p_s}{m} \tag{12.4}$$

where p_s is the detector pixel pitch and m is the magnification factor, which is equal to the ratio between the source-to-detector distance (SDD) and the source-to-object distance

(SOD), shown in Figure 12.1. Each voxel is characterised by a specific grey value that represents a measure of the local X-ray attenuation. When the 3D reconstruction is completed, and before conducting the required analyses, the obtained CT volume can be subjected to pre-processing steps to enhance the data, for example by applying filters to reduce noise or to obtain global or local smoothing (Heinzl et al. 2017).

The next step, which is required for analysing the internal and external surfaces of the reconstructed object, consists of determining the interfaces between material and background (for example air) or between different materials with different attenuation coefficients (i.e. with different densities and atomic numbers). This can be done by using specialised algorithms capable of locating the surface boundaries. Depending on the algorithms used, the surface boundaries can be determined with voxel resolution or with sub-voxel resolution. Sub-voxel resolution is obtained by intra-voxel interpolation, which leads to a more accurate approximation of the actual surface. Figure 12.4 shows an example of a pore, the surface of which was determined with both voxel and sub-voxel resolutions. In addition, different approaches can be adopted for the surface determination. For example, global and local adaptive approaches are commonly based on the 'iso50' threshold value (Otsu 1979), defined as the mean value between background and material (or between two different materials, in the case of multi-material objects). In the case of a global surface determination approach, a global iso50 threshold value is used without distinction within the entire volume. However, such an approach often leads to erroneous surface models due to image artefacts (see Section 12.3.1), irregularities in the scan process, inhomogeneity of the sample material and numerous other effects. Alternatively, local adaptive methods can be used to optimise the surface contour location in each voxel, hence improving the accuracy of the analysis (Kruth et al. 2011, Torralba et al. 2018).

The final analysis steps can be conducted directly on the voxel-based model or on a surface model which is extracted from the determined surface and is normally available as a triangulated model in STL format or as a point cloud. In particular, surface models and their representations as STL files or point clouds are normally needed for the following

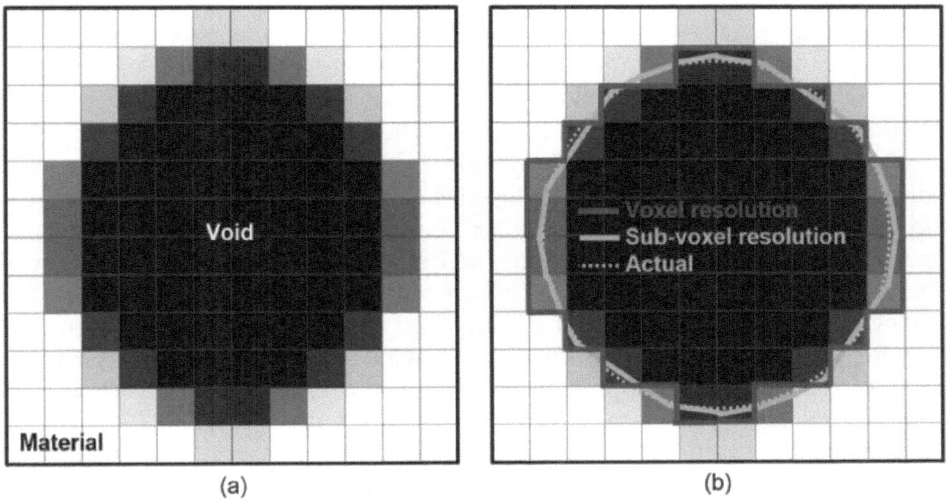

FIGURE 12.4
Example of a pore detected by CT. (a) Voxel-based representation of a 2D slice of the pore. (b) Representation of pore surface determined with voxel resolution (stepped solid line) and sub-voxel resolution (smooth solid line); the actual pore surface is also shown (dotted line).

applications: reverse engineering, rapid prototyping from reverse engineering data, surface topography characterisation and simulations (such as finite element analyses). Other applications, such as 3D defect analysis and 3D powder measurement, are commonly conducted directly on voxel-based models. Dimensional verifications (including comparisons with computer-aided design [CAD] models and thickness analyses) are commonly performed either on voxel-based models or on point clouds/STL models. Examples of the above-mentioned applications are provided in Section 12.4.

12.3 Measurement Errors and Traceability

When industrial CT systems are used to perform dimensional and geometrical measurements, the traceability of measurements to the unit of length (the metre) must be established (see Chapter 10). However, establishment of traceability in X-ray CT can be challenging, due to numerous and complex factors influencing the results of CT measurements (Zanini and Carmignato 2019a). The complexity of some of these factors is further magnified when measuring AM objects; for example, the influence of AM surface texture on CT dimensional measurements can be relevant (Carmignato et al. 2017b). In addition, CT results can be affected by the typical presence of image artefacts, i.e. systematic discrepancies causing shading, rings, streaks or bands that show up in the reconstructed volume, even if they are not actually present in the real object (Kruth et al. 2011). The main error sources and image artefacts are briefly addressed in Section 12.3.1. The subsequent sections discuss three prerequisites for traceability: metrological performance verification (Section 12.3.2), task-specific measurement uncertainty determination (Section 12.3.3) and reference objects that can be used for testing and calibration (Section 12.3.4).

12.3.1 Main Error Sources and CT Image Artefacts

The main error sources influencing the uncertainty of CT dimensional measurements are reviewed in the German guideline VDI/VDE 2630-1.2 (2008) and summarised in Figure 12.5, where they are subdivided into five groups: CT system, workpiece, environment, data processing and operator. Concerning the CT system, the characteristics and performance of its main components (X-ray source, detector and manipulator – see Section 12.2.2) directly influence the measurement uncertainty. In particular, the X-ray source performance is affected, for example, by the accelerating voltage, the filament current, the focal spot size and target characteristics. The detector performance depends on many characteristics, including the pixel pitch, which can directly influence the spatial resolution and signal-to-noise ratio (SNR) of the acquired X-ray projections (Carmignato et al. 2017a). In addition, possible misalignments and instabilities of the relative position and orientation of the hardware components can significantly affect the measurement accuracy (Dewulf et al. 2018).

The characteristics of the workpiece itself have a strong effect on the CT measurement performance. For example, the maximum length of the workpiece penetrated by the X-rays during a scan should allow for an acceptable X-ray transmission in order to produce images with sufficient contrast at the detector (De Chiffre et al. 2014). Moreover, surface texture can cause significant deviations in the dimensional measurements performed by different measuring techniques (for example, CT, tactile and optical CMSs – see Chapter 10)

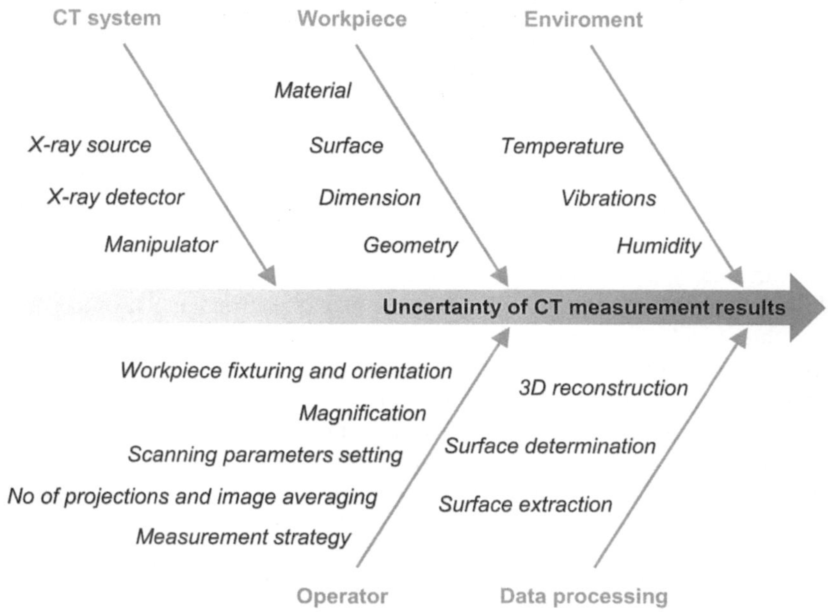

FIGURE 12.5
Ishikawa diagram representing the main error sources contributing to the uncertainty of CT measurements.

due to the different measuring principles producing different surface filtering effects (Carmignato and Savio 2011).

Environmental factors are also important. In particular, thermal gradients can lead to dimensional changes in both the workpiece and the CT system during the scan, with significant effects on the CT scan quality. The temperature and humidity in the cabinet and in the room where the CT system is installed should be maintained within the manufacturer specifications.

The data processing steps described in Section 12.2.3 (i.e. reconstruction, pre-processing, surface determination and STL/point cloud extraction) have a relevant influence, as they all contribute to the determination of the final results and can be implemented using different methods and parameters (see Section 12.2.3).

Finally, the operator is responsible for many choices during the entire CT measurement workflow. For example, the operator can select the workpiece orientation inside the measuring volume and the scanning parameters. In addition, the operator has to choose the fixture to be used to stably mount the workpiece during the scan, without disturbing the acquisition of the scanned object (often low absorption materials such as polymeric foams are chosen for the fixture, but they can suffer from material relaxation and other instabilities). Besides the error sources discussed above, the CT measurement results can be affected and sometimes impaired by image artefacts, which are unwanted effects that may generate erroneous representations of the actual workpiece. One of the most relevant image artefacts is due to the so-called 'beam hardening' effect, which can be understood by recalling that industrial CT systems are in most cases characterised by the emission of polychromatic radiation (i.e. the X-ray beam is composed of X-rays with different energies). Depending on the material and on the thickness to be penetrated, the low energy X-rays (called 'soft' X-rays) can be completely absorbed after a short travelling distance into absorbing material and only high energy X-rays (called 'hard' X-rays) can fully penetrate the object and reach the detector. This effect causes the attenuation of the X-ray beam to become non-linear. However, the reconstruction algorithms are normally based on the incorrect assumption of a linear attenuation. As a result, the outer skin of the object will have a different grey value range with respect to the core, even if the object is made from a single material (see Figure 12.6). Beam hardening can be reduced or eliminated by introducing physical filters of a certain thickness and material (typically Al, Cu or Sn) between the X-ray source and workpiece, to immediately filter out the low energy X-rays; or by software corrections based, for example, on linearisation techniques (Lifton and Carmignato 2017). If not completely eliminated, beam hardening can affect, for example, the analysis of material defects and dimensional measurements. In the first case, moving from the core to the skin of the object, the contrast between the grey values of internal defects and grey values of the material will be different, decreasing the quality of the analysis, since it is normally based on a global threshold value. In the second case, beam hardening may not be a problem for the edge detection of external surfaces, as the contrast between skin and background is increased (see Figure 12.6); however, if internal geometries are present (for example, produced by AM), the contrast will worsen, leading to an inaccurate edge detection (Dewulf et al. 2012). Local adaptive surface determination algorithms can be used in this case to reduce the influence of beam hardening. Another typical effect leading to an image artefact similar to beam hardening is caused by the fact that the detector can also

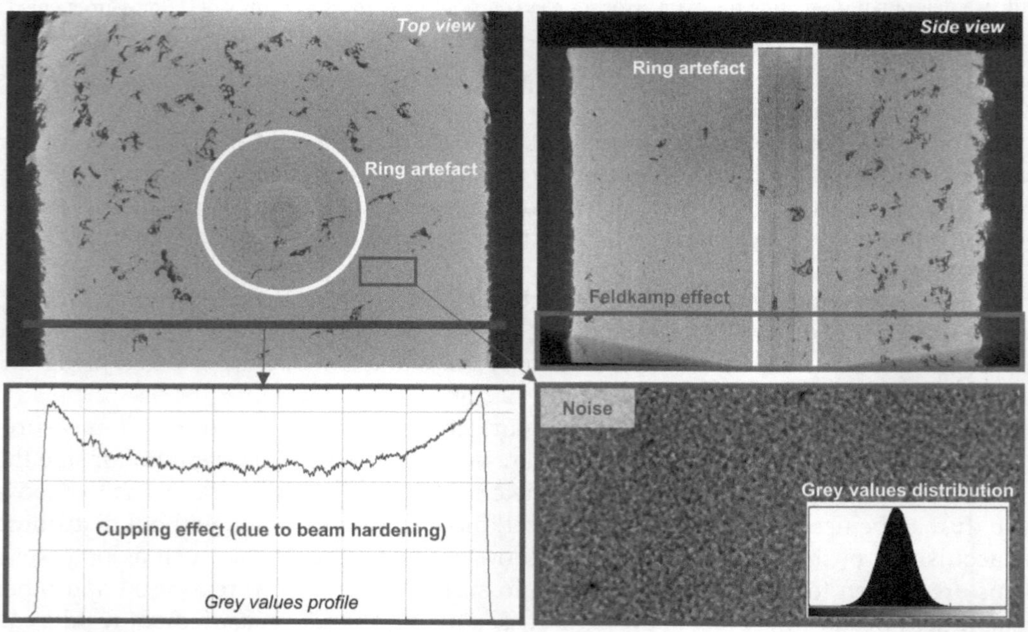

FIGURE 12.6
CT reconstruction of a Ti6Al6V part produced by laser PBF. The reconstruction is affected by several image artefacts. Ring artefacts can be observed on both top and side views. The beam hardening leads to the so-called 'cupping effect', visible from the analysis of grey value profiles (see, for example, the profile computed in correspondence of the horizontal line on the top view). The noise content is related to the standard deviation of the grey value distribution. The Feldkamp effect (visible on the lower part of the side view) is present only on the bottom end of the sample as it was scanned near to the detector border, while the top part of the sample is not affected by such artefact because it was positioned centrally with respect to the detector field of view. (Courtesy of the University of Padova.)

be reached by secondary radiation, known as *scatter*, generated when the emitted X-rays interact with the workpiece material or with components in the surrounding environment (for example, the CT system's cabinet).

A significant artefact that can appear only in the case of cone-beam CT configurations is the so-called 'cone beam artefact' or 'Feldkamp effect' (because it is related to the FDK algorithm [Feldkamp et al. 1984], explained in Section 12.2.3). The generated artefact, caused by the elliptical trajectory (projected on the detector) of object's points that rotate far from the central detector plane and from the central detector axis, is significant especially on surfaces parallel to the X-ray beam, and it becomes more pronounced for surfaces that are close to the top and bottom edges of the detector, as shown in Figure 12.6 (Xue et al. 2015). The effect can be reduced or eliminated with correct orientation of the workpiece inside the field of view. However, if complete elimination is not possible, internal defect analysis and dimensional measurements should be carried out only in regions and surfaces not affected by cone-beam artefacts.

The improper correction of non-ideal or defective ('bad') detector pixels will cause erroneous signal detection at the same location in each projection, leading to the formation of ring artefacts (i.e. rings of sharp contrast concentric to the centre of rotation; see Figure 12.6) in the CT reconstructed volume. The volume regions affected by such artefacts should not be considered when conducting an internal defect analysis, as many fake voids will be found in those regions.

In general, CT projections (and consequently the reconstructed volumes) are inherently characterised by the presence of noise (see Figure 12.6) caused by quantisation of X-ray photons or by detector electronic noise. Noise can be reduced by proper parameter settings (for example, increasing the exposure time and the number of frames to generate a projection) or by software filtering techniques. In the latter case, as a drawback, the filtering may decrease the scan resolution. The presence of noise can affect the quality of internal defect analysis (for example, speckle noise can erroneously be identified as small voids) and the quality of the determined surface (with consequent accuracy issues for dimensional analysis, surface topography measurements and STL model/point cloud extraction).

12.3.2 Metrological Performance Verification

The ISO 10360 series of standards specifies the characteristics and procedures to be used for testing the metrological performance of CMSs (see Chapter 10). However, the relevant ISO technical committee (ISO TC 213 working group 10) is still developing the draft of ISO 10360 part 11, which will address the performance verification of CMSs based on CT (Bartscher et al. 2017). At the time of writing, while the ISO standard is still not available, the most commonly used document for verifying the metrological performance of industrial CT systems is the German guideline VDI/VDE 2630-1.3 (2012), which is explicitly based on ISO 10360-2 (2009), with adaptations to CT systems. In particular, the procedures proposed in the German guideline are aimed at verifying the performance of a CT system using two main types of tests: the local behaviour is verified by evaluating probing errors of form and of size on calibrated spheres, while the global behaviour is verified by means of length measurement errors. An additional metrological characteristic that should be tested is the metrological structural resolution (MSR), which is defined as the size of smallest structure that can still be measured dimensionally. At present, a number of methods have been proposed for MSR evaluation (see, for example, Bartscher et al. 2012, Illemann et al. 2014, Zanini et al. 2017a), but a standard procedure is still a subject of some debate (Bartscher et al. 2017).

12.3.3 Uncertainty Determination

Assessing the task-specific measurement uncertainty is a fundamental step for achieving the traceability of measurements (see Chapter 10) and for taking decisions on the conformance or non-conformance of products with respect to specifications (Wilhelm et al. 2001). However, the presence of multiple complex factors influencing CT results (see Section 12.3.2) makes the uncertainty determination a difficult task for CT dimensional measurements. Due to the difficulty of modelling all the influence factors, analytical methods for uncertainty determination according to the ISO Guide to the Expression of Uncertainty in Measurement (GUM) (BIPM 2008), as well as simulation approaches based on the Monte Carlo method, are difficult to implement (Ferrucci 2017). A feasible experimental approach is represented by the substitution method, which is described for tactile coordinate measuring machines in the standard ISO 15530-3 (2011) and in Chapter 10, and adapted for CT in the German guideline VDI/VDE 2630-2.1 (2015). The application of this guideline requires that a calibrated object similar to the actual workpiece (in terms of material, size and geometry) is available or producible with sufficiently low calibration uncertainty (see, for example, Schmitt and Niggemann 2010 and Müller et al. 2014). An important limit for the application of the substitution method is that in many cases, appropriate reference objects are not available or cannot be calibrated with a sufficiently low uncertainty. Other approaches have been proposed for the uncertainty evaluation of CT measurements of specific parts. For example, an approach was used by Zanini et al. (2018) for determining the uncertainty of CT wear measurements performed on acetabular cups. In this case, the CT system repeatability and reproducibility were found to be the main contributions to the measurement uncertainty. Another method was proposed by Dewulf et al. (2013), based on the concept that CT dimensional measurements are described by the product of the voxel size and the number of voxels. The relative uncertainty was then considered starting from the uncertainty of the voxel size and the uncertainty of the edge detection.

The relevance of uncertainty determination of CT measurements is highlighted by the effort of several research groups internationally and by the results of inter-laboratory comparisons specifically organised for CT dimensional measurements. For example, the *CT Audit* project, coordinated by the University of Padova, was the first international intercomparison of CT systems for dimensional metrology (Carmignato 2012). Subsequently, the Technical University of Denmark coordinated the *CIA-CT* (Angel and De Chiffre 2014) and the *InteraqCT* (Stolfi and De Chiffre 2018) inter-comparisons. The results obtained from the above-mentioned inter-comparisons helped the understanding of the major challenges connected to the uncertainty determination of CT dimensional measurements and promoted the development of dedicated research projects and international standardisation initiatives. However, as discussed in Chapter 10 for optical CMSs, AM objects bring their own complications to uncertainty analysis due to their potentially complex surface form and texture (see Section 12.4). Recently, du Plessis et al. (2019a) presented the results of a round robin test conducted by ten laboratories using X-ray CT to analyse three metal AM parts; the sources of deviation in the results of the different laboratories were discussed and categorized.

12.3.4 Reference Objects

Different types of reference objects (calibrated dimensional artefacts – Carmignato et al. 2020) are commonly used for different roles along the CT measurement workflow, for example, for CT system qualification, for metrological performance verification or for task-specific

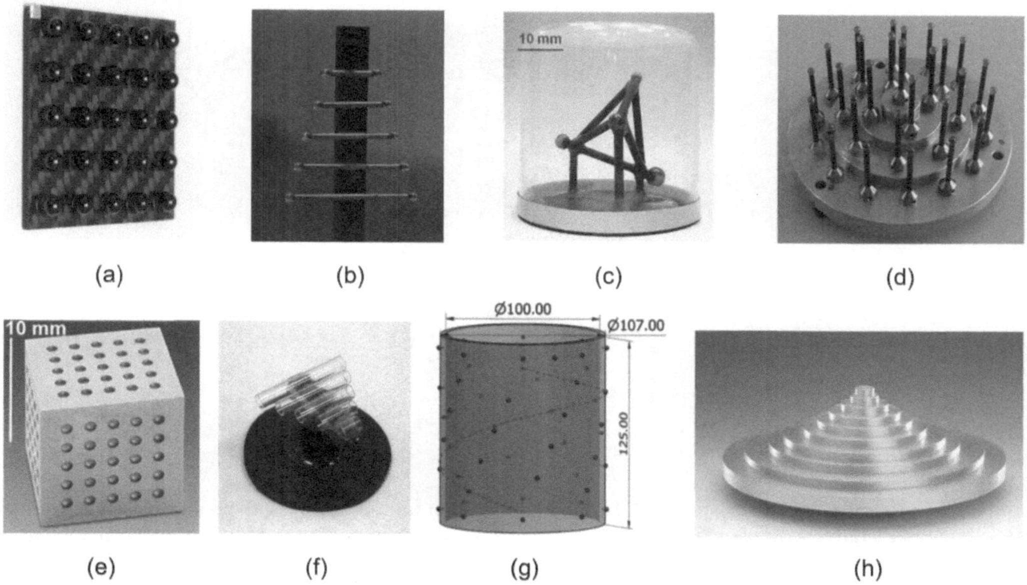

FIGURE 12.7
Examples of reference objects for qualification and testing of CT systems: (a) Ball plate (see Müller 2013) (courtesy of DTU); (b) CT-tree (see Müller 2013) (courtesy of DTU); (c) CT-tetrahedron (see Carmignato 2012) (courtesy of the University of Padova); (d) 27 spheres standard (courtesy of the University of Padova); (e) Calotte cube (courtesy of the PTB); (f) Pan flute gauge (see Carmignato 2012) (courtesy of the University of Padova); (g) CT^2 (see Hermanek et al. 2017b) (courtesy of the University of Padova); (h) Step cylinder (courtesy of the PTB).

uncertainty determination using the substitution method (see Section 12.3.3). Such reference objects must fulfil several requirements. Some requirements are similar to those of the reference objects used for other CMSs, as for example long-term dimensional stability and sufficiently low calibration uncertainty. Other requirements are specifically needed for CT, for instance, the compatibility of the X-ray attenuation coefficient of the material and of the overall dimensions with an acceptable X-ray penetration length. In particular, reference objects used for CT are commonly made of relatively low-absorbing materials, such as aluminium, titanium, ceramics and aluminium oxide (synthetic ruby), possibly interconnected or supported by carbon fibre–reinforced polymer frames, which have good dimensional stability, low coefficient of thermal expansion and low X-ray attenuation coefficient (Bartscher et al. 2017). Figure 12.7 shows examples of reference objects commonly adopted for the qualification and verification of CT systems, as well as for the evaluation of specific influence factors.

Other reference objects have been developed to assess the accuracy of specific types of CT-based analysis (see Leach et al. 2019), such as measurements of gaps in mono- and multi-material parts (Figure 12.8a), porosity analysis (Figure 12.8b) and surface topography characterisation of metal AM parts (Figure 12.8c).

12.4 Applications of CT Metrology for AM

Figure 12.9 illustrates the main possible applications of CT measurements in the field of AM. Although the scheme in Figure 12.9 is specific for powder bed fusion (PBF) processes,

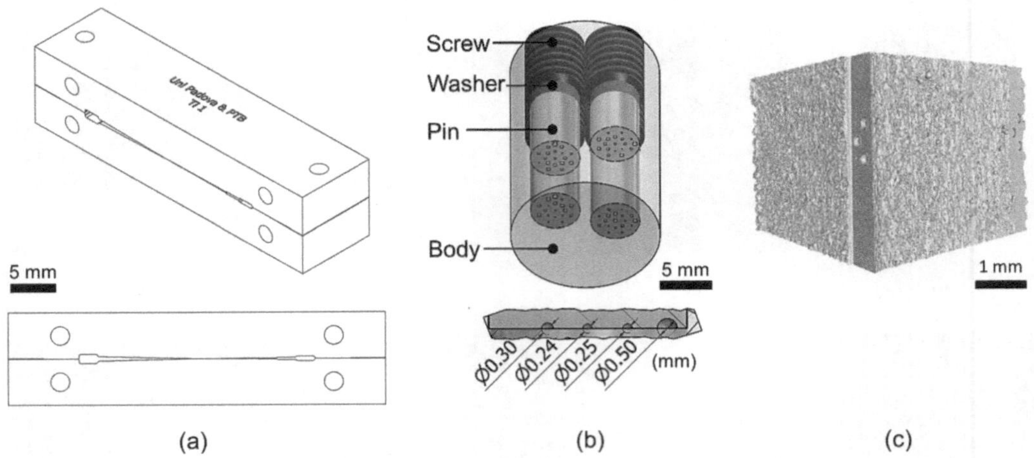

FIGURE 12.8
Examples of reference objects for specific CT-based analysis: (a) multi-material gap standard (see Hermanek et al. 2017c); (b) dismountable standard for the accuracy assessment of porosity analysis (see Hermanek et al. 2017d); (c) AM reference object for the traceability establishment of CT-based roughness measurements of AM surfaces (see Zanini et al. 2019a). (Courtesy of the University of Padova.)

the same concepts hold for other AM processes (except for the powder selection step). In general, CT can be used for quality control at different points of the AM process chain, from selection of the feedstock material to verification of the final product (see 'Quality control by CT' in Figure 12.9). In addition, CT can provide feedback for AM product development and process optimisation (see the 'feedback' arrow in Figure 12.9). Moreover, CT can be used as a reverse engineering tool for acquiring the sample geometry to generate the model of the part to be printed (normally translated into the STL format, which is the most commonly used file format for AM). In this latter case, the reverse engineering step substitutes the design step (see 'Reverse Engineering by CT' in Figure 12.9). Currently, there are no CT systems available for in-process AM inspection, due to the insufficient scanning speed and the technical difficulties of installing a CT device inside an AM machine. However, CT can be employed for comparing the defects detected during the process, for example using on-machine high-speed optical or thermal devices (see Chapter 13), with defects detected by CT after the process end. This comparison is useful to improve the understanding of defect generation mechanisms and of the process itself. For example, some defects arising on a specific layer may close up as the process continues, for instance due to re-heating when producing the successive layer, and such an evolution can be understood by comparing on-machine gathered data with CT data. Besides the possibility of performing porosity analyses, AM parts can be analysed by CT for dimensional and geometrical verification and for surface characterisation. However, it is important to highlight that CT should not be seen merely as a tool for final quality inspection of AM products. CT, instead, is particularly useful for obtaining holistic data that can provide fundamental feedback to improve the process and its preparation, including optimised AM part design, enhanced build set-up (for example, orientation of the part to be built), selection of optimal process parameters (for example, laser scanning speed, scanning strategy, laser power and layer thickness) and selection of optimised feedstock material.

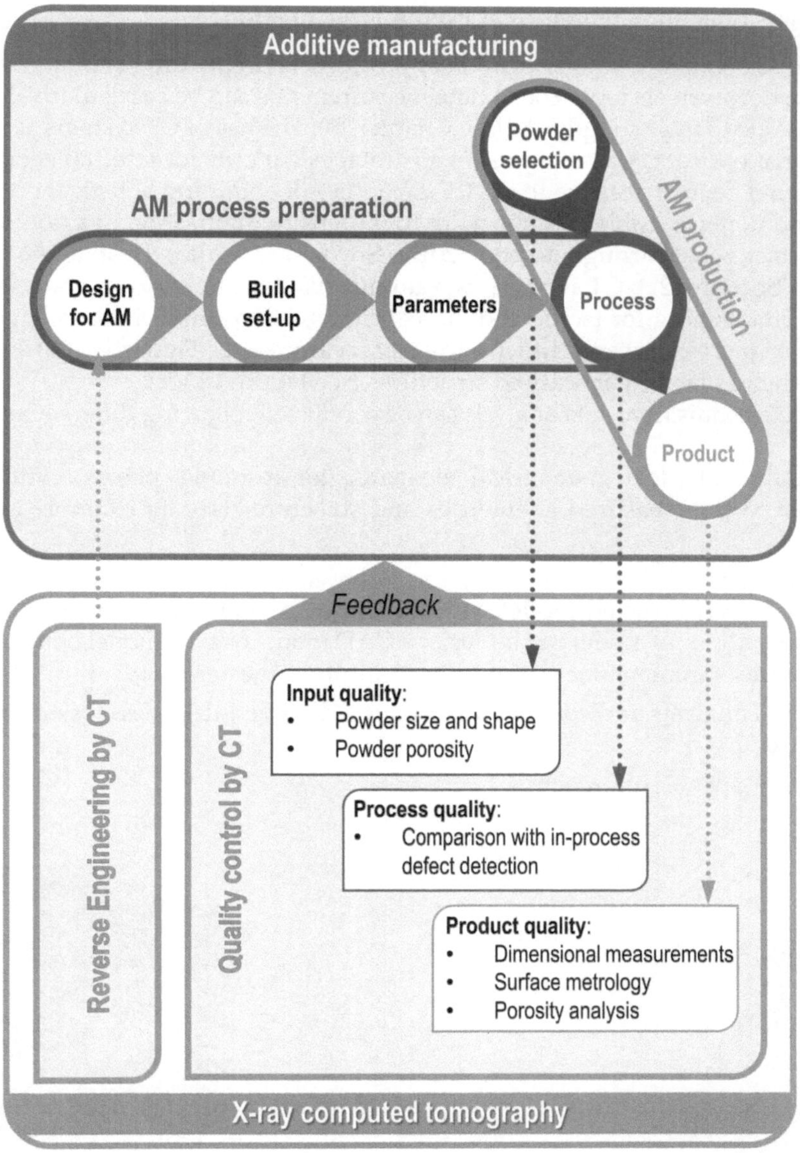

FIGURE 12.9
Schematic representation of the possible applications of X-ray CT in the field of AM, with reference to metal PBF processes.

Possible applications of CT in the AM field, as summarised in Figure 12.9, are presented by means of examples in the following sections. Section 12.4.1 deals with dimensional and geometrical product verification. Section 12.4.2 is dedicated to internal defect analysis. Section 12.4.3 covers the characterisation of AM surface topography. Section 12.4.4 discusses the characterisation of the powder feedstock properties. Finally, Section 12.4.5 addresses the possibilities offered by CT for providing feedback for product development and process optimisation.

12.4.1 Dimensional and Geometrical Product Verification

Dimensional and geometrical verification of products is highly important in manufacturing industry; however, accurate coordinate measurements can be particularly challenging for AM parts due to their complexity (see Chapter 10). The use of CT systems as an alternative to tactile or optical CMSs can present advantages not only for internal geometries but also for external geometries, due to the CT capability of obtaining holistic models, of both accessible and non-accessible features, with high-density digitisation in a non-destructive and non-contact way (Carmignato et al. 2017a, Snyder et al. 2015). Despite the limitations described in Section 12.3, CT represents a unique solution to perform analyses in cases that are too challenging for tactile and optical CMSs, for example, internal channels (see Figure 12.10a), micro-holes and difficult-to-access features (see Figure 12.10b) and intricate geometries such as lattices or scaffold structures (see Figure 12.10c).

After CT data acquisition, data analysis can be conducted following different approaches:

1. Association of ideal geometrical elements (for example, planes, cylinders or spheres) to the measured geometries and structures (see an example in Figure 12.11a).
2. Nominal–actual comparison (see, for example Figure 12.11b), i.e. comparison of the CT reconstructed model with a nominal model (after proper alignment), where the nominal model can be either a CAD model or a 3D model obtained from another measurement (for example, by scanning a master part).
3. Thickness analysis and comparison with the designed thickness (see an example in Figure 12.11c).
4. Other task-specific approaches.

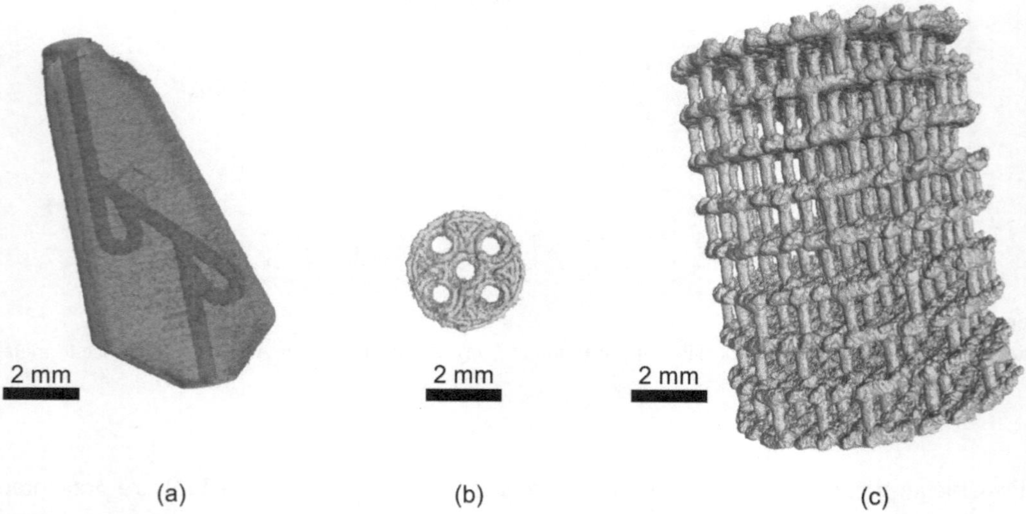

FIGURE 12.10
Examples of CT reconstructed volumes of AM metal parts that are difficult to measure with other CMSs: (a) Ti4Al6V micro-valve with internal channels produced by laser PBF; (b) NiTi micro-component with internal micrometric holes produced via micro-direct metal laser sintering (see Khademzadeh et al. 2018); (c) Ti4Al6V lattice structure produced by laser PBF and including features that are difficult to access from the outside (see Sbettega et al. 2018). (Courtesy of the University of Padova.)

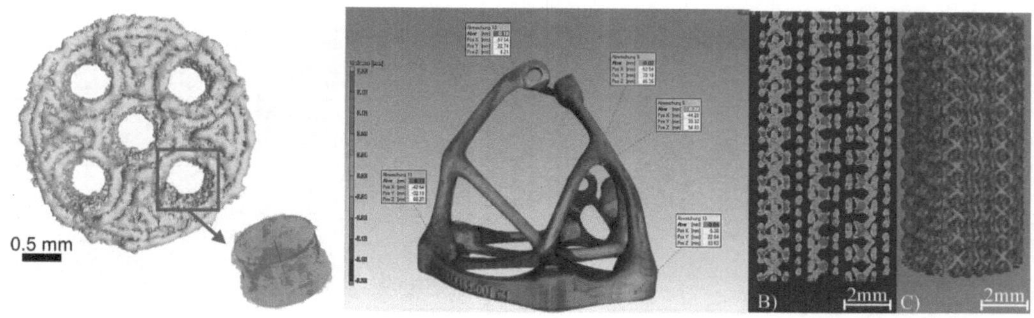

(a) (b) (c)

FIGURE 12.11
Examples of CT data analysis: (a) ideal cylindrical element associated to the inner feature of a NiTi sample produced by AM (from Khademzadeh et al. 2018; with permission); (b) CAD comparison of an aerospace bracket (adapted from Orme et al. 2018; with permission); (c) thickness analysis of a lattice structure (from Van Bael et al. 2011; with permission).

The first approach, through association of ideal geometrical features, uses the common approach adopted with contact CMSs. For example, when measuring the diameter of a cylindrical feature, an ideal cylinder is associated with the extracted points (Figure 12.11a), and its diameter is then computed. In the case of AM parts, the complex shape and high levels of surface texture can significantly influence the comparability of the results obtained from different CMSs due to the different measuring principles and their different filtering effects (Carmignato et al. 2017b). This must be taken into consideration when comparing CT measurement results with reference values obtained, for example, using tactile CMSs.

The second approach determines the deviations of the fabricated part through comparison with a nominal model or other 3D models to be used as a reference. Such deviations can be annotated and displayed directly on the measured surface, as shown for example in Figure 12.11b (Orme et al. 2017, Villarraga-Gómez et al. 2015, du Plessis and le Roux 2018). This type of outcome is useful to gain information about common intrinsic errors of AM parts which cause overall deviations, such as shrinkage and overbuild (du Plessis et al. 2018a), bending and warping or creeping and delamination. This approach, although often common practice, may suffer from the possible low accuracy of the alignment procedure due to the typical presence of significant structural deviations, especially in the case of AM lattice structures (Sbettega et al. 2018). In such cases, other approaches are preferred.

The third approach is based on thickness analysis. Different from the previous case, this approach can eliminate the influence of the alignment to a nominal model. In fact, depending on the specific thickness analysis, it can be performed without reference to a given nominal or CAD model. For example, in the case of a cylindrical structure, the strut thickness can be assessed by defining a search angle centred on the lines perpendicular to the two opposite surfaces considered for the starting point and the end point. In any case, the thickness analysis has been proven to suffer from the influence of high levels of surface texture (Sbettega et al. 2018).

Other task-specific approaches can be implemented using CT data, depending on the requirements of specific applications. In the case of lattice structures, for example, CT can provide a holistic geometrical analysis of strut dimensions, connections at nodal points, structural distortions and flaws within the struts (Pyka et al. 2013, Bauza et al. 2014, Wang et al. 2018, Chang et al. 2019). When lattice structures are produced for applications in the field of energy storage or energy production (Figure 12.12a), for instance, as

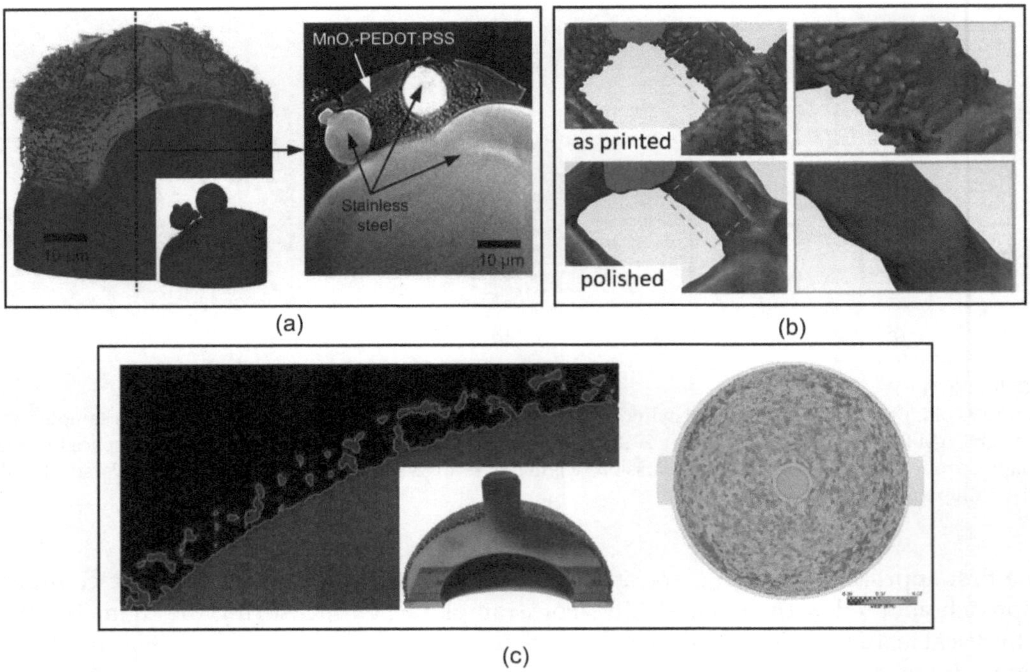

FIGURE 12.12
Examples of porous and scaffold structures from different fields of applications. (a) High resolution CT of a multi-material porous array electrode of a pseudo capacitor, showing the AM stainless-steel structures in grey and the deposited MnOx-PEDOT:PSS layer (from Liu et al. 2016; with permission); (b) 3D rendering of a BCC lattice structure as printed and after polishing, revealing homogenised thickness of the struts (from Chang et al. 2019; with permission); (c) CT segmentation of the porous structures of an acetabular hip prosthesis cup and analysis of the mean local thickness (adapted from Kourra et al. 2018. With permission).

capacitor materials or active layers for fuel cells needing high surface-to-volume ratios, CT-based analyses can be performed to quantify the surface-to-volume ratio, or to measure the structure of porous surfaces and the potential remaining un-melted or partially sintered powder particles (Figure 12.12b) (Liu et al. 2016, Barui et al. 2017, Chang et al. 2019). Additionally, CT is increasingly applied to measure lattice scaffold-like structures for biomedical applications (Figure 12.11c) (Taniguchi et al. 2016, Kourra et al. 2018). For example, the ingrowth of bone on and into the structures can be dimensionally evaluated by CT, enabling a direct comparison between different structure types with respect to their effectiveness. Other examples of task-specific dimensional analyses are in the field of assemblies. CT can be used to measure each individual component in the assembled state and to identify and measure the contact surfaces. This is applied, for example, for the analysis of dental implants that are currently being produced using metal AM (Traini et al. 2008) and for which the dimensional specifications on the contact surfaces are important to avoid bacteria infiltration and implant failure (see, for example, Figure 12.13).

12.4.2 Internal Defect Analysis

Metal AM technologies are capable of producing high-density parts close to the nominal density; however, fully dense parts cannot be produced yet due to the inherent presence of internal voids, typically generated by process instabilities, gas bubbles, oxides

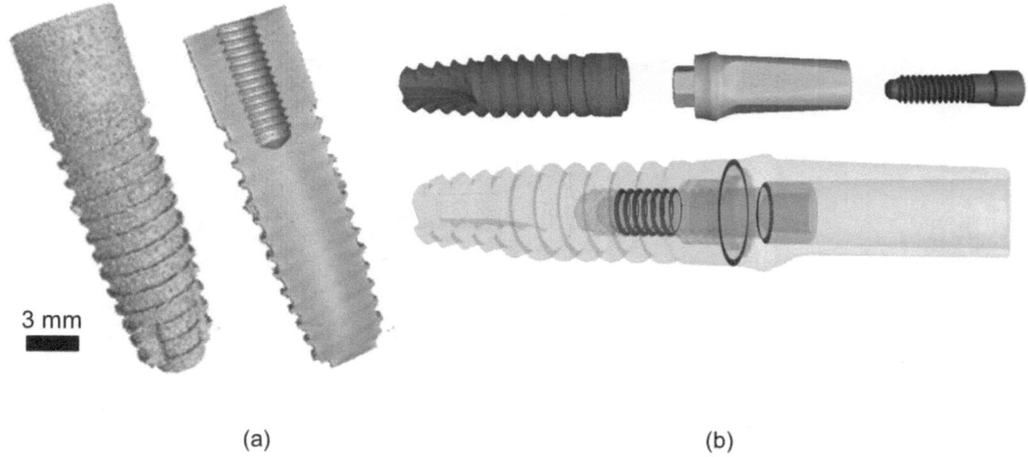

FIGURE 12.13
Examples of CT-based dimensional analyses on dental implants: (a) CT reconstruction of a dental implant produced by laser PBF of Ti4Al6V with micro-machined internal thread (see Zanini and Carmignato 2019b); (b) evaluation of the contact surfaces between the components of a dental implant assembly (see Zanini et al. 2017b). (Courtesy of the University of Padova.)

inclusions and, in the case of powder bed processes, entrapped non-totally melted particles (see Chapter 4). Internal voids and inclusions may act as nuclei for cracks, leading to possible reduced mechanical properties (Kruth et al. 2007, Schmidt et al. 2017). Besides their dimensions, the shape of pores is an important characteristic to be investigated, as it is key information for understanding both the cause of the pore generation and the influence of porosity on the performance of the final product. For example, relatively spherical pores are usually an indication of entrapped gas, typically due to local overheating, while irregular and elongated pores are an indication of the presence of non-totally melted particles within a single layer or between consecutive layers, typically due to insufficient energy (Gong et al. 2013). In addition, the spatial distribution and location of pores within the part volume are useful indications for the quality assessment of an AM product, for improving the process itself and for performing accurate post-process surface refinement. In fact, pores near to the external surface can emerge on the external surface after being machined, for example, by a milling operation (Wits et al. 2016).

As shown in the example in Figure 12.14, CT is capable of providing a complete analysis of internal defects (voids and inclusions). In particular, CT can evaluate defect characteristics, such as volume, size (for example, the equivalent diameter of a sphere with the same volume of the defect), surface area, shape (a commonly used shape indicator is for example, the defect sphericity, calculated as the ratio between the surface area of a sphere with the same volume of the defect and the surface area of the defect itself), elongation (for example, computed form sizes and areas projected on the planes of a reference Cartesian coordinate system) and the spatial distribution of defects within the analysed volume. Due to these capabilities, CT can overcome the limitations of traditional methods commonly used for porosity/density assessment (Wits et al. 2016). For example, the Archimedes method (based on the difference in buoyancy of an object's mass measured in air and submerged into a fluid) has been shown to provide reliable and repeatable results at a relatively low cost and short time (ASTM D2734 2009). However, the Archimedes method only provides an average density or porosity value; it cannot give information about the size, shape and spatial distribution of defects. Moreover, as the relative density is computed with reference

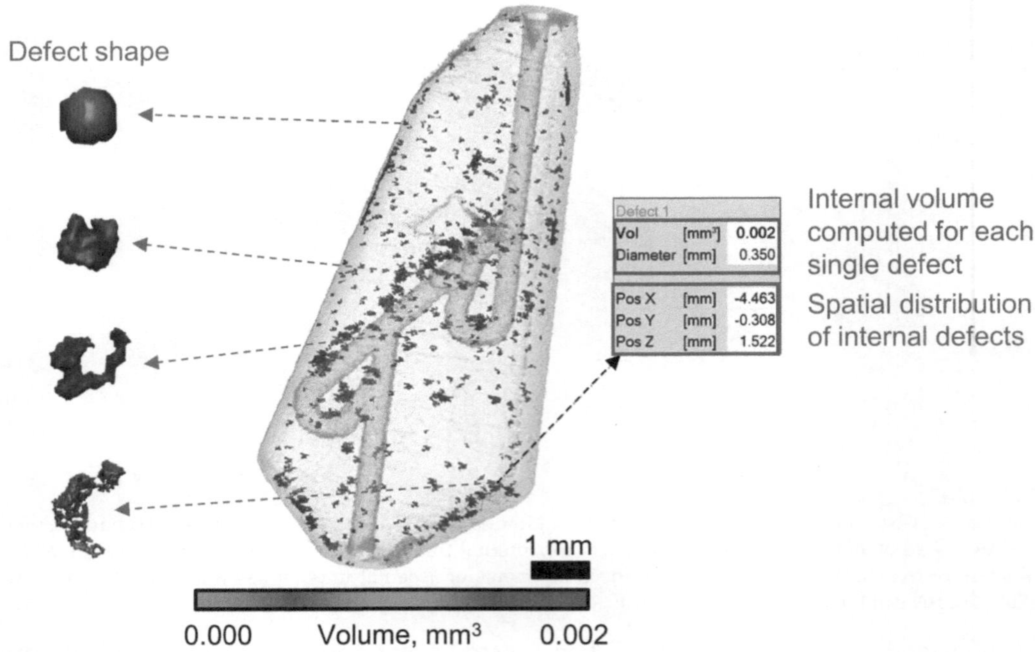

FIGURE 12.14
Example of porosity analysis on a CT reconstruction of a micro-valve produced by laser PBF of Ti6Al4V. The pores are analysed in terms of size, shape and spatial distribution within the entire volume. (Courtesy of the University of Padova.)

to the nominal bulk density of the part material, the measurement uncertainty is increased when the tested parts have non-homogeneous material characteristics, which is often the case with AM components (for example, due to heating gradients occurring during the process or in the case of functionally graded materials). Alternative non-destructive defect analyses are ultrasonic testing and helium pycnometry (see Chapter 9). In ultrasonic techniques, the extraction of 3D information about internal voids/porosity is complex and the accuracy of the results may be influenced by the non-homogeneous size and distribution of defects (Slotwinski et al. 2014). Pycnometry can measure the part volume by gas displacement and is used together with a high-precision balance to measure the part mass that, together with the volume information, is needed to compute the part density. Helium pycnometry has been found not to be accurate enough to measure the porosity of high-density metal AM components (Wits et al. 2016). Metallography is another technique often used for AM parts, based on cutting, epoxy-resin embedding and polishing operations followed by optical analysis, in most cases operated through an optical microscope (Wits et al. 2016, Khademzadeh et al. 2016). Metallography is destructive, time consuming, highly dependent on the quality of sample preparation and limited to 2D analyses of cut sections. Moreover, the limited number of analysed cut sections can, in the case of non-homogeneously distributed defects, result in incorrect global porosity content assessment.

Although CT has important advantages with respect to the methods presented above, the dimensional accuracy of CT-based defect measurements may be affected by the influencing factors already listed and described in Section 12.3. For example, voxel size errors and the surface determination procedure may influence the measured size of individual defects as well as the overall defect content. Moreover, the CT metrological structural resolution, which depends on focal spot size, voxel size and other factors, limits the minimum defect size that can be reliably detected and accurately measured (Hermanek et al. 2019). For establishing the traceability of CT porosity measurements, the comparison with traceable reference measurements is fundamental. In several studies, CT was compared with the Archimedes method and metallography, and it was found that CT tends to produce lower porosity values (Spierings et al. 2011, Wits et al. 2016, Slotwinski et al. 2014). However, neither the Archimedes method nor metallography can be taken as a traceable reference for 3D measurement of pores due to the limitations of these two methods, discussed above. Therefore, to establish the traceability of CT porosity measurements, Hermanek et al. (2019) developed a method based on a dismountable reference object (see Figure 12.8b) with artificial hemispherical defects calibrated before the assembly by accurate coordinate measurements.

12.4.3 Surface Topography Characterisation

Surface metrology of AM parts is covered in Chapter 11. This section examines in more detail the use of CT for AM surface topography characterisation. One of the main reasons why CT is applied for surface metrology is that it allows non-accessible surfaces to be measured (for example, the surfaces of an internal geometry) in a non-destructive way (Townsend et al. 2017, Thompson et al. 2018, Zanini et al. 2019a). Moreover, currently available micro- and nano-focus CT systems are capable of reconstructing micro-scale re-entrant surface features, with appropriate metrological structural resolution (Zanini et al. 2017a). AM metal parts are inherently characterised by complex surface topographies with high levels of texture and high degrees of irregularity at different scales of observation (Townsend et al. 2016, Thompson et al. 2017, Chapter 11). In particular, metal surfaces produced by PBFs can include re-entrant features, such as overhangs and undercuts, characterised by two or more z height values for an (x,y) position (see Figure 12.15, where an AM surface profile is shown as an example). Re-entrant features are mainly caused by the presence of non-totally melted powder particles spattered and attached to the surface, and by the presence of sub-surface defects that may open up to the surface. Such complexity, dominated by re-entrant features and high levels of texture, is undesired for applications requiring good surface finish (Gibson et al. 2015) but is desired for those applications where an increase in the total surface area becomes a functional advantage, such as for battery and capacitor plate designs, for cooling and fluid flow, for adhesive behaviour and for orthopaedic and dental implants (Liu et al. 2016, Townsend et al. 2019). In general, the

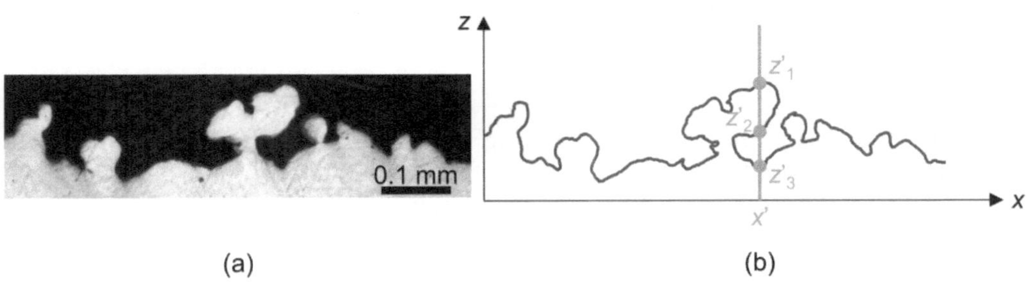

FIGURE 12.15
Example of a metal AM surface containing re-entrant features: (a) cut-section of the metal surface (see Zanini et al. 2019a); (b) corresponding profile showing that the surface portions interested by the presence of re-entrant features are characterised by more than one z height value for an (x,y) position). (Courtesy of the University of Padova.)

surface characteristics of as-build AM products are strongly related to the process parameters; hence, accurate surface characterisation is necessary to improve the process itself, if the objective is either to improve the surface quality or to extend the total surface area.

In order to properly extract areal surface texture data from CT reconstructed volumes, Townsend et al. (2017) proposed a methodology consisting in (i) cropping the data around the surface of interest, (ii) creating and exporting an STL file from the CT surface points, (iii) cleaning the mesh by removing all re-entrant features, (iv) converting the obtained surface points to a height map, and (v) computing the areal surface parameters as defined in ISO 25178-2 (2012). The removal of re-entrant features is necessary to compute conventional areal and profile surface texture parameters (ISO 25178-2 2012, ISO 4288 1998) from CT data. Hence, the limit of this methodology is that a number of surface features acquired by CT are filtered out, hence discarding potentially relevant surface information (see the example reported in Figure 12.16, where a profile with re-entrant features is compared with the same profile after removal of re-entrant features).

In order to include re-entrant features in the topography characterisation of metal AM surfaces, a generalisation of surface texture parameters to be applied on CT data without removing re-entrant surface features was proposed by Pagani et al. (2017, 2018). In particular, the proposed approach is based on the parameterisation of the surface or profile containing re-entrant features. For example, in the case of a surface profile, the parameterisation is conducted by computing the infinitesimal arc length elements of consecutive segments along the investigated profile. If no re-entrant features are present, then the generalised parameters coincide with the corresponding conventional ones. The importance of including the re-entrant surface features on the computation of surface texture parameters was illustrated by Zanini et al. (2019b), comparing the parameters calculated using the generalised parameter definitions (re-entrant features not removed) with the corresponding conventional parameters (re-entrant features removed). Moreover, Khademzadeh et al. (2019) found that the differences between measurements performed on the same surfaces by focus variation microscopy (see Chapter 11) and CT correlated with the percentage of re-entrant features. Accuracy and traceability challenges were studied by Zanini et al. (2019b) using calibrated profiles extracted from cross-sections of an AM reference object (Figure 12.8c).

A different approach to analyse the surface, not based on statistical parameters, is the so-called feature-based analysis (Senin et al. 2018), targeting spatter formations, weld tracks and weld ripples and their geometrical attributes (see Chapter 11). This analysis – which has been implemented for optical measurements so far (Senin et al. 2018) – is in principle applicable when using CT data (see the example in Figure 12.16), but often spatial resolution may not be sufficient for measuring the smallest surface characteristics.

12.4.4 Powder Feedstock Characterisation

The quality of the feedstock material, which might be metal powder in the case of PBF processes, has a direct influence on the quality of metal AM products. The powder characteristics of interest are the size distribution (a wide size range facilitates better particle packing and a higher final part density), the morphology (for which it is recommended to use almost spherical shapes for better flowability and uniformity of powder layers and density) and the possible presence of internal defects (to be avoided) (Gibson et al. 2015). AM powders are commonly characterised using different instruments and approaches, such as sieving, microscopy, spectroscopy or laser diffraction (Sutton et al. 2017, see Chapter 6). The use of multiple methods is typically necessary for determining the different properties

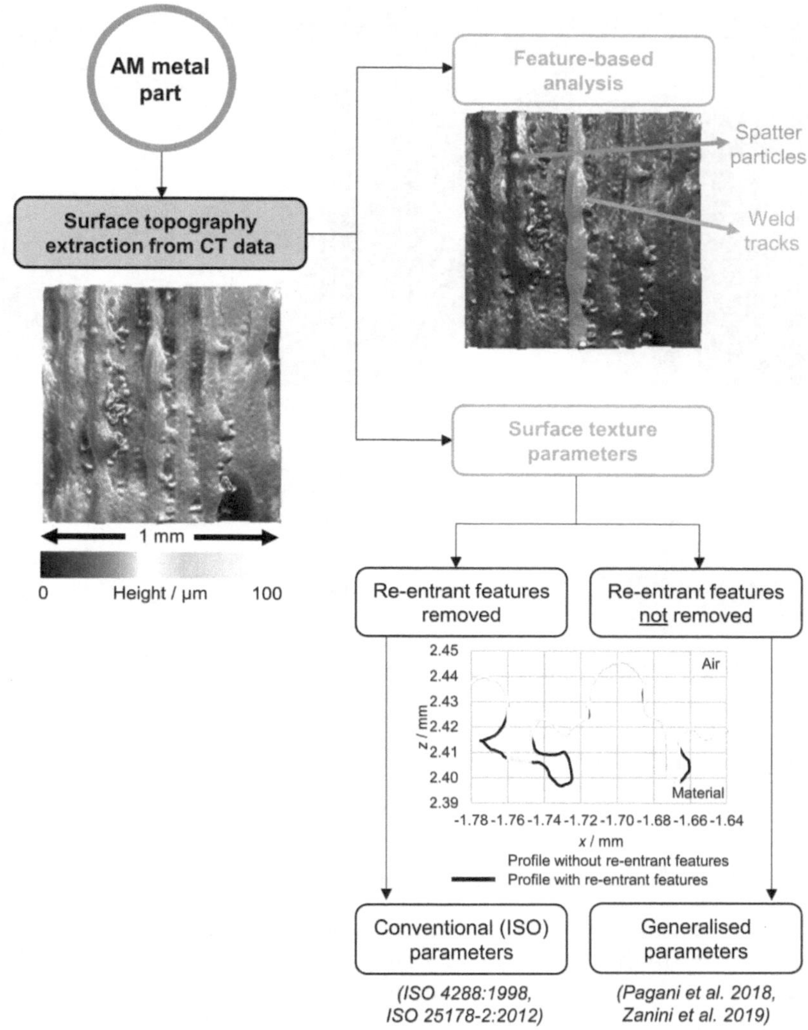

FIGURE 12.16
Schematic representation of possible methods for CT-based surface topography of AM metal parts. After the extraction of the surface topography as a height map, two different types of analysis can be performed: feature-based analysis and characterisation based on surface texture parameters. In the latter case, the evaluation can be done using the conventional (ISO) parameters (when the re-entrant features are previously removed) or using generalised parameters (when the re-entrant features are not removed).

of powders, as most techniques can only determine one single characteristic (Sutton et al. 2017). Recently, CT was successfully proposed to perform a complete quality assessment of powder particles, determining all the characteristics mentioned above (du Plessis and le Roux 2018). Figure 12.17 shows an example of CT analysis of powder, where the size distribution and morphology were determined. Being a new field for CT, different approaches are currently being tested to further improve the significance of the determined feedstock properties. For example, sample preparation and segmentation of the single particles are the main challenges to fully exploit the possibilities of CT within the field of powder characterisation (Heim et al. 2016, Bernier et al. 2018, Sinico et al. 2018, du Plessis et al. 2018a).

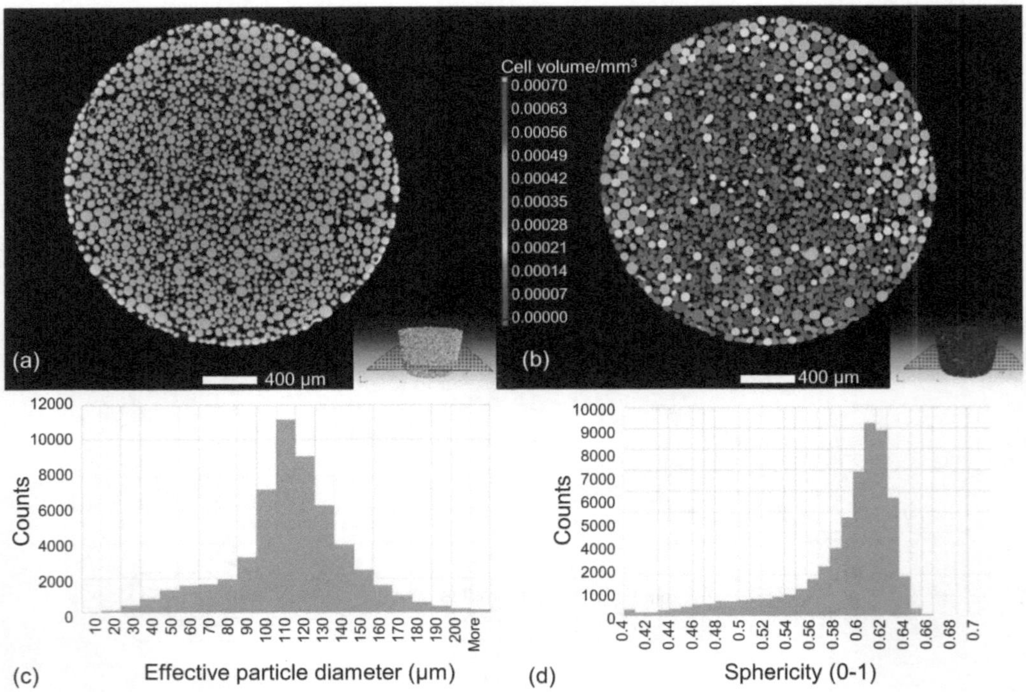

FIGURE 12.17
Example of metal powder analysis conducted by CT: (a) CT slice image of a metal powder sample, (b) analysis of the size of each powder particle on the same slice, (c) extracted information on particle size distribution and (d) sphericity distribution. (Adapted from du Plessis et al. 2018c.)

12.4.5 Product Development and Process Optimisation

In AM, product development and process optimisation are strongly correlated. This is due to the fact that a change in the product design can have a significant effect on the build of the part itself and on the different deviations or defects that can arise during the process. In this context, the accurate acquisition of holistic 3D models representing the geometry of actual products and their characteristics (for example, dimensions, structural deviations, surface defects and internal porosity) is necessary to improve the quality of the manufacturing technology and the final product. Holistic models are also needed to improve process simulations (for example, structural simulations that can be used to improve the part design or process simulations that are used to tune the process parameters for improving the final product properties) and process modelling, which are – together with AM itself – enabling requirements to realise Industry 4.0 (see Chapters 3 and 4). As specified in the previous sections, CT allows the reconstruction of 3D high-density digitisation models of AM parts, including both accessible and non-accessible geometries and features; hence, it is an optimal tool to be used directly by product designers and engineers in the above-depicted framework.

12.4.5.1 CT for Product Development

The product development process can benefit from the combination of AM and CT, because designers are no longer constrained by the limitations of conventional manufacturing processes and conventional measuring techniques. In fact, AM allows the fabrication of parts with highly complex shapes and structures that can subsequently be verified by CT. One of the main advantages of CT for product development is the possibility of obtaining holistic information on external as well as internal structures, and subsequently to export the scanned volume in a manner which enables direct use by product designers and engineers, for example, to re-engineer the manufactured parts (Akdogan et al. 2019). The use of AM and CT for reverse engineering purposes is another relevant aspect to be taken into account when dealing with product development (Favata et al. 2018). The CT reconstructed volume of a real part – for which the nominal model or drawings are not available – can be converted into an STL model and used to print a prototype using an AM machine. Both the CT virtual model and AM physical prototype can be used as starting points to improve the design of that part. Several design iterations can be based on this approach: re-designed parts can be printed again using AM and scanned with CT to be compared with the initial model in order to investigate experimentally the improvements. For example, Bauer et al. (2019) showed a full process flow for the reverse engineering of an AM turbine foil, using CT for acquiring the inner features in combination with optical CMS for measuring the external surface.

Product design can also be improved using CT models to feed software for mechanical/structural simulations, with the benefit of basing the simulations not on an ideal geometry but on the actual one (including discrepancies with respect to the nominal model, structural deviations and internal defects). For example, Dallago et al. (2019) used CT models of as-built lattice structures as input for finite element analyses of the stress distribution at the strut junctions. They found that as-built junctions induce a higher stress concentration with respect to the nominal one, with results that are closer to those obtained from experimental analyses.

Another example of CT-based improvement of AM design is the use of CT for scanning biological samples to learn from nature how to build light but stable and rigid structures or other biologically inspired functional designs. This approach is already being applied for many AM parts, and the number of publications combining AM and design, inspired by nature, is rapidly increasing (du Plessis and Broeckhoven 2019, du Plessis et al. 2019b, du Plessis et al. 2018b).

12.4.5.2 CT for Process Optimisation

CT can be used to provide feedback to the AM process, for example to adapt the printing parameters based on the results obtained from CT measurements of several AM samples (Xu et al. 2019). One of the most common approaches for AM process development and improvement is the design and printing of benchmark artefacts (Leach et al. 2019, see Chapter 5). Benchmark artefacts are meant to be used by customers to test different processes and/or machines to find the most suitable solution for their specific needs. For this reason, numerous benchmarks have been proposed over the years (Moylan et al. 2014). The main requirement is that a benchmark should contain various features, such as holes, tubes, ramps, overhangs, angles, thin walls, fine features and freeform structures. Another important requirement is that the benchmark design must take into account the measurability aspect, i.e. design for metrology (Rivas Santos et al. 2019). In fact, the use of a benchmark can be effective only if the fabricated features are measurable with an acceptable accuracy. Consequently, the use of CT allows new design possibilities for benchmarks, particularly the possibility to include intricate and internal geometries to be measured non-destructively (hence without modifying the part from its original state). Another example of direct feedback from CT to the process is the evaluation of internal porosity (see Section 12.4.2). For example, when the AM laser parameters are optimised for a specific material, the porosity is in general more spherical and smaller compared to the case of non-optimal laser parameters, which can lead to keyhole porosity or lack of fusion porosity, depending on the energy density (King et al. 2014). Furthermore, the distribution of pores inside the sample's volume can provide information about the laser tracks and the proper sintering of the powder. For example, too little overlap between the contour hatch and the bulk hatch can lead to a lack of fusion and thus accumulation of porosity between the two hatch strategies (Baier et al. 2019). In order to maximise the effectiveness of CT analyses, the gathered information must be combined with other available sources of information, for example AM process parameters, AM process simulations (which should be validated through experimental analyses) and in-process monitoring of melt pool, powder bed and flow of shielding gas (see Chapter 13).

12.5 Conclusion and Future Look

This chapter has examined the key roles of X-ray CT in the field of metal AM. The most relevant advantage of CT is its capability of analysing both accessible and non-accessible geometries and features in a non-destructive and non-contact manner, overcoming the limitations of conventional measuring devices such as tactile and optical CMSs. The benefits are particularly relevant for metal AM product verification, where advanced measuring solutions are required for the analysis of parts that are often characterised by highly complex external and internal geometries, and by the presence of multiple types of defects (for example, surface defects, internal voids and inclusions, and geometrical or dimensional errors). However, CT should not be seen merely as a tool for quality inspection of final AM products. Rather, CT is a powerful enabling technology for enhancing AM quality, for example through improved product development and process optimisation. In fact, the unique capability of CT in obtaining holistic and information-rich 3D reconstructions can be fully exploited when such 3D reconstructions are used, for instance, as a basis for developing, optimising or validating numerical models and digital twins that are required in the Industry 4.0 context (see Chapters 3 and 4).

Future developments are needed to overcome the current limitations of industrial CT. For example, CT results are still highly dependent on the user's experience; hence, advanced training and guidance is required. In addition, CT scans are relatively slow, especially when high resolutions and accuracy are required. The development of robust solutions for reducing the required expert user input and automating the CT processing pipeline is still beyond the state of the art (Körner et al. 2019). Moreover, the establishment of measurement traceability is challenging in many cases, due to the numerous and complex error sources affecting the measurement process (Leach et al. 2019). The full potential of X-ray CT as an enabling technology for the smart factory of the future will be reached by increasing the automation, speed, measurement accuracy and integration with AM production lines.

References

Akdogan, A., Vanli, A. S., and Durakbasa, N. 2019. Re-engineering of manufacturing parts by computed tomography data. *Proc. 12th Int. Conf. on Measurement and Quality Control – Cyber Physical Issue*, Springer International Publishing, pp. 114–121.

Ametova, E., Probst, G., and Dewulf, W. 2017. X-ray computed tomography devices and their components. In: Carmignato, S., Dewulf, W., and Leach, R. K. (eds.), *Industrial X-Ray Computed Tomography*, 69–98. Springer: Berlin.

Angel, J., and De Chiffre, L. 2014. Comparison on computed tomography using industrial items. *Ann. CIRP* 63:473–476.

ASTM D 2734-09. 2009. *Standard Test Methods for Void Content of Reinforced Plastics*. ASTM International: West Conshohocken.

Baier, M., Zanini, F., Savio, E., and Carmignato, S. 2018. A new conversion approach between different characterization methods to measure the spot size of micro computed tomography systems. *Proc. euspen*, Venice, Italy, Jun. 445–446.

Baier, M., Sinico, M., Paggi, U., Witvrouw, A., Thijs, L., Dewulf, W., and Carmignato, S. 2019. Improving metal additive manufacturing part design and final part precision using feedback from X-ray computed tomography. *Proc. euspen: Advancing Precision in Additive Manufacturing*, Nantes, France, Sep.

Bartscher, M., Bremer, H., Birth, T., Staude, A., and Ehrig, K. 2012. The resolution of dimensional CT – An edge-based analysis. *Proc. 4th Conf. on Industrial Computed Tomography*, Wels, Austria, Sep. 191–201.

Bartscher, M., Neuschaefer-Rube, U., Illemann, J., Borges de Oliveira, F., Stolfi, A., and Carmignato, S. 2017. Qualification and testing of CT systems. In: Carmignato, S., Dewulf, W., and Leach, R. K. (eds.), *Industrial X-Ray Computed Tomography*, 185–228. Springer: Berlin.

Barui, S., Chatterjee, S., Mandal, S., Kumar, A., and Basu, B. 2017. Microstructure and compression properties of 3D powder printed Ti-6Al-4V scaffolds with designed porosity: Experimental and computational analysis. *Mater. Sci. Eng. C* 70:812–823.

Bauer, F., Schrapp, M., and Szijarto, J. 2019. Accuracy analysis of a piece-to-piece reverse engineering workflow for a turbine foil based on multi-modal computed tomography and additive manufacturing. *Precis. Eng.* 60:63–75.

Bauza, M. B., Moylan, S. P., Panas, R. M., Burke, S. C., Martz, E., Taylor, J. S., Knebel, R. H., Bhogaraju, R., and Grove, M. 2014. Study of accuracy of parts produced using additive manufacturing Carl Zeiss Industrial Metrology truss. *Proc. ASPE: Dimensional Accuracy and Surface Finish in Additive Manufacturing*, Berkeley, CA, USA, Apr.

Bernier, F., Tahara, R., and Gendron, M. 2018. Additive manufacturing powder feedstock characterization using X-ray tomography. *Met. Powder Rep.* 73:158–162.

BIPM JCGM 100. 2008. Evaluation of measurement data – Guide to the expression of uncertainty in measurement. International Organization for Standardization: Geneva.

Buzug, T. M. 2008. *Computed Tomography: From Photon Statistics to Modern Cone-Beam CT*. Springer: Berlin.

Carmignato, S., and Savio, E. 2011. Traceable volume measurements using coordinate measuring systems. *Ann. CIRP* 60:519–522.

Carmignato, S. 2012. Accuracy of industrial computed tomography measurements: Experimental results from an international comparison. *Ann. CIRP* 61:491–494.

Carmignato, S., Dewulf, W., and Leach, R. K. 2017a. *Industrial X-ray Computed Tomography*. Springer: Berlin.

Carmignato, S., Aloisi, V., Medeossi, F., Zanini, F., and Savio, E. 2017b. Influence of surface roughness on computed tomography dimensional measurements. *Ann. CIRP* 66:499–502.

Carmignato, S., De Chiffre, L., Bosse, H., Leach, R. K., Balsamo, A., and Estler, W. T. 2020. Dimensional artefacts to achieve metrological traceability in advanced manufacturing. *Ann. CIRP* 69: 1–25.

Chang, S., Liu, A., Ong, C. Y. A., Zhang, L., Huang, X., Tan, Y. H., Zhao, L., Li, L., and Ding, J. 2019. Highly effective smoothening of 3D-printed metal structures via overpotential electrochemical polishing. *Mater. Res. Lett.* 7:282–289.

Cormack, A. M. 1963. Representation of a function by its line integrals, with some radiological applications. *J. Appl. Phys.* 34:2722–2727.

Dallago, M., Winiarski, B., Zanini, F., Carmignato, S., and Benedetti, M. 2019. On the effect of geometrical imperfections and defects on the fatigue strength of cellular lattice structures additively manufactured via selective laser melting. *Int. J. Fatigue* 124:348–360.

De Chiffre, L., Carmignato, S., Kruth, J.-P., Schmitt, R., and Weckenmann, A. 2014. Industrial applications of computed tomography. *Ann. CIRP* 63:655–677.

Dewulf, W., Tan, Y., and Kiekens, K. 2012. Sense and non-sense of beam hardening correction in CT metrology. *Ann. CIRP* 61:495–498.

Dewulf, W., Kiekens, K., Tan, Y., Welkenhuyzen, F., and Kruth, J. P. 2013. Uncertainty determination and quantification for dimensional measurements with industrial computed tomography. *Ann. CIRP* 62:535–538.

Dewulf, W., Ferrucci, M., Ametova, E., Heřmánek, P., Probst, G., Boeckmans, B., Craeghs, T., and Carmignato, S. 2018. Enhanced dimensional measurement by fast determination and compensation of geometrical misalignments of X-ray computed tomography instruments. *Ann. CIRP* 67:523–526.

du Plessis, A., and le Roux, S. G. 2018. Standardized X-ray tomography testing of additively manufactured parts: A round robin test. *Addit. Manuf.* 24:125–136.

du Plessis, A., Yadroitsev, I., Yadroitsava, I., and le Roux, S. G. 2018a. X-Ray microcomputed tomography in additive manufacturing: A review of the current technology and applications. *3D Print. Addit. Manuf.* 5:227–247.

du Plessis, A., Broeckhoven, C., Yadroitsev, I., Yadroitsava, I., and le Roux, S.G. 2018b. Analyzing nature's protective design: the glyptodont body armor. *J. Mech. Behav. Biomed. Mater.* 82:218–223.

du Plessis, A., Sperling, P., Beerlink, A., du Preez, W., and le Roux, S. G. 2018c. Standard method for microCT-based additive manufacturing quality control 4: Metal powder analysis. *MethodsX* 5:1336–1345.

du Plessis, A., and Broeckhoven, C. 2019. Looking deep into nature: A review of micro-computed tomography in biomimicry. *Acta Biomaterialia* 85:27–40.

du Plessis, A., le Roux, S. G., Waller, J., Sperling, P., Achilles, N., et al. 2019a. Laboratory X-ray tomography for metal additive manufacturing: Round robin test. *Addit. Manuf.* 30:100837.

du Plessis, A., Broeckhoven, C., Yadroitsava, I., Yadroitsev, I., Hands, C. H., Kunju, R., and Bhate, D. 2019b. Beautiful and functional: A review of biomimetic design in additive manufacturing. *Addit. Manuf.* 27:408–427.

Favata, J., and Shahbazmohamadi, S. 2018. Realistic non-destructive testing of integrated circuit bond wiring using 3-D X-ray tomography, reverse engineering, and finite element analysis. *Microelectron. Reliab.* 83:91–100.

Ferrucci, M., Leach, R. K., Giusca, C., Carmignato, S., and Dewulf, W. 2015. Towards geometrical calibration of X-ray computed tomography systems – A review. *Meas Sci Technol.* 26:092003.

Ferrucci, M. 2017. Towards traceability of CT dimensional measurements. In: Carmignato, S., Dewulf, W., and Leach, R. K. (eds.), *Industrial X-Ray Computed Tomography*, 229–266. Springer: Berlin.

Feldkamp, L. A., Davis, L. C., and Kress, J. W. 1984. Practical cone-beam algorithm. *J. Opt. Soc. Am. A* 1:612–619.

Fessler, J. A. 2000. Statistical image reconstruction methods for transmission tomography. *Handbook of Medical Imaging*, vol. 2., 1–70. SPIE Press: Bellingham, WA.

Flisch, A., Wirth, J., Zanini, R., Breitenstein, M., Wendt, F., Mnich, F., and Golz, R. 1999. Industrial computed tomography in reverse engineering applications. *Comput. Tomogr. Image Process.* 45–53.

Gibson, I., Rosen, D. W., and Stucker, B. 2015. *Additive Manufacturing Technologies*. Springer: Berlin.

Gong, H., Rafi, K., Starr, T., and Stucker, B. 2013. The effects of processing parameters on defect regularity in Ti-6Al-4V parts fabricated by selective laser melting and electron beam melting. *Proc. 24th SFF*, Austin, Aug. 12–14.

Heim, K., Bernier, F., Pelletier, R., and Lefebvre, L. P. 2016. High resolution pore size analysis in metallic powders by X-ray tomography. *Case Stud. Nondestruct. Test. Eval.* 6:45–52.

Heinzl, C., Amirkanov, A., and Kastner, J. 2017. Processing, analysis and visualization of CT data. In: Carmignato, S., Dewulf, W., and Leach, R. K. (eds.), *Industrial X-Ray Computed Tomography*, 99–142. Springer: Berlin.

Hermanek, P., Rathore, J. S., Aloisi, V., and Carmignato, S. 2017a. Principles of X-ray computed tomography. In: Carmignato, S., Dewulf, W., and Leach, R. K. (eds.), *Industrial X-Ray Computed Tomography*, 25–67. Springer: Berlin.

Hermanek, P., Ferrucci, M., Dewulf, W., and Carmignato, S. 2017b. Optimized reference object for assessment of computed tomography instrument geometry. *Proc. 7th Conf. Industrial Computed Tomography*, Leuven, Belgium, Feb.

Hermanek, P., Borges de Oliveira, F., Carmignato, S., Bartscher, M., and Savio, E. 2017c. Experimental investigation on multi-material gap measurements by computed tomography using a dedicated reference standard. *Proc. euspen*, Hannover, Germany, May 381–382.

Hermanek, P., and Carmignato, S. 2017d. Porosity measurements by X-ray computed tomography: Accuracy evaluation using a calibrated object. *Prec. Eng.* 49:377–387.

Hermanek, P., Zanini, F., and Carmignato, S. 2019. Traceable porosity measurements in industrial components using X-ray computed tomography. *J. Manuf. Sci. Eng., Trans ASME* 141:051004.

Hounsfield, G. 1976. Historical notes on computerized axial tomography. *J. Can. Assoc. Radiol.* 27:135–142.

Hsieh, J. 2015. *Computed Tomography: Principles, Design, Artifacts, and Recent Advances*. SPIE Press: Bellingham, WA.

Illemann, J., Bartscher, M., Jusko, O., Hartig, F., Neuschaefer-Rube, U., and Wendt, K. 2014. Procedure and reference standard to determine the structural resolution in coordinate metrology. *Meas. Sci. Technol.* 25:064015.

ISO 4288. 1998. Geometric product specification (GPS) – Surface texture – Profile method: Rules and procedures for the assessment of surface texture, International Organization for Standardization: Geneva.

ISO 10360 part 2. 2009. Geometrical product specifications (GPS) – Acceptance and reverification tests for coordinate measuring machines (CMM) – Part 2: CMMs used for measuring size, International Organization for Standardization: Geneva.

ISO 15530 part 3. 2011. Geometrical product specifications (GPS) – Coordinate measuring machines (CMM): Technique for determining the uncertainty of measurement – Part 3: Use of calibrated workpieces or measurement standards, International Organization for Standardization: Geneva.

ISO 25178 part 2. 2012. Geometrical product specifications (GPS) – Surface texture: Areal – Part 2: Terms, definitions and surface texture parameters, International Organization for Standardization: Geneva.

Jiménez, R., Ontiveros, S., Carmignato, S., and Yagüe-Fabra, J. A. 2013. Fundamental correction strategies for accuracy improvement of dimensional measurements obtained from a conventional micro-CT cone beam machine. *Ann. CIRP* 6:143–148.

Khademzadeh, S., Carmignato, S., Parvin, N., Zanini, F., and Bariani, P. F. 2016. Micro porosity analysis in additive manufactured NiTi parts using micro computed tomography and electron microscopy. *Mater. Des.* 90:745–752.

Khademzadeh, S., Zanini, F., Bariani, P. F., and Carmignato, S. 2018. Precision additive manufacturing of NiTi parts using micro direct metal deposition. *Int. J. Adv. Manuf. Technol.* 96:3729–3736.

Khademzadeh, S., Zanini, F., Rocco, J, Brunelli, K., Bariani, P. B., and Carmignato, S. Quality enhancement of microstructure and surface topography of NiTi parts produced by laser powder bed fusion. *CIRP J. Manuf. Sci. Technol.*, in press.

King, W. E., Barth, H. D., Castillo, V. M., Gallegos, G. F., Gibbs, J. W., Hahn, D. E., Kamath, C., and Rubenchik, A. M., 2014. Observation of keyhole-mode laser melting in laser powder-bed fusion additive manufacturing. *J. Mater. Process. Technol.* 214:2915–2925.

Körner, L., Lawes, S., Bate, D., Newton, L., Senin, N., and Leach, R. K. 2019. Increasing throughput in X-ray computed tomography measurement of surface topography using sinogram interpolation. *Meas. Sci. Technol.* 30:125002.

Kourra, N., Warnett, J. M., Attridge, A., Dibling, G., McLoughlin, J., Muirhead-Allwood, S., King, R., and Williams, M. A. 2018. Computed tomography metrological examination of additive manufactured acetabular hip prosthesis cups. *Addit. Manuf.* 22:146–152.

Kruth, J. P., Levy, G., Klocke, F., and Childs, T. H. C. 2007. Consolidation phenomena in laser and powder-bed based layered manufacturing. *Ann. CIRP* 56:730–759.

Kruth, J. P., Bartscher, M., Carmignato, S., Schmitt, R., De Chiffre, L., and Weckenmann, A. 2011. Computed tomography for dimensional metrology. *Ann. CIRP* 60:821–842.

Leach, R. K., Bourell, D., Carmignato, S., Donmez, A., Senin, N., and Dewulf, W. 2019. Geometrical metrology for metal additive manufacturing. *Ann. CIRP* 68:677–700.

Lifton, J. J., and Carmignato, S. 2017. Simulating the influence of scatter and beam hardening in dimensional computed tomography. *Meas Sci Technol*. 28:104001.

Liu, X., Jervis, R., Maher, R. C., Villar-Garcia, I. J., Naylor-Marlow, M., Shearing, P. R., Ouyang, M., Cohen, L., Brandon, N. P., and Wu, B. 2016. 3D-Printed structural pseudocapacitors. *Adv. Mater. Technol.* 1:1–7.

Lopez, A., Bacelar, R., Pires, I., Santos, T. G., Sousa, J. P., and Quintino, L. 2018. Non-destructive testing application of radiography and ultrasound for wire and arc additive manufacturing. *Addit. Manuf.* 21:298–306.

Müller, P. 2013. Coordinate metrology by traceable computed tomography. *PhD Thesis*. Department of Mechanical Engineering, Technical University of Denmark: Lyngby, Denmark.

Müller, P., Hiller, J., Dai, Y., Andreasen, J. L., Hansen, H. N., and De Chiffre, L. 2014. Estimation of measurement uncertainties in X-ray computed tomography metrology using the substitution method. *Ann. CIRP*. 7:222–232.

Moylan, S., Slotwinski, J., Cooke, A., Jurrens, K., and Donmez, M. A. 2014. An additive manufacturing test artifact. *J. Res. Natl. Inst. Stand. Technol.* 119:429–459.

Orme, M., Gschweitl, M., Ferrari, M., Vernon, R., Yancey, R., Mouriaux, F., and Madera, I. 2017. A holistic process-flow from concept to validation for additive manufacturing of light-weight, optimized, metallic components suitable for space flight. *Proc. 58th AIAA/ASCE/AHS/ASC Struct. Struct. Dyn. Mater. Conf.* vol. 1–15., 1540.

Orme, M., Madera, I., Gschweitl, M., and Ferrari, M. 2018. Topology optimization for additive manufacturing as an enabler for light weight flight hardware. Designs 2:51.

Otsu, N. 1979. A threshold selection method from gray-level histogram. *IEEE Trans. Systems. Man. Cybernetics* 9:62–66.

Pagani, L., Qi, Q., Jiang, X., and Scott, P. J. 2017. Towards a new definition of areal surface texture parameters on freeform surface. *Measurement* 109:281–291.

Pagani, L., Zanini, F., Carmignato, S., Jiang, X., and Scott, P. J. 2018. Generalization of profile texture parameters for additively manufactured surfaces. *J. Phys. Conf. Ser.*, 1065:212019.

Panetta, D. 2016. Advances in X-ray detectors for clinical and preclinical computed tomography. *Nucl. Instrum. Methods Phys. Res. Sect. A.* 809:2–12.

Pyka, G., Kerckhofs, G., Papantoniou, I., Speirs, M., Schrooten, J., and Wevers, M. 2013. Surface roughness and morphology customization of additive manufactured open porous Ti6Al4V structures. Materials 6:4737–4757.

Radon, J. 1986. On the determination of functions from their integral values along certain manifolds. *IEEE Trans. Med. Imaging* 5:170–176.

Rivas Santos, V. M., Thompson, A., Sims-Waterhouse, D., Maskery, I., Woolliams, P., and Leach, R. K. 2019. Design and characterisation of an additive manufacturing benchmarking artefact following a design-for-metrology approach. *Addit. Manuf.* 32:100964.

Sbettega, E., Zanini, F., Savio, E., Benedetti, M., and Carmignato, S. 2018. X-ray computed tomography dimensional measurements of powder bed fusion cellular structures. *Proc. euspen*, Venice, Italy, Jun. 465–466.

Schmidt, M., Merklein, M., Bourell, D., Dimitrov, D., Hausotte, T., Wegener, K., Overmeyer, L., Vollertsen, F., and Levy, G. N. 2017. Laser based additive manufacturing in industry and academia. *Ann. CIRP* 66:561–583.

Schmitt, R., and Niggemann, C. 2010. Uncertainty in measurement for x-ray-computed tomography using calibrated work pieces. *Meas. Sci. Tech.* 21:054008.

Senin, N., Thompson, A., and Leach, R. K. 2018. Feature-based characterisation of signature topography in laser powder bed fusion of metals. *Meas. Sci. Tech.* 29:045009.

Sinico, M., Ametova, E., Witvrouw, A., and Dewulf, W. 2018. Characterization of AM metal powder with an industrial microfocus CT: Potential and limitations. *Proc. ASPE: Advancing Precision in Additive Manufacturing*, Berkeley, CA, USA, Jul. 286–291.

Slotwinski, J. A., Garboczi, E. J., and Hebenstreit, K. M. 2014. Porosity measurements and analysis for metal additive manufacturing process control. *J. NIST* 119:494.

Snyder, J. C., Stimpson, C. K., Thole, K. A., and Mongillo, D. J. 2015. Build direction effects on microchannel tolerance and surface roughness. *J. Mech. Design* 137:111411.

Spierings, A. B., Schneider, M., and Eggenberger, R. 2011. Comparison of density measurement techniques for additive manufactured metallic parts. *Rapid Proto. J.*17:380–386.

Stolfi, A., and De Chiffre, L. 2018. Interlaboratory comparison of a physical and a virtual assembly measured by CT. *Prec. Eng.* 51:263–270.

Sun, W., Brown, S., Flay, N., McCarthy, M., and McBride, J. 2016. A reference sample for investigating the stability of the imaging system of x-ray computed tomography. *Meas Sci Technol.* 27:085004.

Sutton, A. T., Kriewall, C. S., Leu, M. C., and Newkirk, J. W. 2017. Powder characterisation techniques and effects of powder characteristics on part properties in powder-bed fusion processes. *Virtual Phys. Prototyping* 12:3–29.

Taguchi, K., and Iwanczyk, J. S. 2013. Vision 20/20: Single photon counting X-ray detectors in medical imaging. *Med. Phys.* 40(10).

Taniguchi, N., Fujibayashi, S., Takemoto, M., Sasaki, K., Otsuki, B., Nakamura, T., Matsushita, T., Kokubo, T., and Matsuda, S. 2016. Effect of pore size on bone ingrowth into porous titanium implants fabricated by additive manufacturing: An in vivo experiment. *Mat. Sci. Eng. C* 59:690–701.

Thompson, A., Senin, N., Giusca, C., and Leach, R. K. 2017. Topography of selectively laser melted surfaces: A comparison of different measurement methods. *Ann. CIRP* 66:543–546.

Thompson, A., Senin, N., Maskery, I., Körner, L., Lawes, S., and Leach, R. K. 2018. Internal surface measurement of metal powder bed fusion parts. *Addit. Manuf.* 20:126–133.

Torralba, M., Jiménez, R., Yagüe-Fabra, J. A., Ontiveros, S., and Tosello, G. 2018. Comparison of surface extraction techniques performance in computed tomography for 3D complex micro-geometry dimensional measurements. *Int. J. Adv. Manuf. Technol.* 97:441–453.

Townsend, A., Senin, N., Blunt, L., Leach, R. K., and Taylor, J. S. 2016. Surface texture metrology for metal additive manufacturing: A review. *Prec. Eng.* 46:34–47.

Townsend, A., Pagani, L., Scott, P. J., and Blunt, L. 2017. Areal surface texture data extraction from X-ray computed tomography reconstructions of metal additively manufactured parts. *Prec. Eng.* 48:254–264.

Townsend, A., Pagani, L., Scott, P. J., and Blunt, L. 2019. Introduction of a surface characterization parameter Sdr prime for analysis of re-entrant features. *J. Nondestr. Eval.* 38:58.

Traini, T., Mangano, C., and Sammons, R. I. 2008. Direct laser metal sintering as a new approach to fabrication of isoelastic functionally graded material for manufacture of porous titanium dental implants. *Dent. Mater.* 24:1525–1533.

Van Bael, S., Kerckhofs, G., Moesen, M., Pyka, G., Schrooten, J., and Kruth, J. P. 2011. Micro-CT-based improvement of geometrical and mechanical controllability of selective laser melted Ti6Al4V porous structures. *Mat. Sci. Eng. A* 528:7423–7431.

VDI/VDE 2630 part 1.2. 2008. Computed tomography in dimensional measurement – Influencing variables on measurement results and recommendations for computed tomography dimensional measurements. Verein Deutscher Ingenieure e.V.: Düsseldorf.

VDI/VDE 2630 part 1.3. 2012. Computed tomography in dimensional measurement – Guideline for the application of DIN EN ISO 10360 for coordinate measuring machines with CT sensors. Verein Deutscher Ingenieure e.V.: Düsseldorf.

VDI/VDE 2630 part 2.1. 2015. Computed tomography in dimensional measurement – Determination of the uncertainty of measurement and the test process suitability of coordinate measurement systems with CT sensors. Verein Deutscher Ingenieure e.V.: Düsseldorf.

Villarraga-Gómez, H., Lee, C., Tarbutton, J., and Smith, S. T. 2015. Dimensional metrology of complex inner geometries built by additive manufacturing. *Proc. ASPE: Achieving Precision Tolerances in Additive Manufacturing*, Raleigh, NC, USA, Apr. 164–169.

Wang, H., Su, K., Su, L., Liang, P., Ji, P., and Wang, C. 2018. The effect of 3D-printed Ti6Al4V scaffolds with various macropore structures on osteointegration and osteogenesis: A biomechanical evaluation. *J. Mech. Behavior Biomed. Mat.* 88:488–496.

Wilhelm, R. G., Hocken, R., and Schwenke, H. 2001. Task specific uncertainty in coordinate measurement. *Ann. CIRP* 50:553–563.

Wits, W. W., Carmignato, S., Zanini, F., and Vaneker, T. H. J. 2016. Porosity testing methods for the quality assessment of selective laser melted parts. *Ann. CIRP* 65:201–204.

Xu, D., Cheng, F., Zhou, Y., Matalaray, T., Xian Lim, P., and Zhao, L. 2019. Process optimization: Internal feature measurement for additive-manufacturing parts using x-ray computed tomography. *Proc. SPIE* 11053:110530H.

Xue, L., Suzuki, H, Ohtake, Y., Fujimoto, H., Abe, M., Sato, O., and Takatsuji, T. 2015. Numerical analysis of the Feldkamp-Davis-Kress effect on industrial x-ray computed tomography for dimensional metrology. *J. Comp. Inf. Sci. Eng.* 15.

Zanini, F., and Carmignato, S. 2017a. Two-spheres method for evaluating the metrological structural resolution in dimensional computed tomography. *Meas. Sci. Technol.* 28:114002.

Zanini, F., Carmignato, S., and Savio, E. 2017b. Assembly analysis of titanium dental implants using X-ray computed tomography. *Proc. euspen*, Hannover, Germany, May 489–490.

Zanini, F., Carmignato, S., Savio, E., and Affatato, S. 2018. Uncertainty determination for X-ray computed tomography wear assessment of polyethylene hip joint prostheses. *Prec. Eng.* 52:477–483.

Zanini, F., Sbettega, E., Sorgato, M., and Carmignato, S. 2019a. New approach for verifying the accuracy of X-ray computed tomography measurements of surface topographies in additively manufactured metal parts. *J. Nondestr. Eval.* 38:12.

Zanini, F., Pagani, L., Savio, E., and Carmignato, S. 2019b. Characterisation of additively manufactured metal surfaces by means of X-ray computed tomography and generalised surface texture parameters. *Ann. CIRP* 68:515–518.

Zanini, F., and Carmignato, S. 2019a. X-ray computed tomography for dimensional metrology. In: Gao, W. (eds.), *Metrology. Precision Manufacturing*, 537–584. Springer: Singapore.

Zanini, F., and Carmignato, S. 2019b. X-ray computed tomography for dimensional measurements of threaded parts. *Proc. euspen*, Bilbao, Spain, Jun. 489–490.

13

On-Machine Measurement, Monitoring and Control

Bianca Maria Colosimo and Marco Grasso

CONTENTS

13.1 Introduction

Previous chapters presented methods and instruments for post-process measurements of dimensional characteristics, surface texture and internal features of additively manufactured parts (see Chapters 10, 11 and 12). This chapter has a different perspective, not only for the qualification of additive manufacturing (AM) processes and products, but also for the development of novel generations of intelligent AM systems. This chapter focuses on on-machine measurements, i.e. measurements that can be performed while the part is being produced by exploiting the layerwise production paradigm that potentially makes a large amount of data and information available. On-machine gathered data present several benefits that have been motivating continuously increasing research studies and industrial development efforts. First, such data enable the potential to anticipate quality measurements and qualification operations during the process itself, reducing the need for difficult and expensive post-process quality inspection. Indeed, additively produced parts characterised by novel and complex geometries (for example, lattice structures, topologically optimised shapes, internal channels) represent a challenge for traditional measurement and quality control tools, as the continuously increasing design freedom is counterbalanced by inspection issues. Internal features and defects may be difficult to inspect via non-destructive testing (for example, in the presence of large and high-mass components where internal flaw detection and high resolution is needed), and the lack of standard qualification methodologies in the presence of such complex shapes may be a barrier for the industrial implementation of AM technologies.

The layerwise production paradigm allows acquisition of a large amount of information about the quality of the part and the stability of the process during the process itself. This represents the key issue in the shift from product quality analysis, based on traditional methods and instruments, to on-machine process sensing and monitoring, where several quantities of interest can be measured track by track and layer by layer.

In addition, a fast and robust detection of defects during the process allows implementation of reactive, adaptive or corrective strategies. The simplest strategy consists of stopping the process once a defect or a process-unstable condition has been detected, avoiding time and resource wastage. More advanced and cutting-edge solutions include either closed-loop control methods to adapt, in real-time, the process parameters to prevent or mitigate the effects of non-optimal processing conditions; or in-process defect correction approaches to remove the defect once it has been detected. The final aim is to achieve zero defects and first-time-right production capabilities by means of intelligent combinations of on-machine monitoring and control. On-machine measurement and monitoring could also be used as technological enablers for the development of new materials, thanks to the capability of providing detailed information about the material-specific effects on the process quality and stability during the entire process.

Motivated by all these potential benefits, most metal AM system developers are investing novel solutions to gather on-machine data and to take the greatest advantage from them. Indeed, several commercial toolkits (see Section 13.8) that have become commercially available in the last few years. However, the ease with which product and process measurements can be made available on-machine and in-process depends on the AM technology. Powder bed fusion (PBF) and directed energy deposition processes provide several opportunities for on-machine sensing at different levels and with different methodologies. On the other hand, other metal AM processes based on material extrusion and binder jetting are not yet industrially mature enough and are more limited in terms of ease of on-machine measurements. Among metal AM technologies, PBF processes are the most widely used in industry for the production of functional components, and they are typically characterised by the most stringent specifications and quality constraints, thanks to their capacity of achieving highest accuracy, resolution and functional performances (see Chapter 3). Because of this, most off-the-shelf sensing and monitoring solutions have been made available in recent years for this category of AM processes. For the same reason, this chapter mainly focuses on metal PBF processes.

Despite several benefits enabled by on-machine measurements, it is worth pointing out that not all defects and processing errors can be detected by exploiting the track-by-track and layer-by-layer analysis of the gathered data. Indeed, some defects do not originate in a single layer or they may originate underneath the visible layer. Furthermore, there are defects/errors arising only after the last layer was produced, for example warpage due to thermal and residual stresses, possibly occurring after removal of the produced part from the build platform.

This chapter first clarifies the basic definitions and terminology, with particular attention to the difference between on-machine sensing, on-machine measurement, on-machine monitoring and process control (Section 13.2). This chapter presents the main types of defects that may occur in PBF processes and the different categories of on-machine measurable quantities related to those defects (Section 13.3), and the available sensing methods and architectures (Section 13.4). Finally, this chapter provides an overview of on-machine measurement (Section 13.5), monitoring (Section 13.6) and control methods (Section 13.7) with a summary of current research and a forward look at the most promising future industrial developments (Section 13.8).

13.2 Basic Definitions and Terminology

The term *on-machine* associated with data gathering and measurements in AM processes is used to indicate that the information of interest is recorded *on* the AM system where the process is taking place. The term *in-process* indicates that the measurement is carried out during the process itself. In PBF processes, data can be recorded while the laser or the electron beam scans the part and/or after the PBF process of the current layer has been completed and before the PBF of the next layer starts. Additional definitions are introduced hereafter and are based on those put forward by Gao et al. (2019).

- *On-machine sensing* – refers to the acquisition of data by means of sensors placed on the AM system. Data are usually available in raw formats (the specific format depends on the type of sensor used), which need to be processed in order to compute the actual features to be monitored with time.

- *On-machine measurement* – refers to the possibility of using on-machine gathered data to measure quality characteristics of the part to anticipate and support post-process quality inspection and measurement. The output of on-machine measurement methods consists of a measurement of some product properties, including geometrical, volumetric and microstructural characteristics. Such measurements can be compared with specification and tolerance limits for on-machine part quality assessment.
- *On-machine monitoring* – refers to the possibility of performing on-machine and/or in-process statistical process monitoring (also known as *statistical process control* – SPC), that is, checking for unnatural process conditions, also referred to as *out-of-control* conditions. On-machine monitoring entails the design of a procedure to issue an alarm rule, usually referred to as a *control chart* (Montgomery 2009), see Section 13.6.
- *Process control* – refers to a closed-loop control of process parameters at different temporal scales, for example from point to point, from track to track or from layer to layer. The closed-loop control is activated by a deviation from the target behaviour that can be observed or predicted. The main objective of process control is to act on the process parameters to bring back the process to the target state.

13.3 Defects and Their Fingerprint in PBF Processes

13.3.1 Causes of Defects

The defects and process errors in PBF can be grouped into four major categories, depending on their root causes (Mani et al. 2017, Everton et al. 2016, Grasso and Colosimo 2017): design and job preparation, feedstock material, equipment and process conditions.

13.3.1.1 Defects Induced by Feedstock Material

Different defects may affect the metal powder used in PBF, ranging from irregular morphologies to the presence of powder-entrapped porosity, satellited or broken particles, etc.

Dimensional and morphological issues affect the apparent density and flowability of the powder, which have a direct impact on the quality and performance of the final part. In addition, metallic powders may be contaminated by moisture, organics, adsorbed gases, oxides and nitride films on particle surfaces (Das 2003). Such contamination, which may vary and increase along consecutive power reuse cycles, may degrade the mechanical properties and the geometrical accuracy of the consolidated component in PBF processes (Das 2003, Tang et al. 2015, Seyda et al. 2012, Hann 2016).

13.3.1.2 Equipment-Induced Defects

Equipment-induced defects include various types of flaws and process errors that may originate as a result of improper or degraded performance of critical machine components and sub-systems, for example the beam scanning/deflection apparatus, the build chamber environmental control and the powder handling and deposition equipment. In laser powder bed fusion (L-PBF), deviations from the nominal beam profile properties, together with out-of-calibration laser scanner conditions and imperfections or contamination of lenses and mirrors, may result in parts with inaccurate final dimensions, internal defects and other location effects that reduce the part-to-part reproducibility (Moylan et al. 2014b, Foster et al. 2015, Chapter 8). In electron beam powder bed fusion (EB-PBF), sources of process errors and defects include worn cathodes, inaccurate beam calibration settings and electromagnetic interference. Other defects are related to the control of the chamber environment in both L-PBF and EB-PBF. The flow rate and laminarity of the inert gas in L-PBF are known to have a direct impact on the part quality, including porosity and geometrical accuracy (Ferrar et al. 2012). The oxygen content and the chamber pressure also influence the interaction between the laser and the material, the surface chemical properties of the solidified material and the process by-products that may induce further unstable conditions and powder bed contamination (Spears and Gold 2016).

Another critical part of the equipment that can induce and propagate defects within the build area is the powder handling and deposition system. A worn recoating system (Foster et al. 2015), and the occurrence of impacts between the recoating system and super-elevated edges of the printed parts, may yield an inhomogeneous powder bed and uneven layer thickness with consequent internal, geometrical and surface defects in the part. Although recoating errors are, in most cases, a consequence of other issues and powder bed contamination, the linear motion of the recoating system causes a propagation of defects originated in some area of the layer along the recoating direction. This is particularly critical in industrial processes where several parts are stacked in the same build volume to optimise production time and costs.

13.3.1.3 Defects due to Improper Design or Job Preparation

Despite the continuous improvements of software tools for AM that support the designer's decisions and drive optimised build design and preparation settings, most design and process choices are still based on the operator's experience. Moreover, the high levels of complexity of additively produced parts often impose compromises or trial-and-error choices that can result in defects and poor part quality. One critical source of defects can be the improper design of the supports, which are needed to avoid geometrical distortion due to thermal stresses and to optimise cooling rates in overhangs or bridges (see Chapter 3). Another possible source of defect can be an improper part orientation within the build volume.

13.3.1.4 Process Setting–Induced Defects

Defects in PBF can often be due to the incorrect selection of process parameters and scanning strategies. Indeed, process parameters (for example, laser power, scan speed and layer thickness) and scan strategies (for example, hatching scanning path and contour scanning path) are not only material dependent but also geometry dependent, and complex products manufactured via PBF are usually characterised by critical geometrical features, such as overhang regions, thin walls, acute corners and other complex features, which would require locally variant process parameter settings. However, almost all commercial PBF systems allow the process parameters to be varied from one part to another, but not within the same part. This makes the selection of optimal parameters and scan strategies a compromise choice that typically does not guarantee first-time-right production. The scan strategy influences the temperature distribution over the slice, and inadequate strategies may increase residual stresses and porosity. In L-PBF, improper scan strategies may also inflate the generation of super-elevated edges and affect the microstructural properties of the part. Experimental 'process mapping' is usually carried out to identify the processability window, i.e. the set of appropriate process parameters and the scan strategy that corresponds to appropriate part density and quality for a given material. Such windows are usually quite narrow, which highlights the difficulty in keeping optimal process conditions for any possible geometry.

The energy density plays an important role in determining the residual stresses in the part and the powder wettability, together with the number and properties of process by-products, i.e. spatter and plume emissions. Spatter is either powder particles blown away during the laser scan of the part or liquid material ejected from the melt pool as a result of unstable solid–liquid transitions (Liu et al. 2015, Khairallah et al. 2016, Ly et al. 2017). The plume is formed by the partial material vaporisation (King et al. 2014) and differs from the surrounding atmosphere in terms of chemical composition, temperature and pressure. When spatter falls on the powder bed, it can produce contamination, as it is characterised by larger size, more irregular shapes and different chemical compositions compared to loose powder particles. When the beam passes over the deposited spatter, the larger size of the spatter may prevent complete melting, which can result in formation of voids within the part (Khairallah et al. 2016). On the other hand, if the spatter is displaced by the recoating device, particle dragging may occur, with a consequent irregularity of the powder bed deposition (Foster et al. 2015). Besides spatter, the material vaporisation and plume formation can interfere with the optical properties of the processing beam path by altering the beam profile and the local energy density. EB-PBF differs from L-PBF both in terms of controllable parameters and process by-products. Also, in EB-PBF, non-optimal process parameters may lead to unstable process conditions and various kinds of defects in the final product.

13.3.2 Types of Defects

Different defects that may originate during PBF processes due to one or both of the causes discussed in Section 13.3.1, leading to undesired waste of material and resources, and limiting the reproducibility and performance capabilities. The following sections review the main types of defects in PBF processes.

13.3.2.1 Porosity

Porosity is the type of defect that has attracted the largest attention in the mainstream literature. Indeed, it is particularly critical for most metal AM applications because it strongly influences both mechanical and fatigue performances, and it affects the crack growth characteristics of the part (Edwards et al. 2013). Porosity consists of voids that can be found in different locations of the part, between adjacent layers (intra-layer pores), close to the surface (under-skin pores) or on the external surface of the part (surface pores). Pores may have different sizes, shapes and spatial distributions as shown in Figure 13.1, depending on their root cause (Maamoun et al. 2018). A distinction commonly used in the literature is between spherical pores (Figure 13.1 b and d) and non-spherical pores (Figure 13.1 a and c) (Sharratt 2015, Maamoun et al. 2018). Intra-layer pores usually belong to the non-spherical category and are also referred to as *acicular pores* by some authors (Smith et al. 2016). Spherical pores are typically produced by gas entrapment into the melt pool as a consequence of convective displacement of molten material in the presence of excessive energy density, or gas entrapment in the feedstock powder particles. Non-spherical pores

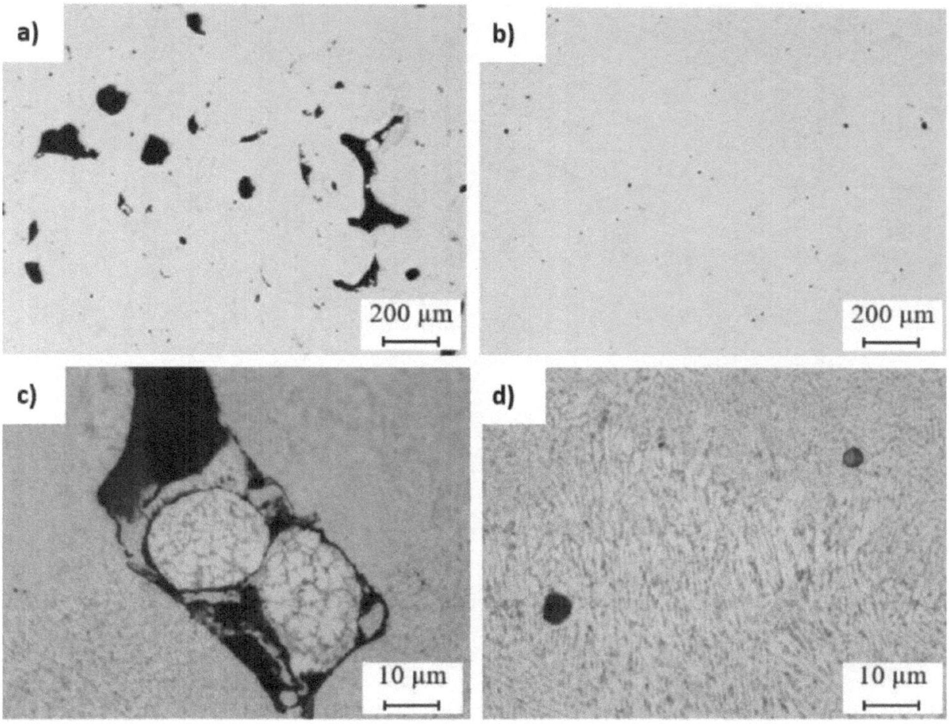

FIGURE 13.1
Example of pores in as-built AlSi10Mg sample fabricated with different L-PBF process parameters: (a) lack-of-fusion pores, (b) spherical gas pores, (c) detail of lack-of-fusion pore, (d) detail of spherical gas pore. (From Maamoun et al. 2018. With permission.)

are typically caused by the opposite condition known as *lack-of-fusion*, that is, an insufficient energy density to achieve proper melting and solidification conditions. Although in various applications, hipping treatments are applied, lack-of-fusion pores are particularly critical due to their large size, irregular shapes and inclusion of un-melted powder particles (Benedetti et al. 2018).

13.3.2.2 Residual Stresses, Cracks and Delamination

Although EB-PBF allows for the mitigation and almost avoidance of residual stresses due to the very high and homogeneous temperature at which the build is kept for the entire process, they still represent one of major sources of defects in L-PBF. Residual stresses in L-PBF may arise from the thermal gradient and the cool-down phase of molten top layers (Mercelis and Kruth 2006). Connected to residual stresses, cracks may occur and propagate during the process as a consequence of stress relief through fracturing when the tensile stress exceeds the ultimate tensile strength of the solid material at a given point and temperature (Harrison et al. 2015, Zhong et al. 2005, Dye et al. 2001). An example of crack caused by the build-up of residual stresses in L-PBF is shown in Figure 13.2 (Parry et al. 2016). Delamination is a particular case of cracking where the onset of cracks is located between adjacent layers and/or between the part and baseplate. Delamination occurs when the residual stresses exceed the binding ability between the top layer and the previous layer.

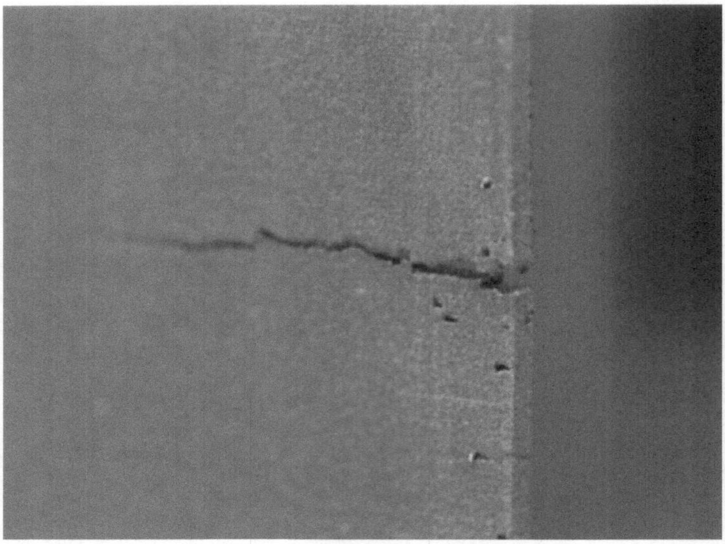

FIGURE 13.2
Crack formation in a Ti6Al4V thin wall produced via L-PBF caused by the build-up of residual stresses. (From Parry et al. 2016. With permission.)

In addition to internal cracks and delamination, detachment from the supports is a typical event in the presence of strong residual stresses in the part.

13.3.2.3 Microstructural Inhomogeneity and Impurities

PBF processes involve highly localised high-heat inputs during very short beam-material interaction times that will, therefore, significantly affect the microstructure of the part (Thijs et al. 2010). Microstructural inhomogeneities or non-equilibrium microstructures may have a detrimental effect on the mechanical and functional performance of the part. Inhomogeneities of the microstructure include impurities and inhomogeneities of grain size characteristics and crystallographic textures (Sharratt 2015). Microstructure customisation via process parameter selection can be considered a viable solution not only to avoid inhomogeneities and part-to-part variability, but also to customise the microstructure depending on the target performance of each local feature of the printed part.

Impurities in the material include inclusions, contamination from other materials and formation of surface oxides (Niu and Chang 1999, Huang et al. 2016). Figure 13.3 shows an example of inclusions and impurities in L-PBF caused by local oxide contaminations (Figure 13.3 a) and partially molten particles (Figure 13.3 b) (Casati et al. 2016).

13.3.2.4 Balling

The balling phenomenon consists of melt-ball formation, i.e. a solidification of the molten material into spheres instead of solid layers (Kruth et al. 2004). The surface tension, combined with low energy inputs, drives the balling phenomenon by preventing the molten material from wetting the powder and the underlying layer. The result is a rough and bead-shaped surface that produces an irregular layer deposition, with detrimental effects on the density of the part and a severe impediment to inter-layer connection (which may

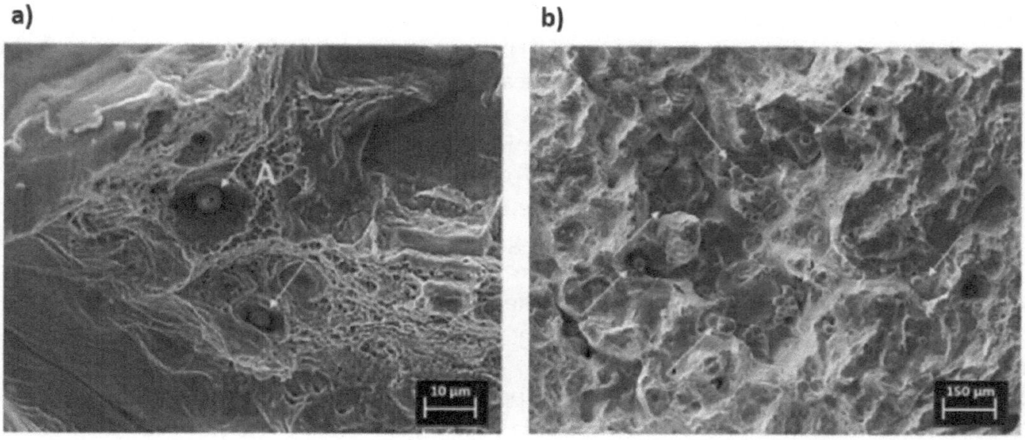

FIGURE 13.3
Example of (a) inclusions and impurities in L-PBF caused by local oxide contaminations and (b) partially molten particles. (From Casati et al. 2016. With permission.)

lead to crack propagation and delamination). Li et al. (2012) discussed three disadvantageous effects of the balling phenomenon in L-PBF: i) it increases the surface texture, ii) it increases the porosity as a large number of pores can be formed between the discontinuous metallic balls and iii) in cases of very severe balling, super-elevated spheres may interfere with the movement of the powder deposition system. Figure 13.4 shows an example of balling effects at different scan speeds in L-PBF (Li et al. 2012).

13.3.2.5 Dimensional and Geometrical Deviations

Dimensional deviations in PBF can be classified into (i) shrinkage and oversize effects, (ii) warping and curling, (iii) dross formation at down-facing surfaces, (iv) super-elevated

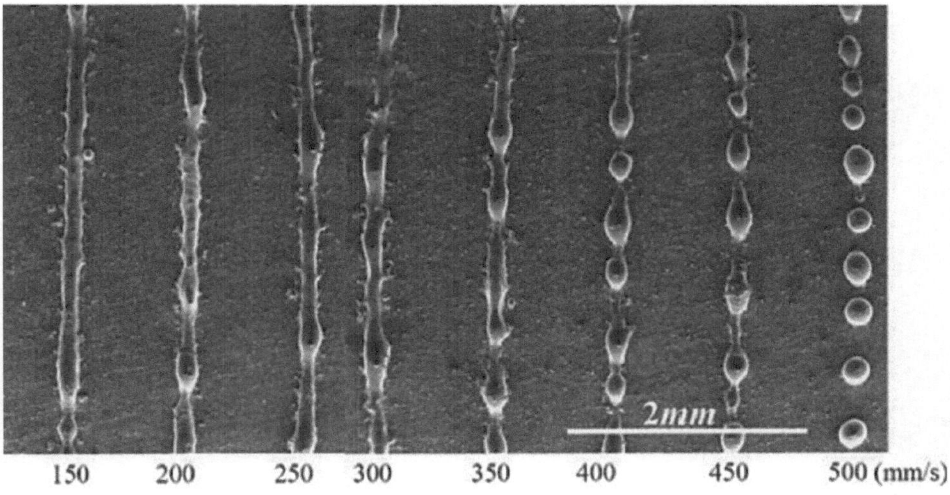

FIGURE 13.4
Example of increasing balling severity at increasing scan speeds in L-PBF. (From Li et al. 2012. With permission.)

edges and (v) other in-plane geometrical distortions. Regarding the size of the part, shrinkage has been reported by different authors (Thomas 2009, Sharratt 2015), even though the opposite effect (i.e. parts that are systematically larger than the nominal) may occur in practice, depending on dimensional compensation settings applied to the job file. Warping is mainly due to the heat dissipation mechanisms and to thermal stresses (Sharratt 2015). The so-called curling phenomenon is one type of warping effect caused by non-uniform thermal expansion and part contraction, usually associated with an uneven shrinkage between the top and the bottom of overhanging areas (Gibson et al. 2010, Mousa 2016). A combination of shrinking and warping effects yield curved profiles of down-facing surfaces intended to be flat. In the presence of the down-facing surfaces, dross formation and defective contours are known to be caused by a lack of, or improper design of, supporting structures.

Super-elevated edges represent another type of out-of-plane geometrical distortion, featured by elevated ridges of solidified material. Super-elevated edges not only affect end-part quality but also induce the propagation of defects due to possible interference with the powder recoating system.

Other distortions affect critical features such as thin walls, overhang areas and acute corners. In correspondence of these features, the melt pool is largely surrounded by loose powder, which has a lower conductivity than the solid material. The diminished heat flux yields local over-heating phenomena that may deteriorate the geometric accuracy. Figure 13.5 shows two examples of geometrical distortions occurred in the presence of overhang acute corners in parts produced via L-PBF (Figure 13.5 a, Grasso et al. 2018b) and EB-PBF (Figure 13.5 b).

13.3.2.6 Surface Defects

In PBF processes, the surface texture is affected by a stair-stepping effect caused by the layer-wise production and by the presence of surface contaminations, irregularities, dross formations and pores. The surface texture depends on the process parameters, the contour scan strategy and the size of the powder particles (see Chapter 11). Surface texture also depends on the surface orientation with respect to the build direction (Fox et al. 2016,

FIGURE 13.5
Two examples of geometrical distortions occurred in the presence of overhang acute corners in parts produced via (a) L-PBF (Grasso et al. 2018; with permission) and (b) EB-PBF.

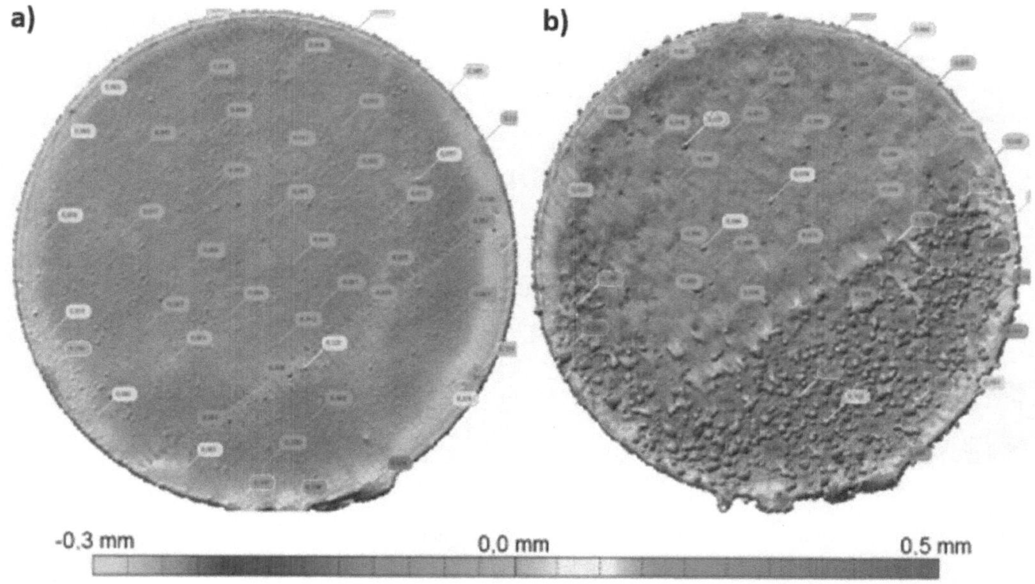

FIGURE 13.6
Examples of surface textures of parts produced via L-PBF and measured via confocal microscopy: surface textures were produced either in the presence of proper (a) or improper (b) inert gas flow conditions. (From Ladewig et al. 2016. With permission.)

Strano et al. 2013). In particular, downward- and upward -facing surfaces are known to have considerably different texture properties (Triantaphyllou et al. 2015). Although in most cases, PBF-produced parts are post-processed (for example, surface and thermal treatments), the surface texture is functionally relevant as it has an effect on the fatigue performance of the part. Because of this, poor texture characteristics have been investigated by different authors (Senin et al. 2017, Sames et al. 2016, Fox et al. 2016, see Chapter 11). Figure 13.6 shows examples of good (Figure 13.6 a) and poor (Figure 13.6 b) surface textures measured via confocal microscopy on parts produced either in the presence of good and improper gas flow conditions in L-PBF (Ladewig et al. 2016). Surface defects are also increased by the balling effect (see Section 13.3.2.4), which prevents the achievement of smooth surfaces.

13.4 On-Machine Sensing Methods and Architectures

13.4.1 Basic Principles

Most common methods suitable for on-machine sensing in PBF can be classified mainly in terms of the electromagnetic range (from visible to infrared bands) and the nature of the measurement output (i.e. spatially integrated or spatially resolved measurements). The basic principles related to these classifications are introduced in Sections 13.4.1.1 –13.4.1.3.

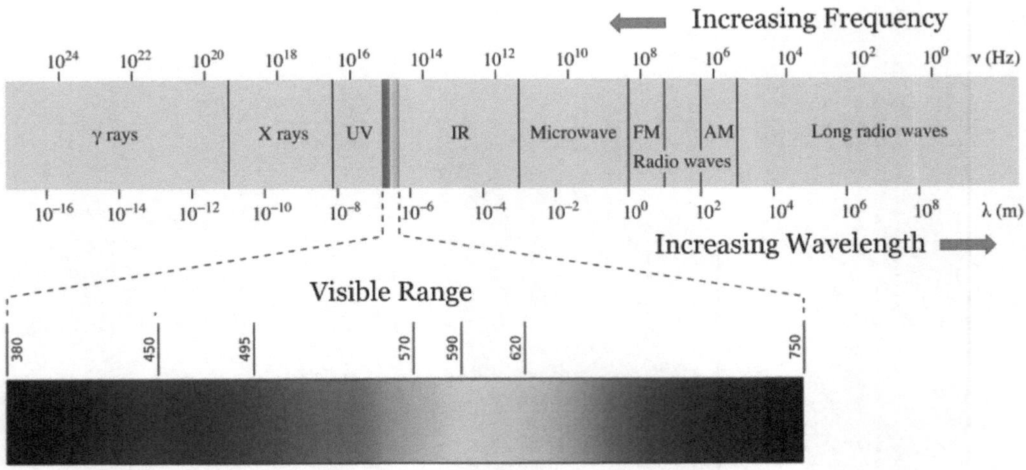

FIGURE 13.7
Spectrum of the electromagnetic radiation with a detail of the visible range.

13.4.1.1 Electromagnetic Spectral Ranges for On-Machine Measurements

The electromagnetic spectrum, which plays a relevant role for on-machine measurements in AM, can be described considering the following ranges (Figure 13.7):

- Visible: 0.4 µm–0.8 µm.
- Near infrared: 0.7 µm–~1 µm.
- Short wave infrared: ~0.9 µm–1.7 µm, or ~0.9 µm–2.5 µm.
- Medium wave infrared: 2 µm–5 µm.
- Long wave infrared: 7.5 µm–14 µm or more.

Electromagnetic radiation emitted by an object, as a consequence of its absolute temperature, mostly occurs in a part of the ultraviolet band, but also from the visible to the infrared range, depending on its absolute temperature. The Planck's law describes the spectral distribution of the intensity of the emitted electromagnetic radiation for a blackbody, i.e. a theoretical object absorbing all the incoming radiation (Williams 2009). According to the Planck's law, the radiance of a blackbody is computed as follows

$$E_{BB}(\lambda, T) = \frac{2hc^2}{\lambda^5}\left(e^{\frac{hc}{\lambda kT}} - 1\right)^{-1} \tag{13.1}$$

where h is Planck's constant, c is the speed of light in a vacuum, λ is the wavelength, k is the Boltzmann constant and T is the absolute temperature in kelvin. A graphical representation of the blackbody radiance as a function of the wavelength for different temperatures is shown in Figure 13.8, where the visible, short-wavelength (SW), mid-wavelength (MW) and long-wavelength (LW) infrared bands are highlighted.

The radiance in each band (also known as *in-band radiance*) that a detector can capture and convert into a signal corresponds to the surface under the curve within the corresponding spectral band.

Figure 13.8 shows that, at low to medium temperatures, for example from 0° C (273 K) to 300 °C (573 K), no significant radiance is measured in the visible ranges. At these

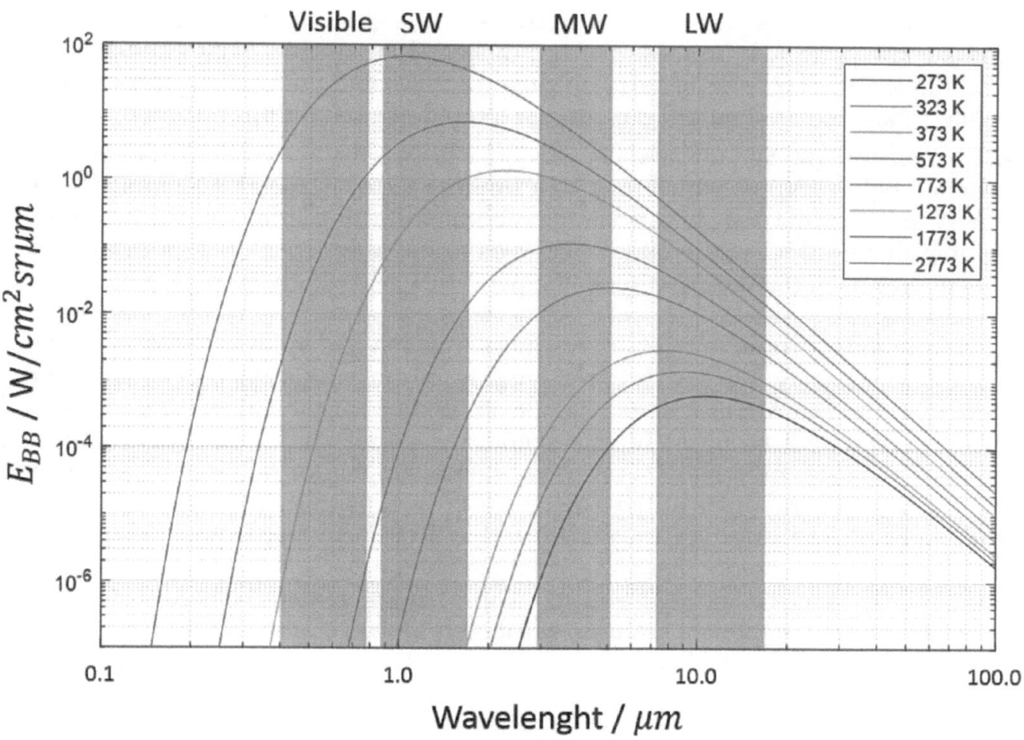

FIGURE 13.8
Planck's curves showing the blackbody radiance as a function of the wavelength for different temperatures.

temperatures, radiance measurement is available only in the infrared range. Only objects at high temperature emit in the visible range. For human eyes, the sensitivity threshold for incandescent emissions is about 525 °C. At high temperatures, for example 1500 °C (1773 K), in-band radiance in the visible range is of the same order of magnitude as for the MW and LW infrared ranges. At these high temperatures, the highest in-band radiance is achieved in the MW infrared range. i.e. $\sum_{MW} E_{BB}\left(\lambda, 1773\,\mathrm{K}\right) = 3.827\,\mathrm{W/cm^2} \cdot \mathrm{sr}$, whereas in the LW infrared range it is $\sum_{LW} E_{BB}\left(\lambda, 1773\,\mathrm{K}\right) = 0.613\,\mathrm{W/cm} \cdot \mathrm{sr}$.

Generally speaking, different factors should be taken into account to determine whether a given spectral range is appropriate for on-machine measurement (Williams 2009).

- *In-band radiance.* The higher the in-band radiance at a given temperature, the higher the signal-to-noise ratio. At low and medium temperatures, no radiance measurements are available from the visible to the SW infrared range. At low temperatures, the in-band radiance is higher at LWs than at MWs, whereas the opposite occurs at high temperatures.
- *Dynamic range.* The relative radiance change caused by a temperature variation is another relevant factor. This change is much higher in the visible range and at short infrared wavelengths than in medium and long infrared bands, which may be a drawback for dynamic measurements. Indeed, large temperature variations, typical

of metal AM processes, may rapidly cause sensor signal saturation when measurements are performed in these spectral ranges. One possible way to mitigate this saturation effect consists of narrowing the spectral bands with filters. Near-infrared filters mounted on standard cameras are commonly used for this purpose.

- *Calibration range*. SW infrared sensors offer good sensitivity but cannot be calibrated below 350 °C, which may be a limiting factor in metal AM applications. MW and LW infrared sensors can be calibrated starting from low temperatures, for example -20 °C, which make them the most appropriate choice for measuring temperature variations from room temperature to very high temperatures.

Medium and long infrared ranges are the most common and appropriate spectral bands used for thermography. The MW infrared range is the most utilised for on-machine monitoring applications in AM as it offers high flexibility in terms of temperature ranges with very high sensitivity, even at very high temperatures. Standard cameras with near infrared filters offer a low-cost alternative to infrared thermography for radiance measurements. However, such cameras come at the expense of a low sensitivity (i.e. the smallest difference of temperature that can be measured) and lack of linear response, with greater risk for saturation in the presence of high thermal gradients.

Although the Planck's law is useful to compare measurement responses at different wavelengths and different temperatures, a real body always emits less than a blackbody at the same temperature. For real bodies, the radiance is the sum of absorption, transmission and reflection. The ratio between the radiance of a body at a given temperature and the radiance of a blackbody at the same temperature is the emissivity, given by

$$\varepsilon = \frac{E_{\lambda}(\lambda, T)}{E_{BB,\lambda}(\lambda, T)}. \tag{13.2}$$

Thus, if the emissivity of a body is known, its temperature can be determined by measuring its spectral radiance. However, in most practical cases, a material emissivity is a function of several factors including temperature, wavelength, observation angle and surface conditions, which makes accurate temperature measurements by means of on-machine thermography challenging or even impossible. This is further complicated by the fact that, during metal AM processes, the material undergoes phase changes, from solid powder particles to molten material and from liquid to solidified bulk material, with the possible formation of metal vapour and plasma.

One way to obtain temperature estimates without the need to know the material emissivity consists in measuring the ratio R of the radiances ($E_{\lambda 1}$ and $E_{\lambda 2}$) emitted by the same object at two separate wavelengths, i.e. (Vinson et al. 2017)

$$R = \frac{E_{\lambda 1}(\lambda, T)}{E_{\lambda 2}(\lambda, T)}. \tag{13.3}$$

Assuming that the dependence of the material emissivity on the wavelength is negligible at all temperatures, i.e. $\varepsilon_{\lambda 1} = \varepsilon_{\lambda 2}$, the ratio R can be computed as

$$R = \frac{\lambda_2^5 \varepsilon_{\lambda 1}}{\lambda_1^5 \varepsilon_{\lambda 2}} \left[\frac{e^{\frac{hc}{\lambda_2 kT}} - 1}{e^{\frac{hc}{\lambda_1 kT}} - 1} \right]. \tag{13.4}$$

Therefore, the material temperature can be estimated from

$$T = \frac{hc}{k}\left[\frac{\dfrac{1}{\lambda_2} - \dfrac{1}{\lambda_1}}{\ln(R) - \ln\left(\dfrac{\lambda_2^5}{\lambda_1^5}\right)}\right]. \tag{13.5}$$

Thermography measurement methods based on this principle are also called *ratio-thermography, dual-bandwidth* or *two-colour* thermography. Additional details are discussed when spatially integrated sensors are introduced in Section 13.4.1.2.

13.4.1.2 Spatially Integrated Sensors

The electromagnetic radiation emitted and reflected by the material can be measured by means of either spatially integrated or spatially resolved sensors. The former integrates the total amount of radiation focused onto the detector to generate a signal that is an integral value. The latter exploits an image sensor that converts the incoming detected radiation into an image.

Spatially integrated sensors are pyrometers, i.e. non-contact instruments that measure the radiance of an object and, if properly calibrated, its temperature. Pyrometers may include different types of detectors. The most common type for on-machine monitoring applications is the photodiode, which converts the incoming electromagnetic radiation into an electrical current signal, which is proportional to the radiation intensity.

Pyrometers may cover a wide range of wavelength bands, depending on the measurement needs. The output signal is proportional to the integral of the incoming radiation captured within a given field of view. Since the output, at a given point in time, is a scalar measurement, pyrometers enable high-speed data acquisition on the order of tens or hundreds of kilohertz. However, the provided information content from pyrometers is much lower than that gathered via spatially resolved sensors.

Single wavelength pyrometers are affected by the calibration issues discussed in Section 13.4.1.1 because of the temperature measurement dependence on the material emissivity. So-called 'ratio pyrometers' allow the avoidance of this issue by measuring the ratio of two radiances at separate wavelengths. Two types of ratio pyrometers are available: two-colour and dual-wavelength pyrometers.

- *Two-colour pyrometers.* These use a 'sandwich detector', i.e. a detector where two wavelength filters are laid one on top of the other. One wavelength is a broad wave band and the other wavelength is a narrower band that is a subset of the broader one.
- *Dual-wavelength pyrometers.* These use two separate and distinct wavelength sets. Because the sensor design allows for separate wavelengths, they can be independently selected and combined, depending on the specific measurement application.

Dual-wavelength pyrometers are less affected by inference such as flames, vapour clouds and plasma emissions, which make them more appropriate than two-colour pyrometers for on-machine monitoring of AM processes. Moreover, two-colour pyrometers are designed to take an average temperature measurement of the radiation captured within the field of view, whereas dual-wavelength pyrometers yield a measurement that is more heavily weighted to the hottest temperature measured in the field of view, which makes them more appropriate for on-machine measurements at the melt-pool level as the contribution of cold areas in the field of view is reduced.

For certain materials and under certain conditions, the emissivity of the material varies with the wavelength, which reduces the accuracy of temperature measurements via ratio thermography. This may occur as a consequence of changes in the alloy, its surface texture, surface oxidations and other parameters. Due to the various phase changes and surface modifications in AM that metal materials undergo during the process, the assumption of constant emissivity at different wavelengths can be violated. This makes accurate temperature measurements in metal AM applications a challenging task. However, depending on the specific application, approximate temperatures may be sufficient. Moreover, as far as on-machine monitoring applications are concerned, where the aim is to detect a change from a target, in-control, radiance pattern may be sufficient without the need for temperature estimates.

13.4.1.3 Spatially Resolved Sensors

The second category of sensors is spatially resolved sensors. These are imaging sensors, i.e. cameras, that can be classified depending on the wavelength range in which they are sensitive, the type of detector, the spatial and temporal resolution and other parameters.

Standard cameras convert the incoming electromagnetic radiation in the visible range into an electric charge through an array of photodiodes to generate an image. Two types of sensors can be used in standard cameras, i.e. charge-coupled device (CCD) and complementary metal-oxide-semiconductor (CMOS) sensors.

- *CCD sensor.* In a CCD sensor, every pixel's charge is transferred sequentially to a common output structure to be converted to voltage.
- *CMOS sensor.* In a CMOS sensor, each pixel has its own charge-to-voltage conversion, and the sensor often includes amplifiers, noise-correction and digitisation circuits.

Each type of sensor has its own positive and negative characteristics that mainly depend on the type of application. Generally speaking, CMOS sensors offer some advantages in terms of higher sampling frequency and monolithic integration in compact volumes that make them particularly suitable for many on-machine monitoring applications and for integration into existing AM machines.

In terms of resolution, the performance of spatially resolved sensors is a compromise between spatial and temporal resolution. The larger the spatial resolution, the lower the maximum image acquisition speed. For the same spatial resolution, a higher temporal resolution can be achieved by reducing the field of view via image cropping. Moreover, depending on the optics mounted on the camera, high spatial resolutions are difficult to achieve over large fields of view, which introduces a further compromise, depending on the application of interest.

Moving from the visible range to the near-infrared range, the sensitivity of the sensor increasingly reduces, but it is usually sufficient to generate a signal. CCD sensors are more sensitive in the near-infrared range, but it is possible to design CMOS sensors with enhanced near-infrared sensitivity for special application in this wavelength band.

If the wavelength range of interest is the infrared, different types of detectors should be used. Cameras for thermography measurements in the infrared range are called *thermal imaging cameras* or simply *thermal cameras.*

Infrared detectors can be classified into thermal types and quantum types.

- *Thermal detectors.* These are characterised by a sensitivity that does not depend on the wavelength and they do not need cooling. However, they are characterised by a slower response time and lower measurement performance.
- *Quantum detectors.* These allow a faster response with high measurement performance, but their sensitivity depends on the wavelength. Quantum detectors also need cooling.

Most thermal cameras on the market are single wavelength cameras, i.e. they allow acquisition of video imaging streams by using a sensor, filters and optics that are calibrated in a specific wavelength band. Dual-wavelength thermal imaging devices are also available but are typically complex and expensive instruments with limitations in terms of temporal resolution. Alternatively, it is possible to combine spatially integrated and spatially resolved thermal measurements or thermal imaging from two sensors calibrated in different wavelength bands to obtain more robust temperature measurements (Peralta et al. 2016, Marshall et al. 2015).

13.4.2 Data Gathering Levels

On-machine measurable quantities are usually referred to as *process signatures,* as they represent signatures of the part quality or proxies of the process condition over time. The signatures that can be measured during the PBF process may be classified into different categories depending on the spatial or temporal scale to which they belong and the type of information they enclose. One possible way to classify the relevant process signatures into distinct observation levels is shown in Figure 13.9.

- *Level 0.* This consists of signals that are already available from the system (for example, from chamber control sensors, from the programmable logic controller [PLC] of the machine). Some of these signals include relevant information to determine the quality and stability of the process. Indeed, the major advantage of Level 0 observation is the possibility of monitoring the process without the need for external or additional sensors. The main disadvantage is the larger distance from the actual process dynamics related to the material/energy source interaction compared with other ad-hoc sensors. This monitoring level is particularly attractive in EB-PBF, where many signals are often freely available from embedded sensors (Grasso et al. 2018b). Although the use of embedded signals still represents a relatively unexplored path in the literature devoted to PBF process monitoring, it may offer interesting opportunities for future industrial developments.
- *Level 1.* This includes process signatures that can be measured to detect and localise flaws within the powder bed and/or in-plane and out-of-plane anomalies in the printed slice. The on-machine determination of the powder bed homogeneity is important to determine recoating errors (for example, local lack of powder), detect rippling caused by recoater bouncing effects and/or rectilinear grooves generated either by particles dragging or other recoating system damage. In-plane geometry

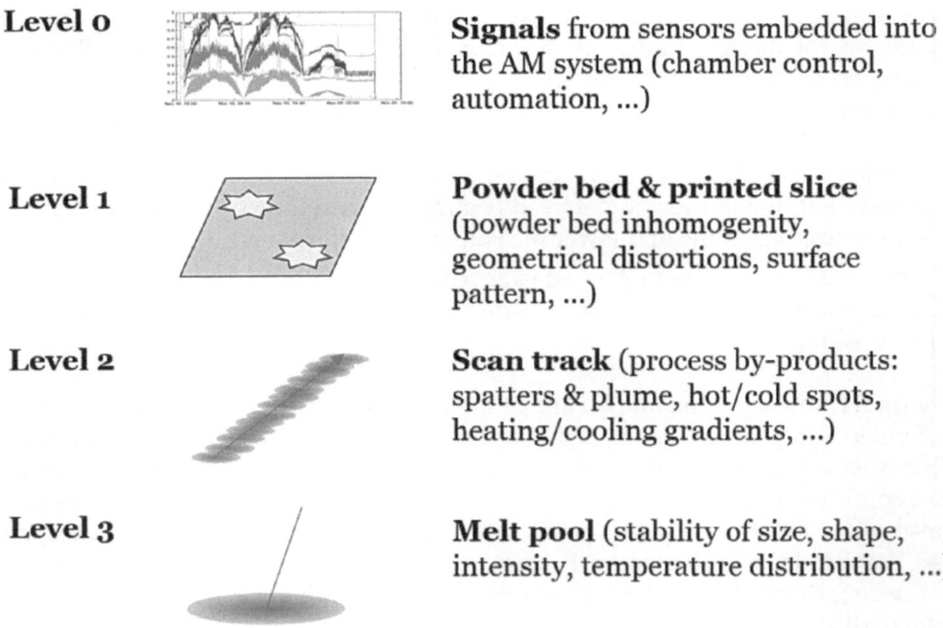

FIGURE 13.9
Data gathering levels for on-machine process signature measurement in PBF processes.

errors consist of deviations from the nominal shape on a layerwise basis, whereas out-of-plane distortions include anomalous surface patterns and so-called super-elevated edges that cause the surface topography of the printed slice to deviate from flat.

- *Level 2.* This is the measurement of quantities that are representative of the process quality and stability while the beam scans the tracks within the print area. In particular, three types of quantities can be measured: i) the geometry of the track, ii) the temperature profile over the track, including the occurrence of hot and cold spots and iii) the process by-products. The geometry and the temperature profile of the track are relevant to detect the onset of balling phenomena, lack-of-fusion or local over-heating conditions, surface and geometric errors and porosity formation. In addition, spatter and plume emissions–related quantities can be measured in this level as proxies of the process stability and final quality of the part. Quantities of possible interest include the number of spatter particles, their area, temperature, orientation, speed, etc. (Repossini et al. 2018, Andani 2017a, Andani 2017b, Ye et al. 2018, Zheng et al. 2018, Anwar and Pham 2018, Zhang et al. 2018).
- *Level 3.* This is the measurement of signatures representative of the highest level of detail at which the PBF process can be observed, i.e. the melt pool. The melt pool is known to be a primary feature of interest in any process that involves a beam–material interaction aimed at achieving a local fusion of the material. Indeed, the stability, dimensions and behaviour of the melt pool determine to a great extent the quality of the part and stability of the process (Craeghs et al. 2010, Craeghs et al. 2012, Clijsters et al. 2014, Berumen et al. 2010, Lott et al. 2011, Doubenskaia et al. 2012, Doubenskaia et al. 2015, Thijs et al. 2010). The signatures of major interest at the melt-pool level include the melt-pool size (i.e. the area or the diameter),

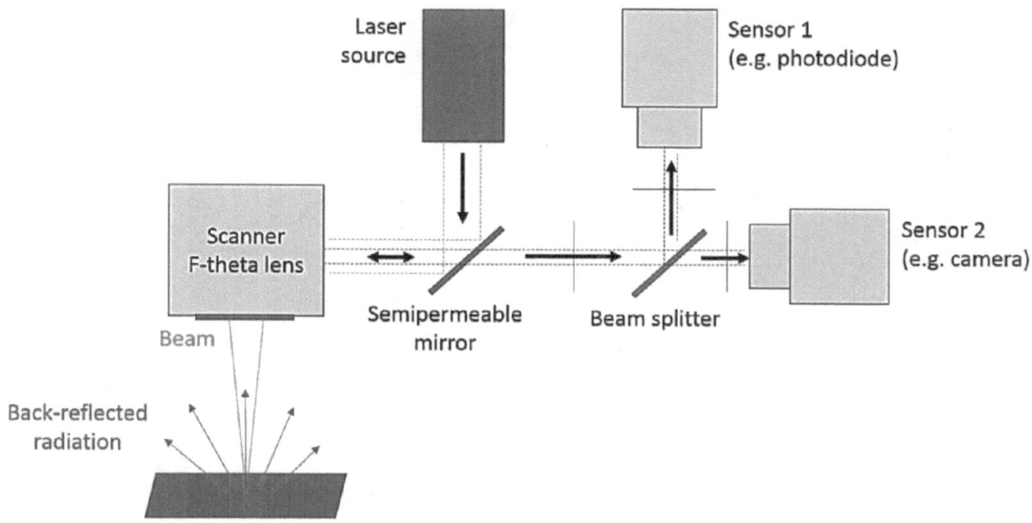

FIGURE 13.10
Example of co-axial sensing architecture in L-PBF.

its shape and the temperature intensity/profile of the melt pool itself. These signatures and their stability over time determine the geometrical accuracy of the track and surface and geometrical properties of the final part.

13.4.3 On-Machine Sensing Architectures

As far as data gathering levels from 1 to 3 are concerned, on-machine sensing architectures applicable in L-PBF can be classified into *co-axial* and *off-axis* architectures (Figure 13.10 and Figure 13.11, respectively), whereas in EB-PBF, only off-axial sensing architectures are applicable. In co-axial architectures, the sensors exploit the optical path of the laser to measure the signatures of interest. In off-axis architectures, the sensors are placed outside the optical path, which allows measurement of additional quantities by means of a larger field of view and on different spatial/temporal scales. Co-axial architectures are not possible in EB-PBF because the energy is provided to the material by means of an electron beam displaced by electromagnetic coils, and hence there is no optical path suitable for sensor integration.

13.4.3.1 Co-Axial Sensing

L-PBF enables the possibility of exploiting the back-reflected radiation from the melt pool and the surrounding area, through the optical path of the laser, to measure different quantities of interest representative of the melt-pool behaviour (monitoring level 3). One example of a co-axial sensing architecture is shown in Figure 13.10. This architecture was first developed by the Katholieke Universiteit Leuven and licensed by Concept Laser (Berumen et al. 2010, Clijsters et al. 2014, Craeghs et al. 2010, Craeghs et al. 2012, Kruth et al. 2007, Demir et al. 2018). The optical setup includes a high-speed near-infrared camera and photodiode. The laser beam is reflected by a partially reflective mirror towards the scanner, and the same mirror enables the radiation emitted by the melt pool to be captured by the two sensors. This radiation is then split towards the two sensors. Different co-axial sensing

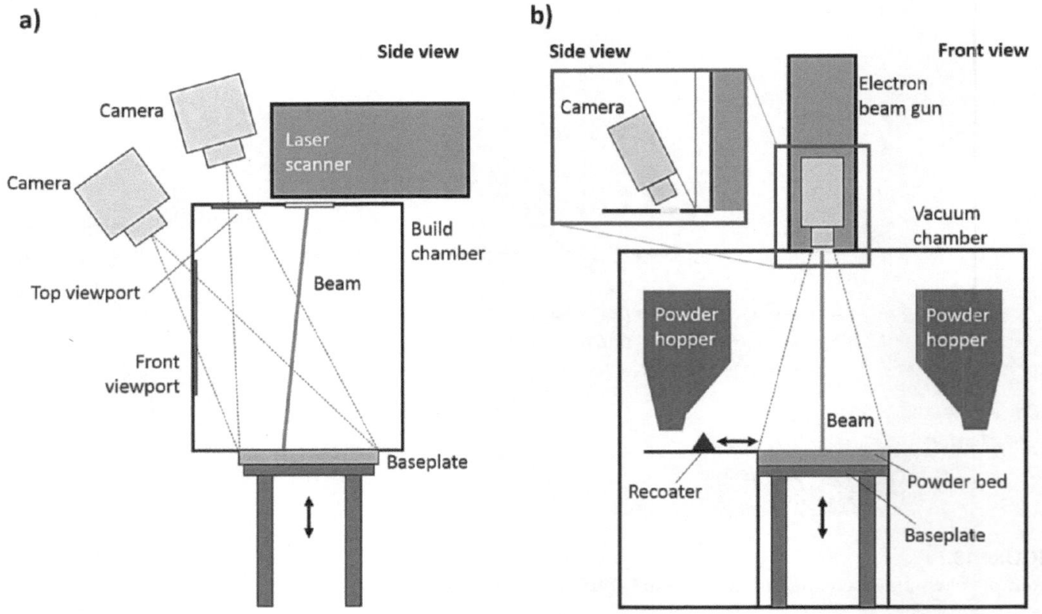

FIGURE 13.11
Example of off-axis sensing architectures in (a) L-PBF and (b) EB-PBF.

architectures may include multiple pyrometers, dual-wavelength or multiple-wavelength pyrometers and different kinds of imaging sensors in the visible or infrared range.

Due to the spatial integration, a high sampling frequency can be achieved by coaxially mounted photodiodes (for example, greater than 50 kHz). The field of view is typically small enough (for example 0.5 mm × 0.5 mm) to enhance the signal-to-noise ratio and avoid capturing radiation coming from regions of the powder bed that surround the melt pool.

Variations of the melt pool size and brightness/temperature may result in variations of the pyrometer signal, which allows monitoring of the stability of the melt pool and possibly anomalous signal changes.

Melt-pool imaging via spatially resolved sensors provides richer information as it allows characterisation of the size and shape of the melt pool together with its spatial brightness or temperature map. The richer information is usually achieved at the expense of sampling frequency, although high-speed video imaging combined with a compromise between field-of-view and spatial resolution (for example, approximately 10 μm/pixel–20 μm/pixel over an area of 1 mm × 1 mm) allow sampling frequencies of up to 10 kHz in most co-axial configurations.

As an example, if the scan speed is 1000 mm/s, a sampling frequency of 10 kHz yields two consecutive images of the melt pool that are spaced apart by 100 μm along the scanning path, which approximately corresponds to the size of the melt pool, leading to a limited loss of information in terms of data sub-sampling.

In some cases, the use of external illumination combined with co-axial sensing may help to improve the signal-to-noise ratio and enhance the output signal quality (Mazzoleni et al. 2019). Indeed, illuminators combined with corresponding filters allow measurement of the radiation emissions at one pre-defined wavelength of interest, cutting out all the information that is deemed not relevant for the given on-machine monitoring application.

13.4.3.2 Off-Axis Sensing

In both L-PBF and EB-PBF, off-axis sensing is suitable to gather information i) during the scanning of each track, ii) once the scan of an entire layer has been concluded or iii) before starting the process in the current layer, after the powder bed has been deposited (Figure 13.11). The latter two levels require the capability to acquire one single image (or multiple images with different illumination settings) just after the powder bed deposition or after the layer scan. The measurement of process signatures while the slice is being melted and solidifies, instead, implies the need to capture fast phenomena and transient events that occur while the beam scans the tracks. In this case, a compromise between spatial resolution, field of view and sampling frequency is needed to minimise information loss and to enhance the signal quality.

Conventional cameras in the visible range are commonly used in L-PBF for imaging the powder bed before and after the laser scan (monitoring level 1). In this case, the field of view is larger than or equal to the build area, but perspective image distortion correction operations are needed. There is no specific requirement on the temporal resolution, as one or a small number of images are acquired at the end of each layer, but the spatial resolution should be high enough (usually in the order of tens of micrometres/pixel) to detect small in-plane and out-of-plane distortions in the powder bed.

Figure 13.12 shows two off-axis powder bed images in L-PBF acquired after the powder recoating operation and before the laser scan. Off-axis imaging in the visible range for level 1 data gathering strongly relies on the illumination condition, as both the surface pattern of the printed slice and the powder bed and the contrast between the background (powder) and foreground (printed slice) areas depend on the intensity, type and relative angle of the illumination source. Non-uniform illumination conditions may mask actual variations in the powder bed homogeneity, and they can make image segmentation for surface pattern and geometry analysis poorly effective. The relative angle between the camera and the illumination source also plays a relevant role (Caltanissetta et al. 2018). When the majority of light is not reflected towards the camera, so-called dark-field illumination is obtained; this may be particularly suitable for image segmentation due to an

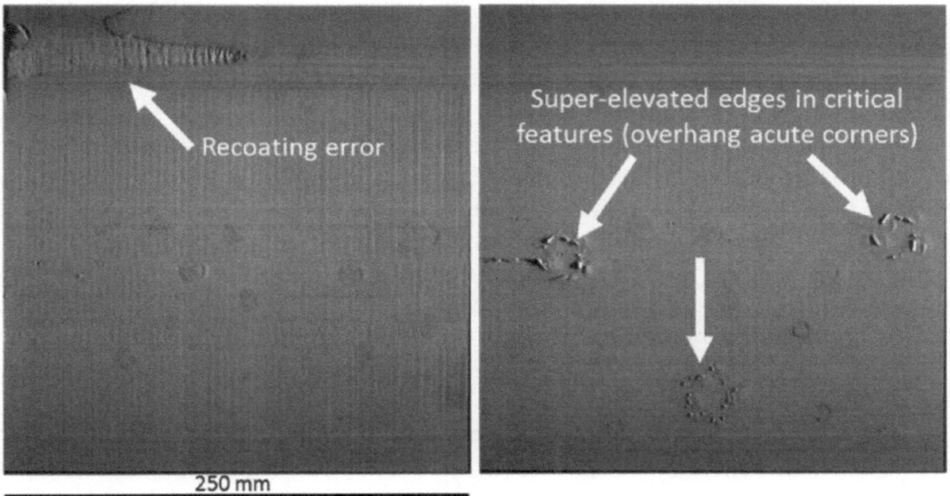

FIGURE 13.12
Example of powder bed imaging in L-PBF.

enhanced contrast between foreground and background regions, but it masks most information about surface irregularities. When the majority of light is reflected towards the camera, bright-field illumination is obtained, which can be more appropriate to capture surface pattern characteristics. Multiple images can be acquired under different illumination conditions and then fused together (Abdelrahman et al. 2017). This fusion allows a reduction of the loss of information and acquisition of a more robust characterisation of the features of interest.

A different approach is to exploit structured illumination via fringe projection combined with monocular or stereo vision for topographical characterisation of the height map of the powder bed and the printed areas. This approach is described in Section 13.5.2, where on-machine metrology methods are introduced.

Layerwise imaging can be used for different on-machine measurement and monitoring purposes, but it lacks the capability to capture dynamic and transient phenomena. Off-axis high-speed cameras in the visible range may be used to address this lack as far as monitoring level 2 is concerned, including the measurement of laser-heated zone size and shape, spatter and plume emissions, hot-spot occurrences, etc. The sampling frequency must be high enough to capture the phenomena on the desired temporal scale to avoid information loss (for example, up to 1 kHz to 10 kHz). On the other hand, the field of view and/or the spatial resolution must be limited to achieve the necessary sampling frequency, but they should be high enough to properly capture all the events of interest.

Figure 13.13 shows an example of video frames acquired by means of a high-speed video imaging system (frame rate equal to 300 Hz) installed off-axially outside the viewport of an L-PBF machine (Grasso et al. 2016, Colosimo and Grasso 2018). The video refers to the L-PBF of a complex shape, where an anomalous heat accumulation occurred in correspondence to an overhang acute corner, leading to a hot-spot event (indicated by the arrow). Since these kinds of hot spots are generated after the laser scan and they remain visible after the laser spot has moved to other locations, their identification may be difficult or even impossible by using co-axial sensors. In general, off-axis video imaging enables the detection of transient phenomena whose measurement and characterisation is not possible by limiting the field of view to the melt-pool area alone.

Video imaging in the visible range is affected by several limitations discussed in Section 13.4.1. Since PBF processes entail wide temperature variations, and the final quality and performance of the product may be strongly affected by the thermal history of the process, off-axis thermal imaging is particularly suitable for on-machine measurement and monitoring.

Thermal cameras are used in both L-PBF and EB-PBF processes to acquire the thermal map of the process during the scan (monitoring level 2) or layerwise (monitoring level 1) to reconstruct the thermal history in local features of the part, and to detect events that are difficult or impossible to measure in the visible range.

Thermal video imaging during the additive production of a part may provide relevant information about the spatial and temporal temperature gradients in the printed area, but it may also provide information about process by-products in L-PBF, that is, spatter and plume emissions (for an example of on-machine monitoring application based on thermal imaging of plume emissions, see Section 13.6.2.2).

As far as off-axis sensing architectures are concerned, other sensing methods have been proposed and tested in the recent years by various authors and, in some cases, patented by PBF system developers. Rather than measuring electromagnetic radiation emissions by means of pyrometers and imaging sensors, other quantities can be measured during the PBF process. These include high-frequency elastic energy releases produced by the

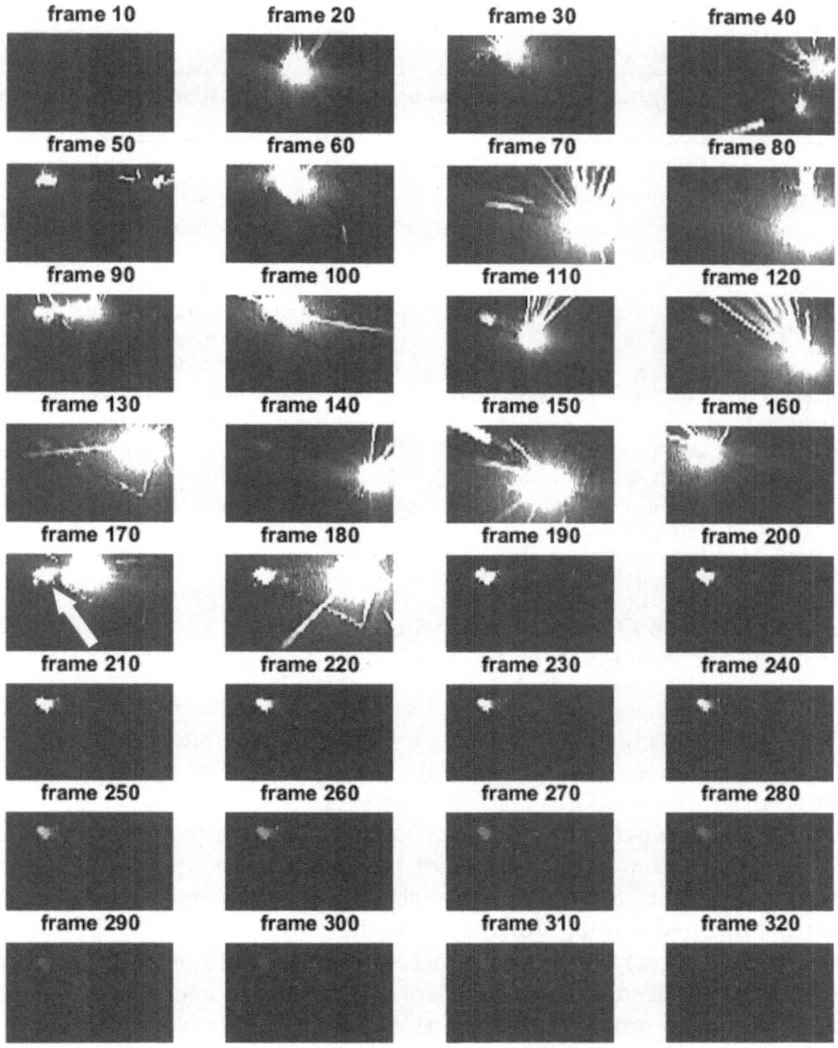

FIGURE 13.13
Examples of video frames acquired by means of an off-axis high-speed camera in the visible range in an L-PBF process (the arrow indicates the hot-spot location). (From Colosimo et al. 2019. With permission.)

formation of micro-cracks, delamination and detachment from the supports, which can be detected by means of acoustic emission or ultrasound sensors (Wasmer 2018, Wasmer et al. 2019).

In EB-PBF, the specific nature of process by-products resulting from the atomic interaction between the electron beam and the material may be used for on-machine sensing and monitoring purposes. In particular, EB-PBF process by-products include back-scattered secondary electrons that can be collected by means of insulated metal plates surrounding the build volume and used to generate on-machine electronic images, in a similar manner to scanning electron microscopy. This yields a layerwise imaging method that enables the characterisation of the powder bed homogeneity and the surface pattern of the printed slice (Arnold et al. 2018, Wong et al. 2019).

TABLE 13.1

Differences between Spatially Integrated and Spatially Resolved Sensors

Output signal	Spatially integrated sensors integral signal	Spatially resolved sensors image
Data gathering level	Mainly level 3	Levels 1 and 2
Sensing architecture	Mainly co-axial	Both co-axial and off-axis
Wavelength	From visible to long wavelength infrared	From visible to long wavelength infrared
Spatial resolution	-	From tens of μm/pixels in case of high-spatial resolution imaging and limited field of view, to hundreds of μm/pixels for high-speed video imaging
Sampling frequency (temporal resolution)	Possibly larger than 50 kHz	Usually limited to 1 kHz to 10 kHz for on-machine monitoring applications
Field of view	Very low for most common applications – melt-pool monitoring (level 3), for example about 0.5 mm × 0.5 mm	Larger than or equal to build area for powder bed monitoring (level 1); smaller than build area for track-level monitoring (level 2).

13.4.4 Mapping between On-Machine Sensing, Process Signatures and Process Defects

This section will summarise the main differences between different sensing methods and their suitability for on-machine detection of process signatures and defects. Table 13.1 first recaps the main differences between spatially integrated and spatially resolved sensing methods for on-machine monitoring of AM processes.

Table 13.2 provides an overview of the process signatures that can be measured on-machine to detect and characterise different types of defects, together with the corresponding sensing methods. Some of these relationships are currently under investigation and need to be confirmed through further research.

Regarding level 1 process signatures, a lack of powder bed homogeneity may change the local layer thickness, leading to possible volumetric and geometrical defects caused by improper energy density provided to the part. Errors in the powder recoating of the slice can also lead to poor welding between one layer and the following layer, with consequent risk of delamination, together with possible geometrical distortion in the presence of severe recoating errors and contamination. The characterisation of the surface pattern and surface topography of the printed slice in each layer provides information not only about possible surface defects but also about the possible occurrence of pores, balling effects and delamination, which are affected by uneven patterns in the layer. Eventually, the on-machine reconstruction of the layerwise slice geometry is relevant to detect geometrical distortion.

Regarding level 2 process signatures, the detection of hot spots may be suitable to identify geometrical distortion, as excessive heat accumulation may cause super-elevated edges in critical geometrical features. The detection of cold spots, on the other hand, may be suitable to identify lack-of-fusion conditions that may produce lack-of-fusion porosity and, if sufficiently severe, also balling and delamination. Cooling history and the temperature profiles measured during the process can provide information about variations in the microstructure of the part and thermal stress accumulation related to improper heat exchanges. Process by-products, such as spatter and plume emissions in L-PBF, can

TABLE 13.2

Mapping between On-Machine Measurable Signatures, Sensing Methods and Process Defects in PBF

			Defects					
Level	**Process signature**	**On-machine sensing method**	**Porosity**	**Residual stresses, cracks and delaminations**	**Micro-structural inhomogeneity and impurities**	**Balling**	**Geometrical/ dimensional distortions**	**Surface defects**
1 (powder bed)	Powder bed homogeneity	Off-axis imaging, visible range	(X)	X			X	
	Slice geometry	Off-axis imaging, visible range					X	
	Slice surface pattern	Off-axis imaging, visible range, fringe projection	(X)	X		X		X
2 (track)	Hot and cold spots	Off-axis video imaging, visible or infrared range	X	X		X	X	
	Temperature profile/ cooling history	Off-axis thermal imaging		X	X			
	Process by-products	Off-axis video imaging, visible or infrared range	X		(X)			
3 (melt pool)	Size	Co-axial video imaging, visible or infrared range	X	X		X		X
	Shape	Co-axial video imaging, visible or infrared range	X	X		X		X
	Average intensity	Co-axial pyrometry	X	X	(X)	X		X
	Intensity profile	Co-axial video imaging, visible or infrared range	X	X	(X)	X		X

Note: An 'X' is shown in correspondence of known relationship while (x) is used to represent second-order, indirect relationships.

be used as potential proxies of volumetric defects. The spatter behaviour is related to the ejection of material from the melt pool and to the formation of denudation zones around the melt pool, which may influence the formation of pores. Large and intense plume emissions may partially absorb and deflect the laser beam, reducing the energy input provided to the part, with consequent lack-of-fusion porosity.

As far as melt-pool signatures (level 3) are concerned, relevant information can be gathered with respect to the possible formation of volumetric defects, both key-hole and lack-of-fusion porosity, thermal stress accumulation because of insufficient heat dissipation, balling phenomena caused by lack-of-fusion conditions and surface defects related to the solidification properties of scanned tracks.

13.5 On-Machine Measurement

On-machine measurement refers to the use of on-machine gathered data about quality characteristics of the part to anticipate and support post-process quality inspection and measurement. For an introduction to general metrological concepts together with specific methods and instruments, the reader is referred to Chapters 10, 11 and 12. In this chapter, the possibility of using some of those methods on-machine, i.e. while metal parts are produced via PBF processes, is discussed. Section 13.5.1 presents on-machine geometry reconstruction, while other methods are briefly presented in Section 13.5.2.

13.5.1 On-Machine Topography Reconstruction

Off-axis layerwise imaging via high-resolution cameras allows identification of the contours of the printed slice in each layer. In principle, this enables the reconstruction of the part on a layerwise basis for the characterisation of possible topographical and dimensional deviations from the reference model. It is worth noting that the part dimensions and its topography measured on-machine are not representative of the final dimensions and topography of the as-built part because of shrinkage and thermal stress–induced distortions, which cannot be captured on a layer-by-layer basis. However, if a major departure from the expected topography is observed in one layer, it is worth signalling it as soon as possible, since it may be representative of a defect that cannot be recovered as the process goes on. This paves the way for metrological use of on-machine gathered data aimed at part topography reconstruction.

In order to reconstruct the topography of the slice in one layer, an on-machine contour identification operator is needed. Figure 13.14 shows an example of a layerwise image acquired in L-PBF by combining an off-axis camera with dark-field illumination (left panel) and the corresponding on-machine identification or slice contours (right panel) (Caltanissetta et al. 2018). A camera perspective correction operation was applied to enable on-machine topography reconstruction.

Several image-segmentation and edge-detection techniques are available for contour detection (for a general overview, see Gonzalez and Woods 2002). The specific nature of layerwise images, characterised by noisy patterns and non-homogeneous pixel intensities within both the background (powder) and foreground (slice) regions, together with badly defined edges, imposes some challenges for many commonly used techniques. One category of methods that has been showed to be relatively effective in segmenting this type of image is the so-called 'active contours method' (Chan and Vese 2001). Active contours

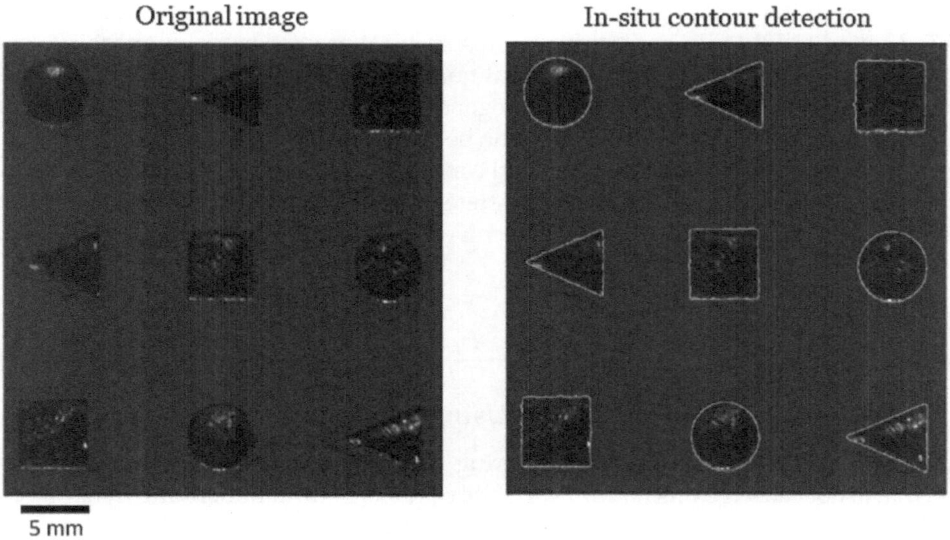

FIGURE 13.14
Example of layerwise image acquired in L-PBF (left panel) and corresponding identification of slice contours (right panel).

are implemented in many computer vision libraries (for example, OpenCV) and image processing toolboxes (for example, Matlab® Image Processing Toolbox).

Despite being a simple and easy-to-implement procedure, on-machine layerwise geometry reconstruction is still missing in industrial PBF systems. One main challenge is the achievement of accurate contour identification. Seminal studies have investigated the measurement bias and variability, as well as the influence of illumination conditions and the sensitivity to segmentation algorithm settings (Caltanissetta et al. 2018, Aminzadeh 2016, Aminzadeh and Kurfess 2016). However, further developments are needed to make this kind of on-machine measurement application industrially available.

13.5.2 Other Methods

Rather than measuring the topography of each printed slice, it could be possible to exploit on-machine gathered data to characterise the volumetric and microstructural properties of the part.

The term *optical tomography* is used to indicate the possibility of using layerwise and on-machine measurement to reconstruct a 3D map of internal flaws (Bamberg et al. 2016). This kind of application is still under study, as there is no assessed method to identify and characterise individual pores on a layer-by-layer basis. Indeed, the few methods proposed so far have attempted to indirectly identify the presence and severity of a pore by looking at some porosity-related quantities, for example, cold-spot regions as proxies of lack-of-fusion pores (Bamberg et al. 2016) or temperature profiles of individual pixels as proxies of over-heating porosity (Schilp et al. 2014).

A different application is on-machine microstructural prediction via on-machine sensing. The layerwise thermal maps based on infrared video imaging could be used to estimate the thermal gradient and solid-liquid interface velocity, which are two indices that can be used to determine the grain orientation within the part, for example columnar against equiaxed grains. The feasibility of this approach has been demonstrated in recent studies (Raplee et al. 2017) devoted to EB-PBF. On-machine microstructural prediction via on-machine sensing could be used to support the microstructural qualification of the part. On the one hand, it allows signalling when and where non-homogeneous microstructural properties have originated within the part. On the other hand, it can be used to support production of functionally graded materials where custom variations of the microstructure are imposed by locally changing the process parameters and/or scanning strategy. In this latter case, on-machine measurements can be used to keep under control the achievement of the target microstructural pattern.

13.6 Statistical Process Monitoring Using On-Machine Sensing

One possible aim of an on-machine monitoring tool consists of signalling a change in the process that is not caused by its natural variability and its natural underlying dynamics. This entails the need for an automated alarm rule to identify the onset of a defect, an unstable process condition or any departure from the in-control pattern of measured variables. The simplest, but also less robust and effective, approach consists of setting heuristic thresholds for each monitored variable in order to signal an alarm whenever a threshold violation is recorded. This is commonly carried out in industrial practice, where the operator is asked to set one, or more, thresholds depending on his/her knowledge of the process. The use of heuristic thresholds can be a cumbersome activity and can produce detrimental effects (increases in the false positive and false negatives rates), especially when moving to multivariate settings, where hundreds of features have to be monitored with time. The detrimental effect of false positives (false alarms) is especially evident when in-process monitoring is assumed.

Statistical process monitoring (SPM), also known in the literature as *statistical process control* (SPC), is a consolidated framework for the design for process monitoring solutions that rely on the characterisation of the natural variability of the (multiple) variables of interest. Section 13.6.1 first introduces the basic principles of SPM. Section 13.6.2 then presents some examples of SPM methods applied to on-machine gathered data in PBF processes. For a more exhaustive and general introduction to SPM and SPC, see Montgomery (2009).

13.6.1 Basic Principles of SPM

13.6.1.1 False Alarms and False Negatives: Type I and Type II Errors

Any type of process, including an AM process, can be labelled as *in statistical control* or simply *in-control* when it is operating with only random causes of variation present, where random causes are inherently part of the process itself (Montgomery 2009). Sources of variability that are not part of the random causes are referred to as *assignable* causes of variation, which include onsets of defects, process errors and process unstable conditions. A process that is operating is the presence of assignable causes is labelled as *out-of-control.*

An SPM problem can be regarded as a statistical hypothesis testing problem that is sequentially repeated over time as new observations become available. In the SPM framework, a null hypothesis and an alternative hypothesis are defined as follows:

H_0: process is in-control (null hypothesis);

H_1: process is out-of-control (alternative hypothesis).

The aim of an SPM method is to properly identify whether the null hypothesis H_0 should be rejected or not, based on available data. Therefore, two errors need to be defined: Type I error α, also called *false alarm* or *false positive* rate; and Type II error β, also called *false negative* rate. These two errors are defined as follows:

$$\alpha = P(\text{Type I error}) = P(\text{reject } \mathrm{H}_0 \mid H_0 \text{ is true});$$

$$\beta = P(\text{Type II error}) = P(\text{fail to reject } \mathrm{H}_0 \mid H_0 \text{ is false}).$$

Type I error is the probability of signalling an alarm, i.e. rejecting the in-control hypothesis, when the process is actually in-control. Type II error is the probability of signalling no alarm, i.e. failing to reject the in-control hypothesis when the process is out-of-control.

As in traditional hypothesis testing, Type I error is a design parameter, whereas a Type II error curve is drawn by computing the Type II error as a function of the 'magnitude' of the shift from the in-control state, to determine the performance of the monitoring procedure.

The statistical instruments used to detect process shifts are called *control charts* (see Section 13.6.1.2). Although control charts are used in traditional quality control applications (Montgomery 2009), where the quality characteristics of interest are usually measured on the product rather than on the process features, they represent general-purpose tools that can also be applied to on-machine monitoring applications (Colosimo 2018, Colosimo et al. 2018).

13.6.1.2 Control Charts

A traditional control chart is a graphical tool which includes i) a centre line that represents the average value of the quality characteristics under in-control process conditions, and ii) one or two control limits chosen such that, if the process is in-control, only 100 × α% of the measured values of the quality characteristics will fall outside these limits. Every measured value of the quality characteristic is plotted onto the control chart in temporal sequence. An alarm is signalled whenever one limit is violated, but additional alarm rules can be defined to signal non-random patterns in the temporal sequence of quality characteristic values plotted onto the control chart, as such patterns may be symptoms of a change in the process behaviour. Indeed, if the process is in-control, all the plotted points should have a random distribution.

The use of control charts for statistical process monitoring involves two sequential phases, denoted by Phase I, also known as the *training* phase; and Phase II, also known as the *monitoring* phase. During the training phase, a dataset gathered under in-control process conditions, and representative of the natural process behaviour, is collected and used to estimate the parameters of the variables to be used for process monitoring and to design the control chart, i.e. choosing the control limits. During the training phase, the collected data can be plotted onto the designed control chart in order to determine whether the process was actually in-control or not during this phase. If out-of-control conditions are signalled and corresponding assignable causes are found by means of a retrospective investigation, an iterative procedure is applied to remove out-of-control data and finally design a control chart that is representative of the actual in-control behaviour of the process. During the following phase, additional measurements of the variable of interest are acquired and their values are plotted onto the previously designed control chart, and an alarm is signalled whenever an out-of-control shift is detected.

13.6.2 Examples of SPM

13.6.2.1 On-Machine Monitoring Example, Level 1

The first example is the design of a simple automated alarm rule to detect inhomogeneous patterns in the powder bed, layer by layer, with a specific focus on the identification of areas not properly covered by the powder, also called *super-elevated edges*. The example is taken from zur Jacobsmühlen et al. (2013). The method works as follows:

- *Training phase.* A set of images of in-control powder beds in L-PBF is recorded (for example, collected during previous builds), and the sample mean and sample standard deviation of the pixel intensities in the presence of correct recoatings are estimated. By using the notation of zur Jacobsmühlen et al. (2013), the mean and standard deviation are labelled as μ and σ, respectively. An upper control limit (*UCL*) for the detection of critical regions is chosen as $UCL = \mu + k\sigma$, where $k = 3$, which corresponds to a Type I error $\alpha = 0.0027$ if the pixel intensity values follow a normal distribution.
- *Monitoring phase.* For each newly acquired powder bed image, the intensity of each pixel is compared with the control limit *UCL*. In order to avoid signalling false alarms when even one or a small number of sparse pixels violate the control limit, the alarm rule envisages a second step to signal an alarm only when a cluster of connected pixels with a critical dimension has intensities above the limit. With this aim, a morphological operation is applied to generate connected regions based on pixels whose intensities are larger than *UCL*. For each connected region, the area A_R and the mean pixel intensity μ_R are computed, and an alarm is signalled if

$$A_R \cdot \mu_R > T \tag{13.6}$$

where T is a threshold that can be estimated based on a set of reference images where critical super-elevated edges where observed.

The method described by zur Jacobsmühlen et al. (2013) is a simple, though effective, on-machine monitoring method that combines two alarm rules to reduce the number of false alarms and tune the out-of-control detection based on available information about the critical size of the defect.

13.6.2.2 On-Machine Monitoring Example, Level 2

A second example is a multivariate control charting method for on-machine monitoring of L-PBF based on the analysis of the plume-emission stability over time. This example is taken from Grasso et al. (2018a). In this case, an off-axis thermal camera was used to acquire video images of the L-PBF process with a frame rate of 50 Hz. Each frame was segmented by using image processing techniques to isolate the region of interest corresponding to the plume emission. Two different descriptors were estimated to characterise the plume property: the area A_j and the mean temperature intensity $\overline{I_j}$ of the plume in the jth frame, with $j = 1, 2, \ldots, n$ (where n is the number of frames). Indeed, the larger and hotter the plume, the larger the material vaporisation, which may interfere with the energy input provided to the material, leading to possible volumetric defects and detrimental properties of the final part. The on-machine monitoring method works as follows.

- *Training phase*. A set of video sequences acquired during some initial layers of the current process is recorded and used to design a multivariate T^2 control chart (Montgomery 2009). An unstable melting condition is expected to be a departure from the in-control pattern that grows over time, layer by layer. Initial layers can be used as a reference to determine whether the process remains stable over time or not. The T^2 statistic is computed by considering the bivariate data vector $\mathbf{x}_j = \left[A_j, \overline{I_j}\right]^T$. The sample mean and the sample variance–covariance matrix are estimated based on plume area and plume mean intensities recorded in this phase, and the upper control limit *UCL* is chosen by setting a target Type I error value.
- *Monitoring phase*. During the production of each subsequent layer, the thermal video is recorded and the plume area and plume mean intensities are computed in each frame, resulting in a new observation of the T^2 statistic in each newly acquired video frame. If the T^2 value violates the control limit, an alarm is signalled. This is an example of a traditional multivariate control chart scheme, combined with on-machine image processing for the automated detection of process instability.

Figure 13.15 shows the resulting control charts in three scenarios (Grasso et al. 2018a). Scenario 1 was representative of an in-control L-PBF process, with a fully dense produced specimen. Scenarios 2 and 3 were representative of out-of-control L-PBF processes, where unstable conditions grew layer by layer, leading to excessive heat accumulation followed by a partial disintegration of the specimen. Figure 13.15 shows that, in out-of-control scenarios, several violations of the control limits occur as a consequence of increasingly unstable plume emissions, characterised by anomalous properties in terms of area and intensity, which finally led to defective parts. This example shows that an on-machine monitoring tool can automatically anticipate the detection of an unstable process, enabling its anticipated interruption and, as a consequence, avoidance of time and resource waste.

13.7 Process Control

On-machine monitoring methods allow implementation of novel 'intelligent' capabilities in AM systems. Indeed, such methods enable the automated detection of out-of-control changes in the process behaviour and the onset of defects in the layer, which can be used to stop the process avoiding further waste of time and resources and the production of scrap. Process interruption is the simplest type of action that can be taken as a consequence of an alarm. However, in order to achieve first-time-right and zero-defect AM capabilities, more sophisticated levels of process control and adjustment are needed. In the current industrial practices, processability windows are experimentally identified for each material of interest. This leads to sets of material-dependent process parameters that are applied to any shape produced with a given material. However, optimal process parameters are not only material-dependent; different geometrical features, different part orientations in the build and different designs of supports would require different process parameters to properly take into account local and global heat accumulation and dissipation effects. To

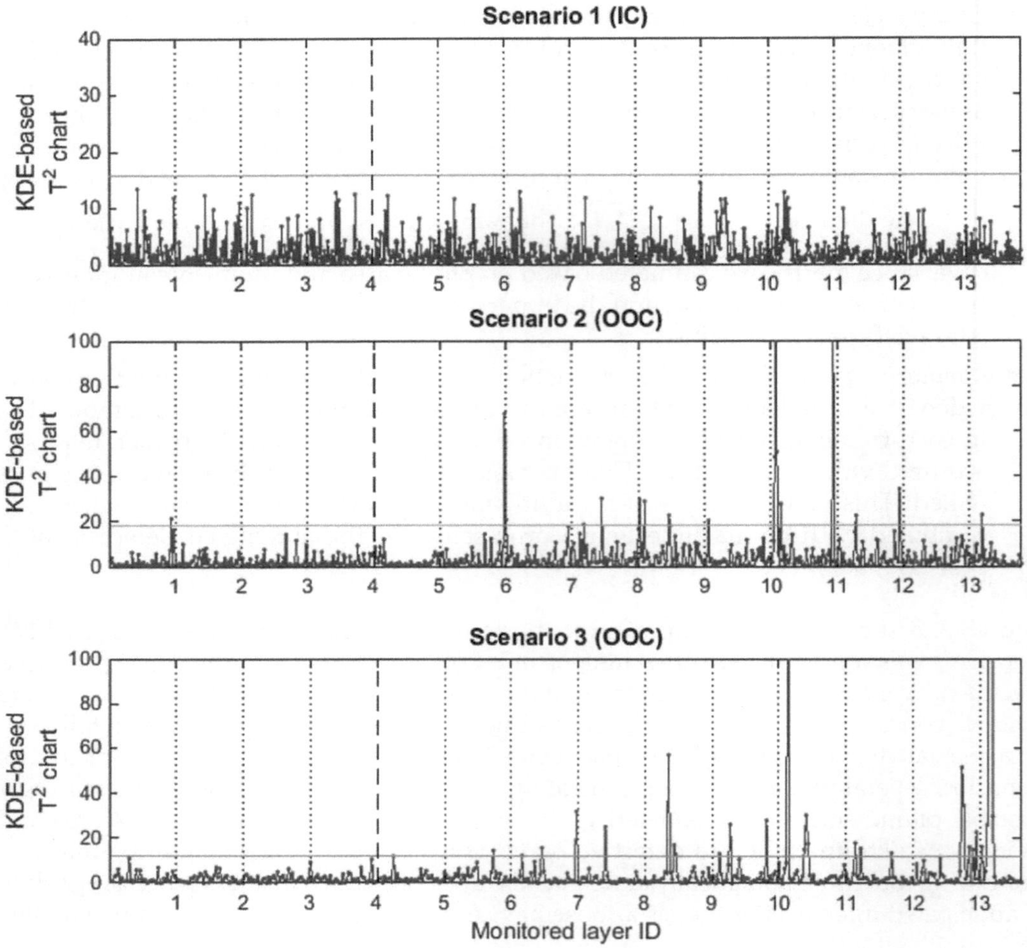

FIGURE 13.15
Plume-based T^2 control charts for on-machine monitoring of the L-PBF in the presence of in-control and out-of-control process conditions; the vertical solid dashed line separates the training phase from the monitoring phase. (From Grasso et al. 2018. With permission.)

this aim, two process control strategies are available, namely feedback and feedforward control strategies. These strategies are introduced in Sections 13.7.1 and 13.7.2 respectively, with a number of examples taken from the literature.

13.7.1 Feedback Control

A feedback control method envisages the closed-loop adaptation of a pre-defined process parameter based on real-time on-machine sensor recording of a quantity that should be kept as close as possible to a target reference. The controller compares the current sensor reading with the reference, also known as the *set point*, and adapts the value of the controlled variable such that the difference from the set point is kept as low as possible. The most commonly used type of closed-loop feedback controller is the proportional, integral, derivative (PID) controller, which adapts the controlled output based on (i) the instantaneous difference between the current signal and the set point (proportional action), (ii) the past difference values (integral action) and (iii) how fast the difference changes over time (derivative action). In most cases, not all of the three actions are combined together, but just a subset of them.

In PBF applications, due to the high laser scan speed, a high temporal resolution of the sensor reading is needed to obtain an effective variation of the process parameter. This can be achieved only by considering a level 3 data gathering approach, where the melt-pool intensity is measured at high speed (for example, between 10 kHz and 20 kHz and higher) by means of co-axial spatially integrated sensing. Variations of melt-pool properties from a target reference indicate a departure from an in-control and stable process state. Process parameters that can be adapted in real time by the feedback controller include the laser power and the scan speed. As an example, a local increase of the melt-pool intensity may be caused by diminished heat dissipation in the presence of overhang regions. This can be mitigated by reducing the laser power or by increasing the scan speed. Figure 13.16 shows a schema of a closed-loop feedback control application in L-PBF.

One example of closed-loop control in L-PBF was presented by Kruth et al. (2007) and Craeghs et al. (2010), who demonstrated the suitability of a feedback control method to adapt the laser power in real time, based on the estimated area of the melt pool. A co-axial pyrometer was used to measure the melt-pool intensity, and an experimental model was used to estimate the melt-pool area based on its measured intensity. A PID controller was implemented, and the set point was determined experimentally for given combinations of process parameters.

In the example presented by Kruth et al. (2007), the experimental tests in L-PBF were performed with a maximum scan speed of 300 mm/s. Current state-of-the-art L-PBF systems use much higher scanning speeds; this makes feedback control implementation in L-PBF processes still a challenging task, as the time constant for the response of the melt pool to change in power or speed can be relatively slow. The larger the scan speed and the larger the distance between optimal process parameters and current process parameters, the lower the expected effectiveness of closed-loop feedback control schemes in L-PBF. The industrial implementation of such feedback control schemes is further complicated by the proprietary nature of controls implemented by different L-PBF system

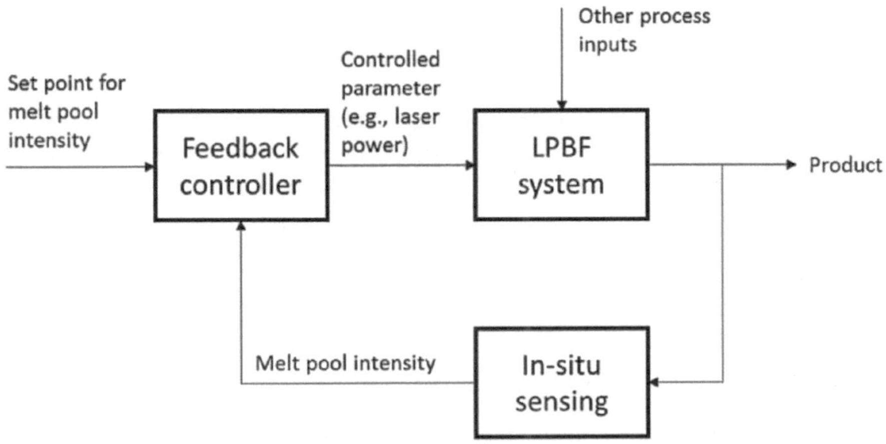

FIGURE 13.16
Example of feedback control scheme in L-PBF.

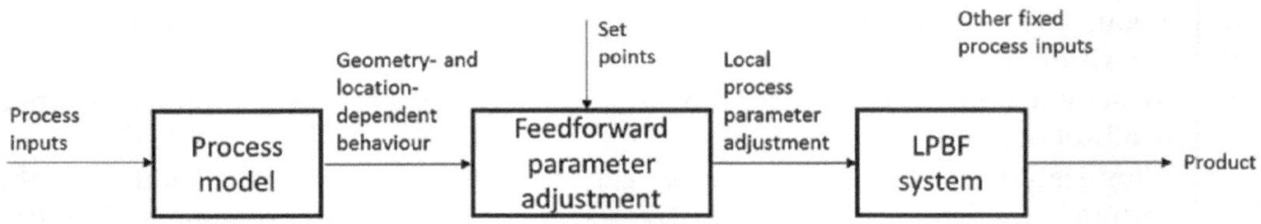

FIGURE 13.17
Example of feedforward control scheme in PBF.

developers and by the evolution of L-PBF machines towards multi-beam systems. Some recent studies (Renken et al. 2019) showed new results that demonstrate the feasibility of feedback control methods in L-PBF, but further research and industrial implementation are needed.

13.7.2 Feedforward Control

Rather than exploiting real-time sensor signals to adapt the process parameters, feedforward control methods use an adjustment of process parameters relying on model-based predictions of local departures from a target behaviour. As an example, in the presence of a bridge, such as that discussed by Kruth et al. (2007), an increase of the melt-pool intensity can be easily predicted in advance. This would allow modification ex-ante of the scanning or exposure strategy in a geometry-dependent way, leading to a mitigation of the detrimental effects caused by inappropriate process parameters. Figure 13.17 shows a simplified schema of a feedforward control methodology in PBF processes.

Feedback and feedforward control methods can be combined in order to overcome the limitations of each single approach and to achieve more robust results. Renken et al. (2019) compared different control strategies in L-PBF to reduce variability of the melt-pool intensity when an unsupported overhanging bridge was produced. Also, in this case, the controlled parameter was the laser power.

Renken et al. (2019) showed that the combination of feedback and feedforward methods minimises the variability of the pyrometer signal when printing overhangs. Indeed, the feedforward control enables the energy input to be set closer to the optimal value, which makes the real-time implementation of closed-loop feedback controllers more effective, leading to final performances that could not possibly be achieved by using just one of these approaches.

The effectiveness of feedforward methods relies on the development of sufficiently accurate and fast predictive models to compensate for the local and systematic variability of physical and functional properties of additively produced parts, which is still an open topic of research in PBF processes.

13.8 Current Research and Future Look

Considerable effort has been devoted so far in the scientific and industrial communities to understanding the nature and the source of defects in PBF processes, their effects on product quality and how they can be mitigated or avoided by acting on controllable parameters.

Indeed, the lack of repeatability and stability of PBF processes, together with several possible sources of defects, have been widely pointed out as major issues that deserve further technological advances to meet challenging industrial requirements (Colosimo et al. 2018, Grasso and Colosimo 2017, Tapia and Elwany 2014, Mani et al. 2017, Spears and Gold 2016, Everton et al. 2016). The development and implementation of on-machine measurement, monitoring and control solutions represents a priority in pushing forward the industrial breakthrough of metal AM systems.

Table 13.3 and Table 13.4 present summaries of the scientific literature devoted to on-machine measurement and on-machine monitoring in PBF processes.

The research in this field is growing and evolving very fast. The first seminal studies were mainly aimed at demonstrating the feasibility of on-machine sensing methods and characterising specific process phenomena with the support of on-machine gathered data. More recent studies have been proposing, testing and demonstrating on-machine measurement and monitoring methodologies. An increasing interest has also been devoted to the use of machine learning and artificial neural network techniques to make sense of large on-machine data streams for robust and reliable identification of defects and process errors (Kwon at al. 2018, Gobert et al. 2018, Okaro et al. 2019, Wasmer et al. 2019, Scime and Beuth 2018).

Recent studies also proposed novel on-machine sensing solutions or the combination of multiple sensors to achieve better on-machine measurement and monitoring performance (Tan Phuc and Seita 2019, Barrett et al. 2018a, Barrett et al. 2018b).

As far as the industrial implementation of these methods is concerned, it is worth noting that most PBF system developers have equipped their systems with on-machine sensing and monitoring modules and toolkits (Table 13.5). Most of these tools are mainly used to

TABLE 13.3

Summary of the Literature Devoted to On-Machine Metrology in PBF

Measured signature		On-machine sensing (main categories)			
		Imaging (visible to near infrared)	**Imaging + fringe projection**	**Thermal imaging (MW to LW infrared)**	**Other**
Layerwise geometry and surface characterization	Geometrical distortions	Foster et al. 2015, Aminzadeh and Kurfess 2016, Aminzadeh 2016, Caltanissetta et al. 2018,		Ridwan et al. 2014	
	Surface topography	Kleszczynski, et al. 2012, zur Jacobsmühlen et al. 2013, Imani et al. 2018, Scime and Beuth 2018, Gobert et al. 2018, Foster et al. 2015, Abdelrahman et al. 2017, Li et al. 2012, Tan-Puc and Seita 2019	Land et al. 2015, Zhang et al. 2016, Dickins et al. 2018		Erler et al. 2014 (2D laser displacement sensor)
Optical tomography		Bamberg et al. 2016, Schilp et al. 2014, Abdelraham et al. 2017		Ridwan et al. 2014	
Micro-structural properties				Raplee et al. 2017	

TABLE 13.4

Mapping of the Literature on On-Machine Process Monitoring in PBF with Respect to the Monitored Signatures and the Sensing Approach

Monitored signature		On-machine sensing method		
		Pyrometry	Imaging (visible to near infrared)	Thermal imaging (MW to LW infrared)
Level 1 Powder bed	Powder bed homogeneity		Foster et al. 2015, zur Jacobsmühlen et al. 2013, Land et al. 2015, Zhang et al. 2016, Tan-Puc and Seita 2019	
Level 2 Track	Hot and cold spots	Bayle and Doubenskaia 2008, Thombansen et al. 2015	Grasso et al. 2016, Colosimo and Grasso 2018	
	Temperature profile/cooling history			Krauss et al. 2012, Krauss et al. 2014, Lane et al. 2015, Bayle and Doubenskaia 2008, Gong et al. 2013, Price et al. 2012, Schilp et al. 2014,
	Process by-products	Bayle and Doubenskaia 2008	Repossini et al. 2018, Andani et al. 2017a, Andani et al. 2017b, Ye et al. 2018	Grasso et al. 2018, Grasso and Colosimo 2019, Bayle and Doubenskaia 2008, Lane et al. 2015
Level 3 Melt pool	Size	Clijsters et al. 2014, Craeghs et al. 2010, Craeghs et al. 2011	Craeghs et al. 2010, Craeghs et al. 2012, Clijsters et al. 2014, Berumen et al. 2010, Kruth et al. 2007, Van Gestel 2015, Demir et al. 2018, Fisher et al. 2018, Kolb et al. 2019, Kwon et al. 2018	
	Shape		Craeghs et al. 2011, Berumen et al. 2010, Van Gestel 2015, Kruth et al. 2007, Kwon et al. 2018	Doubenskaia et al. 2015
	Temperature/ brightness intensity	Craeghs et al. 2011, Berumen et al. 2010, Chivel 2013, Clijsters et al. 2014, Doubenskaia et al. 2012, Pavlov et al. 2010, Thombansen et al. 2015, Demir et al. 2018, Kolb et al. 2019, Okaro et al. 2019	Berumen et al. 2010, Van Gestel 2015, Yadroitsev et al. 2014, Chivel 2013	
	Temperature/ brightness profile		Doubenskaia et al. 2012	Gong et al. 2013, Price et al. 2012

Note: Italic font used for co-axial set-ups and roman font for off-axial set-Ups.

TABLE 13.5

Commercial Toolkits for PBF Process Monitoring

Toolkit name	Developer	Monitored quantity	On-machine sensing
QM meltpool 3D	Concept Laser	Melt pool (area and intensity)	Co-axial photodiodes (co-axial camera also available in research version)
QM coating	Concept Laser	Powder bed	Off-axial camera
EOSTATE MeltPool	EOS	Melt pool	Co-axial and off-axial sensors
EOSTATE PowderBed	EOS	Powder bed	Off-axial camera
EOSTATE Exposure OT	EOS	Thermal map over the entire powder bed	Off-axis camera
Melt Pool Monitoring (MPM) system	SLM Solutions	Melt pool	Co-axial pyrometer
Layer Control System (LCS)	SLM Solutions	Powder bed	Off-axial camera
MeltVIEW	Renishaw	Melt pool	Co-axial photodiodes
	Aconity3D	Melt pool	Co-axial photodiodes and CMOS camera
Truprint Monitoring	Trumpf	Melt pool	Co-axial photodiodes (beta version)
Truprint Monitoring	Trumpf	Powder bed and part geometry	Off-axial camera
	SISMA	Powder bed	Off-axial camera
LayerQam	Arcam	Intensity map in the slice as a proxy of internal porosity	Off-axial camera
XQam	Arcam	Back-scattered X-ray emission	X-ray detector
PrintRite3D	B6 Sigma, Inc.	Different monitoring equipment	Set of co-axial and off-axial sensors available

collect data during the process and provide the user with some post-process data reporting and/or datasets to support the investigation of specific problems and defects. Further development efforts are still needed to implement analytical tools that are able to quickly make sense of gathered data during the process and automatically signal the onset of defects and process instabilities. In addition to the on-machine sensing tools made available by PBF systems manufacturers, third party in-process quality assurance toolkits for metal AM processes have recently been made commercially available (see the system developed by B6 Sigma, Inc. in Table 13.5).

Despite continuous and fast technological developments related to on-machine sensing equipment, on-machine data processing and monitoring methods, together with the first seminal attempts to implement closed-loop control solutions, several challenges and open issues must be faced to develop new generations of smart PBF machines able to achieve first-time-right and zero-defect production capabilities. Some of the most challenging issues, also related to the most promising future research directions, can be summarised as follows (Colosimo et al. 2018).

- *Limits of the layerwise monitoring paradigm.* Although the layerwise production paradigm enables several kinds of on-machine measurements that are not possible in other processes, it also gives rise to some limitations. As an example, looking at the current layer prevents the gathering of information about physical phenomena that are occurring below the layer, involving partial re-melting, heat accumulation and dissipation, and consequent effects on volumetric, micro-structural and thermal stress properties of the material. Although this is an intrinsic limitation imposed by the layerwise production paradigm, sub-surface characterisation methods could be considered in future studies for on-machine applications. Some seminal studies have started investigating new on-machine sensing and measurement techniques (for example Lewis 2019, Smith et al. 2016), but some types of defects still remain not measurable as discussed in Section 13.1. Of course, if it proves possible to predict volumetric features just by looking at the layer being processed, then simpler measurement approaches may be possible, but such correlations have only just started to be investigated (for example, see Thompson et al. 2018).

- *Robust on-machine porosity detection methods (optical tomography).* Volumetric defects are particularly critical in many industrial applications, but accurate methods – so called 'optical tomography' – for their robust identification by means of on-machine sensors are still missing. This is partially related to the above-mentioned limitation of the layerwise paradigm, as pores may generate below the surface and/or re-melting steps may close surface pores identified in previous layers. Several process signatures can be used as proxies of either lack-of-fusion or keyhole porosity, but further research efforts are needed to achieve robust on-machine porosity measurement and monitoring capabilities. The correlation between drivers of defects measured in-process and flaws identified via post-process inspections, for example via X-ray computed tomography (Chapter 12), is still an open issue despite a few seminal studies, for example Yoder et al. 2018 in EB-PBF and Bamberg et al. 2016 in L-PBF.
- *Big data management.* Every data analytics technique applied to on-machine gathered data in AM applications will face the issues related to the handling of very large amounts and fast data streams. Several gigabytes of data may be generated from on-machine sensors during the production of a part, and this pushes the need for computationally efficient methodologies for on-machine and in-process data processing. Only a small number of solutions proposed in the literature were actually implemented and tested on integrated devices with real-time data handling functionalities. Further development efforts are needed in this field.
- *Training.* Since several metal AM applications involve one-of-a-kind productions and geometries that vary along the process from one layer to another, one challenge is the definition of training phases. Indeed, the training dataset must be representative of in-control process conditions, but the underlying dynamics of the process may vary from one layer to another and from one part to another.
- *Transfer learning.* Another issue is how to transfer knowledge and empirical models gathered on one part by using one AM system to other parts produced with the same machine model but with different machines. As an example, it would be relatively convenient to carry out experimental conditions in a limited and controlled set of process conditions and to transfer the acquired knowledge to other conditions, reducing experimental costs and time-to-market. However, this is still an open issue, inflated by the large system-to-system and lab-to-lab variability that characterises metal AM applications. Only a small number of seminal studies have investigated the application of transfer learning methods to AM (Sabbaghi and Huang 2018).

- *Link between process signatures, powder properties and atmospheric conditions.* The literature also lacks methods to integrate and correlate information related to conditions that may vary from one build to another with an impact on the final part quality and performance, but also on the on-machine measurable quantities. One example involves the powder properties, which may change over time as a consequence of powder reuse. Another example is the build chamber atmospheric conditions, which may change during the build and from one build to another. A better understanding of these aspects and the influence of properties of practical interest is still needed.
- *Cyber-physical approach.* Process simulations have a great potential as technological enablers of novel enhanced AM performance and zero-defect production capabilities. As an example, simulations enable feedforward control strategies for local process parameter adjustment, but they also allow the development of on-machine monitoring methods augmented by process simulations, and vice versa. The combination of real data with process simulation is a field that deserves research effort.
- *Effective and robust process control strategies.* Despite seminal studies on closed-loop control in L-PBF and a few recent developments, a wide gap still needs to be filled in order to make intelligent control solutions industrially available. Rather than adapting the process parameters based on model outputs or real-time sensor signals, other on-machine defect mitigation or defect correction solutions have been proposed in the literature (Mireles et al. 2015a, Heeling and Wegener 2018, Grasso et al. 2019, Colosimo et al. 2019). On-machine defect correction represents a further research field that may contribute to the development of novel generations of smart AM systems, passing from highly sensorised machines to intelligent machines that are able to autonomously identify the defect and remove it to produce any part first-time-right.

Acknowledgements

The authors would like to thank the European Institute of Innovation and Technology (EIT), a body of the European Union under the Horizon 2020 (the EU Framework Programme for Research and Innovation) for funding their activity in the framework of the LILIAM (Lifelong Learning in Additive Manufacturing) project.

References

Abdelrahman, M., Reutzel, E. W., Nassar, A. R., Starr, T. L., 2017. Flaw detection in powder bed fusion using optical imaging. *Addit. Manuf.* 15: 1–11.

Aminzadeh, M., 2016. A machine vision system for in-situ quality inspection in metal powder-bed additive manufacturing, Doctoral dissertation, Georgia Institute of Technology.

Aminzadeh, M., Kurfess, T., 2016. Vision-based inspection system for dimensional accuracy n powder-bed additive manufacturing, 2016. In *ASME 2016 11th International Manufacturing Science and Engineering Conference*, V002T04A042–V002T04A042.

Andani, M. T., Dehghani, R., Karamooz-Ravari, M. R., Mirzaeifar, R., Ni, J., 2017a. Spatter formation in selective laser melting process using multi-laser technology. *Mater. Des.*, 131: 460–469.

Andani, M. T., Dehghani, R., Karamooz-Ravari, M. R., Mirzaeifar, R., Ni, J., 2017b. A study on the effect of energy input on spatter particles creation during selective laser melting process. *Addit. Manuf.*, 20: 33–43.

Anwar, A. B., Pham, Q. C., 2018. Study of the spatter distribution on the powder bed during selective laser melting. *Addit. Manuf.*, 22: 86–97.

Arnold, C., Pobel, C., Osmanlic, F., Körner, C., 2018. Layerwise monitoring of electron beam melting via backscatter electron detection. *Rapid Prototyp. J.*, 24: 1401–1406.

Bamberg, J., Zenzinger, G., Ladewig, A., 2016. In-process control of selective laser melting by quantitative optical tomography. In *19th World Conference on Non-Destructive Testing*, Munich, Germany, June.

Barrett, C., Carradero, C., Harris, E., McKnight, J., Walker, J., MacDonald, E., Conner, B., 2018a. Low cost, high speed stereovision for spatter tracking in laser powder bed fusion. In *Proceedings 29th Annual International Solid Freeform Fabrication Symposium*, Austin, Texas, USA, August.

Barrett, C., MacDonald, E., Conner, B., Persi, F., 2018b. Micron-level layer-wise surface profilometry to detect porosity defects in powder bed fusion of Inconel 718. *JOM*, 70: 1844–1852.

Bayle, F., Doubenskaia, M., 2008. Selective laser melting process monitoring with high speed infra-red camera and pyrometer. In *Fundamentals of Laser Assisted Micro-and Nanotechnologies*. Springer.

Benedetti, M., Fontanari, V., Bandini, M., Zanini, F., Carmignato, S., 2018. Low- and high-cycle fatigue resistance of Ti-6Al-4V ELI additively manufactured via selective laser melting: Mean stress and defect sensitivity. *Int. J. Fatigue*, 107: 96–109.

Berumen, S., Bechmann, F., Lindner, S., Kruth, J. P., Craeghs, T., 2010. Quality control of laser-and powder bed-based additive manufacturing (AM) technologies. *Phys. Procedia*, 5: 617–622.

Caltanissetta, F., Grasso, M., Petrò, S., Colosimo, B. M., 2018. Characterization of in-situ measurements based on layerwise imaging in laser powder bed fusion. *Addit. Manuf.*, 24: 183–199.

Casati, R., Lemke, J., Vedani, M., 2016. Microstructure and fracture behavior of 316L austenitic stainless steel produced by selective laser melting. *J. Mater. Sci. Technol.* 32: 738–744.

Chan, T.F., Vese, L.A., 2001. Active contours without edges. *IEEE Trans. Image Process.* 10: 266–277.

Chivel, Y., 2013. Optical in-process temperature monitoring of selective laser melting. *Phys. Procedia*, 41: 904–910.

Clijsters, S., Craeghs, T., Buls, S., Kempen, K., Kruth, J. P., 2014. In situ quality control of the selective laser melting process using a high-speed, real-time melt pool monitoring system. *Int. J. Adv. Manuf. Technol.*, 75: 1089–1101.

Colosimo, B. M., 2018. Modeling and monitoring methods for spatial and image data. *Qual. Eng.*, 30: 94–111.

Colosimo, B. M., Grasso, M., 2018. Spatially weighted PCA for monitoring video image data with application to additive manufacturing. *J. Qual. Technol.*, 50: 391–417.

Colosimo, B. M., Grossi, E., Caltanissetta, F., Grasso, M., 2019. PENELOPE: A new solution for in-situ monitoring and defect removal in L-PBF. In *Proceedings 30th Annual International Solid Freeform Fabrication Symposium*, Austin, Texas, USA, August.

Colosimo, B. M., Huang, Q., Dasgupta, T., Tsung, F., 2018. Opportunities and challenges of quality engineering for additive manufacturing. *J. Qual. Technol.*, 50: 233–252.

Craeghs, T., Bechmann, F., Berumen, S., Kruth, J. P. 2010. Feedback control of layerwise laser melting using optical sensors. *Phys. Procedia*, 5: 505–514.

Craeghs, T., Clijsters, S., Kruth, J. P., Bechmann, F., Ebert, M. C. 2012. Detection of process failures in layerwise laser melting with optical process monitoring. *Phys. Procedia*, 39: 753–759.

Craeghs, T., Clijsters, S., Yasa, E., Bechmann, F., Berumen, S., Kruth, J. P. 2011. Determination of geometrical factors in layerwise laser melting using optical process monitoring. *Opt. Lasers Eng.*, 49: 1440–1446.

Das, S. 2003. Physical aspects of process control in selective laser sintering of metals. *Adv. Eng. Mater.*, 5: 701–711.

Demir, A. G., De Giorgi, C., Previtali, B. 2018. Design and implementation of a multisensor coaxial monitoring system with correction strategies for selective laser melting of a maraging steel. *J. Manuf. Sci. Eng.*, 140: 041003.

Dickins, A., Widjanarko, T., Lawes, S., Stravroulakis, P., Leach, R. K., 2018. Design of a multi-sensor in-situ inspection system for additive manufacturing. In *Proceedings ASPE/euspen Advancing Precision in Additive Manufacturing*, Berkeley, California, USA, June.

Doubenskaia, M., Pavlov, M., Grigoriev, S., Tikhonova, E., Smurov, I. 2012. Comprehensive optical monitoring of selective laser melting. *J. Laser Micro Nanoeng.*, 7: 236–243.

Doubenskaia, M. A., Zhirnov, I. V., Teleshevskiy, V. I., Bertrand, P., Smurov, I. Y. 2015. Determination of true temperature in selective laser melting of metal powder using infrared camera. *Mater. Sci. Forum*, 834: 93–102.

Dye, D., Hunziker, O., Reed, R. C. 2001. Numerical analysis of the weldability of superalloys. *Acta Mater.*, 49: 683–697.

Edwards, P., O'Conner, A., Ramulu, M. 2013. Electron beam additive manufacturing of titanium components: Properties and performance. *J. Manuf. Sci. Eng.*, 135: 061016.

Erler, M., Streek, A., Schulze, C., Exner, H. 2014. Novel machine and measurement concept for micro machining by selective laser sintering. In *Proceedings 25th Annual International Solid Freeform Fabrication Symposium*, Austin, Texas, USA, August.

Everton, S. K., Hirsch, M., Stravroulakis, P., Leach, R. K., Clare, A. T. 2016. Review of in-situ process monitoring and in-situ metrology for metal additive manufacturing. *Mater. Des.*, 95: 431–445.

Ferrar, B., Mullen, L., Jones, E., Stamp, R., Sutcliffe, C. J. 2012. Gas flow effects on selective laser melting (SLM) manufacturing performance. *J. Mater. Process. Technol.*, 212: 355–364.

Fisher, B. A., Lane, B., Yeung, H., Beuth, J. 2018. Toward determining melt pool quality metrics via coaxial monitoring in laser powder bed fusion. *Manuf. Lett.*, 15: 119–121.

Foster, B. K., Reutzel, E. W., Nassar, A. R., Hall, B. T., Brown, S. W., Dickman, C. J. 2015. Optical, layerwise monitoring of powder bed fusion. In *Proceedings 26th Annual International Solid Freeform Fabrication Symposium*, Austin, Texas, USA, August.

Fox, J. C., Moylan, S. P., Lane, B. M. 2016. Effect of process parameters on the surface roughness of overhanging structures in laser powder bed fusion additive manufacturing. *Procedia CIRP*, 45: 131–134.

Gao, W., Haitjema, H., Fang, F. Z., Leach, R. K., Cheung, C. F., Savio, E., Linares, J. M. 2019. On-machine and in-process surface metrology for precision manufacturing. *CIRP Annals*, 68: 843–866.

Gibson, I., Rosen, D. W., Stucker, B. 2010. *Additive Manufacturing Technologies*. New York: Springer.

Gobert, C., Reutzel, E. W., Petrich, J., Nassar, A.R., Phoha, S., 2018. Application of supervised machine learning for defect detection during metallic powder bed fusion additive manufacturing using high resolution imaging. *Addit. Manuf.* 21: 517–528.

Gong, X., Cheng, B., Price, S., Chou, K. 2013. Powder-bed electron-beam-melting additive manufacturing: Powder characterization, process simulation and metrology. In *Proceedings of the ASME District F Early Career Technical Conference*, Birmingham, Alabama, USA, November, 59–66.

Gonzalez, R. C., Woods, R. E. 2002. *Digital Image Processing*. Upper Saddle River, NJ: Prentice Hall.

Grasso, M., Caltanissetta, F., Colosimo, B. M., 2019. A novel self-repairing additive manufacturing system for in-situ defects detection and correction. In *Proceedings Euspen*, Bilbao, Spain, June.

Grasso, M., Colosimo, B. M., 2017. Process defects and in-situ monitoring methods in metal powder bed fusion: A review. *Meas. Sci. Technol.*, 28: 1–25.

Grasso, M., Colosimo, B. M. 2019. A statistical learning method for image-based monitoring of the plume signature in laser powder bed fusion. *Rob. Comput. Integr. Manuf.*, 57: 103–115.

Grasso, M., Demir, A. G., Previtali, B., Colosimo, B. M. 2018a, In-situ monitoring of selective laser melting of zinc powder via infrared imaging of the process plume. *Rob. Comput. Integr. Manuf.*, 49: 229–239.

Grasso, M., Gallina, F., Colosimo, B. M. 2018b, Data fusion methods for statistical process monitoring and quality characterization in metal additive manufacturing. In *15th CIRP Conference on Computer Aided Tolerancing*, Milan, Italy.

Grasso, M., Laguzza, V., Semeraro, Q., Colosimo, B. M. 2016. In-process monitoring of selective laser melting: Spatial detection of defects via image data analysis. *J. Manuf. Sci. Eng.*, 139: 051001-1–051001-16.

Hann, B. A. 2016. Powder reuse and its effects on laser based powder fusion additive manufactured alloy 718. *SAE Int. J. Aerosp.*, 9.

Harrison, N. J., Todd, I., Mumtaz, K. 2015. Reduction of micro-cracking in nickel superalloys processed by selective laser melting: A fundamental alloy design approach. *Acta Mater.*, 94: 59–68.

Heeling, T., Wegener, K. 2018. The effect of multi-beam strategies on selective laser melting of stainless steel 316L. *Addit. Manuf.*, 22: 334–342.

Huang, Q., Hu, N., Yang, X., Zhang, R., Feng, Q. 2016. Microstructure and inclusion of Ti-6Al-4V fabricated by selective laser melting. *Front. Mater. Sci.*, 1–4.

Imani, F., Gaikwad, A., Montazeri, M., Yang, H., Rao, P., 2018. Layerwise in-process quality monitoring in laser powder bed fusion. *ASME Paper No. MSEC*, 6477.

Khairallah, S. A., Anderson, A. T., Rubenchik, A., King, W. E. 2016. Laser powder-bed fusion additive manufacturing: Physics of complex melt flow and formation mechanisms of pores, spatter, and denudation zones. *Acta Mater.*, 108: 36–45.

King, W. E., Barth, H. D., Castillo, V. M., Gallegos, G. F., Gibbs, J. W., Hahn, D. E., Rubenchik, A. M. 2014. Observation of keyhole-mode laser melting in laser powder-bed fusion additive manufacturing. *J. Mater. Process. Technol.*, 214: 2915–2925.

Kleszczynski, S., Zur Jacobsmühlen, J., Sehrt, J. T., Witt, G. 2012. Error detection in laser beam melting systems by high resolution imaging. In *Proceedings 23th Annual International Solid Freeform Fabrication Symposium*, Austin, Texas, USA, August.

Kolb, T., Mahr, A., Huber, F., Tremel, J., Schmidt, M. 2019. Qualification of channels produced by laser powder bed fusion: Analysis of cleaning methods, flow rate and melt pool monitoring data. *Addit. Manuf.*, 25: 430–436.

Krauss, H., Eschey, C., Zaeh, M. 2012. Thermography for monitoring the selective laser melting process. In *Proceedings 23rd Annual International Solid Freeform Fabrication Symposium*, Austin, Texas, USA, August.

Krauss, H., Zeugner, T., Zaeh, M. F. 2014. Layerwise monitoring of the selective laser melting process by thermography. *Phys. Procedia*, 56: 64–71.

Kruth, J. P., Froyen, L., Van Vaerenbergh, J., Mercelis, P., Rombouts, M., Lauwers, B. 2004. Selective laser melting of iron-based powder. *J. Mater. Process. Technol.*, 149: 616–622.

Kruth, J. P., Mercelis, P., Van Vaerenbergh, J., Craeghs, T. 2007. Feedback control of selective laser melting. In *Proceedings of the 3rd International Conference on Advanced Research in Virtual and Rapid Prototyping*, 521–527.

Kwon, O., Kim, H. G., Ham, M. J., Kim, W., Kim, G. H., Cho, J. H., ... and Kim, K. 2018. A deep neural network for classification of melt-pool images in metal additive manufacturing. *J. Intell. Manuf.*, 1–12.

Ladewig, A., Schlick, G., Fisser, M., Schulze, V., Glatzel, U. 2016. Influence of the shielding gas flow on the removal of process by-products in the selective laser melting process. *Addit. Manuf.*, 10: 1–9.

Land, W. S., Zhang, B., Ziegert, J., Davies, A. 2015. In-situ metrology system for laser powder bed fusion additive process. *Procedia Manuf.*, 1: 393–403.

Lane, B., Moylan, S., Whitenton, E. P., Ma, L. 2015. Thermographic measurements of the commercial laser powder bed fusion process at NIST. In *Proceedings 26th Annual International Solid Freeform Fabrication Symposium*, Austin, Texas, USA, August.

Lewis, A. D. 2019. Application of optical coherence tomography for improved in-situ flaw detection in nylon 12 selective laser sintering, Doctoral dissertation, The University of Texas at Austin.

Li, R., Liu, J., Shi, Y., Wang, L., Jiang, W. 2012. Balling behavior of stainless steel and nickel powder during selective laser melting process. *Int. J. Adv. Manuf. Technol.*, 59: 1025–1035.

Liu, Y., Yang, Y., Mai, S., Wang, D., Song, C. 2015. Investigation into spatter behavior during selective laser melting of AISI 316L stainless steel powder. *Mat. Des.*, 87: 797–806.

Lott, P., Schleifenbaum, H., Meiners, W., Wissenbach, K., Hinke, C., Bültmann, J. 2011. Design of an optical system for the in situ process monitoring of selective laser melting (SLM). *Phys. Procedia*, 12: 683–690.

Ly, S., Rubenchik, A. M., Khairallah, S. A., Guss, G., Matthews, M. J. 2017. Metal vapor micro-jet controls material redistribution in laser powder bed fusion additive manufacturing. *Sci. Rep.*, 7: 4085.

Maamoun, A., Xue, Y., Elbestawi, M., & Veldhuis, S. 2018. Effect of selective laser melting process parameters on the quality of al alloy parts: Powder characterization, density, surface roughness, and dimensional accuracy. *Materials*, 11(12): 2343.

Mani, M., Lane, B. M., Donmez, M. A., Feng, S. C., Moylan, S. P. 2017. A review on measurement science needs for real-time control of additive manufacturing metal powder bed fusion processes. *Int. J. Prod. Res.*, 55: 1400–1418.

Marshall, G., Young II, W. J., Shamsaei, N., Craig, J., Wakeman, T., Thompson, S. M. 2015. Dual thermographic monitoring of Ti-6Al-4V cylinders during direct laser deposition. In *Proceedings 26th Annual International Solid Freeform Fabrication Symposium*, Austin, Texas, USA, August.

Mazzoleni, L., Caprio, L., Pacher, M., Demir, A. G., Previtali, B. 2019. External illumination strategies for melt pool geometry monitoring in SLM. *JOM*, 1–10.

Mercelis, P., Kruth, J. P. 2006. Residual stresses in selective laser sintering and selective laser melting. *Rapid Prototyp. J.*, 12: 254–265.

Mireles, J., Ridwan, S., Morton, P. A., Hinojos, A., Wicker, R. B. 2015a. Analysis and correction of defects within parts fabricated using powder bed fusion technology. *Surf. Topogr. Metrol. Prop.*, 3: 034002.

Montgomery, D. C. (2009). *Statistical Quality Control (Vol. 7)*. New York: Wiley.

Mousa, A. A. 2016. Experimental investigations of curling phenomenon in selective laser sintering process. *Rapid Prototyp. J.*, 22: 405–415.

Moylan, S. P., Drescher, J., Donmez, M. A., 2014b Powder bed fusion machine performance testing. In *Proceedings of the 2014 ASPE Spring Topical Meeting – Dimensional Accuracy and Surface Finish in Additive Manufacturing*, Berkeley, CA, USA.

Niu, H. J., Chang, I. T. H. 1999. Instability of scan tracks of selective laser sintering of high speed steel powder. *Scr. Mater.*, 41: 1229–1234.

Okaro, I. A., Jayasinghe, S., Sutcliffe, C., Black, K., Paoletti, P., Green, P. L. 2019. Automatic fault detection for laser powder-bed fusion using semi-supervised machine learning. *Addit. Manuf.*, 27: 42–53.

Parry, L., Ashcroft, I. A., Wildman, R. D. 2016. Understanding the effect of laser scan strategy on residual stress in selective laser melting through thermo-mechanical simulation. *Addit. Manuf.*, 12: 1–15.

Pavlov, M., Doubenskaia, M., Smurov, I. 2010. Pyrometric analysis of thermal processes in SLM technology. *Phys. Procedia*, 5: 523–531.

Peralta, A. D., Enright, M., Megahed, M., Gong, J., Roybal, M., Craig, J. 2016. Towards rapid qualification of powder-bed laser additively manufactured parts. *Integr. Mater. Manuf. Innov.*, 5: 154–176.

Price, S., Cooper, K., Chou, K. 2012. Evaluations of temperature measurements by near-infrared thermography in powder-based electron-beam additive manufacturing. In *Proceedings 23rd Annual International Solid Freeform Fabrication Symposium*, Austin, Texas, USA, August.

Raplee, J., Plotkowski, A., Kirka, M. M., Dinwiddie, R., Okello, A., Dehoff, R. R., Babu, S. S. 2017. Thermographic microstructure monitoring in electron beam additive manufacturing. *Nature Sci. Rep.*, 7: 43554.

Renken, V., von Freyberg, A., Schünemann, K., Pastors, F., Fischer, A., 2019. In-process closed-loop control for stabilising the melt pool temperature in selective laser melting. *Progr. Addit. Manuf.*, 1–11.

Repossini, G., Laguzza, V., Grasso, M., Colosimo, B. M., 2018. On the use of spatter signature for in-situ monitoring of laser powder bed fusion. *Addit. Manuf.*, 16: 35–48.

Ridwan, S., Mireles, J., Gaytan, S. M., Espalin, D., Wicker, R. B. 2014. Automatic layerwise acquisition of thermal and geometric data of the electron beam melting process using infrared thermography. In *Proceedings 25th Annual International Solid Freeform Fabrication Symposium*, Austin, Texas, USA, August.

Sabbaghi, A., Huang, Q. 2018. Model transfer across additive manufacturing processes via mean effect equivalence of lurking variables. *Ann. Appl. Stat.*, 12: 2409–2429.

Sames, W. J., List, F. A., Pannala, S., Dehoff, R. R., Babu, S. S. 2016. The metallurgy and processing science of metal additive manufacturing. *Inter. Mater. Rev.*, 1–46.

Scime, L., Beuth, J., 2018. Anomaly detection and classification in a laser powder bed additive manufacturing process using a trained computer vision algorithm. *Addit. Manuf.* 19: 114–126.

Schilp, J., Seidel, C., Krauss, H., Weirather, J. 2014. Investigations on temperature fields during laser beam melting by means of process monitoring and multiscale process modelling. *Adv. Mech. Eng.*, 6: 217584.

Senin, N., Thompson, A., Leach, R. K. 2017. Characterisation of the topography of metal additive surfaces features with different measurement technology. *Meas. Sci. Technol.* 28: 095003.

Seyda, V., Kaufmann, N., Emmelmann, C. 2012. Investigation of aging processes of Ti-6Al-4V powder material in laser melting. *Phys. Procedia*, 39: 425–431.

Sharratt, B. M. 2015. Non-destructive techniques and technologies for qualification of additive manufactured parts and processes. A literature Review. *Contract Report DRDC-RDDC-2015-C035*, Victoria, BC.

Smith, R. J., Hirsch, M., Patel, R., Li, W., Clare, A. T., Sharples, S. D. 2016. Spatially resolved acoustic spectroscopy for selective laser melting. *J. Mater. Process. Technol.*, 236: 93–102.

Spears, T. G., Gold, S. A. 2016. In-process sensing in selective laser melting (SLM) additive manufacturing. *Integr. Mater. Manuf. Innov.*, 5: 1.

Strano, G., Hao, L., Everson, R. M., Evans, K. E. 2013. Surface roughness analysis, modelling and prediction in selective laser melting. *J. Mater. Process. Technol.*, 213: 589–597.

Tan Phuc, L., Seita, M. 2019. A high-resolution and large field-of-view scanner for in-process characterization of powder bed defects during additive manufacturing. *Mater. Des.*, 164: 107562.

Tang, H. P., Qian, M., Liu, N., Zhang, X. Z., Yang, G. Y., Wang, J. 2015. Effect of powder reuse times on additive manufacturing of Ti-6Al-4V by selective electron beam melting. *JOM*, 67: 555–563.

Tapia, G., Elwany, A. 2014. A review on process monitoring and control in metal-based additive manufacturing. *J. Manuf. Sci. Eng.*, 136: 060801.

Thijs, L., Verhaeghe, F., Craeghs, T., Van Humbeeck, J., Kruth, J. P. 2010. A study of the microstructural evolution during selective laser melting of Ti-6Al-4V. *Acta Mater.*, 58: 3303–3312.

Thomas, D. 2009. The development of design rules for selective laser melting. Doctoral dissertation, University of Wales.

Thombansen, U., Gatej, A., Pereira, M. 2015. Process observation in fiber laser–based selective laser melting. *Opt. Eng.*, 54: 011008–011008.

Thompson A., Tammas-Williams S., Leach R. K., Todd I. 2018. Correlating volume and surface features in additively manufactured parts. In *Proceedings ASPE/euspen Advancing Precision in Additive Manufacturing*, Berkeley, USA, July, 116–120.

Triantaphyllou, A., Giusca, C. L., Macaulay, G. D., Roerig, F., Hoebel, M., Leach, R. K., Milne, K. A. 2015. Surface texture measurement for additive manufacturing. *Surf. Topogr. Metrol. Prop.*, 3: 024002.

Van Gestel, C. 2015. Study of physical phenomena of selective laser melting towards increased productivity. PhD Dissertation, École Polytechnique Fédérale De Lausanne.

Vinson, K., Maxwell, J., Hooper, R. J., Allen, J., Thompson, G. B. 2017. Dual-wavelength in situ pyrometry during additive formation of fibers by laser-induced deposition. *JOM*, 69: 2314–2319.

Wasmer, K., 2018. High-speed x-ray imaging for correlating acoustic signals with quality monitoring: A machine learning approach. In *MSandT 2019 Conference*, Portland, OR.

Wasmer, K., Le-Quang, T., Meylan, B., Shevchik, S. A. 2019. In situ quality monitoring in AM using acoustic emission: A reinforcement learning approach. *J. Mater. Eng. Perform.*, 28: 666–672.

Williams, T. 2009. *Thermal Imaging Cameras: Characteristics and Performance*. Boca Raton, FL: CRC Press.

Wong, H., Neary, D., Jones, E., Fox, P., Sutcliffe, C. 2019. Pilot capability evaluation of a feedback electronic imaging system prototype for in-process monitoring in electron beam additive manufacturing. *Int. J. Adv. Manuf. Technol.*, 100: 707–720.

Yadroitsev, I., Krakhmalev, P., Yadroitsava, I. 2014. Selective laser melting of Ti6Al4V alloy for biomedical applications: Temperature monitoring and microstructural evolution. *J. Alloys Compd.*, 583: 404–409.

Ye, D., Fuh, J. Y. H., Zhang, Y., Hong, G. S., Zhu, K. 2018. In situ monitoring of selective laser melting using plume and spatter signatures by deep belief networks. *ISA Trans.*, 81: 96–104.

Yoder, S., Morgan, S., Kinzy, C., Barnes, E., Kirka, M., Paquit, V., Babu, S. S. 2018. Characterization of topology optimized Ti-6Al-4V components using electron beam powder bed fusion. *Addit. Manuf.*, 19: 184–196.

Zhang, B., Ziegert, J., Farahi, F., Davies, A. 2016. In situ surface topography of laser powder bed fusion using fringe projection. *Addit. Manuf.*, 12: 100–107.

Zheng, H., Li, H., Lang, L., Gong, S., Ge, Y. 2018. Effects of scan speed on vapor plume behavior and spatter generation in laser powder bed fusion additive manufacturing. *J. Manuf. Proc.*, 36: 60–67.

Zhong, M., Sun, H., Liu, W., Zhu, X., He, J. 2005. Boundary liquation and interface cracking characterization in laser deposition of Inconel 738 on directionally solidified Ni-based superalloy. *Scr. Mater.*, 53: 159–164.

Zhang, Y., Hong, G. S., Ye, D., Zhu, K., Fuh, J. Y. 2018. Extraction and evaluation of melt pool, plume and spatter information for powder-bed fusion AM process monitoring. *Mater. Des.*, 156: 458–469.

Zur Jacobsmühlen, J., Kleszczynski, S., Witt, G., Merhof, D. 2013. Elevated region area measurement for quantitative analysis of laser beam melting process stability. In *Instrum. Meas. Technol. Conf. I2MTC*, 707–712.

Index

D

E

I

K

L

M

N

O

Q

R

T

U

V

W

X

Y

Z